SQUARE ROOTS

n	$\sqrt{n}$	n	$\sqrt{n}$	n	$\sqrt{n}$	n	$\sqrt{n}$
1	1.000	51	7.141	101	10.050	151	12.288
2	1.414	52	7.211	102	10.100	152	12.329
3	1.732	53	7.280	103	10.149	153	12.369
4	2.000	54	7.348	104	10.198	154	12.410
5	2.236	55	7.416	105	10.247	155	12.450
6	2.449	56	7.483	106	10.296	156	12.490
7	2.646	57	7.550	107	10.344	157	12.530
8	2.828	58	7.616	108	10.392	158	12.570
9	3.000	59	7.681	109	10.440	159	12.610
10	3.162	60	7.746	110	10.488	160	12.649
11	3.317	61	7.810	111	10.536	161	12.689
12	3.464	62	7.874	112	10.583	162	12.728
13	3.606	63	7.937	113	10.630	163	12.767
14	3.742	64	8.000	114	10.677	164	12.806
15	3.873	65	8.062	115	10.724	165	12.845
16	4.000	66	8.124	116	10.770	166	12.884
17	4.123	67	8.185	117	10.817	167	12.923
18	4.243	68	8.246	118	10.863	168	12.961
19	4.359	69	8.307	119	10.909	169	13.000
20	4.472	70	8.367	120	10.954	170	13.038
21	4.583	71	8.426	121	11.000	171	13.077
22	4.690	72	8.485	122	11.045	172	13.115
23	4.796	73	8.544	123	11.091	173	13.153
24	4.899	74	8.602	124	11.136	174	13.191
25	5.000	75	8.660	125	11.180	175	13.229
26	5.099	76	8.718	126	11.225	176	13.266
27	5.196	77	8.775	127	11.269	177	13.304
28	5.292	78	8.832	128	11.314	178	13.342
29	5.385	79	8.888	129	11.358	179	13.379
30	5.477	80	8.944	130	11.402	180	13.416
31	5.568	81	9.000	131	11.446	181	13.454
32	5.657	82	9.055	132	11.489	182	13.491
33	5.745	83	9.110	133	11.533	183	13.528
34	5.831	84	9.165	134	11.576	184	13.565
35	5.916	85	9.220	135	11.619	185	13.601
36	6.000	86	9.274	136	11.662	186	13.638
37	6.083	87	9.327	137	11.705	187	13.675
38	6.164	88	9.381	138	11.747	188	13.711
39	6.245	89	9.434	139	11.790	189	13.748
40	6.325	90	9.487	140	11.832	190	13.784
41	6.403	91	9.539	141	11.874	191	13.820
42	6.481	92	9.592	142	11.916	192	13.856
43	6.557	93	9.644	143	11.958	193	13.892
44	6.633	94	9.695	144	12.000	194	13.928
45	6.708	95	9.747	145	12.042	195	13.964
46	6.782	96	9.798	146	12.083	196	14.000
47	6.856	97	9.849	147	12.124	197	14.036
48	6.928	98	9.899	148	12.166	198	14.071
49	7.000	99	9.950	149	12.207	199	14.107
50	7.071	100	10.000	150	12.247	200	14.142

INTERMEDIATE ALGEBRA

INTERMEDIATE ALGEBRA

Evan W. Thweatt
American River College

WEST PUBLISHING COMPANY
St. Paul   New York   Los Angeles   San Francisco

Copyeditor: Linda L. Thompson
Interior Design: Lucy Lesiak Design
Composition: The Clarinda Company
Artwork: Editing, Design & Production, Inc.

Library of Congress Cataloging in Publication Data

Thweatt, Evan W.
    Intermediate algebra.

    Includes index.
    1. Algebra.   I. Title.
QA154.2.T47   1984   512.9   83–21923
ISBN 0–314–77832–2

COPYRIGHT © 1984 By WEST PUBLISHING CO.
                    50 West Kellogg Boulevard
                    P.O. Box 43526
                    St. Paul, Minnesota 55164
All rights reserved
Printed in the United States of America

CONTENTS

Preface ix

CHAPTER 1 Sets and the Real Number System 1

1–1 Sets and Set Operations 2
1–2 The Real Number System 6
1–3 Order of Operations and Grouping Symbols 11
1–4 Axioms of Equality for The Real Number System 15
1–5 Axioms of Inequality for The Real Number System 17
1–6 Other Properties of The Real Numbers 21
1–7 Addition and Subtraction of Signed Numbers 29
1–8 Multiplication and Division of Signed Numbers 34

CHAPTER 2 Special Products and Factoring 44

2–1 Properties of Exponents 45
2–2 Adding and Subtracting Polynomials 52
2–3 Multiplying Polynomials 58
2–4 Removing the Greatest Common Factor 64
2–5 Factoring Trinomials 68
2–6 Special Types of Factoring 74

CHAPTER 3 Rational Expressions 86

- 3–1 Basic Properties of Fractions 87
- 3–2 Reducing Rational Expressions 94
- 3–3 Multiplication and Division of Rational Expressions 97
- 3–4 Least Common Multiple 103
- 3–5 Addition and Subtraction of Rational Expressions 106
- 3–6 Complex Fractions 111
- 3–7 Division of Polynomials 115
- 3–8 Synthetic Division 118

CHAPTER 4 First-Degree Equations and Inequalities 127

- 4–1 First-Degree Equations 128
- 4–2 Fractional Equations 134
- 4–3 Applications 138
- 4–4 First-Degree Inequalities 153
- 4–5 Absolute Value Equations 160
- 4–6 Absolute Value Inequalities 164
- 4–7 Changing the Subject of an Equation 167

CHAPTER 5 Exponents, Roots, and Radicals 174

- 5–1 Integral Exponents 175
- 5–2 Rational Exponents 180
- 5–3 Simplifying Radicals 187
- 5–4 Operations with Radicals 193
- 5–4 Introduction to Radical Equations 198

CHAPTER 6 Second-Degree Equations and Inequalities 208

- 6–1 Finding Solution sets for Quadratic Equations by Factoring 209
- 6–2 Completing the Square 214

Contents

- 6–3 Complex Numbers 218
- 6–4 The Quadratic Formula 225
- 6–5 Radical Equations 230
- 6–6 Equations that are Quadratic in Form 233
- 6–7 Quadratic Inequalities 236
- 6–8 Applications 241

CHAPTER 7 Relations and Functions: Part I 255

- 7–1 Ordered Pairs, Relations, and Solutions of Equations in Two Variables 256
- 7–2 Functions 260
- 7–3 Distance Formula and Slope 265
- 7–4 The Linear Function 273
- 7–5 Linear Inequalities 282

CHAPTER 8 Relations and Functions: Part II 294

- 8–1 Quadratic Functions and Relations 295
- 8–2 The Circle 304
- 8–3 The Ellipse and Hyperbola 310
- 8–4 Quadratic Inequalities in Two Variables 317
- 8–5 Ratio and Proportion 319
- 8–6 Variation 324
- 8–7 Inverse of a Function 330

CHAPTER 9 Exponential and Logarithmic Functions 343

- 9–1 The Exponential Function 344
- 9–2 Logarithmic Functions 349
- 9–3 Properties of Logarithms 354
- 9–4 Common Logarithms and Calculations 357
- 9–5 Exponential and Logarithmic Equations 366
- 9–6 Applications 369

CHAPTER 10 Systems of Equations 380

- 10–1 Linear Systems of Two Equations 381
- 10–2 Linear Systems of Three Equations In Three Variables 388
- 10–3 Determinants 392
- 10–4 Higher Order Determinants (Optional) 397
- 10–5 Solution of Systems in Three Variables by Determinants (Optional) 402
- 10–6 Quadratic Systems in Two Variables 405
- 10–7 Applications 409

CHAPTER 11 Sequences and Series 423

- 11–1 Sequences 424
- 11–2 Sigma Notation and Series 428
- 11–3 Arithmetic Progressions 432
- 11–4 Geometric Progressions 437
- 11–5 Infinite Geometric Series 442
- 11–6 The Binomial Theorem 446

Answers to Odd-Numbered Exercises, Review Problems, and Cumulative Tests 459

Index 487

PREFACE

KEY FEATURES AND ORGANIZATION

Intermediate Algebra will give the student a thorough background in the essential topics of algebra that are necessary for any advancement that requires these topics as prerequisites.

The first chapter provides a thorough review of the real number system and its properties. After studying this chapter, a student will know the main subsets of the real number system and how they are formed.

Chapter 2 reviews the properties of whole number exponents and factoring. Special types of factoring are presented, including the use of substitution. Many applications from a wide variety of subjects are given at the end of most sections of this chapter and the chapters that follow. These exercises will let the student see how the abstract concepts being learned are applied in other areas of study.

Chapter 3 presents the concept of a fraction and rational expressions. Most of the effort in this chapter is spent on algebraic expressions rather than number fractions. The assumption is made that the student already knows the rules relating to number fractions. Number fractions are presented solely for the purpose of showing that the rules applied to them are identical to the rules applied to algebraic fractions. A separate section exists on finding the least common multiple. This section prepares the student to find the least common denominator in the next section dealing with adding and subtracting rational expressions. The chapter concludes with a section on synthetic division. This important topic was not put in an appendix, where it is often overlooked. In many colleges, the intermediate algebra student is expected to learn this topic. In those schools that require a student to also take college algebra, synthetic division could be delayed until that course.

Equations are delayed until chapter 4 so that a more complete presentation can be made. Many texts rush to put equations in as early as chapter 2. The author's observation is that students have much more difficulty with factoring and fractions than they do with equations. If these two topics are covered first, then they can be used again in the setting of solving equations. This will help to reinforce what the student has learned about them earlier. In this chapter, fractional equations can now be made to flow naturally as another type of equation rather than as a disconnected topic usually tacked on to the chapter on rational expressions. Also, a much more meaningful and wider variety of word problems can be given at this point than can be given before fractions are covered. Absolute value equations and inequalities are covered as the last part of chapter 4.

Chapter 5 briefly reviews whole number exponents and then moves into negative integer exponents and rational exponents. The student is shown the relationship between a rational exponent and a radical. The familiar topics of simplification and operations on radicals are covered in chapter 5.

Chapter 6 covers the three ways of solving a quadratic equation—by factoring, by completing the square, and by using the quadratic formula. A section on complex numbers is introduced before the section on the quadratic formula. Once complex numbers are introduced, the student is expected to find all solutions to the equations that follow, including complex solutions.

Chapters 7 and 8 cover the topics of relations and functions. A function is defined in terms of a relation using ordered pairs of real numbers. The student is then shown what is meant by the domain and range of a function. Chapter 8 develops the ideas of the conic sections. Graphs of the conic sections are carefully drawn and explained. The student is also taught how to recognize a conic section from its equation. Chapter 8 concludes with a section on the inverse of a function, which sets the stage for exponential and logarithmic functions in the next chapter.

Chapter 9 is written so that logarithms can be taught either by using a calculator or by using tables and interpolation.

Chapter 10 gives several methods of solving systems of equations, including Cramer's rule. Higher-order determinants, as well as their use in solving systems of equations, are introduced but are left as optional topics.

Sequences are developed in chapter 11. Once the student understands a sequence, the transition is made into series using sigma notation. Arithmetic and geometric sequences and series are then developed. Chapter 11 concludes with a section on the binomial theorem.

A few exercises are given in this section in which the exponent on the binomial is not a whole number.

PEDAGOGICAL FEATURES

Chapter objectives are stated at the beginning of each chapter.

The student is encouraged to learn to use a **calculator** by working certain exercises marked with a calculator symbol.

Definitions and **rules** are named and boxed for easy identification and reference. An example usually follows immediately so that the student can see how the definition or rule is applied.

Word problems are presented early in the text, including a great variety from other areas of study.

Chapter summaries and **review problems** are given for each chapter. A **cumulative test** is given for each chapter beginning with Chapter 2.

Answers for the odd-numbered exercises, all chapter review problems, and all the cumulative test exercises are given in an **appendix**.

The even-numbered exercises parallel the odd-numbered exercises for greater **assignment flexibility.**

Tables containing logarithms, squares and square roots, and formulas from geometry are presented on the endpapers.

ANCILLARY AIDS

A **student study guide** can be obtained that provides a self-test for each section and a cumulative test for all sections in each chapter. **Other ancillary aids are available for the instructor.**

COURSE COVERAGE

The length of the text makes it suitable for a one-semester or one-quarter course. For a short course, chapters 1 and 11, as well as sections 10–4 and 10–5, could be omitted. However, the student must be familiar with the basic concepts of a set that are presented in chapter 1.

ACKNOWLEDGMENTS

A great many people have contributed to the development of this text with their meticulous scrutiny and suggestions. I wish to thank Jesús Carreón, Mesa Community College; Richard R. Conley, Ashland Community College; Arthur P. Dull, Diablo Valley College; William Edgar, American River College; Edward Harper, American River College; Michelle Harrell, Miami Dade Community College; Wei-Jen Harrison, American River College; Kathleen B. Keller, Los Angeles Harbor College; David A. Lunsford, Grossmont College; Maryjean Riel, Truckee Meadows Junior College; and Peter Tannenbaum, University of Arizona.

The editorial staff of West Publishing was especially helpful, providing expert suggestions and guidance: Denise Simon, acquiring editor; Barbara Fuller, production editor; and Kristen McCarthy, marketing coordinator. I also wish to express appreciation to Linda Thompson, copyeditor.

I wish to give special thanks to my family, Norma, Terri, and Steve, and to my parents Earl and Gladys Thweatt, for their help and encouragement during the time I was writing, reading, and editing the manuscript.

Evan Thweatt

INTERMEDIATE ALGEBRA

A Brief History of Algebra

Algebra is an old and honored discipline. Much of the body of Computer Science and Mathematics that we study today had its beginning centuries ago in ancient Egypt, Greece, and India. Although some of the earliest works on algebra were from Egypt, the Egyptians apparently did not develop it to a very high form. The Greeks, more interested in geometry than algebra, developed only enough algebra to explain their geometric ideas. No significant contribution to algebra was made by the Greeks until A.D. 300 or 400. Several gifted men of India and Persia made significant progress in the subject between A.D. 500 and 1200.

One of the most influential men of this period was Al-Khwarizmi, a teacher in Bagdad. Around 825 he wrote "Kitab al jabr w'al-mugabala," which is translated as "the science of redintegration and equation." A corruption of the title of his work gives us the Arabic word *algebra*. One of the earliest uses of the word *algebra* appeared in the Italian language around 1200.

The subject that we today call modern algebra was developed primarily from the sixteenth century to the present. This may not sound very modern, but when compared to the algebra of Egypt some 3500 years ago, this is indeed recent.

We begin our study of algebra with the subjects of sets, the real number system, and the basic properties of the real number system.

CHAPTER 1 | Sets and the Real Number System

Objectives of Chapter 1

To be able to:

- ☐ Explain what a set is and tell if it is well defined. (1–1)
- ☐ Form the intersection and union of two sets. (1–1)
- ☐ Define a subset. (1–1)
- ☐ List the subsets of the real number system. (1–2)
- ☐ Diagram the real number system. (1–2)
- ☐ Simplify expressions containing several operations. (1–3)
- ☐ Work with grouping symbols. (1–3)
- ☐ Identify the axiom of equality applied to a given statement. (1–4)
- ☐ Identify the axiom of inequality applied to a given statement. (1–5)
- ☐ Use the commutative and associative axioms of addition and multiplication. (1–6)
- ☐ Use the identity axioms of addition and multiplication. (1–6)
- ☐ Use the distributive axiom to write products as sums or sums as products. (1–6)
- ☐ Determine if a set is closed for a particular operation. (1–6)
- ☐ Use the equality multiplication and equality addition theorems. (1–6)
- ☐ Add or subtract signed numbers. (1–7)
- ☐ Multiply or divide signed numbers. (1–8)

We begin Chapter 1 with a brief study of sets. The idea of a set is helpful in understanding certain mathematical concepts. For example, we can talk about the solution set of an equation, or the set of numbers for which a rational expression is defined.

We will also take a look at the real number system. This system of numbers will be broken down into several subsets of numbers, and each subset will be studied in detail. The axioms which apply to these sets of numbers will also be examined. These axioms will then be used to explain a few of the "mysteries" of algebra, such as why the product of two negative numbers gives a positive number for an answer.

1–1 Sets and Set Operations

A **set** is a collection of objects. A set is named with a capital letter and is indicated by braces. For example, we can write the set of whole numbers less than 5 as $A = \{0, 1, 2, 3, 4\}$. Every set must be **well defined** to be useful. This means we must be able to tell if an object belongs to a set. For example, the set $\{0, 1, 3\}$ is well defined because we know 3 belongs to it and 5 does not. The set {good automobiles} is not well defined because it is not clear what is meant by a good automobile.

The objects that belong to a set are called the **members** or **elements** of the set. We may indicate a set either by listing its elements or by using set-builder notation. For example, if we want all the whole numbers divisible by 2, we can write

$$\{0, 2, 4, 6, \ldots\} \quad (1)$$

where the three dots show the pattern is continued, or we can write

$$\{x | x \text{ is divisible by 2 and } x \text{ is a whole number}\} \quad (2)$$

The statement in (2) uses set-builder notation and is read:

{ }	variable name	|	a rule or set of instructions.
the set of	all x	such that	x is divisible by 2 and x is a whole number

Example 1 Use set-builder notation to indicate the whole numbers greater than 5 and less than 10.

Solution $\{x|x$ is greater than 5 and less than 10, x a whole number$\}$. If we actually list the elements in the set in example 1, we get $\{6, 7, 8, 9\}$.

The symbol $\in$ is used to indicate that an element belongs to a set. If $A = \{5, 6, 7\}$, then $5 \in A$; this is read "5 is an element of set A." The symbol $\notin$ indicates that an element does not belong to a set. For the set A just given, we see $8 \notin A$.

The **intersection of two sets** is indicated by the symbol $\cap$. We write the intersection of sets A and B as $A \cap B$, where $A \cap B$ is the set that contains all elements belonging to both A and B. This statement is given formally in definition 1–1. Using set-builder notation gives conciseness to the definition.

Definition 1–1

Set Intersection

For sets A and B, $A \cap B = \{x|x \in A \text{ and } X \in B\}$.

Example 2 If $A = \{5, 6, 7, 8\}$ and $B = \{0, 6, 8, 9\}$, find $A \cap B$.

Solution $A \cap B = \{6, 8\}$.

A set is empty if it contains no elements. The **empty set** is specified by the symbol $\emptyset$ or $\{\ \}$.

Example 3 Find the intersection of $A = \{1, 2, 3\}$ and $B = \{4, 5\}$.

Solution The two sets have no elements in common. Hence, $A \cap B = \emptyset$.

Note that $\emptyset \neq \{0\}$. The set $\{0\}$ is not empty because it contains one element, the number zero. Also, the empty set is not given by $\{\emptyset\}$. The symbol $\{\emptyset\}$ denotes a set whose only element is the empty set.

The **union of two sets** A and B, written $A \cup B$, is the set containing all the elements belonging to either A or B or both.

Definition 1–2

Union of Sets

For sets A and B, $A \cup B = \{x|x \in A \text{ or } x \in B\}$.

Example 4 If $A = \{0, 1, 2, 3\}$ and $B = \{2, 3, 4, 5\}$, find $A \cup B$.

Solution $A \cup B = \{0, 1, 2, 3, 4, 5\}$

If every element of set A is an element of set B, then A is a **subset** of B. This is written as $A \subseteq B$, and is read "A is a subset of B." If A is a subset of B and set B contains at least one element that is not contained in set A, then A is called a proper subset of B. This is written as $A \subset B$.

Definition 1–3

Subset of a Set

For sets A and B, $A \subseteq B$ if and only if $x \in A$ implies $x \in B$.

The phrase *if and only if* in definition 1–3 means two things.

1. If $A \subseteq B$, then each element of A also belongs to B.
2. If each element of A also belongs to B, then $A \subseteq B$.

The empty set is a subset of every set. To see why, suppose for a moment the empty set is not a subset of some set, call it set A. This means the empty set must contain at least one element not contained in set A. This is impossible because the empty set, by definition, has no elements. Hence the empty set is a subset of set A. Because set A could be any set, the empty set is a subset of every set.

Example 5 Is $A = \{1, 2, 3\}$ a subset of $B = \{1, 2, 5, 7\}$?

Solution No, because $3 \in A$ but $3 \notin B$. ■

Two sets are **equal** if each is a subset of the other.

Definition 1–4

Set Equality

If A and B are two sets, $A = B$ if and only if $A \subseteq B$ and $B \subseteq A$.

Example 6 If $A = \{1, 2, 3\}$ and $B = \{2, 1, 3\}$, is $A = B$?

Solution Yes, because each set is a subset of the other. ■

Example 7 If $A = \{1, 2, 3, 4, 5\}$, $B = \{2, 4, 5, 6\}$, and $C = \{4, 5, 7, 9\}$, find:
(a) $(A \cap B) \cup C$;   (b) $(A \cup B) \cap (B \cap C)$.

Solution (a) We first find $A \cap B$.

$$A \cap B = \{1, 2, 3, 4, 5\} \cap \{2, 4, 5, 6\} = \{2, 4, 5\}$$

Thus,

$$(A \cap B) \cup C = \{2, 4, 5\} \cup \{4, 5, 7, 9\} = \{2, 4, 5, 7, 9\}$$

Sets and The Real Number System   5

(b) First find $A \cup B$ and $B \cap C$.

$$A \cup B = \{1, 2, 3, 4, 5\} \cup \{2, 4, 5, 6\} = \{1, 2, 3, 4, 5, 6\}$$
$$B \cap C = \{2, 4, 5, 6\} \cap \{4, 5, 7, 9\} = \{4, 5\}$$

Thus,

$$(A \cup B) \cap (B \cap C) = \{1, 2, 3, 4, 5, 6\} \cap \{4, 5\}$$
$$= \{4, 5\}$$

1–1 Exercises

In exercises 1–13, find the indicated sets if $A = \{0, 1, 2, 3, 4, 5\}$, $B = \{2, 3, 4, 5, 6, 7\}$, $C = \{3, 4, 5, 6, 7, 8\}$, and $D = \{9, 10\}$. *(See examples 1–4 and 7.)*

1. $A \cap B$ *2,3,4,5*
2. $A \cup C$ *0,1,2,3,4,5,6,7,8*
3. $(A \cap B) \cup C$ *2,3,4,5,6,7,8*
4. $(A \cup B) \cap C$ *2,3,4,5,6,7,8*
5. $(A \cap B) \cap C$ *2,3,4,5,6,7,8*
6. $(A \cap D) \cap B$
7. $(A \cup B) \cap (A \cap C)$
8. $(A \cap C) \cup (C \cap D)$
9. Show that $(A \cap B) \cap C = A \cap (B \cap C)$. (This is called the *associative property of set intersection*.)
10. Show that $A \cap B = B \cap A$. (This is called the *commutative property of set intersection*.)
11. Show that $(A \cup B) \cup C = A \cup (B \cup C)$. (This is called the *associative property of set union*.)
12. Show that $A \cap (B \cup C) = (A \cap B) \cup (A \cap C)$ This is called the *distributive property of set intersection over set union*.)
13. Show that $A \cap (A \cup B) = A$ (This is called the *property of absorption*.)

For exercises 14–22, answer each question if $A = \{1, 2\}$, $B = \{1, 2, 3, 4, 5\}$, $C = \{5, 8\}$, and $D = \{8, 9, 10\}$. *(See example 5.)*

14. Is $A \subset B$?
15. Is $A \in B$?
16. Is $3 \in (A \cap B)$?
17. Is $(A \cap C) \subseteq B$?
18. Is $(\emptyset \cap A) \subseteq A$?
19. Is $(D \cup C) \subset (A \cup B)$?
20. Is $(C \cap D) \in B$?
21. Is $3 \in (C \cup D)$?
22. Is $(A \cup C) \subset (B \cup D)$?

Determine if the sets in exercises 23–31 are well defined. *(See page 2 for the meaning of a well-defined set.)*

23. $\{5, 9, 10\}$
24. $\{2, 4, 6\}$
25. {all good automobiles}
26. {all living presidents of the United States}
27. {all tall people}

28. {1, 2, 3, *a*, *b*}
29. {all people over 20 feet tall}
30. {all kind people}
31. {all females who have been president of the United States}

In exercises 32–35, let A = {5, 6, 9} and B = {6, 5, 9}.

32. Is $A = B$?
33. Is $A \subset B$?
34. Is $A \subseteq B$?
35. Is $A \cap B = A$?

In exercises 36–40, let A = {5, 6, 7, 10, 11}. Use the list method to find each set.

36. $\{x | x \in A \text{ and } x \text{ is an even number}\}$
37. $\{x | x \in A \text{ and } x \text{ is an odd number}\}$
38. $\{x | x \text{ is } \in A \text{ and } x \text{ is greater than } 12\}$
39. $\{x | x \in A \text{ and } x \text{ is divisible by } 3\}$
40. $\{x | x \in A \text{ and } x \text{ is divisible by } 5\}$

In exercises 41–46, let A = {3, 4, 5, 6, 7} and B = {5, 6, 8, 9}. Use the list method to find each set.

41. $\{x | x \in A \text{ and } x \in B\}$
42. $\{x | x \in B \text{ and } x \text{ is an even number}\}$
43. $\{x | x \in A \text{ and } x \neq 5\}$
44. $\{x | x \in A \text{ and } x \notin B\}$
45. $\{x | x \in A \text{ or } x \in B\}$
46. $\{x | x \notin A \text{ and } x \in B\}$

1–2 The Real Number System

The **real number system** contains five primary subsets of numbers. These five subsets and the letters representing them are:

N = set of **natural numbers** = $\{1, 2, 3, 4, \ldots\}$

W = set of **whole numbers** = $\{0, 1, 2, 3, \ldots\}$

J = set of **integers** = $\{\ldots, -3, -2, -1, 0, 1, 2, 3, \ldots\}$

Q = set of **rational numbers** = $\left\{ \dfrac{a}{b} \;\middle|\; a, b \text{ are integers and } b \neq 0 \right\}$

H = set of **irrational numbers** = {all real numbers that can be represented as nonrepeating, nonterminating decimals}

Sets and The Real Number System

Any of these sets of numbers can be graphed on a number line. For the purpose of drawing the graphs, we define the number that corresponds to a particular point on a number line as the coordinate of that point. For example, on the number line in figure 1–1, 4 is the coordinate of the point A.

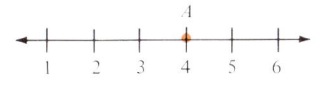

Figure 1–1

The graph of the natural numbers is made by designating some point on a number line with the coordinate 1. Then equal line segments are marked off to the right of 1, and these points will represent the other natural numbers (see figure 1–2).

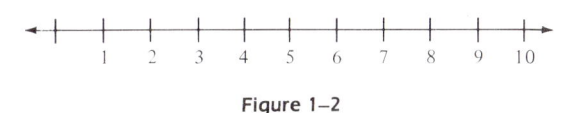

Figure 1–2

The graph of the integers is shown in figure 1–3.

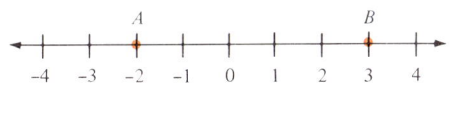

Figure 1–3

The arrows at each end of the line in figure 1–3 indicate the graph continues indefinitely in both directions. The coordinate of point A is -2, read "negative two." The coordinate of point B is 3, read either as "three" or "positive three." The positive sign $(+)$ is usually not written to the left of a positive number. We understand that a number written without a sign in front of it represents a positive number. For example, $7 = +7$.

All the rational numbers cannot be shown on a number line in the way that the integers can. The difficulty lies in the fact that we cannot name the next greater rational number after any given rational number. For example, we cannot give the next rational number greater than $\frac{5}{8}$. No matter what number we name as the next greater, it can be shown that an infinite number of rational numbers lie between it and the given rational number. A graphical representation of the specific rational numbers $\frac{3}{4}$, $-\frac{5}{8}$, 3, and $-\frac{15}{4}$ is shown in figure 1–4.

A rational number can be represented as a terminating decimal or a repeating decimal. For example, the terminating decimal 0.375 is the same as $\frac{3}{8}$. That is, if we divide 3 by 8, the quotient is 0.375. The repeating decimal 0.141414 . . . is the same as $\frac{14}{99}$, which can be verified by dividing 14 by 99. A decimal is a repeating decimal if a block or set of digits repeat endlessly. For example, in 3.5162162 . . ., the digits 162 repeat endlessly.

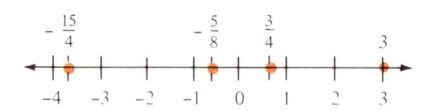

Figure 1–4

Example 1 Write 14/99 as a repeating decimal.

Solution
```
        0.1414 . . .
99 ) 14.0000
     9 9
     ———
       410
       396
       ———
       140
        99
       ———
       410
       396
       ———
        14
```

The irrational numbers are those real numbers that are not rational. They can be represented by nonterminating, nonrepeating decimals. For example,

$$\sqrt{2} = 1.41421356 \ldots, \quad \pi = 3.14159265 \ldots,$$
$$\text{and} \quad 0.1010010001 \ldots$$

are examples of irrational numbers.

There is a one-to-one correspondence between the set of real numbers and the points on a number line. This means that for each real number, there corresponds one and only one point on a number line. Also, for each point on the line, there corresponds one and only one real number.

Figure 1–5 shows the relationship of the set of real numbers to its primary subsets. The set of real numbers is formed by the union of the set of rational numbers and the set of irrational numbers. These two sets are **disjoint,** meaning that no number can be both rational and irrational. Also, observe that sets N, W, and J are subsets of Q, the set of rational numbers.

Example 2 Answer each question; referring to Figure 1–5 on page 9.

(a) Is $N \subset W$?   (b) Is $J \subseteq R$?   (c) Is $W \subseteq N$?

(d) Is N a subset of both W and J?

(e) Do some elements of W not belong to N? Which ones?

Solution (a) Yes   (b) Yes   (c) No   (d) Yes
(e) Yes; the element 0

DIAGRAM OF THE REAL NUMBER SYSTEM

```
                    Real numbers, R
                    /            \
        Irrational numbers, H   Rational numbers, Q
                                     |
                                 Integers, J
                                     |
                                 Whole numbers, W
                                     |
                                 Natural numbers, N
```

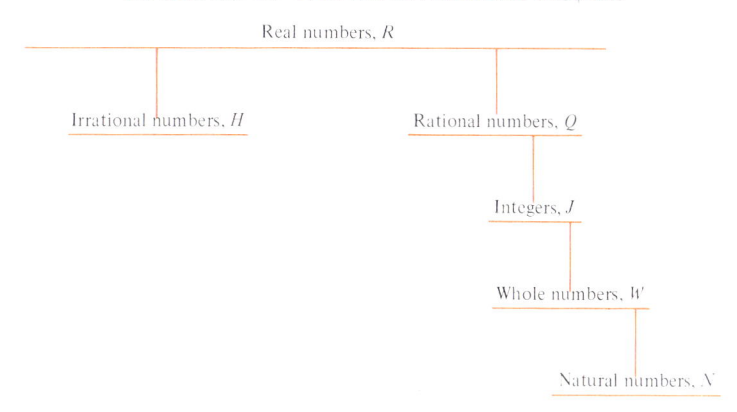

Figure 1-5

1-2 Exercises

1. Is $J \subseteq Q$?
2. Is $H \subset R$?
3. Is every element of Q also an element of J?
4. Are some elements of Q also elements of J?
5. Is it possible for a number to be both rational and irrational?
6. Show that 9 is a rational number.
7. If a number is rational, does it follow that it is an integer?
8. If a number is an integer, does it follow that it is a rational number?
9. Show that 2.5 belongs to the set of rational numbers.
10. Show that $1\frac{2}{3}$ is a rational number and give two other rational number expressions equivalent to it.
11. Which repeating decimal is represented by the fraction 3/11?
12. List three rational numbers between 2 and 3.
13. Is $\frac{5}{0}$ a rational number? Why?
14. Which element of the set of whole numbers is not a natural number?
15. Show that 2.4/3 is a rational number.
16. Give three irrational numbers.
17. Give two numbers that are rational numbers but not integers.

18. Give a number that is a real number but not a whole number or an integer.
19. Show that if $A = \{1, 3, 0.75\}$ and $B = \{0.75, 5, 6\}$, then $A \cap B$ has an element that is a rational number.
20. Show that $N \cap W$ is a subset of J.
21. Show that $N \subset W$.
22. Is the union of the set of natural numbers and their opposites equal to J?

In exercises 23–32, let $A = \{5, -7, 1.1313\ldots, 1.121121112\ldots, 0, 5/8, -6.8, \sqrt{2}/5\}$. List all the elements from set A that satisfy the given conditions.

23. natural numbers
24. whole numbers
25. negative numbers
26. positive numbers
27. integers
28. rational numbers
29. decimal numbers that are rational numbers
30. decimal numbers that are irrational numbers
31. repeating decimals
32. real numbers

In exercises 33–44, answer true or false.

33. Zero is not a real number.
34. Repeating decimals are not rational numbers.
35. Positive odd numbers are natural numbers.
36. Zero is an even number.
37. Decimal numbers are not real numbers.
38. Whole numbers are not integers.
39. The natural numbers do not include zero.
40. The symbols $+7$ and 7 represent the same number.
41. Every point on a real number line corresponds to some real number.
42. If a number is negative, it must be written with a minus sign in front of it.
43. The number zero is neither positive nor negative.
44. $\frac{9}{2}$ is a fraction but not a real number.

In exercises 45–52, divide the numerator of the given fraction by the denominator and determine whether the fraction represents a terminating or repeating decimal.

45. $\frac{5}{8}$ **46.** $\frac{9}{5}$ **47.** $\frac{2}{3}$ **48.** $\frac{5}{12}$

49. $\frac{2}{7}$ **50.** $\frac{2}{9}$ **51.** $\frac{3}{20}$ **52.** $\frac{3}{16}$

1-3 Order of Operations and Grouping Symbols

To simplify $3 \cdot 5 + 8$, we could first multiply the 3 and the 5 to get 15 and then add the 8. If we did this, we would have $3 \cdot 5 + 8 = 15 + 8 = 23$. On the other hand, we might first add the 5 and 8 to get 13 and then multiply 13 by 3. If we worked the problem this way, we would get $3 \cdot 5 + 8 = 3 \cdot 13 = 39$. Notice that we have two different answers. Which one, if either, is correct? An agreement is needed to determine which operations are done first. In working problems with more than one sign of operation (and containing no grouping symbols), we agree to the following.

Rule 1-1

Order of Operations

1. First, evaluate any powers.
2. Second, do all multiplications and divisions from left to right.
3. Third, do all additions and subtractions from left to right.

If this order of operations is applied to $3 \cdot 5 + 8$, we see 23 is the correct answer because the multiplication must be done first followed by the addition.

Multiplication is not always done first just because it is listed first in the order of operations. If a division comes before a multiplication in a problem, the division is done first. Multiplication and division have equal weight, and the one coming first in the problem is done first. A similar statement can be made for the addition and subtraction.

Example 1 Simplify $25 - 4 \cdot 5$.

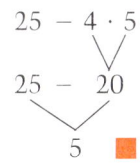

Solution
$25 - 4 \cdot 5$
$25 - 20$
5 ■

Example 2   Simplify $5 - 3 + 4 \cdot 3 + 24 \div 3$.

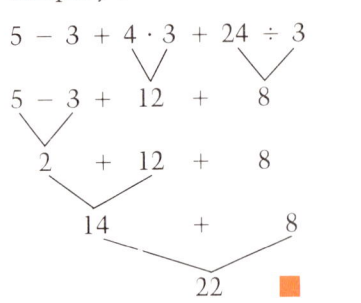

Solution

$$5 - 3 + 4 \cdot 3 + 24 \div 3$$
$$5 - 3 + 12 + 8$$
$$2 + 12 + 8$$
$$14 + 8$$
$$22 \; \blacksquare$$

Example 3   Simplify $2^3 + 10 \div 5 - 3^2$.

Solution

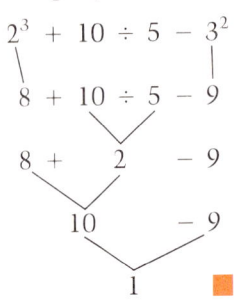

$$2^3 + 10 \div 5 - 3^2$$
$$8 + 10 \div 5 - 9$$
$$8 + 2 - 9$$
$$10 - 9$$
$$1 \; \blacksquare$$

The presence of grouping symbols in an expression containing several operations can make the simplification of that expression a great deal easier. For example, consider the expression $3 \cdot 5 + 8$ mentioned earlier. If we write it as $(3 \cdot 5) + 8$ and agree to do the operation within the parentheses first, we get

$$(3 \cdot 5) + 8 = 15 + 8 = 23$$

For the expression $3(5 + 8)$, we have

$$3(5 + 8) = 3(13) = 39$$

The operations within a set of grouping symbols must be done first, in the order given in rule 1–1. Once the grouping symbols are removed, any further simplifications are done according to the steps given in rule 1–1.

The most commonly used grouping symbols are the following:

parentheses	( )
brackets	[ ]
braces	{ }
bar or vinculum	────

In fractions, we sometimes write the fraction bar as a solidus (slash mark). For example, we could write $\frac{3}{5}$ as 3/5.

Sets and The Real Number System 13

Example 4 Simplify $3[4^2 + 2(5^2 - 4 \cdot 2)] - \frac{3}{3}$.

Solution

$3[4^2 + 2(5^2 - 4 \cdot 2)] - \frac{3}{3}$

$3[4^2 + 2(25 - 8)] - \frac{3}{3}$

$3[4^2 + 2(17)] - \frac{3}{3}$

$3[4^2 + 34] - \frac{3}{3}$

$3[16 + 34] - \frac{3}{3}$

$3[50] - \frac{3}{3}$

$150 - 1$

149

Example 5 Supply grouping symbols to give the indicated answer.
$3 \cdot 4 + 2 - 5 = 13$

Solution For $3 \cdot 4 + 2 - 5$ to equal 13, we must put parentheses around $4 + 2$, which gives

$$3(4 + 2) - 5 = 13$$

1–3 Exercises

In exercises 1–32, simplify, giving a single number for each. *(See examples 1–4.)*

1. $17 - 5 \cdot 2$
2. $24 \cdot 2 - 7 \cdot 4 + 6$
3. $8 - 2 + 12 \div 6$
4. $9 + 5 \div 5 + 18$
5. $2 \cdot 3 \div 6 \cdot 5 - 4 \cdot 5 \div 10$
6. $4^2 - 8 \div 2 + 3^3$
7. $8^2 - 7 \cdot 3 + 5 \cdot 0$
8. $9 \cdot 2 + 10 \div 5 - 3^2$
9. $100 - 20 \div 2 + 9 \cdot 3 + 2^2$
10. $\dfrac{6^2 + 5 \cdot 2 + 4}{2 \cdot 5^2}$
11. $\dfrac{20 \cdot 4 \div 10 + 2}{5} + \dfrac{5^2 + 2 \cdot 5}{3 + 10 \div 5}$
12. $\dfrac{6 \cdot 2 + 3 \cdot 4}{2^2 \cdot 3} + \dfrac{2 \cdot 5^2 + 5 \cdot 10}{8 + 20 \div 10}$
13. $\dfrac{22 \div 11 + 8 + 2 \cdot 5}{7 - 12 \div 6}$
14. $2 \cdot [4 - (6 - 3)]$

15. $86 \cdot 2\{4 \cdot [8 - (5 - 3)] + 8\}$
16. $28 - [(16 + 3) - (8 \cdot 2)]$
17. $\dfrac{29 - (16 + 3) + 27}{(2 \cdot 3)^2 + 1}$
18. $2^3[16 - (4 + 2)]$
19. $8 \cdot 3 + 5^2 - (7 - 3) + (2 \cdot 2^2)$
20. $5 \cdot 8^2 - (7 - 2) + 5$
21. $9 + \{7 + [8 - (7 - 6)] + 2\}$
22. $6^2 + \dfrac{(20 - 4)}{8}$
23. $12 \cdot \dfrac{3^2 + (6 - 3)}{4}$
24. $8^2 + (2^2 + 5^2) - [3^2 - (10 - 4)]$
25. $\dfrac{5^2 + 4^2 - 3^2}{8}$
26. $\dfrac{18 \div 9 \cdot (3^2 + 4)}{13}$
27. $\dfrac{25 - (3^2 + 4^2)}{10}$
28. $14 + [5 + (4 + 3 \cdot 4) - 4^2]$
29. $(20)^2 - [12 \cdot 5 + 4(2 \cdot 3 - 2)]$
30. $\dfrac{(17)^2 + (14 - 3)}{2 \cdot (10)^2 + (6^2 + 8^2)}$
31. $(10)^3 + [5(4 + 12 \div 3) + 7]$
32. $(25)^2 + [7^2 - (7 \cdot 2 - 3)]$

In exercises 33–36, supply grouping symbols in the following exercises to give the indicated answer. (See example 5.)

33. $5 \cdot 18 \div 3 \cdot 2 - 3;\quad 57$
34. $5 \cdot 8 + 3 \cdot 2 + 1;\quad 71$
35. $4 \cdot 9 + 2 \cdot 5 \cdot 3 - 2;\quad 148$
36. $2 \cdot 30 - 2 \cdot 10 \div 2 + 5;\quad 50$
37. Think of any natural number and add 3 to it. Multiply this sum by 6. Divide this product by 2 and then subtract 9. Divide this difference by 3. What is the answer?
38. Think of any natural number and square it. Now add 5. Multiply this result by 4. Now divide by 2 and then subtract 10. Divide this result by 2. Take the positive square root of the resulting quotient. What is the answer?
39. Subtract 5 from your age. Multiply this difference by 16. Divide this product by the cube of 2. Now add 10. Divide this result by 2. What is the answer?

Sets and The Real Number System

40. Three coins have a value of 16¢. One is not a penny and one is not a dime. What are the coins?

In exercises 41–44, simplify each expression.

41. $2.51 + 2(9.16)$
42. $(8.3)^2 + 5(8.3)$
43. $7(5.22) \div 3(9.82)$
44. $\dfrac{9(7.11)^2 + 3(6.21)}{0.54}$

1–4 Axioms of Equality for the Real Number System

Most mathematical systems rest upon one or more assumptions, called axioms. An axiom is a simple, obvious fact that is accepted as being true without proof. Algebra has as its most basic assumptions the following four axioms.

The Reflexive Axiom

If a is any real number, then $a = a$.

The reflexive axiom states that any number is equal to itself.

The Symmetric Axiom

If a and b are real numbers and $a = b$, then $b = a$.

Example 1 If $5 = \frac{10}{2}$, then $\frac{10}{2} = 5$.

To paraphrase the symmetric axiom we might say that the two sides of an equation may be interchanged.

The Transitive Axiom

If a, b, and c are real numbers and $a = b$ and $b = c$, then $a = c$.

Example 2 If $5 = \frac{10}{2}$ and $\frac{10}{2} = \frac{20}{4}$, then $5 = \frac{20}{4}$.

The Substitution Axiom

If a and b are real numbers and $a = b$, then a may be replaced by b or b may be replaced by a within any expression or statement representing a real number without changing the number.

Example 3   If $2b = 6$ and $a = b$, then $2a = 6$.

Example 4   Tell which axiom of equality is used in each of the following statements.

(a) $2c = 2c$
(b) If $a = b$ and if $2a = 5c$, then $2b = 5c$
(c) If $2x = y$ and if $y = 4z$, then $2x = 4z$.
(d) If $4c = \dfrac{p}{2}$, then $\dfrac{p}{2} = 4c$.

Solution   (a) reflexive axiom.   (b) substitution axiom. (a has been replaced by b in $2a = 5c$.)   (c) transitive axiom or substitution axiom. (d) symmetric axiom.

1–4 Exercises

In exercises 1–18, identify the axiom of equality that has been used. *(See examples 1–4.)*

1. If $2 = 2x + 1$, then $2x + 1 = 2$.
2. If $a = 2c$ and $2c = 5$, then $a = 5$.
3. If $x = 3$ and $2x = 6$, then $2 \cdot 3 = 6$.
4. $2x + 1 = 2x + 1$
5. $(a - b)^2 = (a - b)^2$
6. If $6x = 3$ and $4y = 3$, then $6x = 4y$.
7. If $2x + 5 = 4$, then $4 = 2x + 5$.
8. $9x - 2 = 9x - 2$
9. If $x = 4$, $y = 3$, and $3x = 4y$, then $3 \cdot 4 = 4 \cdot 3$.
10. If $c + d = b$ and $b = 4$, then $c + d = 4$.
11. If $2 + a = b$, then $b = 2 + a$.
12. If $2a = -4$, then $-4 = 2a$.
13. $5 = 5$
14. If $5c + 2 = d$ and $c = x$, then $5x + 2 = d$.
15. If $4p = 3t$ and $3t = 9r$, then $4p = 9r$.

Sets and The Real Number System 17

16. If $2a = 5c$ and $c = 3$, then $2a = 15$.
17. If $a + 8 = 3 + c$ and $c = 18$, then $a + 8 = 21$.
18. $3t - 2 = 3t - 2$
19. What result is obtained in the reflexive statement $3a + b = 3a + b$ if a is replaced by 6 and b is replaced by 4?
20. What result is obtained in the reflexive statement $9c - 2b = 9c - 2b$ if $c = 5$ and $b = 3$?
21. Verify that $2a + 3 = c$ and $c = 2a + 3$ give the same results by letting $a = 5$ and $c = 13$. To which axiom must we appeal to know that these results are the same?
22. Is $2c - 4 = 2c - 4$ an illustration of the reflexive axiom?

In exercises 23–31, replace the question mark to make the given statement true, using the property that is given.

Example If $2x = y$, then ___?___ $= 2x$; symmetric axiom of equality.

Solution If $2x = y$, then ___y___ $= 2x$. ■

23. $2x + y =$ ___?___; reflexive axiom of equality.
24. If $2a = 3b$ and $b = c$, then ___?___ $= 3c$; substitution axiom of equality.
25. If $5a = 4$ and $y = a$, then $5y =$ ___?___; substitution axiom of equality.
26. If $9a = 4$ and $4 = 5c$, then ___?___ $=$ ___?___; transitive axiom of equality.
27. If $y = b$, then $6x + y + 3 = 6x +$ ___?___ $+ 3$; substitution axiom of equality.
28. $x^2 + 3 =$ ___?___; reflexive axiom of equality.
29. If $5 = 8 - 3$, then ___?___ $=$ ___?___; symmetric axiom of equality.
30. If $\frac{40}{4} = 10$ and $10 = \frac{70}{7}$, then ___?___ $=$ ___?___; transitive axiom of equality.
31. If $9x = c$ and $x = a^2$; then ___?___ $=$ ___?___; substitution axiom of equality.

1–5 Axioms of Inequality for the Real Number System

Just as we need axioms of equality for our study of algebra, we also need axioms relating to inequalities. The symbol $<$ is used to indicate that one number is less than another. If a is less than b, we write $a < b$ and read "a is less than b." Other symbols used to indicate inequalities and the way they are read are given in the table. An example, and the way it is read, is also shown for each symbol.

Symbol	Read	Example	Read
$>$	is greater than	$5 > 3$	5 is greater than 3.
$\leq$	is less than or equal to	$7 \leq 9$	7 is less than or equal to 9.
$\geq$	is greater than or equal to	$8 \geq 3$	8 is greater than or equal to 3.
$\neq$	is not equal to	$8 \neq 5$	8 is not equal to 5.
$\not>$	is not greater than	$6 \not> 8$	6 is not greater than 8.
$\not\leq$	is not less than or equal to	$7 \not\leq 2$	7 is not less than or equal to 2.

The concept of inequality is defined in definition 1–5.

Definition 1–5

Less Than

If a and b are any two distinct real numbers, then $a < b$ if and only if:

1. a lies to the left of b on a number line or
2. There exists some positive real number c such that $a + c = b$.

The two statements in this definition are equivalent.

Example 1

Consider the number line shown.

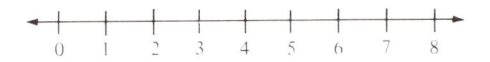

```
←—+——+——+——+——+——+——+——+——+→
   0   1   2   3   4   5   6   7   8
```

We see that 5 is to the left of 8. Thus, by definition, $5 < 8$. We also know that there exists some positive number c such that $5 + c = 8$. Using the number line, we see $5 + 3 = 8$. Because $5 + 3 = 8$ and 3 is positive, we may conclude that $5 < 8$ by part 2 of definition 1–5. ■

The second condition allows us to turn an inequality into an equality by adding the appropriate positive number to the smaller of the two numbers making up the inequality. If $2x + 1 < 8$, then for some positive real number c, $2x + 1 + c = 8$.

Trichotomy Axiom

If a and b are real numbers, then one and only one of the following is true: $a < b$, or $a = b$, or $a > b$.

Transitive Axiom

If a, b, and c are real numbers and $a < b$ and $b < c$, then $a < c$.

Example 2

(a) If $2 < 5$ and $5 < 8$, then $2 < 8$.
(b) If $2x < 5$ and $5 < y$, then $2x < y$.

Example 3

Tell which property of inequality is used.

(a) If $13 < 25$, then $13 + c = 25$ for some $c > 0$.
(b) $18 < 20$
(c) $5 < 8$ and $8 < 10$. Therefore, $5 < 10$.
(d) If $-16 < -10$, then $-16 + c = -10$ for some $c > 0$.

Solution

(a) definition of inequality   (b) definition of inequality; $18 + 2 = 20$ and 2 is positive.   (c) transitive axiom   (d) definition of inequality

1-5 Exercises

In Exercises 1–12, tell which property of inequality is used.

1. Because $7 = 7$, then $7 \not< 7$ and $7 \not> 7$.
2. If $2x < a$ and $a < c$, then $2x < c$.
3. If $2x < y$, then for some positive number c, $2x + c = y$.
4. If $(2 + 5) < 13$ and $13 < 20$, then $(2 + 5) < 20$.
5. $18 < 42$.
6. $(1 + a) < 10$ if $a = 4$.
7. If $(14 + x) < 10$, then $(14 + x) + c = 10$ for some positive number c.
8. If $(a + b) + c = 16$ for some positive number c, then $(a + b) < 16$.
9. If $2x > 4$ and $4 > y$, then $2x > y$.
10. Because $17 \not> 20$ and $17 \neq 20$, then $17 < 20$.
11. If $\frac{20}{2} = 10$, then $\frac{20}{2} \not> 10$ and $\frac{20}{2} \not< 10$.
12. If $2x < 5$ and $5 < c$, then $2x < c$.

In exercises 13–19, replace the question mark with the appropriate number or symbol to make the statement true using the given axiom or definition of inequality.

Example   If $2a < 9$ and $9 < d$, then ___?___ $< d$; transitive axiom of inequality.

Solution If $2a < 9$ and $9 < d$, then __$2a$__ $< d$.

13. If $3a < 5$, then for some positive number c, $3a + c = $ ___?___ ; definition of inequality.
14. If $5 < 9$, then $5 + $ ___?___ $= 9$; definition of inequality.
15. If $5 \not> 7$ and $5 \neq 7$, then 5 ___?___ 7; trichotomy axiom.
16. If $(5 + x) < 3$, then for some positive number c, ___?___ $+ c = 3$; definition of inequality.
17. If $4x < 5$ and $5 < 3b$, then $4x < $ ___?___ ; transitive axiom of inequality.
18. If $a < 5$, then a lies to ___?___ of 5 on the number line; definition of inequality.
19. If $3x + c = 5$ for some positive real number c, then $3x$ ___?___ 5; definition of inequality.

In exercises 20–25, rewrite the given problem as a true statement without using the slash mark.

Example   Rewrite without using a slash mark.
(a) $9 \not> 12$ (b) $t \not< 9$

Solution (a) $9 < 12$ (b) $t \geq 9$

20. $t \not< 12$
21. $x \neq 5$
22. $5 \not> 8$
23. $7 \neq t$
24. $(4 + 8) \not\leq x$
25. $9 \neq (t - 1)$
26. $(t - 3) \not\leq 8$

In exercises 27–42, indicate which definition or axiom of equality or inequality has been used.

27. If $9x = 2y$, then $2y = 9x$.
28. $27 < 32$
29. If $3x < 2y$, then for some positive number c, $3x + c = 2y$.
30. If $3x = 4$ and $4 = y$, then $3x = y$.
31. $2x + 3 = 2x + 3$
32. $(3 + 7) < 22$

Sets and The Real Number System

33. If $5x = 2y$ and $y = 3$, then $5x = 6$.
34. Because $7 < 9$, then $7 \not= 9$.
35. Because $24 > 20$, then $24 \neq 20$.
36. If $2x < 3$ and $3 < 5$, then $2x < 5$.
37. $(18 + 3) < 25$
38. Because $3 + 2 = 5$, then $3 < 5$.
39. If $a = 5b$ and $b = 3$, then $a = 15$.
40. If $2x + 3 = y$, then $y = 2x + 3$.
41. If $ac = b$ and $c = x$, then $ax = b$.
42. If $3x + 5 = 10$, then $3x < 10$.
43. A geologist found that the weight of rocks in box A was heavier than that of rocks in box B. Express this fact in two ways using greater than and less than symbols. Let the weight of box A be s pounds and the weight of box B be t pounds.
44. (a) In exercise 43, suppose the rocks in box A are of uniform size and weight. Suppose those in box B are also of uniform size and weight, but not the same size and weight as those in box A. Let p represent the weight of a rock from box A and q represent the weight of a rock from box B. If two rocks from box A weigh the same as three rocks from box B, express this fact using an equation.
    (b) If $p = 3$ pounds, make a substitution in part (a) to indicate the weight of three rocks from box B.
45. Suppose a particle moves along a straight line according to the equation $d = 5x^2 + 3x$, where d is the distance, measured in feet, from the origin.
    (a) Use a substitution and express d in terms of t if $x = t + 3$.
    (b) Find d if $t = 2$.

1–6 Other Properties of the Real Numbers

In addition to the properties of equality and inequality discussed in the last two sections, the real numbers have other important properties. Some familiarity with these other properties is a necessity if we are to understand why algebra works as it does.

The axioms presented here are based on the assumption that addition, subtraction, multiplication, and division of the real numbers are **binary operations.** By this, we mean that only two real numbers at a time may be added, subtracted, multiplied, or divided. (Try it. If you wanted to add $3 + 5 + 7$, what do you do? You might add 3 and 5 to get 8 and then add this result to 7 to get $8 + 7 = 15$. At each step, you work with only two numbers.)

Closure Axiom for Addition
For any two real numbers a and b,
$$a + b \text{ is a real number.}$$

Closure Axiom for Multiplication
For any two real numbers a and b,
$$ab \text{ is a real number.}$$

These two closure axioms state that if we add or multiply any two real numbers, the result must also be a real number.

Commutative Axiom for Addition
For any real numbers a and b,
$$a + b = b + a$$

Commutative Axiom for Multiplication
For any real numbers a and b,
$$ab = ba$$

Example 1

$$3 + 2 = + 2 + 3 \qquad\qquad 8 \cdot 5 = 5 \cdot 8$$
$$5 = 5 \qquad\qquad\qquad\qquad 40 = 40$$

Associative Axiom for Addition
For any real numbers a, b, and c,
$$(a + b) + c = a + (b + c)$$

Associative Axiom for Multiplication
For any real numbers a, b, and c,
$$(ab)c = a(bc)$$

Example 2

$$(3 + 2) + 6 = 3 + (2 + 6) \qquad\qquad (3 \cdot 4) \cdot 2 = 3 \cdot (4 \cdot 2)$$
$$5 + 6 = 3 + 8 \qquad\qquad\qquad 12 \cdot 2 = 3 \cdot 8$$
$$11 = 11 \qquad\qquad\qquad\qquad 24 = 24$$

Additive Identity Axiom

There is a unique real number 0 such that for any real number a,

$$a + 0 = a \quad \text{and} \quad 0 + a = a$$

Multiplicative Identity Axiom

There is a unique real number 1 such that for any real number a,

$$a \cdot 1 = a \quad \text{and} \quad 1 \cdot a = a$$

The numbers 0 and 1 preserve the identity of the number for addition and multiplication, respectively.

Example 3

$6 + 0 = 6$
$7 \cdot 1 = 7$ ■

Zero is called the **additive identity** and one is called the **multiplicative identity.**

Additive Inverse Axiom

For each real number a, there exists a unique real number $-a$ (called the **additive inverse** or **opposite** of a) such that

$$a + (-a) = 0 \quad \text{and} \quad (-a) + a = 0$$

Example 4

$5 + (-5) = 0$ ■

Multiplicative Inverse Axiom

For each nonzero real number a, there exists a unique real number $1/a$ such that

$$a \cdot \left(\frac{1}{a}\right) = 1 \quad \text{and} \quad \left(\frac{1}{a}\right) \cdot a = 1$$

The symbol $1/a$ is called the **multiplicative inverse** or **reciprocal** of a.

Example 5 $6(\frac{1}{6}) = 1$

> **Distributive Axiom**
> For any real numbers a, b, and c.
> $$a(b + c) = ab + ac$$
> and
> $$(b + c)a = ba + ca$$

Example 6
$5(2 + 3) = 5 \cdot 2 + 5 \cdot 3$
$5(5) = 10 + 15$
$25 = 25$

Example 7 Use the distributive axiom to rewrite $5(2 + b)$.
Solution
$5(2 + b) = 5 \cdot 2 + 5b$
$ = 10 + 5b$

Example 8 Use the distributive axiom to simplify $3x + 5x$.
Solution
$3x + 5x = (3 + 5)x$
$ = 8x$

Facts that can be proven by using previously accepted facts are called **theorems.** Two theorems with their proofs are now given. Reasons are given in each step in the proofs so that you may see how the axioms we have been learning can be applied to prove other statements.

> **Equality Addition Theorem**
> If a, b, and c are real numbers and $a = b$, then $a + c = b + c$.

This theorem states that we may add the same number to both sides of an equation.

Proof

Statements	Reasons
1. a, b, and c are real numbers and $a = b$.	1. Given information.

Sets and The Real Number System

2. $a + c$ is a real number.
3. $a + c = a + c$
4. $a + c = b + c$

2. Closure axiom for addition.
3. Reflexive axiom.
4. Substitution axiom.

In the next theorem, we supply the statements. You should place a piece of paper in the *Reasons* column and fill in the reasons. The proof is very similar to the one just given. The correct reasons are given at the bottom of the next page. Supply your reasons and then check to see how well you did.

Equality Multiplication Theorem

If a, b, and c are real numbers and $a = b$, then $ac = bc$.

This theorem states that we may multiply both sides of an equation by the same number.

Proof

Statements

1. a, b, and c are real numbers and $a = b$.
2. ac is a real number.
3. $ac = ac$
4. $ac = bc$

Reasons

1.
2.
3.
4.

If any real number is multiplied by zero, the product is zero. This fact is stated without proof as the next theorem.

Multiplication by Zero Theorem

If a is any real number, $a \cdot 0 = 0 \cdot a = 0$

The final theorem we give is called the *double-negative theorem*.

Double-Negative Theorem

If a is any real number, then $-(-a) = a$.

Example 9

$-(-8) = 8$; Read "the opposite of negative eight is eight."
If $a = -8$ in the theorem, then

$$-[-(-8)] = -[+8] = -8$$

26   Intermediate Algebra

The essence of the theorem is that the opposite of the opposite of a given number is the given number.

Example 10 Tell which axioms or theorems have been used in the following.

(a) $5 + 3 = 3 + 5$
(b) $3 + (4 + 5) = (3 + 4) + 5$
(c) $3 \cdot 1 = 3$
(d) $3(5 + 2) = 3 \cdot 5 + 3 \cdot 2$
(e) $5 + 3 = 8$
(f) $5 \cdot 0 = 0$

Solution (a) commutative axiom for addition   (b) associative axiom for addition   (c) identity axiom for multiplication   (d) distributive axiom   (e) addition fact   (f) multiplication by zero theorem

1–6 Exercises

In exercises 1–23, tell which axioms or theorems are used. If more than one is used in an exercise, name each one.

1. $9 + 3 = 3 + 9$
2. $a \cdot (3b) = (a \cdot 3) \cdot b$
3. $3 + 0 = 3$
4. $3 \cdot 1 + (5 + 0) = 8$ (*Hint:* Two axioms and one arithmetic fact used.)
5. $a \cdot (3 + 4) = a \cdot 3 + a \cdot 4$
6. $9a \cdot 1 = 9a$
7. $3(4 \cdot a) = a \cdot (3 \cdot 4)$
8. $7a + 5a = a \cdot (7 + 5)$
9. $2x + 3xy = x(2 + 3y)$
10. $ab + 1 = 1 + ab$
11. If $3 = 2x$, then $5 \cdot 3 = 5 \cdot 2x$.
12. If $(4 + 5) = x$, then $(4 + 5) + y = x + y$.
13. $(x + y) + z = (y + x) + z$
14. $a + (b + c) = (b + a) + c$
15. $9c + 9d = 9(c + d)$
16. $(2x + y) + 0 = 2x + y$
17. If $2x = 3y$ and c is any real number, then $2x \cdot c = 3y \cdot c$.

Answers to proof on page 25:
1. Given information   2. Closure axiom for multiplication
3. Reflexive axiom   4. Substitution axiom

Sets and The Real Number System

18. $5 \cdot (a \cdot c) = (5 \cdot a) \cdot c$
19. If $9x - 2 = c$, then $2 \cdot (9x - 2) = 2 \cdot c$.
20. $5 \cdot \dfrac{1}{5} + xy = 1 + xy$
21. $9x + 3y = 3(3x + y)$
22. $5 + (-5) + 8 = 8$
23. $25 + (x \cdot y) = (y \cdot x) + 25$

In exercises 24–27, supply the reason for each step.

Example

$(z + y) + 3 = 3 + (z + y)$ commutative axiom for addition
$ = 3 + (y + z)$ commutative axiom for addition

24. $(d \cdot c + f \cdot e) + 2 = 2 + (d \cdot c + f \cdot e)$
$ = 2 + (c \cdot d + e \cdot f)$
25. $y + (z + x) = y + (x + z)$
$ = (y + x) + z$
$ = (x + y) + z$
26. $3 \cdot y + x \cdot (4 \cdot y) = 3 \cdot y + (x \cdot 4)y$
$ = 3 \cdot y + (4 \cdot x)y$
$ = (3 + 4 \cdot x)y$
27. $2 \cdot 1 + 5ac + 2cb = 2 + 5ac + 2cb$
$ = 2 + 5ac + 2bc$
$ = 2 + (5a + 2b)c$

In exercises 28–40, use the associative axioms to rewrite and simplify each expression.

Example

Use the associative axioms to simplify each.
(a) $3(5x)$ (b) $2 + (5 + c)$

Solution

(a) $3(5x) = (3 \cdot 5)x$ (b) $2 + (5 + c) = (2 + 5) + c$
$ = 15x$ $ = 7 + c$

28. $5(2a)$
29. $9(8c)$
30. $6(5b)$
31. $2(3t)$
32. $5 + (6 + x)$
33. $9 + (12 + t)$
34. $17 + (2 + p)$
35. $23 + (4 + a)$
36. $\dfrac{1}{5}(20c)$
37. $\dfrac{4}{9}(18t)$
38. $\dfrac{5}{6}(12b)$
39. $\dfrac{5}{9} + \left(\dfrac{2}{9} + x\right)$
40. $\dfrac{2}{7} + \left(\dfrac{5}{7} + 3c\right)$

In exercises 41–52, use the distributive axiom to rewrite each expression. Simplify where possible.

Example Use the distributive axiom to rewrite each. Then simplify.
(a) $5(6 + c)$ (b) $6x + 9x$

Solution
(a) $5(6 + c) = 5 \cdot 6 + 5c$
$\qquad\qquad = 30 + 5c$

(b) $6x + 9x = (6 + 9)x$
$\qquad\qquad = 15x$

41. $5(9 + b)$ **42.** $3(10 + c)$ **43.** $7x + 6x$
44. $9a + 7a$ **45.** $(5 + x)3$ **46.** $(12 + c)5$
47. $\frac{1}{5}x + \frac{2}{5}x$ **48.** $\frac{5}{8}(16 + c)$ **49.** $5c + 2c + 3c$
50. $8a + 4a + 6a$ **51.** $5(10 + 2 + c)$ **52.** $6(8 + 4 + t)$

In exercises 53–58, simplify. Name each axiom or theorem used. Refer to the example following exercise 52.

Example Use as many of the axioms mentioned in this section as necessary to simplify $6(3 + \frac{1}{6}a + 5b)$. Name each axiom used.

Solution
$6\left(3 + \frac{1}{6}a + 5b\right)$

$= 6 \cdot 3 + 6\left(\frac{1}{6}a\right) + 6(5b)$ distributive axiom

$= 18 + 6\left(\frac{1}{6}a\right) + 6(5b)$ multiplication fact

$= 18 + \left(6 \cdot \frac{1}{6}\right)a + (6 \cdot 5)b$ associative axiom for multiplication

$= 18 + 1a + (6 \cdot 5)b$ multiplicative inverse axiom

$= 18 + a + (6 \cdot 5)b$ multiplicative identity axiom

$= 18 + a + 30b$ multiplication fact

53. $7\left[5 + \frac{1}{7}(b) + 6c\right]$

54. $9\left[8 + \frac{1}{9}(x) + 5y\right]$

Sets and The Real Number System

55. $9x + 5x + 5y + 6y$
56. $10a + 7a + 5b + 9b + 2$
57. $6c + 7c + 8d + 9d + 5 + 3$
58. $2\left(5b + \dfrac{1}{2}c + 10\right) + 3 \cdot 0$

In exercises 59–66, use the double-negative theorem to simplify.

59. $-(-5)$ **60.** $-(+6)$ **61.** $-[-(6)]$ **62.** $-[-(-8)]$
63. $-(-a)$ **64.** $-(-x)$ **65.** $-[-(c)]$ **66.** $-[-(-c)]$

1–7 Addition and Subtraction of Signed Numbers

Signed numbers are numbers which are either positive or negative. A positive signed number may be written with a plus sign in front of the number, or the plus sign may be omitted. A negative signed number must be written with a minus sign in front of it. The numbers $+9$, 7, $+3$, and 2 are all positive signed numbers, while -6, -10, and -5 are negative signed numbers.

Signed numbers have many applications. For example, sometimes we wish to think of the distance that a number lies from the origin on a number line. For example, 6 lies 6 units to the right of the origin, while -8 lies 8 units to the left of the origin. Because distance is always considered to be positive, we must ignore the minus sign in -8 when we say its distance from the origin is 8 units. To show this, we use two vertical bars to enclose -8 and write $|-8|$. We then think of the notation $|-8|$ as representing 8. That is, $|-8| = 8$. In general, the distance that a number x lies from the origin on a number line is called the **absolute value** of the number and is denoted by $|x|$. The formal definition of the absolute value of a number x is given in definition 1–6.

Definition 1–6

Absolute value

For any real number x,

$$|x| = \begin{cases} x & \text{if } x \geq 0 \\ -x & \text{if } x < 0 \end{cases}$$

The $-x$ in the second part of the definition tells us to form the opposite of x. If x is negative, then $-x$ means the opposite of a negative number, which is a positive number.

Examples

(a) $|3| = 3$
(b) $|-7| = -(-7) = 7$ ■

You probably recall how to add signed numbers. Check your knowledge of adding signed numbers by examining the following four sums:

$(+3) + (+5) = +8$
$(-3) + (+5) = +2$
$(-3) + (-5) = -8$
$(+3) + (-5) = -2$

Now study rule 1–2 and see how each sum relates to either part 1 or part 2 of rule 1–2.

Rule 1–2

Adding Signed Numbers

1. If both numbers have the same sign, add their absolute values. The sum has the same sign as the two numbers.
2. If the numbers have opposite signs, find the absolute values of both numbers. Subtract the lesser absolute value from the greater. The sum has the sign of the number with the greater absolute value.

Example 2

Use rule 1–2 to add the following.
(a) $(+5) + (+2)$ (b) $(-6) + (-3)$ (c) $(-5) + (+8)$

Solution

(a) $(+5) + (+2) = +(|+5| + |+2|) = +(5 + 2) = +7$, or 7
(b) $(-6) + (-3) = -(|-6| + |-3|) = -(6 + 3) = -9$
(c) $(-5) + (+8) = +(|+8| - |-5|) = +(8 - 5) = +3$, or 3 ■

The detailed steps shown in Example 2 are usually not necessary. You will probably soon learn to look at the given problem and calculate the sum mentally.

Example 3

(a) $(-6) + (-8) + (+3) = (-14) + (+3) = -11$
(b) $(-19) + (+20) + (-8) = (+1) + (-8) = -7$
(c) $(-10) + (0) + (-2) = (-10) + (-2) = -12$ ■

Subtraction of two signed numbers a and b follows the rule that $a - b = a + (-b)$. That is, instead of subtracting, we add the opposite, or additive inverse, of the subtrahend, or number being subtracted.

Sets and The Real Number System

This idea is illustrated in the following display. In the first column, the two numbers have been subtracted. In the second column, the opposite of the subtrahend has been added. The results are the same in either case.

$$8 - 3 = 5 \qquad 8 + (-3) = 5$$
$$19 - 16 = 3 \qquad 19 + (-16) = 3$$
$$28 - 8 = 20 \qquad 28 + (-8) = 20$$
$$102 - 84 = 18 \qquad 102 + (-84) = 18$$

Rule 1–3

Subtracting Signed Numbers

To subtract two signed numbers, change the sign of subtraction to addition and change the sign of the subtrahend to the opposite sign. Then add, using the rules of addition of signed numbers.

Example 4

Subtract.

(a) $18 - (-3)$ (b) $-26 - (+4)$

Solution

(a) change this sign to its opposite

$$18 - (-3) = 18 + (+3) = 21$$

change to addition

(b) change this sign to its opposite

$$-26 - (+4) = -26 + (-4) = -30$$

change to addition

You may check your answer by adding: Add your answer to the number being subtracted and it must equal the first number. Make sure you do this in the original subtraction problem and not in the one where you changed signs. In example 4(a), note that

$$18 - (-3) = 18 + (+3) = 21$$

To check the correctness of this answer observe that

$$21 + (-3) = 18 \quad ■$$

Example 5

(a) $17 - (-3) = 17 + (+3) = 20$ *Check:* $20 + (-3) = 17$
(b) $-18 - (-4) = -18 + (+4) = -14$ *Check:* $-14 + (-4) = -18$
(c) $-72 - (+4) = -72 + (-4) = -76$ *Check:* $-76 + (+4) = -72$
(d) $42 - 18 = 42 + (-18) = 24$ *Check:* $24 + 18 = 42$
(e) $-16 - 5 = -16 + (-5) = -21$ *Check:* $-21 + 5 = -16$ ■

1–7 Exercises

Simplify exercises 1–12 until a single integer is obtained. (See examples 2–3.)

1. $(11) + (-3)$
2. $(+17) + (-25) + (+3)$
3. $(-102) + (+63) + 0$
4. $(-776) + (9,600) + (-24)$
5. $(-96) + (79) + 63$
6. $86 + (-84) + 63$
7. $-8 + [(17 + 4) + (-3)]$
8. $7 + 3 + [7 + (-2)] + (3 + 7)$
9. $26 + [(-7 + 5)] + 3$
10. $3 + \{(-8) + (-4 + [-5])\} + (-10)$
11. $36 + (-43) + [-19 + (-5)]$
12. $100 + (-82) + [7 + (-14)] + (-16)$

In exercises 13–19, use the distributive axiom to make the computations easier.

Example $22(8) + 22(-5) = 22[8 + (-5)] = 22(3) = 66$

13. $16(-24) + 16(+26)$
14. $72(102) + 72(-2)$
15. $60\left(\dfrac{3}{4} + \dfrac{2}{12}\right)$
16. $90\left(\dfrac{3}{5} + \dfrac{1}{15}\right)$
17. $26 + 9(26)$
18. $36(42) + 36(21) + 36(37)$
19. $82(125) + 82(-25)$

In exercises 20–44, perform the indicated operations. (See examples 4 and 5.)

20. $74 - (-20)$
21. $-16 - (-14)$
22. $96 - 27$
23. $-82 - (-3)$
24. $-8 - 4$
25. $[-8 - (-2)] - [17 - 5]$
26. $(-2) - (-3) - (-8)$
27. $0 - (-8)$
28. $5 + (-3) - (-4) - (+6)$
29. $19 - (-3) + (-8) - (5 - 3)$
30. $36 - [8 - (-3)]$
31. $54 - (-3) + [9 + (-4) - (-3)]$
32. $75 - [42 + (-16) + (-26)]$
33. $|-4| - (7 - 3)$
34. $|[8 - (-3)] - 15| - (-3)$
35. $-10 - 8 - 3 - 4$
36. $23 - (-8 - 5 - 2)$
37. $63 - 8 - [-7 + (-3)]$
38. $8 + 7 - [4 - (-12) + (-3)]$
39. $\dfrac{28 - (-4) + 16}{16 - (-8)}$
40. $\dfrac{100 - [7 - (-3)]}{20 + [20 - (-5)]}$
41. $\{17 - [9 - (-3)]\}^2$

Sets and The Real Number System

42. $\dfrac{100 - (8 \cdot 3 - 4)}{-8 - (-28)}$

43. $\dfrac{8 \cdot 5 - [2 - (3)] - 1}{10 - (-7)}$

44. $-93 - 4 \cdot 5 - (-8) + (-6)$

The distributive axiom also applies to subtraction. That is, $a(b - c) = ab - ac$. For example, $3(5 - a) = 3 \cdot 5 - 3a = 15 - 3a$. In exercises 45–56, apply the distributive axiom and other axioms, rules and theorems as needed to simplify each expression.

45. $5(3 - a)$
46. $2(5 - x)$
47. $3(6 - 5c)$
48. $10(8 - 2d)$
49. $7(2a - b + c)$
50. $9(6a - 5b - 2c)$
51. $12(6a + 4b + 1)$
52. $c(3a + 2b)$
53. $x(5 + 2y)$
54. $3(3a + |-2| + 6)$
55. $9(2b - |-3| - 2)$
56. $10(|-4| - |-3| + 2c)$

In exercises 57–66, use the distributive axiom to simplify.

Example Use the distributive axiom to simplify $2x - 8x + 3x$.

Solution
$$\begin{aligned} 2x - 8x + 3x &= (2 - 8 + 3)x \\ &= [2 + (-8) + 3]x \\ &= [-6 + 3]x \\ &= -3x \end{aligned}$$

57. $7x - 9x + 2x$
58. $7a + 6a - 2a$
59. $8c - 10c + 5$
60. $2t - 4t + 9$
61. $8c - 4c + 5c + 7d$
62. $3a + 5a - 10a - 2b$
63. $-5x - 6x + 7y + 2y$
64. $8x - 3x + 6y - 4y$
65. $12x - 15x + 10y - 12y$
66. $4b - c + d - 4d$
67. $(-2.31) - (16.23) - (-12.81)$
68. $(26.8) - (-4.8)^2 + (-0.17)$

69. Benjamin is standing at the 7300-foot level on a mountain. He descends (goes down) 731 feet and then ascends (goes up) 562 feet. What is his present altitude?

70. The summit of a mountain is 10,321 feet above sea level. A cabin is located at a point 5,628 feet below the summit. At what altitude is the cabin located?

71. A woman was $9,456 in debt. She sold some stock for $6,332 and applied all the money from the sale of the stock to the debt. How much was she still in debt?

72. A plane is flying 200 feet above the ground over a point that is 260 feet below sea level. The top of a small hill just ahead of the plane has an altitude of 632 feet above sea level. How much altitude must the plane gain if it is to clear the top of the hill by 500 feet?

1-8 Multiplication and Division of Signed Numbers

Multiplication of two signed numbers follows the rule that the product of two positive numbers is a positive number, that the product of a positive number and a negative number is a negative number, and that the product of two negative numbers is a positive number. We accept without proof that the product of two positive numbers is a positive number. We then illustrate that the product of a negative number and a positive number is negative by showing that $(+3)(-2) = -6$.

Show that $(+3)(-2) = -6$.

$(+2) + (-2) = 0$	The sum of two opposites is zero.
$(+3)[(+2) + (-2)] = (+3) \cdot (0)$	Equality multiplication theorem. (We multiplied both sides of the equation by $+3$.)
$[(+3)(+2)] + [(+3)(-2)] = 0$	Distributive property and multiplication by zero theorem.
$(+6) + [(+3)(-2)] = 0$	The product of two positive numbers is a positive number.
$(+3)(-2) = -6$	Every number (in this case, $+6$) has one and only one additive inverse. Because $(+3)(-2)$ and -6 are both additive inverses of $+6$, they must be equal.

Do you see the full impact of the fourth line? Whatever $(+3)(-2)$ is, it must be the opposite of $+6$, because the sum of $+6$ and $(-3)(+2)$ is zero. But the only number we can add to $+6$ to get zero is -6. Thus, we are forced to conclude that $(+3)(-2) = -6$, which is stated in the fifth line.

In a similar fashion, we can easily show that $(-3)(-2) = +6$.

$(+2) + (-2) = 0$	The sum of two opposites is zero.
$(-3)[(+2) + (-2)] = (-3)(0)$	Equality multiplication theorem. (We multiplied both sides of the equation by -3.)
$[(-3)(+2)] + [(-3)(-2)] = 0$	Distributive property and multiplication by zero theorem.
$(-6) + [(-3)(-2)] = 0$	The product of a positive and negative number is a negative number.

Sets and The Real Number System

$$(-3)(-2) = +6$$

Every number (in this case, -6) has one and only one additive inverse. Because $(-3)(-2)$ and $+6$ are both additive inverses of -6, they must be equal.

The conclusion in the last line is arrived at in exactly the same manner as the conclusion in the steps for showing $(+3)(-2) = -6$. In the fourth line, something is added to -6 to give a sum of zero. Because the only number we can add to -6 to give a sum of zero is $+6$, we conclude that $(-3)(-2)$ is the same as $+6$.

These two illustrations are not general proofs for all real numbers; they are specific examples showing how signed numbers are related under multiplication. It is to be hoped that they will convince you if you were a doubter.

In view of this discussion on multiplying signed numbers, we have the following important rule.

Rule 1–4

Multiplying Two Signed Numbers

1. If both numbers have like signs, multiply their absolute values. The product will be positive.
2. If the numbers have unlike signs, multiply their absolute values. The product will be negative.

The rule of signs for dividing two signed numbers is given in rule 1–5.

Rule 1–5

Dividing Two Signed Numbers

1. If both numbers have like signs, find the quotient of their absolute values. The quotient will be positive.
2. If the numbers have unlike signs, find the quotient of their absolute values. The quotient will be negative.

Division may be checked by multiplying the quotient and divisor to see if this gives the dividend. For example, $(-15) \div (+3) = -5$ because $(-5)(+3) = -15$.

Example 1

(a) $(-6)(+3) = -18$
(b) $(-4)(-7) = 28$
(c) $(-17)(-3)(+2) = +102$

(d) $(-3)^3 = (-3)(-3)(-3) = -27$

(e) $-5^2 \cdot (-2)^2 = -(5 \cdot 5) \cdot (-2)(-2) = (-25)(4) = -100$. (*Caution:* $-5^2 \neq (-5)^2$. -5^2 means $-(5)^2 = -25$.)

(f) $(-2)^3(-4) = (-2)(-2)(-2)(-4) = (-8)(-4) = 32$.

(g) $\dfrac{-18}{-2} = +9$

(h) $\dfrac{(-6)^2(2)}{-9} = \dfrac{36(2)}{-9} = \dfrac{72}{-9} = -8$

(i) $\dfrac{-3^2 \cdot 2}{6} = \dfrac{-9 \cdot 2}{6} = \dfrac{-18}{6} = -3$ (*Caution:* $-3^2 \neq (-3)^2$. -3^2 means $-(3)^2 = -9$.)

(j) $\dfrac{(-3)^2 \cdot 2}{6} = \dfrac{9 \cdot 2}{6} = \dfrac{18}{6} = 3$ ■

Example 2 Evaluate $-5a^2 + 3b - c$ if $a = -3$, $b = 4$ and $c = -2$.

Solution
$$\begin{aligned}-5a^2 + 3b - c &= -5(-3)^2 + 3(4) - (-2) \\ &= -5(9) + 12 + (+2) \\ &= -45 + 12 + 2 \\ &= -33 + 2 \\ &= -31 \end{aligned}$$ ■

Example 3 Evaluate $\dfrac{9x - 4y}{-2(-3y - 2) - 2}$ if $x = 4$ and $y = -5$.

Solution
$$\begin{aligned}\dfrac{9x - 4y}{-2(-3y - 2) - 2} &= \dfrac{9(4) - 4(-5)}{-2[-3(-5) - 2] - 2} \\ &= \dfrac{36 - (-20)}{-2(15 - 2) - 2} \\ &= \dfrac{36 + (+20)}{-2(13) - 2} \\ &= \dfrac{56}{-26 - 2} \\ &= \dfrac{56}{-26 + (-2)} \\ &= \dfrac{56}{-28} \\ &= -2 \end{aligned}$$ ■

1–8 Exercises

In exercises 1–30, perform the indicated operations. (See example 1.)

1. $(-9)(-5)$
2. $(-18)(+4)$
3. $-5^2 \cdot (-2)^3$
4. $(-28)(-2)(-1)$
5. $16(-3)(0)$
6. $-3^2(-2)^2$

Sets and The Real Number System

7. $(-3)^2(-2)^2$
8. $(-1)^4(-2)^3$
9. $7(-4)(-1)^3$
10. $16(-2)(3)$
11. $2(-5)(-4)^2$
12. $(-8)(-1)^4(2)^2$
13. $(-2)^4(-3)$
14. $(-2)^3(-3)^2$
15. $-(-8)(-3)$
16. $3[(-2) + (-6)]$
17. $(-4)[(-2) - (-3)]$
18. $\dfrac{-3^2 \cdot 9}{-1}$ [See example 1(i)]
19. $\dfrac{7^2(-2)}{-14}$
20. $\dfrac{8[-4 - (-3)]}{-2}$
21. $\dfrac{16(-3)}{-6}$
22. $\dfrac{(-4)[5 - (6 - 2)]}{-2}$
23. $\dfrac{19(-3)(-4)}{-6}$
24. $\dfrac{18(-3^2 + 5)}{2(-2)}$ [See example 1(i)]
25. $\dfrac{84(-1)}{(-1)^3(21)}$
26. $\dfrac{40[8 - (-9)]}{(-3)(-4) - (-5)}$
27. $-8\{9 - [(-5)(-4) - (-8) \cdot 3]\} - 2$
28. $-4^2 \cdot (-4)^2(-1)^3$
29. $75 - \{25 - [10(-5)]\}$
30. $-[-(-3) + (-4)^3]$

In exercises 31–35, evaluate the expression $2a^2 + 3b - 2c$ for the given values of a, b, and c. *(See examples 2 and 3.)*

31. $a = 3, b = -2, c = 5$
32. $a = -4, b = 7, c = -6$
33. $a = 7, b = 10, c = -8$
34. $a = -4, b = -9, c = -3$
35. $a = 5, b = 0, c = -1$

In exercises 36–41, evaluate the expression $-2[-3(x - 2y) - y^2]$ for the given values of x and y. *(See examples 2 and 3.)*

36. $x = 6, y = 9$
37. $x = 0, y = -2$
38. $x = 0, y = 0$
39. $x = -12, y = -6$
40. $x = 15, y = 7$
41. $x = -8, y = 4$

In exercises 42–45, evaluate the expression $\dfrac{6 - (x - y)^2 - 2z}{-z^2 + 2}$ for the given values of x, y, and z. Some answers will be fractions. *(See examples 2 and 3.)*

42. $x = 1, y = 1, z = 1$
43. $x = 0, y = -2, z = 0$
44. $x = 5, y = -2, z = -3$
45. $x = -2, y = -5, z = 0$
46. Simplify $\dfrac{(7.1) + (8.3)}{-0.02}$.
47. Simplify $(9.1)^2 - (-1.1)^3 + 0.016$.
48. Evaluate the expression $2a^2 - 4b$ if $a = 1.5$ and $b = -6.12$.

Chapter 1 Summary

Vocabulary

absolute value	multiplicative inverse
additive identity	natural number
additive inverse	order of operations
binary operation	parentheses
braces	rational number
brackets	real number
disjoint sets	reciprocal
element of a set	set
empty set	signed number
equal sets	subset
irrational number	theorem
integer	union of sets
intersection of sets	vinculum
member of a set	well-defined set
multiplicative identity	whole number

The Real Number System

The real number system consists of two major subsets, the rational numbers and the irrational numbers. The rational numbers have three subsets, the set of natural numbers, the set of whole numbers, and the set of integers.

DIAGRAM OF THE REAL NUMBER SYSTEM

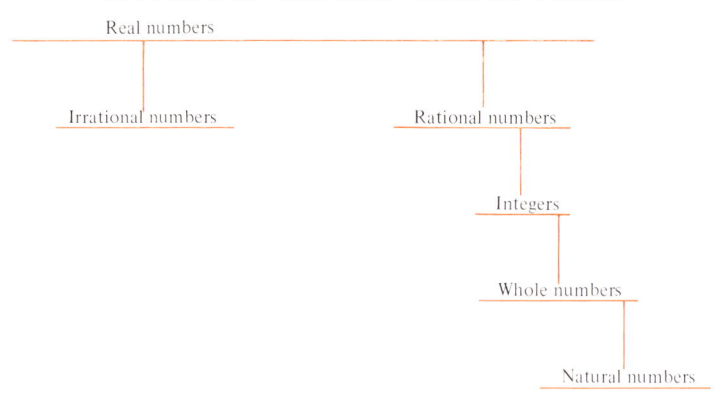

Real numbers
├── Irrational numbers
└── Rational numbers
    └── Integers
        └── Whole numbers
            └── Natural numbers

Axioms

Axioms of Equality

Reflexive Axiom

If a is any real number, then $a = a$.

Symmetric Axiom

If a and b are real numbers and $a = b$, then $b = a$.

Transitive Axiom

If a, b, and c are real numbers and $a = b$ and $b = c$, then $a = c$.

Substitution Axiom

If a and b are real numbers and $a = b$, then a may be replaced by b or b may be replaced by a within any expression or statement representing a real number without changing the number.

Axioms of Inequality

Trichotomy Axiom

If a and b are real numbers, then one and only one of the following is true: $a < b$, or $a = b$, or $a > b$.

Transitive Axiom

If a, b, and c re real numbers and $a < b$ and $b < c$, then $a < c$.

Axioms of Addition

Closure Axiom

For any two real numbers a and b, $a + b$ is a real number.

Commutative Axiom

If a and b are real numbers, then $a + b = b + a$.

Associative Axiom

If a, b, and c are real numbers, then $(a + b) + c = a + (b + c)$.

Axioms of Multiplication

Closure Axiom

For any two real numbers a and b, $a \cdot b$ is a real number.

Commutative Axiom

If a and b are real numbers, then $a \cdot b = b \cdot a$.

Associative Axiom

If a, b, and c are real numbers, then $(a \cdot b) \cdot c = a \cdot (b \cdot c)$.

Identity Axioms

Additive Identity Axiom

There exists a unique real number 0 such that for every real number a,
$$a + 0 = 0 + a = a$$

Multiplicative Identity Axiom

There exists a unique real number 1 such that for every real number a,
$$a \cdot 1 = 1 \cdot a = a$$

Inverse Axioms

Additive Inverse Axiom

For each real number a, there exists a unique real number $-a$ such that $a + (-a) = 0$ and $(-a) + a = 0$.

Multiplicative Inverse Axiom

For each nonzero real number a, there exists a unique real number $1/a$ such that $a \cdot \left(\dfrac{1}{a}\right) = 1$ and $\left(\dfrac{1}{a}\right)a = 1$

Distributive Axiom

For any real numbers a, b, and c, $a(b + c) = ab + ac$ and $(b + c)a = ba + ca$.

Theorems

Theorems of Equality

Equality Addition Theorem

If a, b, and c are real numbers and $a = b$, then $a + c = b + c$.

Equality Multiplication Theorem

If a, b, and c are real numbers and $a = b$, then $ac = bc$.

Multiplication by Zero Theorem

If a is any real number, $a \cdot 0 = 0 \cdot a = 0$.

Double Negative Theorem

If a is any real number, then $-(-a) = a$.

Definitions

1–1 Set Intersection

For sets A and B, $A \cup B = \{x | x \in A \text{ and } x \in B\}$.

1–2 Union of Sets

For sets A and B, $A \cup B = \{x | x \in A \text{ or } x \in B\}$.

1–3 Subset of a Set

For sets A and B, $A \subseteq B$ if and only if $x \in A$ implies $x \in B$.

1–4 Set Equality

For sets A and B, $A = B$ if and only if $A \subseteq B$ and $B \subseteq A$.

1–5 Less Than

If a and b are any two distinct real numbers, then $a < b$ if and only if:
1. a lies to the left of b on the number line or
2. There exists some positive real number c such that $a + c = b$.

1–6 Absolute Value

For any real number x,

$$|x| = \begin{cases} x & \text{if } x \geq 0 \\ -x & \text{if } x < 0 \end{cases}$$

Rules

1–1 Order of Operations

In numerical and algebraic expressions, we agree to:

1. First, evaluate any powers.
2. Second, do all multiplications and divisions from left to right.
3. Third, do all additions and subtractions from left to right.

1–2 Adding Signed Numbers

1. If both numbers have the same sign, add the absolute values of the numbers. The sum has the same sign as the two numbers.
2. If the numbers have opposite signs, find the absolute values of both numbers. Subtract the lesser absolute value from the greater. The sum has the sign of the number with the greater absolute value.

1–3 Subtracting Signed Numbers

To subtract two signed numbers, change the sign of subtraction to addition and change the sign of the subtrahend to the opposite sign. Then add, using the rules of addition of signed numbers.

1–4 Multiplying Two Signed Numbers

1. If both numbers have like signs, multiply their absolute values. The product will be positive.
2. If the numbers have unlike signs, multiply their absolute values. The product will be negative.

1–5 Dividing Two Signed Numbers

1. If both numbers have like signs, find the quotient of their absolute values. The quotient will be positive.
2. If the numbers have unlike signs, find the quotient of their absolute values. The quotient will be negative.

Grouping Symbols

Parentheses, ( ), brackets, [ ], braces, { }, and the fraction bar, ——, are used to enclose numbers or algebraic expressions. If one or more sets of grouping symbols fall within another set of grouping symbols, the indicated operations within the innermost set or sets of grouping symbols are to be done first. A good rule of thumb would be to start from the inside and work towards the outermost set of grouping symbols.

Review Problems

Section 1-1 — *Answer true or false for problems 1–5 if $A = \{3, 5, 7, 9, 10\}$.*

1. $3 \subseteq A$
2. $\emptyset \subseteq A$
3. $9 \in A$
4. $\{7, 9\} \subseteq A$
5. $\{5, 8\} \subset A$

For problems 6–10, use $A = \{1, 2, 3, 4, 5\}$, $B = \{3, 5, 7, 9\}$, and $C = \{2, 3, 5, 8\}$.

6. Find $A \cap B$.
7. Find $A \cup C$.
8. Find $(A \cap B) \cup C$.
9. Find $\emptyset \cup (A \cap B)$.
10. Find $(A \cap B) \cap C$.

Section 1-2

11. Is it true or false that every natural number is also a whole number?
12. The set of rational numbers consists of how many proper subsets, as named in the text?
13. Can a real number be both rational and irrational? Why?
14. Write $\frac{3}{8}$ as a terminating decimal.
15. In an expression containing both addition and multiplication, which is done first if there are no grouping symbols in the expression?
16. In an expression containing only addition and subtraction (but not necessarily in that order) how do we know which to do first?
17. Which of the following does not represent a number: $\frac{8}{0}$, $\frac{0}{8}$?

Section 1-3 — *In problems 18–28, simplify the expressions.*

18. $8 \cdot 3 + 4$
19. $16 \div 4 + 10 \cdot 3$
20. $42 \div 7 \cdot 3 + 5$
21. $19 \cdot 3 - 7$
22. $7 + [3(7 \cdot 2 - 3) + 6]$
23. $100 \div 4 + [5 + (6 - 3)]$
24. $3^2 + 6 - 2^2$
25. $\dfrac{3^4 + 5}{43}$
26. $\dfrac{2^3 + 5^2}{2^3 + 3}$
27. $\dfrac{18 - 3(5 - 2)}{12 - 3}$
28. $(3^2 + 2)(2 + 2^4)$

Section 1-4—1-6 — *In problems 29–35 tell which axiom is used.*

29. $5 + 2 = 5 + 2$
30. $3 \cdot 2 = 2 \cdot 3$
31. If $3a = 2c$ and $c = x - 3$, then $3a = 2(x - 3)$.
32. If $5 < c$ and $c < b$, then $5 < b$.
33. $5 + (3 + 2) = (5 + 3) + 2$

Sets and The Real Number System

34. $5 = \dfrac{10}{2}$ or $5 < \dfrac{10}{2}$ or $5 > \dfrac{10}{2}$

35. $1 \cdot (a + b) = a + b$

Section 1–7

In problems 36–40, add or subtract.

36. $(-8) + (-5)$ **37.** $(-19) - (-5)$
38. $(-8) + (-6) - (-6)$ **39.** $0 - (-10)$
40. $-11 - (-8) + 5$

Section 1–8

In problems 41–45, multiply or divide.

41. $(-8)(-4)$ **42.** $\dfrac{(-9)(-4)(-1)}{-12}$ **43.** $\dfrac{(-3)^2(4)(-1)^3}{(-2)^2}$

44. $\dfrac{16(-2)(-1)}{(-2)^3}$ **45.** $\dfrac{18(-3)(+2)}{(-3)^2}$

Sections 1–7 and 1–8

In problems 46–50, simplify.

46. $\dfrac{3^2 + 5 + (3 + 2) + 1}{(-2)^2 + 2 \cdot 3}$ **47.** $\dfrac{3(5^2 \cdot 2) + 10}{2^2 \cdot 5}$

48. $3\{2[5(9 - 4)]\}$ **49.** $(6^2 - 3)(2^3 - 5)$ **50.** $(9 \cdot 2)[3^2(9 - 2^3)]$

51. Evaluate the expression $\dfrac{(2a^2 - 3b)}{-4}$ if $a = 6$ and $b = -4$.

52. Suppose x and y represent two real numbers whose product is a negative number. Express this fact using variables and mathematical symbols.

CHAPTER 2 | Special Products and Factoring

Objectives of Chapter 2

To be able to:
- ☐ Identify an exponent and base. (2–1)
- ☐ Write repeated factors using exponents. (2–1)
- ☐ Multiply numbers whose bases are the same. (2–1)
- ☐ Divide numbers whose bases are the same. (2–1)
- ☐ Give the meaning of a zero exponent. (2–1)
- ☐ Use the product rule for exponents. (2–1)
- ☐ Raise a fraction to a given power. (2–1)
- ☐ Identify a polynomial by the number of terms that it contains. (2–2)
- ☐ Give the coefficient of a factor or set of factors. (2–2)
- ☐ Add and subtract polynomials. (2–2)
- ☐ Give the degree of a polynomial. (2–2)
- ☐ Evaluate a polynomial at specified values. (2–2)
- ☐ Multiply polynomials. (2–3)
- ☐ Factor by removing the greatest common factor. (2–4)
- ☐ Factor trinomials. (2–5)
- ☐ Factor the difference of two squares. (2–6)
- ☐ Factor the sum or difference of two cubes. (2–6)
- ☐ Factor by grouping. (2–6)

Multiplying two or more numbers together gives an answer we call a **product.** For example, in 3 · 5 = 15, we call 15 the product of 3 and 5. The process can also be reversed. If we write 15 as 3 · 5, we say that 15 has been factored, and 3 and 5 are called **factors** of 15. The process of writing a number as a product of two or more numbers is called *factoring*. In this chapter, we learn how to multiply and factor certain kinds of algebraic expressions. However, we begin the chapter by reviewing properties of exponents.

2–1 Properties of Exponents

If a factor is repeated two or more times, the expression containing the repeated factor can be written using an exponent. For example,

$$x \cdot x \cdot x \cdot x \cdot x$$

is written as x^5, and it is read as "x raised to the fifth power."

In the expression x^5, x is called the **base** and 5 is called the **exponent.** In general, a natural number exponent is a number which indicates the number of times a factor is repeated.

Definition 2–1

Exponent

The nth power of x is written as

$$x^n = \underbrace{x \cdot x \cdot x \cdots x}_{n \text{ factors of } x}$$

In the definition just given, if $n = 1$, then

$$x^1 = x$$

Example 1

(a) $x^2 = x \cdot x$, read "x squared" or "x to the second power."
(b) $x^3 = x \cdot x \cdot x$, read "x cubed" or "x to the third power."
(c) $5y^3 = 5 \cdot y \cdot y \cdot y$ (Note that only y is cubed.)
(d) $(2a)^2 = (2a)(2a) = (2 \cdot 2)(a \cdot a) = 4a^2$
(e) $(-3)^2 = (-3)(-3) = 9$
(f) $-3^2 = -(3 \cdot 3) = -9$ (Be careful with expressions of this type. We must square 3 first to get 9 and then take the opposite of 9 to get -9.)

Expressions with the same base are multiplied by writing the common base and adding the exponents. To see this, note

$$a^2 a^3 = (a \cdot a)(a \cdot a \cdot a)$$
$$= a \cdot a \cdot a \cdot a \cdot a$$
$$= a^5$$

That is, $a^2 a^3 = a^{2+3} = a^5$.

Rule 2–1

Multiplying Numbers That Have the Same Base

If a is any real number and if m and n are natural numbers, then $a^m a^n = a^{m+n}$.

Example 2

(a) $x^5 x^4 = x^{5+4} = x^9$

(b) $(3a^4)(2a^3) = (3 \cdot 2)(a^4 a^3)$
$= 6a^{4+3}$
$= 6a^7$

(c) $x^5 y^6 = x^5 y^6$ (This problem cannot be simplified by using rule 2–1 because the bases are not the same.)

In an expression such as $(x^3)^2$, the base for the exponent 2 is x^3. This means x^3 is used twice as a factor. Then

$$(x^3)^2 = x^3 \cdot x^3$$
$$= x^{3+3} \qquad \text{rule 2–1}$$
$$= x^6$$

This example suggests that if a power is raised to a power, we multiply the two exponents to obtain the final exponent.

Rule 2–2

Raising a Power to a Power

If a is any real number and if m and n are natural numbers, then $(a^m)^n = a^{mn}$.

Example 3

(a) $(5^2)^4 = 5^{2 \cdot 4} = 5^8$

(b) $(p^3)^4 = p^{3 \cdot 4} = p^{12}$

(c) $[(x-3)^3]^2 = (x-3)^{3 \cdot 2} = (x-3)^6$

When expressions which have the same base are divided, the exponents are subtracted. For example,

$$\frac{x^5}{x^2} = \frac{x \cdot x \cdot x \cdot x \cdot x}{x \cdot x} = x \cdot x \cdot x = x^3$$

Special Products and Factoring

That is,
$$\frac{x^5}{x^2} = x^{5-2} = x^3.$$

If the denominator of the expression contains the greater exponent, then
$$\frac{x^2}{x^5} = \frac{x \cdot x}{x \cdot x \cdot x \cdot x \cdot x} = \frac{1}{x \cdot x \cdot x} = \frac{1}{x^3}$$

Throughout this discussion, we assume $x \neq 0$ to eliminate the possibility of division by zero.

Rule 2-3 — Dividing Numbers That Have the Same Base

If a is any nonzero real number and m and n are natural numbers,

$$\frac{a^m}{a^n} = a^{m-n} \quad \text{if } m > n$$
$$\frac{a^m}{a^n} = \frac{1}{a^{n-m}} \quad \text{if } n > m$$
$$\frac{a^m}{a^n} = 1 \quad \text{if } m = n$$

Example 4

(a) $\dfrac{5^6}{5^2} = 5^{6-2} = 5^4$

(b) $\dfrac{3^2}{3^7} = \dfrac{1}{3^{7-2}} = \dfrac{1}{3^5}$

(c) $\dfrac{t^{10}}{t^{10}} = 1$

(d) $\dfrac{(-5)^3}{(-5)^5} = \dfrac{1}{(-5)^{5-3}} = \dfrac{1}{(-5)^2} = \dfrac{1}{25}$

(e) $\dfrac{10x^5}{2x^3} = \dfrac{10}{2} \cdot \dfrac{x^5}{x^3} = 5 \cdot x^{5-3} = 5x^2$

(f) $\dfrac{y^7 z^4}{y^{10} z} = \dfrac{y^7}{y^{10}} \cdot \dfrac{z^4}{z^1} = \dfrac{1}{y^{10-7}} \cdot z^{4-1} = \dfrac{1}{y^3} \cdot z^3 = \dfrac{z^3}{y^3}$ ∎

If an expression containing two or more factors is raised to some given power, then each factor of the expression is raised to that power. For example,

$$\begin{aligned}
(a^3 b^2)^2 &= (a^3 b^2)(a^3 b^2) && \text{definition 2-1} \\
&= (a^3 \cdot a^3)(b^2 \cdot b^2) && \text{associative and commutative axioms} \\
&= a^6 b^4 && \text{rule 2-1}
\end{aligned}$$

That is,

$$(a^3b^2)^2 = (a^3)^2(b^2)^2$$
$$= a^6b^4$$

This example suggests the following rule.

Rule 2–4

Raising a Product to a Power

If a and b are real numbers and if m, n, and p are natural numbers, then

$$(a^m b^n)^p = a^{mp} b^{np}$$

Example 5

(a) $(2^2 \cdot 3)^2 = 2^{2 \cdot 2} \cdot 3^{1 \cdot 2} = 2^4 \cdot 3^2$ (or 144, if simplified)
(b) $(3^2 \cdot a^4)^2 = 3^{2 \cdot 2} \cdot a^{4 \cdot 2} = 3^4 \cdot a^8 = 81a^8$
(c) $(-2x^5)^3 = (-2)^3(x^5)^3 = -8x^{15}$
(d) $(a^5 b^4 c^3)^5 = a^{5 \cdot 5} b^{4 \cdot 5} c^{3 \cdot 5} = a^{25} b^{20} c^{15}$ ■

If a fraction is raised to a power, both the numerator and denominator are raised to that power. For example,

$$\left(\frac{3}{4}\right)^2 = \frac{3}{4} \cdot \frac{3}{4} = \frac{3 \cdot 3}{4 \cdot 4} = \frac{3^2}{4^2} = \frac{9}{16}$$

Rule 2–5

Raising a Quotient to a Power

If a is any real number if b is any nonzero real number, and if m is a natural number,

$$\left(\frac{a}{b}\right)^m = \frac{a^m}{b^m}$$

Example 6

(a) $\left(\dfrac{5}{8}\right)^2 = \dfrac{5^2}{8^2} = \dfrac{25}{64}$

(b) $\left(\dfrac{-3}{2}\right)^3 = \dfrac{(-3)^3}{2^3} = \dfrac{-27}{8}$

(c) $\left(\dfrac{a^3 b^4}{c^5}\right)^2 = \dfrac{(a^3 b^4)^2}{(c^5)^2} = \dfrac{a^6 b^8}{c^{10}}$ ■

We now take a look at what we mean by a zero exponent. Rules 2-1–2-5 were given in terms of natural number exponents only.
Suppose these laws of exponents are also true when any of the

Special Products and Factoring

exponents given by m, n, or p are zero. For example, if we let $m = 0$ in rule 2–1, we get

$$a^0 \cdot a^n = a^{0+n}$$
$$= a^n \qquad (1)$$

However, it is also true that

$$1 \cdot a^n = a^n \qquad (2)$$

From equations (1) and (2), we see that we must define a^0 to be 1. That is, $a^0 = 1$.

Definition 2–2

Zero Exponent

If a is any nonzero real number, then $a^0 = 1$.

If a were zero in definition 2–2, then a^0 could represent such a nonsense statement as $0^5/0^5 = 0^{5-5} = 0^0$. Because division by zero is not possible, 0^0 is meaningless.

Example 7

(a) $5^0 = 1$

(b) $\left(\dfrac{3}{5}\right)^0 = 1$

(c) $(x - 3)^0 = 1$ if $x \neq 3$

(d) $(3a^0)^2 = (3 \cdot 1)^2 = 3^2 = 9$

2–1 Exercises

In exercises 1–47, use the rules of exponents to simplify.

1. $(9z)(8z^2)$
2. $(3x)^2(3x)$
3. $(6x - y)(6x - y)(6x - y)$
4. $5 \cdot x \cdot x + 6 \cdot x \cdot x \cdot x$
5. $-3^3 a^3 (b^2)^3$
6. $-6^2(-3b^3)(4a^3)(-2b)$
7. $5a^2(2a^4)^3(-2a)$
8. $(2x)^2(9x)(-1)^3$
9. $\dfrac{x^9}{x^4}$
10. $\dfrac{20p^2}{4p^7}$
11. $\dfrac{2x^6 y^7}{8x^4 y^2}$
12. $\dfrac{(2x - 3)^2(ab^2)}{(2x - 3)^6(a^2 b^4)}$
13. $\dfrac{12x^0 y^6}{3x^4 y^{10}}$
14. $\dfrac{-(-4)^2(3)^4}{(4^5)(3^8)}$
15. $\dfrac{(2r - 3)^0 (r - 4)^2}{(r - 4)^3}$
16. $\dfrac{(2x)^0(-4)^2}{16x}$
17. $3b^0(b^{2m-3})$
18. $\dfrac{3x^0 \cdot x^5}{2x}$

19. $m^0 + p^0 + 3q^0$

20. $(9x + 2y^2 - z)^0$

21. $\dfrac{[(2x - 3)^2(4x - 3)]^2}{4x - 3}$

22. $\dfrac{(-3xy^2)(7x^3y^2)(-2x^5)}{(-21x^4y^3)(10xy^6)}$

23. $\dfrac{(2m^2)^3(4m)^2}{32m^2}$

24. $\dfrac{[(9a + 2)^2 + (4a - 3)^2]^0}{(6a + 1)^2}$

25. $\dfrac{(-3a)^3(-b^3)}{9b^2}$

26. $\dfrac{(3a)^0(5b)^2}{(5bc^3)(3b^0)}$

27. $\dfrac{(7ac^2)^2(2b)^2}{(7a^2c^3)(2b^2)^0}$

28. $-(5 + 2)^2$ (*Hint:* Add the numbers first and then square.)

29. $[-(3 + 2)^2]^0$

30. $\dfrac{2m - 3}{(2m - 3)^2(-5)^2}$

31. $\dfrac{5^2 \cdot 5^4 \cdot 5^3}{2 \cdot 5^7}$

32. $\dfrac{3^2(-3)^4(5)}{(-3)^3(5^2)}$

33. $\left(\dfrac{3}{8}\right)^2\left(\dfrac{3}{8}\right)$

34. $\left(\dfrac{2}{3}\right)^2\left(\dfrac{5}{8}\right)^0$

35. $\left(\dfrac{5}{8} + \dfrac{3}{8}\right)^{14}$

36. $\left(\dfrac{19}{8} + \dfrac{-11}{8}\right)^2$

37. $\dfrac{17^{12}}{17^{10}}$

38. $\dfrac{2^{20}(3)^5}{2^{16}(3^2)^3}$

39. $-(-5)^0(2a - 3)^5$

40. $\dfrac{[(5a - 2b)^2]^4[(3a - 2b)^3]^2}{[(5a - 2b)(3a - 2b)^2]^2}$

41. $(2a - 3)^m(2a - 3)^2$

42. $2np(2^{3x})$

43. $[(3a^7)^2]^2$

44. $\{[(2a^2)^3]^2\}^2$

45. $\{[(3x^2)^2]^0\}^5$

46. $x^{5.61}x^{9.82}$ (*Hint:* Assume rule 2–1 holds for decimal exponents.)

47. $x^{4.99}x^{5.3}$

48. The distance s that an object will fall in t seconds in a vacuum is given by

$$s = 16t^2$$

where s is measured in feet and t is measured in seconds. Find s for each value of t.

(a) $t = 3$

(b) $t = 9$

(c) How much greater is the distance found in part b than the distance found in part a?

49. The distance an object will move can be expressed in terms of its velocity and acceleration by the equation

$$d = \dfrac{v_f^2 - v_0^2}{2a}$$

where v_f and v_0 are the final and beginning velocities, respectively, of the object and a is the acceleration of the object. Find d if $v_f = 8$, $v_0 = 4$, and $a = 6$. (The symbol v_f is read "v sub f" and v_0 is read "v sub zero.")

50. A calculus student differentiated a function and got

$$y' = -2x^2 + \frac{8}{x^2}$$

Find y' if $x = 4$. (The symbol y' is read "y prime" and it is a symbol indicating the derivative of a function.)

51. The distance s that an object will fall in t seconds is given by

$$s = v_0 t + \frac{1}{2}at^2$$

where v_0 is the initial velocity of the object and a is the acceleration constant. Find s if $v_0 = 15$, $t = 5$, and $a = 32$.

52. The kinetic energy (KE) of an object is given by KE $= \frac{1}{2}mv^2$, where m is the mass of the object and v is the velocity of the object. Find KE if $m = 20$ and $v = 6$.

53. The total surface area of a right circular cylinder is $S = 2\pi rh + 2\pi r^2$, where r is the radius of the cylinder and h is the height. Find the surface area of a cylinder if $r = 7$ and $h = 14$. Use $\frac{22}{7}$ for π.

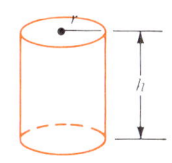

54. A calculus student found the derivative of a function to be

$$y' = \frac{-64}{x^5}$$

Find the value of y' if $x = -2$. (See exercise 50 for an explanation of the symbol y'.)

55. The square of the impedance for a particular electrical circuit is given by

$$Z^2 = R^2 + X_c^2$$

where Z is the impedence and R and X_c represent resistance and capacitive reactance, respectively. Find Z^2 if $R = 20$ and $X_c = 10$.

56. The energy, E, stored in the dielectric field of a capacitor is

$$E = \frac{1}{2}CV^2$$

where C is the capacitance and V is the voltage across the capacitor. Find E if $C = 0.000004$ and $V = 1000$.

57. Einstein's famous equation for energy is

$$E = mc^2$$

where m is a mass measured in grams and c is the speed of light measured in centimeters per second. Find E if $m = 2$ and $c = 3 \cdot 10^{10}$.

58. A calculus student found the derivative of a function to be
$$y' = 24x(3x^2)^3$$
Find y' if $x = -1$.

2–2 Adding and Subtracting Polynomials

If an algebraic expression consists of a finite number of additions, subtractions, multiplications and divisions, and if the exponents on its variables are whole numbers, the expression is called a **rational expression.** For example,

$$\frac{3x^2 + 2}{5x}, \; 6x^2, \text{ and } \frac{7x^3}{9x - 4}$$

are rational expressions.

A **polynomial** is a rational expression consisting of one or more terms in which no variable occurs in the denominator of the terms. A polynomial of one term is called a **monomial.** A polynomial of two terms is called a **binomial,** and a polynomial with three terms is called a **trinomial.** There is no special name for polynomials having more than three terms. They are called polynomials of four terms, polynomials of five terms, and so forth.

Example 1 Identify each of the following as a monomial, binomial, trinomial, or as a polynomial of more than three terms.

(a) $3x^2 + 2$
(b) $2x^2y$
(c) $2x + 3 - (x^2 + 5)$
(d) $5x^3 - 3x^2 + 8x - 2$
(e) $x^5 - 2x^4 + 5x^3 - 8x^2 + x - 3$

Solution (a) Binomial.
(b) Monomial.
(c) Trinomial, because $(x^2 + 5)$ is one term.
(d) Polynomial of four terms.
(e) Polynomial of six terms.

Special Products and Factoring

If numbers and variables are written as factors, any one or more of the factors is said to be the **coefficient** of the remaining factors. For example, in the monomial $5x^2yz^3$, $5x^2$ is the coefficient of yz^3 and $5yz^3$ is the coefficient of x^2. In particular, 5 is called the **numerical coefficient** of x^2yz^3. In an expression such as xy^2, the numerical coefficient is understood to be 1. When the word *coefficient* is used, it will mean numerical coefficient unless indicated otherwise.

Example 2   In $6x^3yz$, what is the coefficient of $6y$?
Solution x^3z

Example 3   Name the numerical coefficient of $7x^2y$.
Solution   7

In our study of polynomials, we restrict ourselves, for the most part, to polynomials of one variable. The **degree of a monomial** in one variable is the exponent of the variable in that term. If the monomial has more than one variable, then the degree of the monomial in any one of the variables is the same as the exponent of that variable. For example, the monomial $5x^2y^3$ is of degree 3 in y. If the monomial has more than one variable, then the degree of the monomial is the sum of the exponents of all the variables in the monomial. The degree of the monomial $5x^2y^3$ is 5 because it is of degree 2 in x and degree 3 in y.

The **degree of a polynomial** is the degree of the term of highest degree in the polynomial.

Example 4   The monomial $-7x^2y^5$ is of what degree in y?
Solution   The exponent on y is 5. The monomial is of degree 5 in y.

Example 5   Give the degree of the polynomial $7x^2 - 6x^3 + 4x^5 + 2$.
Solution   The term of highest degree is the term $4x^5$. Therefore, the polynomial is of degree 5.

As a special case, any nonzero constant term is said to have degree zero because, for example, 7 could be written as $7x^0$. The constant zero has no degree because 0 can be written as $0x^2$ or $0x^5$ or $0x^0$.

In any polynomial, the preferred order is usually to write the term of highest degree first, then the term with the next-highest degree, and so forth, writing the constant term as the last term. This is called arranging the terms of a polynomial in *descending powers of the variable.*

Example 6   Arrange $7x^2 - 6x^3 + 4x^5 + 2$ in descending powers of the variable.

Solution   If we arrange the polynomial according to the instructions just given, we have $4x^5 - 6x^3 + 7x^2 + 2$.

Two or more terms are said to be **like terms** if they contain corresponding variables raised to corresponding powers. The terms $5x^2y$ and $5xy^2$ are *not* like terms. The variables are the same, but the corresponding exponents are not the same. The terms $3a^2x^2$ and $2a^2y^2$ are not like terms because the variable parts are different. However, $6x^2y$ and $-2x^2y$ are like terms because they have the same variables raised to the same powers.

Like terms may be added or subtracted. The distributive axiom gives a means of doing this.

Example 7   Add: $5x + 3x$.

Solution
$$5x + 3x = (5 + 3)x \qquad \text{distributive axiom}$$
$$= 8x$$

Example 8   Subtract: $2x^2y - 4x^2y$.

Solution
$$2x^2y - 4x^2y = (2 - 4)x^2y \qquad \text{distributive axiom}$$
$$= -2x^2y$$

Generally, like terms are added or subtracted by adding or subtracting the numerical coefficients of the terms and affixing the common letter part to the final coefficient.

Example 9   Combine like terms in $9ab - 2ac + 10ab$.

Solution
$$9ab - 2ac + 10ab = (9ab + 10ab) - 2ac$$
$$= 19ab - 2ac$$

Note that $19ab$ and $2ac$ are not like terms and cannot be combined.

Special Products and Factoring

Example 10 Simplify $3x^2 - 4ab^2 - 5x^2 + 4a^2b + 7x + 7ab^2$.

Solution
$3x^2 - 4ab^2 - 5x^2 + 4a^2b + 7x + 7ab^2$
$= (3x^2 - 5x^2) + (-4ab^2 + 7ab^2) + 4a^2b + 7x$
$= -2x^2 + 3ab^2 + 4a^2b + 7x$
$= -2x^2 + 7x + 4a^2b + 3ab^2$ ■

Example 11 Simplify $(5x - 2z) - (7x - 4z)$.

Solution First remove the parentheses. Then combine like terms.
$(5x - 2z) - (7x - 4z) = 1 \cdot (5x - 2z) - 1 \cdot (7x - 4z)$
$= 5x - 2z - 7x + 4z$
$= (5x - 7x) + (-2z + 4z)$
$= -2x + 2z$ ■

Example 12 Evaluate the rational expression for the given values of the variables.

$$\frac{5s^2 - 4}{2s - p} + p, \quad s = 6, p = 10$$

Solution
$\frac{5s^2 - 4}{2s - p} + p = \frac{5(6)^2 - 4}{2(6) - 10} + 10$
$= \frac{5(36) - 4}{12 - 10} + 10$
$= \frac{180 - 4}{2} + 10$
$= \frac{176}{2} + 10$
$= 88 + 10$
$= 98$ ■

A polynomial can be represented by a variable such as p or q. For example, if we write $p = 3x^2 + 6x + 4$, then p represents the polynomial $3x^2 + 6x + 4$. If a certain value is assigned to x, then the value of the polynomial is given by $p(x)$, read "p of x." The notation $p(x)$ does not mean p times x. If $p(x) = 3x^2 + 6x + 4$, then $p(4)$ is the value of the polynomial when $x = 4$.

Example 13 If $p(x) = 3x^2 + 6x + 4$, find each value.

(a) $p(2)$
(b) $p(3c)$

Solution
(a) $p(2) = 3(2)^2 + 6(2) + 4$
$= 12 + 12 + 4$
$= 28$

The given polynomial has a value of 28 when x has a value of 2.

(b) $p(3c) = 3(3c)^2 + 6(3c) + 4$
$= 3(9c^2) + 18c + 4$
$= 27c^2 + 18c + 4$

2–2 Exercises

In exercises 1–14, identify the polynomials as monomials, binomials, trinomials, or polynomials of more than three terms. Do not change the form of the given polynomial for the purpose of making the identification. (See example 1.)

1. $5x + 2$
2. $4a^2$
3. $9x - 2y$
4. $5x^2 - 5x - 3$
5. $-7x^3 + 2x^2 - x - 1$
6. 8
7. $5(x + 2) + 7$
8. $-2x(x - 2) + 7$
9. $6x + 3 - 2(x - 1)$
10. $9x^5 - 2x^4 - 6x^3 - 2x^2 - x + 1$
11. $5x^0 + 2$
12. $9x(3x - 2) + 5(x + 4)$
13. $2(5y - 3) - 3(2y + 4)$
14. $8c + 5(x - 1) + 2(x + 4)$

In exercises 15–28, apply the distributive axiom and the rules of exponents to simplify each expression. Then identify the simplified expression as a monomial, binomial, trinomial, or a polynomial of more than three terms.

Example

$5(x - 3) + 2(x + 5) = 5x - 5 \cdot 3 + 2x + 2 \cdot 5$
$= 5x - 15 + 2x + 10$
$= (5x + 2x) + (-15 + 10)$
$= 7x - 5$

The simplified expression is a binomial.

15. $5x(x - 1)$
16. $2(3x + 2y - 1)$
17. $3x(x + 4y)$
18. $2(x^3 + 5x^2 + 6x - 1)$
19. $3x(2x - 3) + 5x^3$
20. $5(6x^3 - 2x + 1)$
21. $2(2x^5 - 6x) + 3(x^4 - 2x)$
22. $-3(4x - 3) + 5(x - 2)$
23. $8x + 5(2x - 3) + 6x$
24. $9x^2 + x(x - 3) + 5x$
25. $2(x - 3) + 5(x + 2) - 3(x - 4)$
26. $2x^2 + 6(x - 3) - 2(x^2 - 3)$
27. $3x^3 + 2x^2(x - 4) - 5x^3 - 6x$
28. $-18x + 6(3x - 4) - 5(2x^2 - 3x)$
29. Give the degree of each term of $9x^3 + 6x - 3$.
30. Give the degree of each term of $5x^5 - 2x^3 + 4$.

In exercises 31–42, give the degree of each polynomial. (See example 5.)

31. $8x^3 - 6x^2 + 5$
32. $9x^5 - 6x^4 - 6x$
33. $x^2 - 6x^3 - 9x + 5$
34. $6x^2 + 9x^3 - x$

Special Products and Factoring

35. $18 - 6x^2 + 10x^3$ **36.** $8x - 7 + x^3$
37. $3 - x + 5x^5 + 3x^4$ **38.** $5x$
39. $-6x + 2$ **40.** 5
41. 8 **42.** 0

In exercises 43–52, for the monomial $8x^3y^4z$, give the coefficient of each factor. (See examples 2 and 3.)

43. x^3z **44.** $8x^3$ **45.** z
46. y^4z **47.** $8y^4$ **48.** $8x^3y^4$
49. x^3y^4z (What is the special name for this coefficient?)
50. x^3 **51.** 8 **52.** x^3y^4

In exercises 53–70, simplify by combining like terms. (See examples 7–11.)

53. $3x + 5x - x$ **54.** $19x^2 - 5y^2 + 6x^2$
55. $4a - 5b + 5a - 6b$ **56.** $5a - (4c + 2a)$
57. $9x^2 - (5x^2 - 4y) + (6x^2 - 3y)$ **58.** $6a - [-(3a - 4c)] + 5c$
59. $6x - 3y + (x - 4) + 5y$ **60.** $9x + [x + (3x^2 - x) + 3]$
61. $5a^2x^2 + 7ax^2 - 6a^2x^2 + 2ax^2$ **62.** $7cd - 4cd^2 - 2cd$
63. $[3a^2 - (4xy + 5a^2) - 3] + 3a^2$ **64.** $(2a - c) + (2a - b + c)$
65. $3a^2bc - 4ab^2c - 2abc + 7a^2bc - 2abc$
66. $5a - [-(3a - 6) - (4a - 3)]$
67. $(7x - 4y) - (2y + 3x) - (4x + 2y)$
68. $8a - 2b - (3c + b)$
69. $18ab^2 - 6cd - (7ab^2 - ab)$
70. $s^2 - 2t + 5t - 3 - (8s^2 - s)$

We can use a vertical, or column, arrangement for adding polynomials. Arrange the polynomials in exercises 71–75 in columns of like terms and add. If the polynomials do not agree term for term, leave space for the missing terms.

Example Find $(3a^2b - 2ac - 7a) + (3a^2b + 2ab - 4c)$.

Solution
$$\begin{array}{l}3a^2b \qquad\quad - 2ac - 7a \\ \underline{3a^2b + 2ab \qquad\qquad - 4c} \\ 6a^2b + 2ab - 2ac - 7a - 4c\end{array}$$

71. $(3x^2y - 6x - 4) + (6x^2y - 9x + 5)$
72. $(-2x^2y + 7x - 2) + (8x^2y - 2x - 4)$
73. $(7x^2y - 2xy + 5) + (3x^2y - 2y + 4)$
74. $(9a^3b^2 - 7ab^3 - 2c) + (7a^2b + 4ab^3 + 5)$
75. $6x^2y - 4x + 5) + (7xy^2 - 9x)$

In exercises 76–79, subtract polynomials using columns of like terms. Use zeros to hold the places of missing terms, if desired.

Example

Solution

Subtract: $(6a^3b^2 + 6a^2b - 8ab + 4) - (-2a^3b^2 - 3a^2b + 3ab - 5)$.

$$\begin{array}{r} 6a^3b^2 + 6a^2b - 8ab + 4 \\ \underline{\ominus 2a^3b^2 \oplus 3a^2b \ominus 3ab \oplus 5} \\ 8a^3b^2 + 9a^2b - 11ab + 9 \end{array}$$

Remember, to subtract, change the signs of the bottom terms and add. The circles show the new signs to be used.

76. $(19ac^2 + 7ac - 2c) - (4ac^2 - 5ac + 5c)$

77. $(21x^2y - 5xy + 5) - (8x^2y - 5xy^2 + 3x - 4)$

78. $(3x^2 - 4y + z - 3) - (-3x - 4y + 5)$

79. $(7abc^2 - 7ac + 4) - (6ac^2 - 2ac + 5a)$

In exercises 80–84, evaluate the rational expressions for the given values of the variables. (See example 12.)

80. $\dfrac{9x^2 - 4y}{3x - 2}$; $x = 2, y = -4$

81. $\dfrac{8a - (2c)^2}{3b}$; $a = 5, b = 4, c = -2$

82. $\dfrac{(6x - 3)^2}{9x - (2y + 2)}$; $x = 3, y = 0$

83. $\dfrac{5c - 2d}{3c - d}$; $c = 2, d = 6$

84. $8x^2y - (3x^2 - 4)$; $x = -1, y = 0$

In exercises 85–92, find each value if $p(x) = 2x^2 - 3x + 5$. (See example 13.)

85. $p(2)$ **86.** $p(-3)$ **87.** $p(-6)$ **88.** $p(0)$
89. $p(-1)$ **90.** $p(ab)$ **91.** $p(-2c)$ **92.** $p(2a^2)$

In exercises 93–94, simplify.

93. $(5.09x^2 + 7.51x - 12.02) - (4.23x^2 - 16.54x - 3.03)$

94. $5.3 - [-(9.87x^2 - 19.45)] + (15.4x^2 - 31.02x)$

2–3 Multiplying Polynomials

When multiplying polynomials we need to know how to work with several situations, such as multiplying a monomial by a monomial, a binomial by a trinomial, or a binomial by a binomial, to name a few.

Special Products and Factoring

Regardless of the kinds of polynomials being multiplied, the multiplication is carried out by using the commutative, associative, and distributive axioms, as well as other rules we have studied.

Example 1 Multiply: $(3x^2y^3)(5x^4y^2)$.
Solution $(3x^2y^3)(5x^4y^2) = (3 \cdot 5)(x^2 \cdot x^4)(y^3 \cdot y^2) = 15x^6y^5$ ◼

The middle step in example 1 can be omitted after a little practice.

Example 2 Multiply: $(-6a^3b^0c^2)(7a^6b^2d^3)$.
Solution $(-6a^3b^0c^2)(7a^6b^2d^3) = (-6 \cdot 7)(a^3 \cdot a^6)(b^0 \cdot b^2)(c^2)(d^3)$
$= -42a^9b^2c^2d^3$ ◼

Example 3 Multiply: $(-3x)(5x - 4y)$.
Solution $-3x(5x - 4y) = -3x(5x) - (-3x)(4y)$
$= -15x^2 - (-12xy)$
$= -15x^2 + 12xy$ ◼

In examples of this type, you need to be extremely careful with minus signs. A common mistake is to multiply $-3x(5x - 4y)$ and get $-15x^2 - 4y$ or even $-15x^2 - 12xy$. Both of these answers are wrong. (Why?)

Example 4 Multiply: $(5x - 2)(3x + 4)$.
Solution Apply the distributive axiom first.

$(5x - 2)(3x + 4) = (5x - 2)(3x) + (5x - 2)(4)$
$= 3x(5x - 2) + 4(5x - 2)$ Commutative axiom of multiplication.
$= (3x)(5x) - 3x(2) + 4(5x) - 4(2)$ Distributive axiom.
$= 15x^2 - 6x + 20x - 8$ Multiply the monomials in each term.
$= 15x^2 + 14x - 8$ ◼   Combine like terms.

In the third line in the solution to example 4, each term of the first binomial is multiplied by each term of the second binomial. This observation will allow us to make the multiplications faster.

Example 5 Multiply: $(6x + 5)(3x - 2)$.

Solution
$$(6x + 5)(3x - 2) = 6x(3x) - 6x(2) + 5(3x) - 5(2)$$
$$= 18x^2 - 12x + 15x - 10$$
$$= 18x^2 + 3x - 10$$

Do you see how it was done?

Now let's make another observation. The term $3x$ in the answer in example 5 is the sum of the product of the two inside terms and the product of the two outside terms in the two given binomials. That is, in $(6x + 5)(3x - 2)$, the $3x$ is obtained as shown by the arrows.

$$(6x + 5)(3x - 2)$$

Add:
$$15x$$
$$-12x$$
$$3x$$

Now look at the diagram in example 6 and see how the answer is written without going through any intermediate steps. The circled numbers indicate the order in which the mental multiplications should be made.

Example 6 Multiply: $(2x + 3)(3x + 5)$.

Solution

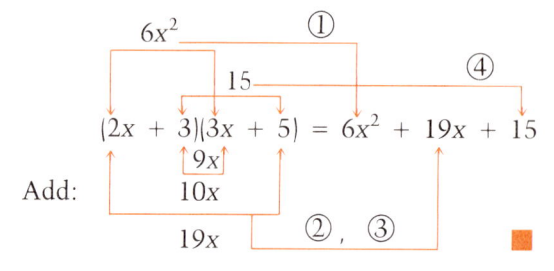

$$(2x + 3)(3x + 5) = 6x^2 + 19x + 15$$

Add:
$$9x$$
$$10x$$
$$19x$$

You should practice this procedure until you can do it perfectly. We cannot stress too much the need to be able to make these mental multiplications.

If binomials have different variables, then use the method shown in example 5. This type of multiplication is illustrated in example 7.

Example 7 Multiply: $(2x + 3y)(2a - 3b)$.

Solution
$$(2x + 3y)(2a - 3b) = 2x(2a) - 2x(3b) + 3y(2a) - 3y(3b)$$
$$= 4ax - 6bx + 6ay - 9by$$

Special Products and Factoring

To square a binomial of the form $a + b$, note that
$$(a + b)^2 = (a + b)(a + b)$$
$$= a^2 + ab + ab + b^2$$
$$= a^2 + 2ab + b^2$$

The steps for squaring a binomial mentally are given in rule 2–6.

Rule 2–6

Squaring a Binomial
$$(a + b)^2 = a^2 + 2ab + b^2$$

1. Square the first term of the binomial to get the first term of the answer.
2. Multiply the two terms of the binomial together and double this product to get the middle term of the answer.
3. Square the last term of the binomial to get the last term of the answer.

Example 8

Square: $(3x + 4)^2$.

Solution

The squaring process is shown with three numbered arrows to correspond to the three steps in rule 2–6.

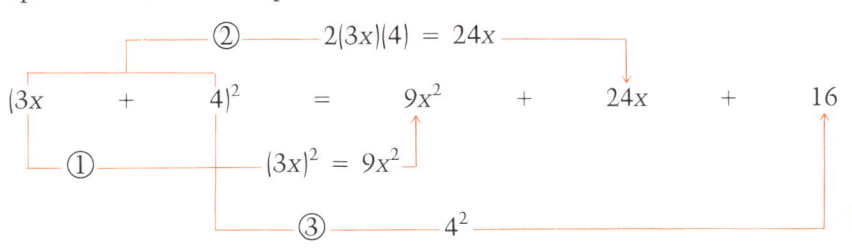

$$(3x + 4)^2 = 9x^2 + 24x + 16$$

① $(3x)^2 = 9x^2$
② $2(3x)(4) = 24x$
③ 4^2

Example 9

Multiply: $(2x - 5)(2x^2 - 6x + 3)$.

Solution

Multiply the first term of the binomial by each term in the trinomial and then multiply -5 in the binomial by each term in the trinomial; then collect like terms.

$$(2x - 5)(2x^2 - 6x + 3) = 2x(2x^2) - 2x(6x) + 2x(3) - 5(2x^2)$$
$$- 5(-6x) - 5(3)$$
$$= 4x^3 - 12x^2 + 6x - 10x^2 + 30x - 15$$
$$= 4x^3 - 22x^2 + 36x - 15$$

Example 10

Multiply the two polynomials in example 9 using a vertical arrangement.

Solution   Place the binomial beneath the trinomial as shown. Then multiply, much as you would with whole numbers.

$$\begin{array}{r} 2x^2 - 6x + 3 \\ 2x - 5 \\ \hline -10x^2 + 30x - 15 \\ 4x^3 - 12x^2 + 6x \\ \hline 4x^3 - 22x + 36x - 15 \end{array}$$

⟵ Multiply -5 by $2x^2 - 6x + 3$.
⟵ Multiply $2x$ by $2x^2 - 6x + 3$ and keep the like terms in columns.
⟵ Add the partial products. ■

The method shown in example 9 is a horizontal arrangement for multiplying polynomials and the method shown in example 10 is a vertical arrangement. You may use whichever method you prefer, unless you are given instructions to use one or the other.

2–3 Exercises

In exercises 1–12, multiply. (See example 3.)

1. $7a^2(6a^2 - 4a - 2b)$
2. $-2x(5x^2 - 3y)$
3. $-6x^2(-2ax + 4y)$
4. $5x^2y^3(6xy^4 - 7x^3y^4)$
5. $8a^2(5a^2b - 2ac)$
6. $-3c(2c^2 - 3c^2d)$
7. $9a^2b(-3a^2b - 2ab^2)$
8. $4p^2q(-q^2 - 3p^2)$
9. $4a^2c(5a^2 - 2c^2 - 3a^2c)$
10. $-5x^2y(x - y - 5xy^2)$
11. $-a(-a - b - c + a^2c)$
12. $-r^2(r^2s - s^2 + 3rs)$

In exercises 13–30, multiply and simplify the given binomials. Try to make as many of the multiplications mentally as possible. (See example 6.)

13. $(3x - 4)(5x - 3)$
14. $(2a - 7)(3a + 4)$
15. $(-3x + 7)(x - 5)$
16. $(-x + 5)(2x - 3)$
17. $(3x + 4)(2x - 5)$
18. $(8x - 4)(2x - 5)$
19. $(9x - 6)(x + 5)$
20. $(3c - 2b)(2c + 3b)$
21. $(x - 2y)(3x + 4y)$
22. $(10x - 1)(2x + 3)$
23. $(2u - 3v)(u + v)$
24. $(5t - 4r)(-2t + 5r)$
25. $(3 - 2x)(5 + 4x)$
26. $(x - 2)(2 + 5x)$
27. $(9 + y)(5y - 5)$
28. $(10 + 2t)(3t + 1)$
29. $(8ab - 2c)(2ab + 5c)$
30. $(xy + 5)(3xy - 8)$

In exercises 31–38, multiply using a horizontal arrangement. (See example 9.)

31. $(2x - 3)(x^2 + 5x - 3)$
32. $(x + 5)(2x^2 - x + 3)$
33. $(x - 6)(x^3 + 7x - 4)$
34. $(2x - 1)(x^3 - 5x + 4)$

Special Products and Factoring

35. $(x - 5)(x^6 - 4x^2 - 2x)$
36. $(9x - 2)(x^3 - 6)$
37. $(2a - 4)(a^2 - 5)$
38. $(2b - 5)(3b^3 - 2b)$

In exercises 39–46, multiply using a vertical arrangement. (See example 10.)

39. $(2x - 3)(x^2 + x - 3)$
40. $(x + 5)(x^2 - 6x + 3)$
41. $(8x + 1)(x^2 - x - 1)$
42. $(x - 3)(5x^2 - 6x + 2)$
43. $(x^2 - 2x - 1)(2x^2 + x - 3)$
44. $(3x^2 - x + 1)(x^2 - 2x + 3)$
45. $(x^3 - 1)(4x^2 - 5x + 1)$
46. $(x^2 + 5)(x^3 + 6x)$

In exercises 47–66, square the binomials, using rule 2–6. (See example 8.)

47. $(2x + 3)^2$
48. $(8x + 1)^2$
49. $(5x - 1)^2$
50. $(9x + 2)^2$
51. $(8x + 5)^2$
52. $(6x - y)^2$
53. $(2x - 3y)^2$
54. $(x - 2y)^2$
55. $(-2x - 3)^2$
56. $(-5x - 2)^2$
57. $(-7y - 2)^2$
58. $(-6y - 5)^2$
59. $(2c - ab)^2$
60. $(3r - 4s)^2$
61. $[5 - (a + b)]^2$
62. $[2 + (x + y)]^2$
63. $[(x + 3y) + 5]^2$
64. $[(2x + y) - 4]^2$
65. $[(2 + 3x) - 3]^2$
66. $[(7 + 2y) + 6]^2$

In exercises 67–78, multiply and simplify the polynomials.

67. $3x(5xy - 2)$
68. $(2a - 3)(4a^2 - 2a + 1)$
69. $(9a + 2)^2$
70. $(6x + 5)(3)$
71. $(x - 1 + y)(2x + 3y + 2)$
72. $(x^2 + x - 3)(5 + x - x^2)$
73. $(2x + y)^3$
74. $(a + 2b)(a + b)^2$
75. $(a^2 - b^2)(a^2 + b^2)$
76. $(a - b)(a^2 + ab + b^2)$
77. $(x - 3)(x + 3)(x^2 + 9)$
78. $(x^2 - 4)(x^2 + 4)(x^4 + 16)$

In exercises 79–86, multiply and simplify.

79. $x^3(x^{2n} - 3)$
80. $x^5(x^{3n-4} - x^2)$
81. $x^n(2x^n - x^m)$
82. $x^{m+1}(x^{2m} - x^3)$
83. $a^n(3a^n - a^{n+1})$
84. $a^{n-1}(a^{n+2} - a^5)$
85. $y^5(2y^{n+1} + 3y^n)$
86. $x^{6+n}(x^n + x^{6+n})$

87. A calculus student worked a calculus problem and got
$$y' = (3x + 2)(2)(9x - 3) + (9x - 3)^2(3)$$
but the student failed to simplify the result by multiplying the polynomials in each term of the expression for y'. Finish simplifying the problem, including collecting like terms.

88. The median for a set of data is a number such that half the data are smaller than the number and half the data are larger than the number; it is given by the equation

$$M = L + a\frac{x_i}{f}$$

Find M if $L = 30$, $a = 4$, $x_i = 12$, and $f = 36$.

89. Find $(7.3x - 2.7)^2$.

90. Find $7.1x(9.3x - 5.4) - (4.3x - 10.4)$.

2–4 Removing the Greatest Common Factor

In this and following sections we essentially reverse the procedure of the last section. In section 2–3 we learned how to multiply $3(x + y)$ to get $3x + 3y$. Now we want to take $3x + 3y$ and write it in factored form instead of as a polynomial of two terms. Since each term of $3x + 3y$ has a common factor of 3, we factor out the 3 and write $3x + 3y = 3(x + y)$.

How would we factor $4x + 12$? Notice that $4x + 12 = 4x + 4 \cdot 3$, and we can see 4 is common to both terms. By the distributive axiom, $4x + 12 = 4x + 4 \cdot 3 = 4(x + 3)$. The monomial factor 4 is customarily not factored further; that is, we would not write $4(x + 3)$ as $2 \cdot 2(x + 3)$.

Although $4x + 12 = 4(x + 3)$, there are an infinite number of ways to factor $4x + 12$. Verify by multiplying that each of the following expressions is equivalent to $4x + 12$.

$$4x + 12 = 4(x + 3)$$
$$4x + 12 = 8\left(\frac{1}{2}x + \frac{3}{2}\right)$$
$$4x + 12 = \frac{1}{2}(8x + 24)$$
$$4x + 12 = 2(2x + 6)$$

Which is correct? They all are correct. However, convention dictates that we factor according to the steps given in procedure 2–1.

Procedure 2–1

Removing the Greatest Common Factor
1. All coefficients must be integers.
2. The monomial that is removed must be as large as possible and still maintain integers for the coefficients of the variable terms.
3. All variables must have whole number exponents.

Thus we would not factor $4x + 12$ as $8(\frac{1}{2}x + \frac{3}{2})$ because the coefficient of the variable term is not an integer. Furthermore, we would not factor $4x + 12$ as $\frac{1}{2}(8x + 24)$ because the monomial that is removed can be made larger and still have an integer for the coefficient of the variable term. We would also not factor $4x + 12$ as $2(2x + 6)$ because the largest integer possible has not been removed. This leaves only $4(x + 3)$ as the factorization meeting the requirements of procedure 2–1.

Finally, notice that $3x - 6 = 3(x - 2) = -3(-x + 2)$. Either factorization is correct.

Example 1 Factor $5x + 10$.
Solution $5x + 10 = 5x + 5 \cdot 2 = 5(x + 2)$ ■

Example 2 Factor $3x^2 + 6x$.
Solution $3x^2 + 6x = (3x) \cdot x + (3x) \cdot 2 = 3x(x + 2)$ ■

Example 3 Factor $7x^2y^3 - 21x^3y^5$.
Solution $7x^2y^3 - 21x^3y^5 = (7x^2y^3) \cdot 1 - (7x^2y^3) \cdot 3xy^2$
$= 7x^2y^3(1 - 3xy^2)$ ■

In example 3, note that because the first term is the same as the quantity factored from the terms, the number 1 must be written inside the parentheses as the first term.

Example 4 Factor $12a^2b - 24a^3b^4$.
Solution $12a^2b - 24a^3b^4 = (12a^2b) \cdot 1 - (12a^2b) \cdot 2ab^3$
$= 12a^2b(1 - 2ab^3)$ ■

Example 5 Factor $(3x - 2)(5a) - (3x - 2)(7b)$.
Solution $(3x - 2)(5a) - (3x - 2)(7b) = (3x - 2)(5a - 7b)$ ■

If you have difficulty with example 5, use a substitution and let $y = 3x - 2$. Then,

$$(3x - 2)(5a) - (3x - 2)(7b) = y(5a) - y(7b)$$
$$= y(5a - 7b) \quad (1)$$

Now substitute $3x - 2$ for y in (1) to get

$$y(5a - 7b) = (3x - 2)(5a - 7b) \quad ■$$

Example 6
Factor $(2x + 3)(a^2b) - (2x + 3)(ab)$.

Solution
The terms in this example have a common binomial factor, $2x + 3$, and a common monomial factor, ab. When both these factors are removed from each term we get

$$(2x + 3)(a^2b) - (2x + 3)(ab) = [(2x + 3)(ab)](a - 1)$$

Example 7
Factor $6x^5 - 3x^b$ if $b > 5$.

Solution
$$6x^5 - 3x^b = 3x^5(2) - 3x^5(x^{b-5})$$
$$= 3x^5(2 - x^{b-5})$$

Example 8
Factor $5x^m + 10x^n$ if $n > m$.

Solution
$$5x^m + 10x^n = 5x^m(1) + 5x^m(2x^{n-m})$$
$$= 5x^m(1 + 2x^{n-m})$$

2–4 Exercises

In exercises 1–12, replace the question mark or marks in each exercise with the correct expression.

Example
(a) $7x - 2ax = ?(7 - 2a)$
(b) $-7ac + 14a = -7a(\ ?\ )$

Solution
(a) $7x - 2ax = x(7 - 2a)$
(b) $-7ac + 14a = -7a(c - 2)$

1. $5a - 5b = ?(a - b)$
2. $5a - 5b = ?(-a + b)$
3. $2ax - bx = x(\ ?\ -\ ?\ )$
4. $9ax - 4z = -1(\ ?\ )$
5. $5a^2b - 10a + 5a^2 = 5a(\ ?\ )$
6. $-ab - 4ac = ?(b + 4c)$
7. $9xy^2 - 6x^2y^2 - 3x = 3x(\ ?\ -\ ?\ -\ ?\ )$
8. $-2bc + 4d^2c - 8c = ?(b - 2d^2 + 4)$
9. $9a^2d - 12ad + 18a^2c = -3a(\ ?\ )$
10. $5x^2y - 2cd = -(\ ?\ +\ ?\ )$
11. $-4x^2 - 5yz = -(\ ?\ )$
12. $-6x^2y - 2x^2y = 1(\ ?\ )$

In exercises 13–40, factor by removing the greatest common factor. If a variable appears as an exponent, assume it represents a natural number. (See examples 1–4.)

13. $6x - 6y$
14. $7x - 14$
15. $6ab + 18a$
16. $12a^2b - 15b$

17. $-5a + 7ab$
18. $-6xy - 4x$
19. $a^4b^6z^7 + a^3b^3z^4 - a^2b^5z^3$
20. $9a^2b - 12b$
21. $13x^2 + 39x^4 - 26x$
22. $10x^2y - 10xy$
23. $12x - 24x^3y + 36x^4$
24. $17x^6y^7 - 51xy^6 - 34x^2y^3$
25. $7a^2 - 6a + 7a^4$
26. $2a - 5b + 4c$
27. $13x^2y + 15xy + 16x$
28. $8a^3 - 12a^3b + 4ab^5$
29. $50a^2b - 25ab^3 + 5ab$
30. $-x^2y - xy^2 - y$
31. $(2x - 3)(5a) - (2x - 3)(7b)$ (See example 5.)
32. $(2x + 7)(x^2) - (2x + 7)(y)$ (See example 5.)
33. $x^2(3x - 4) - 2a(3x - 4)$
34. $(8a - 3b)(6x) - (8a - 3b)(4y)$ (See example 6.)
35. $(8a - b)(7ab) - (8a - b)(6x) + (8a - b)(5y)$
36. $(2x - 3)(6a^2b) + (2x - 3)(5a)$ (See example 6.)
37. $(4y - 3)^2(7b) - (4y - 3)^3(6a)$ (Hint: Let $(4y - 3) = q$. See example 5.)
38. $(7x - 5)^3(7ab) - (7x - 5)^2(8b)$
39. $(2x - 3)(5x - 4)(5c) - (2x - 3)(5x - 4)(7d)$
40. $(7x + 2)^2(3x - 1)(8c) - (7x + 2)(3x - 1)(4)$

In exercises 41–48, factor according to the condition given. *(See examples 7 and 8.)*

41. $3x^5 - 6x^c$, $c > 5$
42. $8a^6 - 7a^m$, $m > 6$
43. $5y^7 + 3y^b$, $b > 7$
44. $2x^8 + 3x^t$, $8 > t$
45. $-3p^n + 6p^m$, $m > n$
46. $-8x^c + 4x^d$, $d > c$
47. $5x^{m+3} - 2x^m$, $m > 3$
48. $3x^{n+5} - 6x^n$, $n > 5$

49. If one circle has a radius of r_1 and a second, larger, circle has a radius of r_2, then the difference in the areas of the two circles is

$$A = \pi r_2^2 - \pi r_1^2$$

Factor the right side of the equation.

50. A calculus student differentiated a function to get

$$y' = 5(3x^2 - 4) + 12x(3x^2 - 4)(5x - 1)$$

Factor the right side of this equation and then find the value of y' by replacing the x with 2 in the factored form of the equation.

51. When a body of volume V_b is heated, the volume will expand according to the equation

$$V_f = V_b + kV_bt$$

where V_f is the final volume, V_b is the beginning volume before the body is heated, k is a constant, and t is the temperature range through which the body was heated. Factor the right side of the equation.

52. A silo has a hemispherical roof, and the roof is covered uniformly with 4 inches of ice. The radius of the hemisphere is 5 feet. Find the volume of the ice by using the formula

$$V = \tfrac{2}{3}\pi r_2^3 - \tfrac{2}{3}\pi r_1^3$$

where r_2 is the radius to the outer edge of the ice and r_1 is the radius of the hemispherical roof. Factor the right side of the equation and then find the volume of the ice in cubic inches. Use 3.14 for π. Give the answer to the nearest whole number.

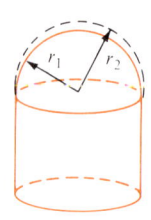

2–5 Factoring Trinomials

The trinomials we consider in this section can be factored into two binomials. The factoring process is mostly a matter of trial and error. However, it is possible to know exactly what signs to place within the binomials. Carefully observe the signs of the following four trinomials and their binomial factors. (We illustrate shortly how to arrive at the factors.)

Trinomial	Binomial Factors
$x^2 + 7x + 12$	$(x + 3)(x + 4)$
$x^2 - 9x + 20$	$(x - 5)(x - 4)$
$x^2 + 4x - 12$	$(x + 6)(x - 2)$
$x^2 - x - 20$	$(x + 4)(x - 5)$

From these four examples, we can give a means of placing the signs within the binomial factors.

Procedure 2–2

Determining Signs of the Binomial Factors When Factoring Trinomials

If the signs of the last two terms of the trinomial are (assuming a positive second-degree term):	Then the signs of the binomial factors will be:
+ +	(+)(+)
− +	(−)(−)
+ −	(+)(−) or (−)(+)
− −	(+)(−) or (−)(+)

Factoring Trinomials Whose Leading Coefficient is 1

If a trinomial is written with descending powers, the first term is called the *leading term* of the trinomial. If the leading term has a coefficient of 1 and if the trinomial has the form

$$x^2 + (a + b)x + ab$$

then the trinomial factors as

$$(x + a)(x + b)$$

Now consider the trinomial $x^2 + 5x + 6$. We know from our discussion on signs that both binomials must have a plus sign. To begin factoring $x^2 + 5x + 6$, we write

$$x^2 + 5x + 6 = (\quad + \quad)(\quad + \quad)$$

Next, we place variables whose product is x^2 within the parentheses. This leads to

$$x^2 + 5x + 6 = (x + \quad)(x + \quad)$$

Next, we need to find numbers whose product is 6 and whose sum is 5. These can only be 2 and 3. This gives

$$x^2 + 5x + 6 + (x + 2)(x + 3)$$

Because multiplication is commutative, we could also write

$$x^2 + 5x + 6 = (x + 3)(x + 2)$$

Even when you are sure you are right with your factoring, you should form the habit of mentally multiplying the two binomials together to see if the product is the trinomial.

Now let us factor $x^2 - 3x - 10$. By procedure 2–2, we write

$$x^2 - 3x - 10 = (\quad + \quad)(\quad - \quad)$$
$$= (x + \quad)(x - \quad)$$

Now we ask what numbers have a product of -10 and a sum of -3. These numbers are -5 and $+2$. Placing these within the parentheses, we get

$$x^2 - 3x - 10 = (x + 2)(x - 5)$$

If you had factored incorrectly as

$$x^2 - 3x - 10 = (x + 5)(x - 2)$$

you would have discovered this when you made your mental multiplication of the factors, for

$$(x + 5)(x - 2) = x^2 + 3x - 10$$

which is incorrect. Always multiply the two binomials. If you make a mistake, you will find it.

Example 1   Factor $x^2 + 10x + 16$.
Solution $x^2 + 10x + 16 = (x + 2)(x + 8)$

Example 2   Factor $p^2 + 4p - 21$.
Solution $p^2 + 4p - 21 = (p + 7)(p - 3)$

Example 3   Factor $t^2 - 8t + 15$.
Solution $t^2 - 8t + 15 = (t - 3)(t - 5)$

Example 4   Factor $3m^2 - 3m - 60$.
Solution
$$3m^2 - 3m - 60 = 3(m^2 - m - 20)$$
$$= 3(m - 5)(m + 4)$$

The monomial factor 3 is removed first.

Factoring Trinomials Whose Leading Coefficient is Not 1

We now consider trinomials where the coefficient of the second-degree term is not 1. The factorization of these trinomials requires more effort, but knowing which signs to place within the binomial factors is a big help.

If we start with $2x^2 + 5x - 12$, then we know the signs are $(\ +\ )(\ -\ )$. Hence, as a start,

$$2x^2 + 5x - 12 = (\ +\ )(\ -\ )$$

Further, the first terms of the binomials must have a product of $2x^2$. This product is obtained by multiplying $2x$ and x. So far, we have

$$2x^2 + 5x - 12 = (2x + \ )(x - \ )$$

or

$$2x^2 + 5x - 12 = (2x - \ )(x + \ )$$

At this point in the factoring process, you may not know whether to try the $2x$ with the plus sign or the minus sign. Don't worry about that now. When we multiply the binomials together, we will know which way it should be. We must now decide what the last two numbers in each binomial should be. Because they have a product of -12, these numbers could be

$-3(4)$ or $3(-4)$ or $6(-2)$ or $2(-6)$ or $12(-1)$ or $1(-12)$

Generally, try the intermediate factors first.

We deliberately try a set that is wrong—-2 and 6—to show what could happen to you.

$$2x^2 + 5x - 12 = (2x - 2)(x + 6)$$

Special Products and Factoring

If we multiply $2x - 2$ and $x + 6$, we get
$$2x^2 + 10x - 12$$
which is wrong. The middle term should be $5x$. Also, the terms of the trinomial have no common factor, but the factor $(2x - 2) = 2(x - 1)$ has a common factor of 2. If the terms of the trinomial do not have a common factor, then neither of the binomial factors will have terms with a common factor either.

Let us try again, using 3 and 4.
$$\begin{aligned} 2x^2 + 5x - 12 &= (2x + 3)(x - 4) \\ &= 2x^2 - 5x - 12 \end{aligned}$$

This is correct except for the sign of the middle term. At this point, we switch the signs within the binomials to obtain
$$2x^2 + 5x - 12 = (2x - 3)(x + 4)$$

Example 5   Factor $6x^2 - 5x - 6$.
Solution $6x^2 - 5x - 6 = (2x - 3)(3x + 2)$ ■

Example 6   Factor $2y^2 + 11y + 12$.
Solution $2y^2 + 11y + 12 = (y + 4)(2y + 3)$ ■

Example 7   Factor $6x^2 + 4x - 16$.
Solution $\begin{aligned} 6x^2 + 4x - 16 &= 2(3x^2 + 2x - 8) \qquad \text{Remove the common factor, 2.} \\ &= 2(3x - 4)(x + 2) \end{aligned}$ ■

Example 8   Factor $4n^3 - 14n^2 - 30n$.
Solution $\begin{aligned} 4n^3 - 14n^2 - 30n &= 2n(2n^2 - 7n - 15) \qquad \text{Remove the common} \\ &= 2n(n - 5)(2n + 3) \qquad\quad\text{factor, } 2n. \end{aligned}$ ■

Example 9   Factor $2(x + 5)^2 - 3(x + 5) - 2$.
Solution   In a problem such as this, let $x + 5 = y$. Then we get

$$2y^2 - 3y - 2 = (2y + 1)(y - 2) \qquad \text{Factor.}$$
$$\begin{aligned} 2(x + 5)^2 - 3(x + 5) - 2 &= [2(x + 5) + 1][(x + 5) - 2] \qquad \text{Substitute} \\ &= [2x + 10 + 1][x + 5 - 2] \qquad\quad x + 5 \text{ for } y. \\ &= (2x + 11)(x + 3) \qquad\qquad\qquad\text{Remove} \end{aligned}$$
parentheses.
Simplify within the brackets.

Example 10 Factor $x^4 - x^2 - 6$.
Solution $x^4 - x^2 - 6 = (x^2 + 2)(x^2 - 3)$

Example 11 Factor $x^{2n} + 5x^n + 6$.
Solution $x^{2n} + 5x^n + 6 = (x^n + 2)(x^n + 3)$

Factoring Perfect-Square Trinomials

The last type of factoring we consider in this section is the factoring of **perfect-square trinomials.** Perfect-square trinomials are formed by squaring a binomial. For example, we know

$$(x + 5)^2 = x^2 + 10x + 25$$

Consequently, $x^2 + 10x + 25$ can be factored as

$$x^2 + 10x + 25 = (x + 5)(x + 5)$$
$$= (x + 5)^2$$

To factor a perfect-square trinomial, take the square roots of the first and last terms to obtain the two terms to form each binomial. The sign in each binomial is the same as the sign of the middle term of the trinomial. The steps for recognizing a perfect-square trinomial are given in procedure 2–3.

Procedure 2–3

Recognizing a Perfect-Square Trinomial
1. The first and last terms are positive and perfect squares.
2. The middle term equals twice the product of the square roots of the first and last terms. The sign of the middle term may be either positive or negative.

NOTE: If $N \geq 0$, $\sqrt{N} = b$ if $b \cdot b = N$. The number represented by $\sqrt{N}$ is called the principal (or non negative) square root of N. In example 12, if $x \geq 0$, $\sqrt{9x^2} = 3x$ because $(3x)(3x) = 9x^2$. We shall generally assume that if we have an expression of the form $(expression)^2$, that the expression represents a positive number, and $\sqrt{(expression)^2}$ = the expression. The topic of roots will be studied in detail in chapter 5.

Example 12 Is $9x^2 + 12x + 4$ a perfect-square trinomial?
Solution Yes, $9x^2$ and 4 are perfect squares and $12x = 2\sqrt{9x^2}\sqrt{4} = 2(3x)(2) = 12x$.

Special Products and Factoring

Example 13   Is $5x^2 + 9x + 4$ a perfect-square trinomial?
Solution   No, 5 is not a perfect square.

Example 14   Is $16x^2 - 20x + 9$ a perfect-square trinomial?
Solution   No, $20x \ne 2\sqrt{16x^2}\sqrt{9} = 2(4 \cdot x)(3) = 24x$.

Example 15   Factor $25x^2 - 20x + 4$.
Solution $25x^2 - 20x + 4 = (5x - 2)^2$

Example 16   Factor $-4x^2 + 20x - 25$.
Solution $-4x^2 + 20x - 25 = -(4x^2 - 20x + 25) = -(2x - 5)^2$

Example 17   Factor $9x^2 - 6x + 4$.
Solution $9x^2 - 6x + 4$ is **prime** because it is not a perfect-square trinomial and cannot be factored as a general trinomial.

Example 18   Factor $x^{2n} + 6x^n y^m + 9y^{2m}$.
Solution $x^{2n} + 6x^n y^m + 9y^{2m} = (x^n + 3y^m)(x^n + 3y^m) = (x^n + 3y^m)^2$

2–5 Exercises

In exercises 1–34, factor the trinomials. If a trinomial cannot be factored, state that it is prime. Remove any common factors first. Any variable used as an exponent represents a natural number. (See examples 1–9.)

1. $x^2 + x - 12$
2. $x^2 - 11x + 30$
3. $t^2 + 3t - 28$
4. $x^2 - 9x - 22$
5. $x^2 + 11x - 12$
6. $3z^2 - 6z - 24$
7. $2u^2 - 8u - 90$
8. $x^2 + 2x + 3$
9. $x^2 + 5x + 7$
10. $w^2 + 6w + 5$
11. $-x^2 + 3x + 40$ *(Hint:* Factor out -1 first.)
12. $-x^2 + 12x - 35$
13. $-2c^2 - c + 21$
14. $-2s^2 - 3s + 27$
15. $3x^2 + 14x - 5$
16. $5x^2 - 14x - 3$
17. $-7x^2 + 29x - 4$
18. $2x^2 - 15x + 28$
19. $7a^2 + 5a - 2$
20. $12p^2 - 12p + 3$
21. $8a^2 + 5a + 7$
22. $8x^2 - 10xy - 3y^2$
23. $3a^2 - 17ab + 10b^2$

24. $2x^2 - bx - 3b^2$

25. $21c^2 - 13cd + 2d^2$

26. $11x^2 - 34xy + 3y^2$

27. $15x^2 - 2x - 8$

28. $24x^2 + 23x - 12$

29. $(x - 6)^2 - 2(x - 6) - 8$ [*Hint:* Let $x - 6 = y$ (see example 9).]

30. $2(x + 4)^2 + 7(x + 4) - 15$

31. $2(a - b)^2 + 7(a - b) + 6$

32. $3(a - 2)^2 + 19(a - 2) + 6$

33. $3(b + 6)^2 - 13(b + 6) - 10$

34. $10(x + 2)^2 + 7b(x + 2) + b^2$

In exercises 35–42, determine if the trinomials are perfect-square trinomials. (See examples 12–14.)

35. $4v^2 + 20v + 25$

36. $t^2 - 14t + 49$

37. $x^2 + 6x + 9$

38. $x^2 + 16x + 64$

39. $5d^2 + 20d + 16$

40. $4A^2 - 10A + 9$

41. $9x^2 - 24xy + 16y$

42. $4x^2 - 16x - 16$

In exercises 43–58, factor the perfect square trinomials. If the trinomial is not a perfect-square trinomial, make a statement to that effect. (See examples 15–18.)

43. $4B^2 - 12B + 9$

44. $4D^2 - 20D + 25$

45. $9x^2 + 24x + 16$

46. $25x^2 + 20x + 4$

47. $9x^2 - 30xy + 25y^2$

48. $25a^2 - 40ab + 16b^2$

49. $16p^2 - 30p + 25$

50. $8q^2 + 20q + 4$

51. $64a^2 - 32a + 4$

52. $72t^2 - 120t + 50$

53. $48a^2 + 24a + 3$

54. $25a^2x + 20ax + 4x$

55. $9y - 24sy + 16s^2y$

56. $-25a + 30ar - 9ar^2$

57. $-9bx^2 - 24bxy - 16by^2$

58. $cx^2 + 3cxy + 9cy^2$

In exercises 59–64, factor. (See example 18.)

59. $x^{2a} + 5x^a + 6$

60. $x^{2b} + 2x^b - 15$

61. $x^4 - x^2 - 12$

62. $x^6 + 2x^3 - 35$

63. $7a^{2m} + 6a^m b^n - b^{2n}$

64. $6x^{2a} - 11x^a - 10$

2–6   Special Types of Factoring

Factoring the Difference of Two Squares

In this section we show special types of polynomials to be factored. The first of these polynomials is referred to as the *difference of two squares.* If we look briefly at the products of polynomials, which we found in section 2–3, we see the following:

1. $(x - 5)(x + 5) = x^2 - 25$
2. $(2x - 3)(2x + 3) = 4x^2 - 9$
3. $(7x - 2)(7x + 2) = 49x^2 - 4$

Notice that the right side of all three equations is the difference of two squares. How do we reverse the process to factor the difference of two squares? The factors on the left in every case represent the product of the sum and difference of the same two numbers. Further, these two numbers are the square roots of the terms on the right. Procedure 2–4 outlines the steps in factoring the difference of two squares.

Procedure 2–4

Factoring the Difference of Two Squares

$$a^2 - b^2 = (a + b)(a - b)$$

Step 1. Form two sets of parentheses, with one set getting a plus sign and the other set a minus sign. This gives
$$a^2 - b^2 = (\quad + \quad)(\quad - \quad)$$

Step 2. Take the principal square root of a^2. This number becomes the first term in both binomials. We now have
$$a^2 - b^2 = (a + \quad)(a - \quad)$$
$$\sqrt{a^2} = a$$

Step 3. Take the principal square root of b^2. This number becomes the last term in both binomials.
$$a^2 - b^2 = (a + b)(a - b)$$
$$\sqrt{b^2} = b$$

Note: a and b represent positive numbers.

The following examples will illustrate how to factor the difference of two squares.

Example 1

Factor $36x^2 - 4$.

Solution

$$36x^2 - 4 = 4(9x^2 - 1) = 4(3x - 1)(3x + 1)$$
$$\sqrt{1} = 1$$
$$\sqrt{9x^2} = 3x$$

Example 2

Factor $x^2 + 16$.

Solution

$x^2 + 16$ is prime. The sum of two squares cannot be factored over the real numbers. ■

Example 3

Factor $(a + b)^2 - 4$.

Solution

$(a + b)^2 - 4 = (a + b + 2)(a + b - 2)$. If you don't see this, let $(a + b) = y$ and write $(a + b)^2 - 4$ as $y^2 - 4$. Then

$$y^2 - 4 = (y + 2)(y - 2)$$
$$(a + b)^2 - 4 = [(a + b) + 2][(a + b) - 2]$$
$$= (a + b + 2)(a + b - 2) \blacksquare$$

Example 4

Factor $(a - b)^2 - (x + 3)^2$.

Solution

First let $(a - b) = m$ and $(x + 3) = n$. We may then write $(a - b)^2 - (x + 3)^2 = m^2 - n^2 = (m - n)(m + n)$. Finally, substitute the values for m and n into this expression.

$$(a - b)^2 - (x + 3)^2 = (m - n)(m + n)$$
$$= [(a - b) - (x + 3)][(a - b) + (x + 3)]$$
$$= (a - b - x - 3)(a - b + x + 3) \blacksquare$$

Example 5

Factor $x^4 - y^4$.

Solution

$$x^4 - y^4 = (x^2 - y^2)(x^2 + y^2)$$
$$= (x - y)(x + y)(x^2 + y^2) \blacksquare$$

Factoring the Sum or Difference of Two Cubes

To see how to factor the sum or difference of two cubes, notice:

1. $(a - b)(a^2 + ab + b^2) = a^3 + a^2b + ab^2 - a^2b - ab^2 - b^3$
   $= a^3 - b^3$
2. $(a + b)(a^2 - ab + b^2) = a^3 - a^2b + ab^2 + a^2b - ab^2 + b^3$
   $= a^3 + b^3$

An expression of the form $a^3 - b^3$ is called the *difference of two cubes*. If it has the form $a^3 + b^3$, it is called the *sum of two cubes*. Both can be factored by reversing steps 1 and 2.

The steps for factoring the sum or difference of two cubes are summarized in rule procedure 2–5.

Procedure 2–5

Factoring the Sum or Difference of Two Cubes

$$x^3 \pm y^3 = (x \pm y)(x^2 \mp xy + y^2)$$

1. The factors will be a binomial and a trinomial.
2. The terms of the binomial factor are the cube roots of the two terms of the original problem. The sign between the two terms in the binomial is the same as the sign between the terms of the original problem.

Special Products and Factoring

3. The first term in the trinomial factor is the square of the first term in the binomial factor.
4. The middle term of the trinomial factor is the opposite of the product of the two terms of the binomial factor.
5. The last term in the trinomial factor is the square of the last term in the binomial factor.

The next few examples show how to factor the sum or difference of two cubes. Notice in particular how procedure 2–5 has been applied to the trinomial factor in the factorization process.

Example 6 Factor $a^3 + 8$.

Solution
$a^3 + 8 = (a + 2)(a^2 - 2a + 4)$
(Note: $\sqrt[3]{N} = b$ if $b \cdot b \cdot b = N$. In Example 6, $\sqrt[3]{a^3} = a$ because $a \cdot a \cdot a = a^3$. In like manner, $\sqrt[3]{8} = 2$ because $2 \cdot 2 \cdot 2 = 8$.)

Example 7 Factor $27x^3 - 8$.

Solution $27x^3 - 8 = (3x - 2)(9x^2 + 6x + 4)$

Example 8 Factor $16x^3 + 2$.

Solution Factor out the common factor of 2 first.
$$16x^3 + 2 = 2(8x^3 + 1)$$
$$= 2(2x + 1)(4x^2 - 2x + 1)$$

Example 9 Factor $(a + b)^3 + 27$.

Solution The factoring in this problem may be simplified by using a substitution and letting $y = a + b$. Doing this gives,
$$(a + b)^3 + 27 = y^3 + 27$$
$$= (y + 3)(y^2 - 3y + 9) \qquad (1)$$
We must now replace y with $a + b$ in (1).
$$(a + b)^3 + 27 = [(a + b) + 3][(a + b)^2 - 3(a + b) + 9]$$
$$= (a + b + 3)(a^2 + 2ab + b^2 - 3a - 3b + 9)$$

Factoring by Grouping

Certain types of expressions can be factored by selectively grouping the terms making up the expression, as shown in example 10.

Example 10 Factor $ac + bd + bc + ad$.

Solution The idea in factoring by grouping is to select terms from the expression with a common factor. In this example, the first and third terms have a common factor of c and the second and fourth terms have a common factor of d. Thus

$ac + bd + bc + ad = (ac + bc) + (bd + ad)$ Group pairs of terms with a common factor.

$\qquad\qquad\qquad\quad = c(a + b) + d(b + a)$ Remove the common factor from each term of the two binomials.

$\qquad\qquad\qquad\quad = c(a + b) + d(a + b)$ Commutative axiom of addition.

$\qquad\qquad\qquad\quad = (a + b)(c + d)$ ■ Factor $(a + b)$ from each term.

Some four-term polynomials can be factored by grouping the four terms as one term and three terms, as in example 11.

Example 11 Factor $x^2 - a^2 - 2ab - b^2$.

Solution Group the last three terms by removing -1 from each of these terms.

$x^2 - a^2 - 2ab - b^2 = x^2 - (a^2 + 2ab + b^2)$
$\qquad\qquad\qquad\quad = x^2 - (a + b)^2$ The right side of the equation is the difference of two squares.

$\qquad\qquad\qquad\quad = [x - (a + b)][x + (a + b)]$
$\qquad\qquad\qquad\quad = (x - a - b)(x + a + b)$ ■

2–6 Exercises

In exercises 1–6, factor as the difference of two squares. (See example 1.)

1. $x^2 - 16$
2. $9 - x^2$
3. $16x^2 - 4y^2$
4. $49a^2 - 4b^2$
5. $x^2y^2 - 4$
6. $36a^2 - 81b^2$

In exercises 7–16, factor as the difference of two squares. (See examples 3 and 4.)

7. $(a + b)^2 - 9$
8. $(2x - y)^2 - 16$
9. $(3x - 2)^2 - (x + 5)^2$
10. $(2x - 5)^2 - (x + 3)^2$
11. $(x + 5)^2 - 9y^2$
12. $(x^2 + 2)^2 - (x + 5)^2$

Special Products and Factoring

13. $(x + y - 1)^2 - (2x + y + 2)^2$
14. $(x + y - 1)^2 - (3x + y + 4)^2$
15. $(x^2 + y + 2)^2 - (x^2 - y - 2)^2$
16. $(2a + b)^2 - (4a - b + 2)^2$

In exercises 17–32, factor as either the sum or difference of two cubes. (See examples 6–9.)

17. $x^3 - y^3$
18. $x^3 + 27$
19. $a^3 + b^3$
20. $2a^3 + 2b^3$
21. $8x^3 + y^3$
22. $3x^3 - 24y^3$
23. $x^3 - 27y^3$
24. $8x^3 - 125y^3$
25. $64x^3 - 27y^3$
26. $8x^3y^3 + 27a^6$
27. $16x^3 + 2y^3$
28. $c^2a^3 - 8c^2y^3$
29. $(x + y)^3 - 8$
30. $(x + 2)^3 - 27$
31. $(2x - 3)^3 + 1$
32. $(3x + 1)^3 + 8$

In exercises 33–50, factor either as the difference of two squares or as the sum or difference of two cubes. (See examples 1–9.)

33. $x^4 - y^4$
34. $x^9 - y^9$
35. $4a^4 - 64y^4$
36. $4x^2 + 16y^2$
37. $x^6 - y^6$ (*Hint:* Factor as the difference of two squares.)
38. $2a^3 - 16b^3y^6$
39. $s^8 - t^8$
40. $b^6 - 8c^3$
41. $x^{10} - y^{10}$
42. $250x^3 + 54y^6$
43. $2x^3 - 16y^3$
44. $-27a^3 + 8y^6$
45. $m^6 + n^3$
46. $-m^9 + n^3$
47. $x^4 - y^2$
48. $9a^4 - 9b^4$
49. $6a^3 - 48y^6$
50. $7a^3 + 56y^6$

In exercises 51–72, factor by grouping. (See example 10.)

51. $ac + ad + 2bc + 2bd$
52. $ax + ay + dx + dy$
53. $2ab - 6b + a - 3$
54. $cx - cy + dx - dy$
55. $a^2 - ac - 2ab + 2bc$
56. $cx^2 + dx^2 - cy^2 - dy^2$
57. $2xy + 8y - 3x - 12$
58. $9a + 9b - ax^2 - bx^2$
59. $a^3 - 4a^2 + 2a - 8$
60. $ax^3 + bx^3 + ay^3 + by^3$
61. $x^2 - xy - xy^2 - y^3$
62. $xy - 4x - 4y + 16$
63. $6c^3 - 9c + 8c^2 - 12$
64. $ab - 6a - 5b + 30$
65. $9ab - 27a - 2b + 6$
66. $2xy + 8x - 3y - 12$
67. $acx - 1 + a - cx$
68. $8ab + 8a + b + 1$
69. $pqt + 1 - pq - t$
70. $2cd - c - 2d + 1$
71. $ac + 2c - ad - 2d + 2a + 4$
72. $ac - 4d + 3a - 4c + ad - 12$

73. The difference in the volumes of two spheres with radii r_2 and r_1 is
$$V = \tfrac{4}{3}\pi r_2^3 - \tfrac{4}{3}\pi r_1^3$$
Completely factor the right side of this equation.

74. An algebra student was given the task of reducing the following fraction:

$$\frac{5a + ax + 5b + bx}{15 + 3x}$$

Before the fraction could be reduced, the student had to first factor the numerator and denominator. Do the factoring for the student. (You do not have to reduce the fraction unless you remember how. Reducing fractions will be studied in the next chapter.)

Chapter 2 Summary

Vocabulary

base
binomial
coefficient
degree of a polynomial
exponent
factors
like terms
monomial

numerical coefficient
perfect-square trinomial
polynomial
prime
product
rational expression
trinomial

Factoring Polynomials

Removing the Greatest Common Factor (see procedure 2–1)

$$5ax^2 + 10a^2y = 5a(x^2 + 2y)$$

Factoring Trinomials

Trinomials with a leading coefficient of one

$$x^2 + (a + b)x + ab = (x + a)(x + b)$$

Trinomials with a leading coefficient not equal to one

$$acx^2 + (ad + bc)xy + bdy^2 = (ax + by)(cx + dy)$$

Perfect square trinomials

$$x^2 + 2xy + y^2 = (x + y)^2$$

Difference of Two Squares (see procedure 2–4)

$$x^2 - y^2 = (x + y)(x - y)$$

Difference of Two Cubes (see procedure 2–5)

$$x^3 - y^3 = (x - y)(x^2 + xy + y^2)$$

Sum of Two Cubes (see procedure 2–5)

$$x^3 + y^3 = (x + y)(x^2 - xy + y^2)$$

Special Products and Factoring

Using Substitution

$(x + y + 3)^2 - 4$; let $z = (x + y + 3)$. Then
$$z^2 - 4 = (z - 2)(z + 2)$$
$$(x + y + 3)^2 - 4 = [(x + y + 3) - 2][(x + y + 3) + 2]$$
$$= (x + y + 1)(x + y + 5)$$

By Grouping

$$ax + ay + bx + by = (ax + ay) + (bx + by) = a(x + y) + b(x + y)$$
$$= (x + y)(a + b)$$

Definitions

2-1 Exponent

The nth power of x is written as

$$x^n = \underbrace{x \cdot x \cdot x \cdots x.}_{n \text{ factors of } x}$$

2-2 Zero Exponent

If a is any nonzero real number, then $a^0 = 1$.

Rules

In rules 2-1–2-3, a is any real number and m, n, and p are natural numbers.

2-1 Multiplying Numbers That Have the Same Base

$$a^m \cdot a^n = a^{m+n}.$$

2-2 Raising a Power to a Power

$$(a^m)^n = a^{mn}$$

2-3 Dividing Numbers That Have the Same Base

If $a \neq 0$,

$$\frac{a^m}{a^n} = a^{m-n} \quad \text{if } m > n$$

$$\frac{a^m}{a^n} = \frac{1}{a^{n-m}} \quad \text{if } n > m$$

$$\frac{a^m}{a^n} = 1 \quad \text{if } m = n$$

2-4 Raising a Product to a Power

If a and b are real numbers and if m, n, and p are natural numbers, then $(a^m b^n)^p = a^{mp} b^{np}$.

2-5 Raising a Quotient to a Power

If a is any real number, if b is any nonzero real number, and if m is a natural number, $\left(\dfrac{a}{b}\right)^m = \dfrac{a^m}{b^m}$.

2-6 Squaring a Binomial

$$(a + b)^2 = a^2 + 2ab + b^2$$

1. Square the first term of the binomial to get the first term of the answer.
2. Multiply the two terms of the binomial together and double this product to get the middle term of the answer.
3. Square the last term of the binomial to get the last term of the answer.

Procedures

2-1 Removing the Greatest Common Factor

1. All coefficients must be integers.
2. The monomial factored out must be as large as possible and still maintain integers for the coefficients of the variable terms.
3. All exponents on the variables must be whole numbers.

2-2 Determining Signs of the Binomial Factors when Factoring Trinomials

If the signs of the last two terms of the trinomial are (assuming a positive second-degree term):	Then the signs of the binomial factors will be:
+ +	(+)(+)
− +	(−)(−)
+ −	(+)(−) or (−)(+)
− −	(+)(−) or (−)(+)

2-3 Recognizing a Perfect-Square Trinomial

1. The first and last terms are positive and perfect squares.
2. The middle term equals twice the product of the square roots of the first and last terms. The sign of the middle term may be either positive or negative.

2-4 Factoring the Difference of Two Squares

$$a^2 - b^2 = (a + b)(a - b)$$

Step 1. Form two sets of parentheses, with one set getting a plus sign and the other set a minus sign. This gives $a^2 - b^2 = (\ \ +\ \ )(\ \ -\ \ )$

Special Products and Factoring   83

Step 2. Take the principal square root of a^2. This number becomes the first term in both binomials. We now have
$$a^2 - b^2 = (a + \quad)(a - \quad)$$
$$\sqrt{a^2} = a$$

Step 3. Take the principal square root of b^2. This number becomes the last term in both binomials.
$$a^2 - b^2 = (a + b)(a - b)$$
$$\sqrt{b^2} = b$$

NOTE: a and b represent positive numbers.

2–5 Factoring the Sum or Difference of Two Cubes

$$x^3 \pm y^3 = (x \pm y)(x^2 \mp xy + y^2)$$

1. The factors will be a binomial and a trinomial.
2. The terms of the binomial factor are the cube roots of the two terms of the original problem. The sign between the two terms in the binomial is the same as the sign between the terms of the original problem.
3. The first term in the trinomial factor is the square of the first term in the binomial factor.
4. The middle term of the trinomial factor is the opposite of the product of the two terms of the binomial factor.
5. The last term in the trinomial factor is the square of the last term in the binomial factor.

Review Problems

Section 2–1   Write in a shorter form, using a base and an exponent.

1. $x \cdot x \cdot x \cdot x$
2. $(3x + 2)(3x + 2)$
3. $7y(7y)^3$
4. $(5x)^2(5x)$

Simplify.

5. $\dfrac{50x^5}{5x^3}$
6. $\dfrac{-4x^6y^5}{2x^7y^2}$
7. $\dfrac{(5x + 3)(9x + 2)(5x + 3)^2}{(5x + 3)^0}$
8. $\dfrac{(-2x)^2(5x^4)}{10x^2}$
9. $\dfrac{(3x^0)(-6x^2y)}{xy}$
10. $\dfrac{5x - 3}{(5x - 3)^4(x + y)^2}$
11. $\dfrac{3^2(-2)^3(5)^5}{3^4(-2)^7(5)^3}$

12. The volume of a sphere is $V = \frac{4}{3}\pi r^3$, where r is the radius of the sphere. Find V if $r = 21$ inches. Use $\frac{22}{7}$ for π.

13. The location of an object from the origin on a number line is given by $d = 2t^2 - 3t$. Find d if $t = 5$.

14. Evaluate the expression $y = \dfrac{-x^2 + 5x}{-5}$ if $x = 5$.

Section 2–2

Simplify by combining like terms.

15. $(5x - 3y) + (2x - 4y)$

16. $(3x^2 + 4x - 5) + (7x^2 - 6x + 3)$

17. $(9x^4 - 7x^2 + 3) - (-x^4 + 7x^3 + 5)$

18. $(7x^3y^2 - 7xy^2 + 6x) - (2x^2y^3 + 3xy^2 - 2x)$

19. Evaluate the expression $(3x^2 + 5) - (x^2 + 3)$ for $x = 2$
(a) Without first simplifying the expression.
(b) By first combining like terms and then evaluating for $x = 2$.

20. If $P(x) = 3x^2 - 4x + 2$, find $P(3)$.

21. If $P(x) = -x^3 - 2x^2 - 3$, find $P(-2)$.

22. If $P(x) = x^2 + 3$, find $P(x + h)$.

Section 2–3

Multiply the following polynomials. Collect like terms if possible.

23. $(x^2 - 6x)(x + 5)$

24. $3x^2y(7xy^2 - 2xy^3 + 5)$

25. $(7x - 3)^2$

26. $(2x - 3y)^2$

27. $(x + 2)(2x^2 - 3x + 1)$

28. $(x - 3)(x^2 - 6x + 1)$

Sections 2–4—2–6

Factor.

29. $7x^5 - 2x^3 + 7x^2$

30. $7x^2y + 21xy^2 + 14xy$

31. $x^2 + 4x - 21$

32. $2x^2 + 13x + 6$

33. $14x^2 + 17x - 6$

34. $x^3 - 8$

35. $9x^2 - 4$

36. $5x^2 - 20$

37. $x^2 + 20x + 100$

38. $4x^2 + 12x + 9$

39. $18x^2 - 48x + 32$

40. $cx^2 - 3cx + dx - 3d$

41. $x^2 - 3x - 2xy + 6y$

42. $y^2 - 9x^2 - 12x - 4$

43. $(3x + 2y - 3)^2 - 16$

44. $(x + 3)^3 - 8$

45. The difference between the areas A_1 and A_2 of two triangles with the same base and different altitudes is given by

$$A_1 - A_2 = \tfrac{1}{2}bH - \tfrac{1}{2}bh$$

where H and h represent the altitudes of the first and second triangles, respectively. Factor the right side of the equation.

Cumulative Test for Chapters 1–2

Chapter 1

1. Name the five main subsets of numbers, as discussed in the text, for the real number system.

If $A = \{2, 5, 8, 9\}$, answer true or false for problems 2–5.

2. $5 \in A$
3. $\{5, 7\} \subset A$
4. $6 \notin A$
5. Set A is well-defined.

Answer true or false for problems 6–8.

6. Every real number may be classified either as an irrational number or a whole number.
7. If a number is a real number, it is an integer.
8. Every whole number is a rational number.
9. Simplify: $\dfrac{3 + 5 - 8 + 10}{1 + 3^2}$.
10. The statement $(a + b) + 5 = 5 + (a + b)$ illustrates which axiom of equality?
11. The statement $1 \cdot 6 + 5 = 6 + 5$ illustrates which axiom of equality?
12. Simplify: $3\{2 - [5 + (-2) - 3] - [9 - (-2)]\}$.
13. Simplify: $\dfrac{-4^2 + 8(-3) - (-2)}{(-3)^2 - 2(-5)}$.
14. Evaluate the expression $\dfrac{5a^2 - 2b}{-4b}$ if $a = 4$ and $b = -5$.

Chapter 2

15. Simplify: $\dfrac{-4^2 a^2 b^4}{8a^3 b}$.
16. The area, A, of a circle is $A = \pi r^2$, where r is the radius of the circle. Find A if $r = 7$ inches. Use $\tfrac{22}{7}$ for π.
17. Simplify by combining like terms: $(9x^2 + 17x - 3) - (4x^2 - 8x - 9)$.
18. Square and simplify: $(4x - 3y)^2$.
19. Factor completely: $15x^3 + 50x^2 - 40x$.
20. Factor completely: $16a^3 b - 2b$.

CHAPTER 3 | Rational Expressions

Objectives of Chapter 3

To be able to:

☐ Recognize a rational expression. (3–1)

☐ Determine what values make the denominator of a fraction zero. (3–1)

☐ Determine when fractions are equal. (3–1)

☐ Apply the fundamental theorem of fractions. (3–1)

☐ Raise fractions to higher terms. (3–1)

☐ Simplify the signs of a fraction. (3–1)

☐ Reduce a rational expression. (3–2)

☐ Multiply and divide rational expressions. (3–3)

☐ Write a fraction as a product of its numerator and the reciprocal of its denominator. (3–3)

☐ Find the least common multiple (LCM) for two or more numbers. (3–4)

☐ Find the least common denominator (LCD) for two or more fractions. (3–5)

☐ Add and subtract rational expressions. (3–5)

☐ Simplify a complex fraction. (3–6)

☐ Divide polynomials. (3–7)

☐ To use synthetic division to divide and factor polynomials. (3–8)

Rational Expressions

In arithmetic it is important to know how to add, subtract, multiply, and divide fractions. We also need to know how to perform these same operations on algebraic fractions.

We begin our discussion with a review of the basic properties of fractions, and then we apply these properties to algebraic fractions.

3-1 Basic Properties of Fractions

A definition of a rational expression was given in section 2-2. For our purposes in this chapter, it will be convenient to think of a rational expression as a quotient of polynomials. For example,

$$\frac{3x}{2x^2 + 5}, \quad \frac{9x^2 - 4x + 3}{x - 1}, \quad \frac{1}{x - 2y}, \quad \text{and} \quad \frac{3x + 5}{1}$$

are rational expressions. We also refer to a rational expression as a fraction, because this is the most common name for a rational expression. A fraction is not defined for a zero denominator, so we must be careful not to give values to the variables that will cause the denominator to become zero. For example, if $x = 3$ in the rational expression $2x/(x - 3)$, then the fraction becomes $6/(3 - 3) = 6/0$, which is meaningless.

Example 1 For what values of x will the denominator of

$$\frac{5x + 2}{2x^2 + 5x - 3}$$

be zero?

Solution The values of x that give a zero denominator can be found if the denominator of the fraction is first factored. Then each factor is set equal to zero.

$$\frac{5x + 2}{2x^2 + 5x - 3} = \frac{5x + 2}{(2x - 1)(x + 3)}$$

If each factor in the denominator is now set equal to zero, we get

$$\begin{array}{ll} 2x - 1 = 0 & \text{or} \quad x + 3 = 0 \\ 2x = 1 & \qquad x = -3 \\ x = \dfrac{1}{2} & \end{array}$$

The fraction will have a zero denominator (and thus be undefined) if $x = \frac{1}{2}$ or $x = -3$. ■

Two fractions are equal if their cross products are equal. If we write $\frac{3}{4} = \frac{6}{8}$ as

$$\frac{3}{4} \diagup\!\!\!\!\!\!= \diagdown\!\!\!\!\!\!\frac{6}{8}$$

the diagonal lines show which numbers form the cross products. Notice that $3 \cdot 8 = 4 \cdot 6$. These products are the same, and $\frac{3}{4}$ is said to be *equal to* (or *equivalent to*) $\frac{6}{8}$. The idea discussed here is given as rule 3–1.

Rule 3–1

Equality of Fractions

$\dfrac{a}{b} = \dfrac{c}{d}$ if and only if $ad = bc$ $b, d, \neq 0$

Example 2   Is $\frac{5}{8} = \frac{45}{72}$?

Solution   Yes, because $5 \cdot 72 = 8 \cdot 45$, or $360 = 360$.

Example 3   Is $\frac{5}{3} = \frac{11}{7}$?

Solution   No, because $5 \cdot 7 \neq 3 \cdot 11$, or $35 \neq 33$.

Example 4   Is $\dfrac{x}{3} = \dfrac{2x}{6}$?

Solution   Yes, because $x \cdot 6 = 3 \cdot 2x$, or $6x = 6x$.

Example 5   Is $\dfrac{x - 4}{3} = \dfrac{x^2 - 16}{3(x + 4)}$?

Solution   Yes, because $3(x - 4)(x + 4) = 3(x^2 - 16)$.

The fractions in example 5 are equal, but only for the permissable replacements for the variable. Notice they are equal for all $x \neq -4$.

The numerator and denominator of a fraction are called **terms of the fraction.** We often need to raise a fraction to higher terms. This is done by making use of the fundamental theorem of fractions. To illustrate this theorem before we formally state it, note that

Rational Expressions

$$\frac{5}{8} = \frac{5}{8} \cdot 1$$
$$= \frac{5}{8} \cdot \frac{2}{2}$$
$$= \frac{5 \cdot 2}{8 \cdot 2}$$
$$= \frac{10}{16}$$

Rule 3–1 can be used to verify that $\frac{5}{8} = \frac{10}{16}$. What did we do to $\frac{5}{8}$ to get $\frac{10}{16}$? Examination of the steps just shown illustrates that we multiplied $\frac{5}{8}$ by $\frac{2}{2}$ (or 1) to get $\frac{10}{16}$. In general, it is always permissible to multiply a fraction by c/c if $c \neq 0$.

Rule 3–2

Fundamental Theorem of Fractions

If a, b, and c are real numbers and $c \neq 0$, then

$$\frac{a}{b} = \frac{ac}{bc} \qquad b \neq 0$$

Example 6 Use rule 3–2 to write $\frac{5}{8}$ as a fraction with a denominator of 24.

Solution $\frac{5}{8} = \frac{5 \cdot 3}{8 \cdot 3} = \frac{15}{24}$ ■

Example 7 Write $3/x$, $x \neq 0$, as a fraction with a denominator of $5x^2$.

Solution $\frac{3}{x} = \frac{3(5x)}{x(5x)} = \frac{15x}{5x^2}$ ■

Example 8 Write $5/(x - 3)$ as a fraction with a denominator of $x^2 - 9$.

Solution In a problem such as this, it is convenient to write

$$\frac{5}{x - 3} \cdot \frac{}{} = \frac{}{x^2 - 9}$$

where the blank holds the space for the factor by which we are going to multiply the numerator and denominator of the given fraction to get a denominator of $x^2 - 9$. To see by what $x - 3$ is multiplied to get $x^2 - 9$, we must factor $x^2 - 9$. Our example now looks like this:

$$\frac{5}{x - 3} \cdot \frac{}{} = \frac{}{(x - 3)(x + 3)}$$

We now observe from this step that $x - 3$ in $5/(x - 3)$ was multiplied by $x + 3$ to obtain the factored denominator. Thus, we must multiply the given fraction by $(x + 3)/(x + 3)$.

$$\frac{5}{x - 3} \cdot \frac{x + 3}{x + 3} = \frac{5(x + 3)}{(x - 3)(x + 3)} = \frac{5x + 15}{x^2 - 9}$$

We now turn our attention to the signs of a fraction. Notice that

$$\frac{-10}{5} = -2, \quad \frac{10}{-5} = -2, \quad \text{and} \quad -\frac{10}{5} = -2$$

In all three cases, it makes no difference if the minus sign is in the numerator, the denominator, or in front of the fraction. The value of the fraction is the same regardless of the position of the minus sign.

Rule 3–3

A Fraction Containing One Minus Sign

If a and b are real numbers and $b \neq 0$,

$$-\frac{a}{b} = \frac{a}{-b} = \frac{-a}{b}$$

Now consider the case where the fraction has exactly two minus signs. Using examples we see,

$$\frac{-10}{-2} = 5 \quad \text{and} \quad -\frac{-10}{2} = -(-5) = 5 \quad \text{and} \quad -\frac{10}{-2} = -(-5) = 5$$

Rule 3–4

A Fraction Containing Two Minus Signs

If a and b are real numbers and $b \neq 0$,

$$\frac{-a}{-b} = \frac{a}{b} \quad \text{and} \quad -\frac{-a}{b} = \frac{a}{b} \quad \text{and} \quad -\frac{a}{-b} = \frac{a}{b}$$

If a fraction has three minus signs, it is the same as if it had only one minus sign. We can illustrate this by noting

$$-\frac{-10}{-5} = -\frac{10}{5}$$

by rule 3–4, and

$$-\frac{10}{5} = -(2) = -2$$

by rule 3–3. The idea expressed here gives rule 3–5.

Rule 3–5

A Fraction Containing Three Minus Signs

If a and b are real numbers and $b \neq 0$,

$$-\frac{-a}{-b} = \frac{-a}{b}$$

If a fraction has exactly one minus sign, it is usually more convenient to place it in the numerator rather than in the denominator. A fraction written as a/b or $-a/b$ is in **standard form** if fully reduced.

Example 9 Simplify: $\dfrac{-(x-y)}{-2}$.

Solution
$$\frac{-(x-y)}{-2} = \frac{x-y}{2} \qquad \text{rule 3–4}$$

Example 10 Simplify: $-\dfrac{-(x-y)}{-2y}$.

Solution
$$-\frac{-(x-y)}{-2y} = \frac{-(x-y)}{2y} \qquad \text{rule 3–5}$$
$$= \frac{y-x}{2y}$$

Example 11 Simplify: $\dfrac{-2h}{3h-2}$.

Solution $\dfrac{-2h}{3h-2}$ cannot be simplified by any direct application of rules 3–3, 3–4, or 3–5. However, if -1 is factored from the denominator, the given problem can be written as

$$\frac{-2h}{3h-2} = \frac{-2h}{-(2-3h)}$$
$$= \frac{2h}{2-3h} \qquad \text{rule 3–4}$$

Example 12 Write $\dfrac{9a-3b}{-2}$ in standard form.

Solution
$$\frac{9a-3b}{-2} = \frac{-(9a-3b)}{2} \qquad \text{rule 3–3}$$
$$= \frac{3b-9a}{2}$$

3–1 Exercises

In exercises 1–16, tell what value or values of the variable would make the denominator zero. (See example 1.)

1. $\dfrac{2}{x+2}$
2. $\dfrac{5x}{x-3}$
3. $\dfrac{2x+4}{x+5}$
4. $-\dfrac{2x}{2x-6}$
5. $\dfrac{5a}{2-a}$
6. $\dfrac{3a}{5-c}$
7. $\dfrac{5x}{(x-2)(x+4)}$
8. $\dfrac{5x-3}{(x-4)(x+3)}$
9. $\dfrac{-5c}{x^2+6x+8}$ (*Hint:* Factor the denominator.)
10. $\dfrac{2c-3}{x^2-7x+10}$
11. $\dfrac{x+5}{x^2-9}$
12. $\dfrac{x+6}{x^2+5x+6}$
13. $\dfrac{5}{x^3-9x}$
14. $\dfrac{3x}{2x^2-x-1}$
15. $\dfrac{9y}{t^2+5t}$
16. $\dfrac{32}{x^3+x^2-16x-16}$

In exercises 17–26, determine if the two fractions are equal by applying rule 3–1 or rule 3–2. (See examples 2–5.)

17. Is $\dfrac{7}{4}=\dfrac{9}{5}$?
18. Is $\dfrac{6}{16}=\dfrac{3}{8}$?
19. Is $\dfrac{6}{22}=\dfrac{3}{11}$?
20. Is $\dfrac{7}{9}=\dfrac{21}{27}$?
21. Is $\dfrac{3}{5}=\dfrac{27}{45}$?
22. Is $\dfrac{6}{8x}=\dfrac{30}{40x}$ $(x \neq 0)$?
23. Is $\dfrac{8}{3c}=\dfrac{21}{9c}$ $(c \neq 0)$?
24. Is $\dfrac{5}{x}=\dfrac{30}{6x}$ $(x \neq 0)$?
25. Is $\dfrac{3}{2x}=\dfrac{12}{9x}$ $(x \neq 0)$?
26. Is $\dfrac{a-2}{3b}=\dfrac{6a-12}{18b}$ $(b \neq 0)$?

In exercises 27–34, write the given fraction in higher terms with the specified numerator or denominator by applying the fundamental theorem of fractions. (See examples 6–8.)

Rational Expressions

27. $\dfrac{3}{8} = \dfrac{?}{32}$

28. $\dfrac{9}{15} = \dfrac{27}{?}$

29. $\dfrac{6a}{4b} = \dfrac{?}{-12b^2}$

30. $\dfrac{8a - 3}{4x - 1} = \dfrac{?}{20x - 5}$

31. $\dfrac{2c + 3}{b + 3} = \dfrac{?}{5b + 15}$

32. $\dfrac{s - t}{s + t} = \dfrac{s^2 - t^2}{?}$

33. $\dfrac{x - 3}{2x + 3} = \dfrac{?}{2x^2 + x - 3}$

34. $\dfrac{x + 5}{x - 2} = \dfrac{?}{3x^2 - 7x + 2}$

In exercises 35–44, use rules 3–3 through 3–5 to simplify the signs in the given fractions. Leave your answer in standard form. (See examples 9–12.)

35. $-\dfrac{-8}{3}$

36. $\dfrac{-10}{-3}$

37. $-\dfrac{-x}{-2y}$

38. $-\dfrac{16}{-17}$

39. $-\dfrac{-(a - b)}{-2c}$

40. $\dfrac{-(2x - 3)}{-4x}$

41. $-\dfrac{-6y + 2}{-(x - y)}$

42. $-\dfrac{-4h + 3}{-5h}$

43. $-\dfrac{6x - y}{-(a + b)}$

44. $\dfrac{-(3a - b)}{-5}$

In exercises 45–54, use the rules of signs for fractions and the fundamental theorem of fractions to replace the question mark with a number or variable expression.

Example

(a) $\dfrac{-5x}{-3} = \dfrac{?}{3}$ 　 (b) $\dfrac{3 - x}{-4} = \dfrac{?}{4}$

Solution

(a) $\dfrac{-5x}{-3} = \dfrac{5x}{3}$ 　 (b) $\dfrac{3 - x}{-4} = \dfrac{-1(3 - x)}{-1(-4)} = \dfrac{x - 3}{4}$ 　 rule 3–2

45. $-\dfrac{-7x}{2} = \dfrac{?}{2}$

46. $\dfrac{-3x}{-4} = \dfrac{3x}{?}$

47. $-\dfrac{-(a + b)}{-4} = -\dfrac{?}{4}$

48. $\dfrac{-3 - 4y}{-2} = \dfrac{3 + 4y}{?}$

49. $\dfrac{-6x + 5}{-a} = \dfrac{?}{a}$

50. $\dfrac{5 - 2x}{6} = \dfrac{2x - 5}{?}$

51. $\dfrac{7 - 2a}{-3} = \dfrac{?}{3}$

52. $-\dfrac{7}{a - 5} = -\dfrac{?}{5 - a}$

53. $-\dfrac{-6}{c - 9} = -\dfrac{?}{9 - c}$

54. $\dfrac{3b + 4}{5} = \dfrac{?}{5}$

55. In electronics, the inductance of a coil can be determined from the equation

$$L = \dfrac{X_L}{2\pi f}$$

What value of f would give the denominator of the fraction on the right side of the equation a value of zero?

56. In the study of sound and the Doppler effect, the equation

$$F = \frac{f \cdot v}{v_0 - v_s}$$

gives the observed frequency, F, of the sound wave as compared to the actual frequency, f, of the sound wave. What must be true of v_0 and v_s if the right side of the equation has a zero denominator?

57. The radius of a circle can be found from the equation

$$r = \frac{C}{2\pi}$$

What must be true about the equation if r is to have a value of zero? Is it possible for the denominator of the fraction on the right side of the equation to have a value of zero?

3–2 Reducing Rational Expressions

Fractions may be reduced by applying rule 3–2, the fundamental theorem of fractions. A fraction is reduced when the numerator and denominator contain no common factors other than 1. For example, by applying the fundamental theorem of fractions;

$$\frac{6}{8} = \frac{3 \cdot 2}{4 \cdot 2} = \frac{3}{4} \cdot \frac{2}{2} = \frac{3}{4} \cdot 1 = \frac{3}{4}$$

This example illustrates the general principle in reducing all fractions, whether they are numerical fractions or rational expressions. Generally, the procedure consists of factoring both the numerator and denominator and then dividing out the common factors in the numerator and denominator of the fraction.

Example 1 Reduce the following by applying the fundamental theorem of fractions.

(a) $\dfrac{16}{24}$ (b) $\dfrac{5x}{6x^2}$ (c) $\dfrac{3x + 3y}{6}$ (d) $\dfrac{(5a^2b)^3}{25a^3b}$ (e) $\dfrac{6}{6x + 6y}$

Solution The slash marks are used to indicate the factors that are divided out.

(a) $\dfrac{16}{24} = \dfrac{\cancel{8} \cdot 2}{\cancel{8} \cdot 3} = \dfrac{2}{3}$

Rational Expressions

(b) $\dfrac{5x}{6x^2} = \dfrac{5\cancel{x}}{6 \cdot \cancel{x} \cdot x} = \dfrac{5}{6x}$

(c) $\dfrac{3x + 3y}{6} = \dfrac{\cancel{3}(x + y)}{\cancel{3} \cdot 2} = \dfrac{x + y}{2}$

(d) $\dfrac{(5a^2b)^3}{25a^3b} = \dfrac{125a^6b^3}{25a^3b} = \dfrac{5a^3b^2(\cancel{25a^3b})}{\cancel{25a^3b}} = 5a^3b^2$

(e) $\dfrac{6}{6x + 6y} = \dfrac{1 \cdot \cancel{6}}{\cancel{6}(x + y)} = \dfrac{1}{x + y}$

You need to be careful with problems similar to the one given in example 1(e). Many students incorrectly work this problem as

$$\dfrac{6}{6x + 6y} = \dfrac{6}{6(x + y)} = x + y$$

There is a great deal of difference between $x + y$ and $1/(x + y)$.

Example 2 $\dfrac{3r^2 + 7r - 6}{2r^2 + 5r - 3} = \dfrac{(3r - 2)\cancel{(r + 3)}}{(2r - 1)\cancel{(r + 3)}} = \dfrac{3r - 2}{2r - 1}$

Example 3 $\dfrac{x^3 - y^3}{5x(x^2 + xy + y^2)} = \dfrac{(x - y)\cancel{(x^2 + xy + y^2)}}{5x\cancel{(x^2 + xy + y^2)}} = \dfrac{x - y}{5x}$

3–2 Exercises

In exercises 1–40, reduce each fraction. (See examples 1–3.)

1. $\dfrac{15}{24}$
2. $\dfrac{-17}{34}$
3. $\dfrac{6x}{3y}$
4. $\dfrac{5xy}{10x^2y}$
5. $\dfrac{12x^2y^3z^5}{4xy^6z^7}$
6. $\dfrac{-18a^2b^3c}{3a^4b^7c^0}$
7. $\dfrac{(-4a^2b^3)^2}{2a^3b^4}$ *(See example 1(d).)*
8. $\dfrac{(-2a^3b^4c)^3}{-4a^6b^7c^3}$ *(See example 1(d).)*
9. $\dfrac{(-3a^2b^4)^3}{(3a^5b^4)^2}$
10. $\dfrac{(7ab^3)^0}{9a^4b^2}$
11. $\dfrac{2x - 2y}{4}$
12. $\dfrac{5x + 5y}{20}$
13. $\dfrac{3x^2 + 6x^3}{5x^2 \cdot 3y^2}$
14. $\dfrac{4w^3 - 2z^2}{2w - 3z}$

15. $\dfrac{5x - 5xy}{3a - 3ay}$

16. $\dfrac{8x^2 - 12x}{10xy - 15y}$

17. $\dfrac{12a^2b - 6a^3c + 18a}{12a}$

18. $\dfrac{-2x^3y^2 + 10x^2y^3 - 14x^4y^2}{10x^2y^2}$

19. $\dfrac{24x^6y^6 - 12x^2y^4 + 36x^4y^4}{-24x^2y^4}$

20. $\dfrac{5x^2y^3 + 7xy - 2xz}{6x^2 + 3x}$

21. $\dfrac{h^2 - k^2}{h + k}$

22. $\dfrac{25x^2 - 9}{5x - 3}$

23. $\dfrac{49x^2 - 4y^2}{7x - 2y}$

24. $-\dfrac{9n^2 - 4}{3n + 2}$

25. $\dfrac{18x^2 - 32y^2}{6x - 8y}$

26. $\dfrac{50x^2 - 2y^2}{15x + 3y}$

27. $\dfrac{x^2 - y^3}{2x^2 + 2xy + 2y^2}$ (See example 3.)

28. $\dfrac{8 - y^3}{4 - 2y}$

29. $\dfrac{27x^3 - 8y^3}{9x^2 + 6xy + 4y^2}$

30. $\dfrac{3x^3 - 3y^3}{6x - 6y}$

31. $\dfrac{ac - ad + bc - bd}{c - d}$

32. $\dfrac{c - d}{cx - dx + cy - dy}$

33. $\dfrac{2x - 2y}{2(x^2 - y^2)}$

34. $\dfrac{m - n}{2m^2 - mn - n^2}$

35. $\dfrac{x^2 + x - 12}{x^2 - x - 20}$

36. $\dfrac{3p^2 - 11p + 6}{3p^2 + 13p - 10}$

37. $\dfrac{x^2 - xy - 2y^2}{x^2 - y^2}$

38. $\dfrac{2x^2 + xy - 3y^2}{4x^2 + 12xy + 9y^2}$

39. $\dfrac{c^4 - d^4}{c^2 - d^2}$

40. $\dfrac{ab + ad + bc + cd}{ax - ay + cx - cy}$

41. A calculus student worked a problem and got

$$y' = \dfrac{6x^2 - 24}{x^4 - 16}$$

The student had to then reduce the right side of the equation and evaluate the reduced fraction for $x = 2$. What should the answer be?

42. An analytic geometry student was given the task of drawing the graph of

$$y = \dfrac{5x + 15}{x^2 - x - 12}$$

Before drawing the graph, the student reduced the right side of the equation. What would the reduced equation be?

43. Reduce: $\dfrac{0.5x^2 - 0.2x}{0.25x^2 - 0.04}$.

44. Reduce: $\dfrac{0.3x - 3}{0.01x^2 + 0.1x - 2}$.

3-3 Multiplication and Division of Rational Expressions

Multiplication of rational expressions is defined to follow the same rule as multiplication of rational numbers. Recall from arithmetic that

$$\frac{3}{4} \cdot \frac{5}{8} = \frac{3 \cdot 5}{4 \cdot 8} = \frac{15}{32}$$

We define multiplication of rational expressions in the same way.

Definition 3-1

Multiplying Rational Expressions

If A/B and C/D are rational expressions and if $B, D \neq 0$, then

$$\frac{A}{B} \cdot \frac{C}{D} = \frac{A \cdot C}{B \cdot D}$$

Example 1

Multiply and simplify: $\dfrac{5x}{3} \cdot \dfrac{9}{2x^2}$

Solution

$$\frac{5x}{3} \cdot \frac{9}{2x^2} = \frac{5x(9)}{3(2x^2)}$$
$$= \frac{45x}{6x^2}$$
$$= \frac{3x(15)}{3x(2x)}$$
$$= \frac{15}{2x} \quad \blacksquare$$

Division of rational expressions is also defined according to the rules for dividing rational numbers. However, we want to explain why we "invert and multiply" when dividing fractions.

First of all, recall that a **complex fraction** is any fraction with a fraction in either the numerator, the denominator, or both. Thus

$$\frac{\frac{3}{4}}{\frac{5}{9}}, \quad \frac{\frac{5}{8}}{3}, \quad \text{and} \quad \frac{1}{\frac{5}{6}}$$

are examples of complex fractions. Also, recall from arithmetic that to simplify $\frac{5}{8} \div \frac{3}{7}$, you did this:

$$\frac{5}{8} \div \frac{3}{7} = \frac{5}{8} \cdot \frac{7}{3} = \frac{35}{24} = 1\frac{11}{24}$$

That is, you inverted the divisor, $\frac{3}{7}$, and multiplied it by $\frac{5}{8}$. Why?

Generally, what we must do is force the denominator to become 1 by using the fundamental theorem of fractions. When the denominator becomes 1 and the numerator is simplified, the division is finished.

In the example just given, notice we can write the following steps.

1. $\dfrac{5}{8} \div \dfrac{3}{7} = \dfrac{\frac{5}{8}}{\frac{3}{7}}$

2. $= \dfrac{\frac{5}{8} \cdot \frac{7}{3}}{\frac{3}{7} \cdot \frac{7}{3}}$ Use the fundamental theorem of fractions; multiply both numerator and denominator by $\frac{7}{3}$, the reciprocal of the denominator.

3. $= \dfrac{\frac{5}{8} \cdot \frac{7}{3}}{1}$ The denominator in (2) simplifies to the number 1.

4. $= \dfrac{\frac{35}{24}}{1}$ Simplify the numerator of (3) to obtain $\frac{35}{24}$.

5. $= \dfrac{35}{24}$ Dividing a number by 1 does not change the number.

Notice the numerator in step 2 is the product $\frac{5}{8} \cdot \frac{7}{3}$, which is the inverted divisor multiplied by the numerator. Because the denominator in step 2 is the product of two reciprocals, it has a value of 1, which is shown in step 3. The denominator is now 1, indicating that the division is complete except for any simplifications. Steps 4 and 5 are simplification steps only. The result shown in step 5 can be left as $\frac{35}{24}$ or it can be written as $1\frac{11}{24}$.

This discussion is formulated as definition 3–2.

Definition 3–2

Division of Rational Expressions
If A/B and C/D are rational expressions, then
$$\frac{A}{B} \div \frac{C}{D} = \frac{A}{B} \cdot \frac{D}{C} \qquad B, C, D \neq 0$$

Rational Expressions

Example 2 Divide and simplify: $\dfrac{3a^2}{8b} \div \dfrac{6}{b}$

Solution
$$\dfrac{3a^2}{8b} \div \dfrac{6}{b} = \dfrac{3a^2}{8b} \cdot \dfrac{b}{6}$$
$$= \dfrac{3a^2 b}{48b}$$
$$= \dfrac{3b(a^2)}{3b(16)}$$
$$= \dfrac{a^2}{16} \blacksquare$$

In simplifying fractions, it is sometimes helpful to know that $A/B = A \cdot (1/B)$ for any rational expressions A and B, where $B \neq 0$. For example,

$$\dfrac{3}{5} = 3 \cdot \dfrac{1}{5} \quad \text{and} \quad \dfrac{2x}{3y} = 2x \cdot \dfrac{1}{3y}$$

Definition 3–3

Rewriting Rational Expressions

For any rational expressions A and B, $B \neq 0$,

$$\dfrac{A}{B} = A \cdot \dfrac{1}{B}$$

Definition 3–3 tells us that instead of dividing by B, we may multiply by its reciprocal, $1/B$, instead. (This idea is the basis for the invert-and-multiply technique just discussed.)

Expressions such as $(A - B)$ and $(B - A)$ are opposites, or additive inverses. This is easy to show by proving that the sum of $(A - B)$ and $(B - A)$ is 0.

$(A - B) + (B - A) = [A + (-B)] + [B + (-A)]$ definition of subtraction

$\qquad\qquad\qquad\quad = [A + (-A)] + [B + (-B)]$ associative and commutative axioms for addition

$\qquad\qquad\qquad\quad = 0 + 0$ additive inverse axiom

$\qquad\qquad\qquad\quad = 0$

Thus $(A - B)$ and $(B - A)$ are opposites because their sum is zero. Two opposites have a quotient equal to -1. This fact can be most helpful in reducing algebraic fractions. For example, $(A - B)/(B - A) = -1$ (see example 6).

Example 3 Multiply: $\dfrac{6}{8} \cdot \dfrac{12}{18}$.

Solution $\dfrac{6}{8} \cdot \dfrac{12}{18} = \dfrac{6 \cdot 12}{8 \cdot 18} = \dfrac{\cancel{6} \cdot 1 \cdot \cancel{4} \cdot \cancel{3}}{\cancel{4} \cdot 2 \cdot \cancel{6} \cdot \cancel{3}} = \dfrac{1}{2}$ ■

The efficient way to multiply two or more fractions is to first factor the numerators and denominators. Then divide out all common factors in the numerators and denominators. Multiply the remaining factors to obtain the final answer.

Example 4 Divide: $\dfrac{8}{9} \div \dfrac{24}{27}$.

Solution $\dfrac{8}{9} \div \dfrac{24}{27} = \dfrac{8}{9} \cdot \dfrac{27}{24} = \dfrac{\cancel{8} \cdot 1 \cdot \cancel{9} \cdot \cancel{3}}{\cancel{9} \cdot 1 \cdot \cancel{8} \cdot \cancel{3}} = 1$ ■

Example 5 Divide: $\dfrac{x^2 - y^2}{x^2 + 5x + 6} \div \dfrac{x + y}{x^2 + x - 6}$.

Solution
$\dfrac{x^2 - y^2}{x^2 + 5x + 6} \div \dfrac{x + y}{x^2 + x - 6}$
$= \dfrac{x^2 - y^2}{x^2 + 5x + 6} \cdot \dfrac{x^2 + x - 6}{x + y}$
$= \dfrac{(x - y)\cancel{(x + y)}\cancel{(x + 3)}(x - 2)}{\cancel{(x + 3)}(x + 2)\cancel{(x + y)}}$
$= \dfrac{(x - y)(x - 2)}{x + 2}$

The final answer may be left in factored form. ■

Example 6 Multiply: $\dfrac{4x^2 - 9}{x^2 + 2x - 3} \cdot \dfrac{x^2 + 7x + 12}{10(3 - 2x)}$.

Solution
$\dfrac{4x^2 - 9}{x^2 + 2x - 3} \cdot \dfrac{x^2 + 7x + 12}{10(3 - 2x)} = \dfrac{\overset{-1}{\cancel{(2x - 3)}}(2x + 3)\cancel{(x + 3)}(x + 4)}{\cancel{(x + 3)}(x - 1)10\cancel{(3 - 2x)}}$
$= \dfrac{-1(2x + 3)(x + 4)}{10(x - 1)}$
$= \dfrac{-(2x + 3)(x + 4)}{10(x - 1)}$ ■

Rational Expressions

In example 6, notice the factor $3 - 2x$ was divided into the factor $2x - 3$, giving a quotient of -1.

3–3 Exercises

In exercises 1–6, an expression of the form A/B is given. Write the given expression in the form $A \cdot (1/B)$.

Example Write $\dfrac{2a}{x + y}$ in the form $A \cdot (1/B)$.

Solution $\dfrac{2a}{x + y} = 2a \cdot \dfrac{1}{x + y}$

1. $\dfrac{2x}{3y}$
2. $\dfrac{3x - 4}{2x}$
3. $\dfrac{5v - 2}{v - 3}$
4. $\dfrac{6x + 1}{2x + 3}$
5. $\dfrac{(3x - 2)(x + 4)}{x - 1}$
6. $\dfrac{(7x + 5)(x - 3)}{(x + 2)(x - 4)}$

In exercises 7–12, write the opposite (additive inverse) of the given problem. Write your final answer without any grouping symbols.

7. $7x - 2$
8. $-6c + 2$
9. $-7x - 3 + 2y$
10. $-[-(3x + 2)]$
11. $-(2 - t)$
12. $-(-4s + 3)$

In exercises 13–43, multiply or divide the given rational expressions. Leave all answers in reduced form by dividing out common factors in the numerator and denominator. (See examples 1–6.)

13. $\dfrac{9}{16} \cdot \dfrac{30}{21}$
14. $\dfrac{5x^2 y^3}{10xy} \cdot \dfrac{20x^6}{-10x^2 y^4}$
15. $\dfrac{10a^3 b^5 c^2}{-2ab^7 c^3} \cdot \dfrac{-4a^5 b^6 c}{5a^4 b^3 c^0}$
16. $\dfrac{(2x^3 y^2)^2}{(-4ab)^2} \cdot \dfrac{4a^3 b^4}{(2xy)^3}$
17. $\dfrac{2x^3}{5y^2} \cdot \dfrac{-10z^3}{4x^2 y} \cdot \dfrac{7y^5}{21x^3 z}$
18. $18x^2 y \cdot \dfrac{5}{-9x^4 y^5}$
19. $\dfrac{7a^3 b^5}{9x^3 b^2} \div \dfrac{-28a^6 b^7}{27x^4 b^5}$
20. $\dfrac{-20x^3 y}{3x^4 y} \cdot \dfrac{12x^3 y}{16x^5 y^2} \cdot \dfrac{-8x^6 z}{4xz^5}$
21. $\left(\dfrac{9x^2 y}{-7xz^4} \cdot \dfrac{28xz^5}{8x^2 y} \right) \div \dfrac{3xz^4}{4x^5 y}$
22. $\left(\dfrac{16a^3 b^2}{-2x^3 y} \div \dfrac{8a^4 b^2}{6xy} \right) \cdot \dfrac{2ax}{5a^2 y}$
23. $\dfrac{8x^4 + 10x^2 - 6x}{2x}$ (*Hint:* Write the given problem as $\dfrac{8x^4}{2x} + \dfrac{10x^2}{2x} - \dfrac{6x}{2x}$ and reduce each term.)
24. $\dfrac{24x^5 - 18x^4 + 6x^2}{-6x^2}$
25. $\dfrac{17x^2 y^3 - 34xy^4}{-17xy}$

26. $\dfrac{19a^2b^2 - 38ab^3 - 57ab}{-19ab}$

27. $\dfrac{14x^2y^3 - 7axy^2 + 21xy}{7x^2y^2}$

28. $\dfrac{12c^2d^2 + 24cd^3 - 8cd^5}{8c^3d^2}$

29. $\dfrac{15a^3b^4 - 30ab^3 + 20a}{-25ab}$

30. $\dfrac{x^2y - xy^2}{-xy}$

31. $\dfrac{9x - 12y}{3x} \cdot \dfrac{9}{4y - 3x}$ (See example 6.)

32. $\dfrac{6x - 3y}{5x^2} \div \dfrac{y - 2x}{10x^6}$

33. $\dfrac{x^2 - 4y^2}{x^2 + 5xy + 6y^2} \cdot \dfrac{7x - 14y}{x^2 - xy - 2y^2}$

34. $\dfrac{3m^2 - 14m + 8}{m^2 + m - 20} \cdot \dfrac{2m^2 + 11m + 5}{3m^2 + m - 2}$

35. $\dfrac{5n^2 + 9n - 2}{n^2 + 4n + 4} \div \dfrac{5n^2 + 19n - 4}{n^2 + 6n + 8}$

36. $\dfrac{d^3 - 8}{d^2 - d - 20} \div \dfrac{d^2 + 2d + 4}{d^2 + 2d - 8}$

37. $\dfrac{x^2 - y^2}{2x^2 + 7x - 15} \cdot \dfrac{3 - 2x}{x + 5} \cdot \dfrac{3x^2 + 13x - 10}{x^2 - 2xy + y^2}$

38. $\dfrac{2x^2 - 7x + 6}{2x^2 + 9x - 5} \cdot \dfrac{x^2 + 9x + 20}{x^2 - 6x + 8} \cdot \dfrac{2x^2 - 7x + 3}{2x + 5x - 12}$

39. $\left(\dfrac{6x^2 + 11x - 2}{x^2 - 3x - 4} \cdot \dfrac{x^2 - 1}{6x^2 - 7x + 1} \right) \div \dfrac{x + 2}{x^2 - 6x + 8}$

40. $\dfrac{a^2 - b^2}{ac + ad - bc - bd} \cdot \dfrac{cx - cy + dx - dy}{ax + bx - ay - by}$

41. $\dfrac{ax + ay + cx + cy}{2ax + 2ay - bx - by} \div \dfrac{cx + dx + cy + dy}{2ac + 2ad - bc - bd}$

42. $\dfrac{3x^2 - 11x + 6}{x^2 + x - 2} \cdot \dfrac{2x + 4}{6 - 11x + 3x^2}$

43. $\dfrac{x^2 - 3xy + 2y^2}{x^2 - 2xy - 3y^2} \cdot \dfrac{3y^2 - 7xy + 2x^2}{x^2 - xy - 2y^2} \cdot \dfrac{x^2 + 2xy + y^2}{x^2 + xy - 2y^2}$

In exercises 44–48, $P(x) = \dfrac{x}{x^2 - 4}$, $Q(x) = \dfrac{(x - 2)(2x + 5)}{2x}$, and $R(x) = \dfrac{6x^3}{2x^2 - 4x}$. Find each value.

44. $P(x) \cdot Q(x)$

45. $P(x) \div R(x)$

46. $Q(x) \cdot R(x)$

47. $Q(x) \div R(x)$

48. $[P(x) \cdot Q(x)] \div R(x)$

3-4 Least Common Multiple

Before fractions can be added or subtracted, a common denominator must be found. Further, it is usually most convenient if the common denominator is the least common denominator. The **least common denominator (LCD)**, as the name implies, is the smallest number that two or more denominators will divide exactly.

The LCD is actually the least common multiple of the denominators. The smallest number into which two or more numbers will divide exactly is called the **least common multiple (LCM)** for those numbers.

Before we see how to find the LCM for two or more numbers, we need a few definitions.

Any natural number greater than one whose only integral factors are 1 and itself is called a **prime number.** The first four prime numbers are 2, 3, 5, and 7.

Any natural number greater than one which is not a prime number is called a **composite number.** The first four composite numbers are 4, 6, 8, and 9.

Any composite number can be written as the product of two or more prime numbers. This product of primes is referred to as the **prime factorization** of the composite number. For example, the prime factorization of 42 is $2 \cdot 3 \cdot 7$. The **prime factors** of 42 are 2, 3, and 7.

Now let us find the LCM for 10 and 15. First, write the prime factorizations for both numbers.

$$10 = 2 \cdot 5$$
$$15 = 3 \cdot 5$$

Now write a product containing each prime factor appearing in the prime factorizations for both numbers. We write

$$2 \cdot 3 \cdot 5 = 30$$

Thus the LCM for 10 and 15 is 30.

Now let us find the LCM for 18 and 135. We proceed as before, but this time we write repeated prime factors using exponents.

$$18 = 2 \cdot 3 \cdot 3 = 2 \cdot 3^2$$
$$135 = 3 \cdot 3 \cdot 3 \cdot 5 = 3^3 \cdot 5$$

The LCM is the product that contains each factor raised to the larger power of that factor in the prime factorizations for the two numbers. Thus, the LCM for 18 and 135 is $2 \cdot 3^3 \cdot 5 = 270$.

The procedure for finding the LCM is stated as procedure 3–1.

Procedure 3-1

Finding the LCM

To find the LCM of two or more numbers, write a product that contains each factor raised to the largest power found on that factor in any of the prime factorizations for the given numbers.

Example 1

Find the LCM for 35 and 40.

Solution

$35 = 5 \cdot 7$
$40 = 2^3 \cdot 5$
LCM $= 2^3 \cdot 5 \cdot 7 = 280$

Example 2

Let X and Y represent numbers whose prime factorizations are given. Find the LCM for X and Y.

$$X = 2^4 \cdot 5^2 \cdot 7^4$$
$$Y = 2^3 \cdot 7^3 \cdot 11$$

Solution

LCM $= 2^4 \cdot 5^2 \cdot 7^4 \cdot 11$

The LCM can be quite large, and it is often left in factored form. This is particularly true in algebraic expressions.

Example 3

Find the LCM for $6x$ and $2x - 6$.

Solution

$6x = 2 \cdot 3 \cdot x$
$2x - 6 = 2(x - 3)$
LCM $= 2 \cdot 3 \cdot x(x - 3)$

Notice that the LCM for algebraic expressions is found in exactly the same way we find the LCM for numbers.

Example 4

Find the LCM for $9x^2 - 4$, $3x^2 + x - 2$, and $2x^2 - x - 3$.

Solution

$9x^2 - 4 = (3x - 2)(3x + 2)$
$3x^2 + x - 2 = (3x - 2)(x + 1)$
$2x^2 - x - 3 = (x + 1)(2x - 3)$
LCM $= (3x - 2)(3x + 2)(x + 1)(2x - 3)$

Rational Expressions

Example 5

Suppose X, Y, and Z are three completely factored algebraic expressions, as shown. Find the LCM for X, Y, Z.

$$X = (3x + 1)^2(x - 3)$$
$$Y = (3x + 1)(x - 3)^3(x - 1)$$
$$Z = 6(3x - 1)(x + 2)$$

Solution

LCM $= 6(3x + 1)^2(x - 3)^3(x - 1)(3x - 1)(x + 2)$

Clearly, we want to leave the LCM in its factored form. Also, it is not necessary to factor 6 as $2 \cdot 3$. ■

3–4 Exercises

In exercises 1–27, find the LCM for the given numbers or algebraic expressions. Leave the LCM in factored form. (See examples 1–4.)

1. 16 and 24
2. 25 and 35
3. 30 and 21
4. 20, 30, and 40
5. 16, 24, and 35
6. 20, 32, and 45
7. 18, 27, 81, and 90
8. 22, 30, 42, and 50
9. $6a$ and $9b^2$
10. $10a^2$ and $15ab$
11. $16a^2b$ and $22a^3b^2$
12. $20x^3y^4$ and $63x^2yz$
13. $11x^3yz$, $22xyz^2$, and $42xy$
14. $25xyz$, $30x^2yz$, and $45x$
15. $8xy$, $72x^2y$, and $80x^3$
16. $9x$, $28x^4y$, and $30y^5$
17. $5x$ and $10x^2 + 15x$
18. $6x^2$ and $24x - 30$
19. $2x - 6$ and $7x - 21$
20. $5x - 10$ and $10x + 30$
21. $x - y$ and $x^2 - y^2$
22. $2x - 3$ and $4x^2 - 9$
23. 7, $12x - 9$, and $16x^2 - 9$
24. $x^2 - 5x - 6$ and $2x^2 + 5x + 3$
25. $3a^2 - 5a + 2$ and $a^2 - 2a + 1$
26. $t^3 - 8$ and $5t^2 + 10t + 20$
27. $2x^2 - 7x + 6$, $4x^2 - 12x + 9$, and $x^2 + 3x + 2$

In exercises 28–34, find the LCM for X, Y, and Z. (See example 5.)

28. $X = 5(x - 3)^3(x + 4)$; $Y = 10(x - 3)(x - 5)$; $Z = 20(x + 4)^2$
29. $X = 3x^2(6x - 1)^3$; $Y = 24x(7x - 3)(6x - 1)$; $Z = 42(2x - 1)$
30. $X = x^3y(3x + 2)(x - 1)^4$; $Y = xy^3(x - 1)(x + 5)^3$; $Z = x(x + 5)^4(x + 2)$
31. $X = 5x^2(x - 3)^2(x + 5)(x - 2)$; $Y = 6x^2(x - 2)^4(x + 5)^2$; $Z = 24(x + 5)(x - 3)^2$
32. $X = -7x(x + 5)^2(x - 3)$; $Y = x^3(x + 6)^3(x - 1)$; $Z = (x - 3)(x + 5)(x + 6)^3$
33. $X = -11x^3(8x - 1)^2(2c + 3)^2$; $Y = 22x(3x + 1)(3c - 1)$; $Z = (3x - 1)^3$
34. $X = (2x - 1)(x + 4)$; $Y = (4x - 1)(2x - 1)^2$; $Z = 3(x + 4)^5$

3–5 Addition and Subtraction of Rational Expressions

Before two rational expressions or numbers can be added or subtracted, they each must have the same denominator. If they already have the same denominator, then we can write them as

$$\frac{x}{y} + \frac{z}{y} = x \cdot \frac{1}{y} + z \cdot \frac{1}{y} \quad \text{definition 3–3}$$

$$= \frac{1}{y} \cdot (x + z) \quad \text{commutative and distributive axioms}$$

$$= \frac{x + z}{y} \quad \text{definition 3–3}$$

That is, to add two numbers whose denominators are the same, we add the numerators and put this sum over the common denominator.

If the denominators are not the same, we must find a common denominator. Suppose we are given fractions

$$\frac{a}{b} + \frac{c}{d}, \quad b, d \neq 0$$

We must make the denominators the same. To do this, we use the fundamental theorem of fractions to write

$$\frac{a}{b} = \frac{a \cdot d}{b \cdot d} \quad \text{and} \quad \frac{c}{d} = \frac{c \cdot b}{d \cdot b} = \frac{b \cdot c}{b \cdot d}$$

Then

$$\frac{a}{b} + \frac{c}{d} = \frac{a \cdot d}{b \cdot d} + \frac{b \cdot c}{b \cdot d}$$

$$= (a \cdot d)\frac{1}{b \cdot d} + (b \cdot c)\frac{1}{b \cdot d} \quad \text{definition 3–3}$$

$$= \frac{1}{b \cdot d}(a \cdot d + b \cdot c) \quad \text{commutative and distributive axioms}$$

$$= \frac{ad + bc}{bd} \quad \text{definition 3–3}$$

We summarize this discussion as rule 3–6.

Rule 3–6

Adding Fractions

1. $\dfrac{x}{y} + \dfrac{z}{y} = \dfrac{x + z}{y} \quad y \neq 0$

2. $\dfrac{a}{b} + \dfrac{c}{d} = \dfrac{ad + bc}{bd} \quad b, d \neq 0$

Rational Expressions

Example 1   Add: $\dfrac{7}{8} + \dfrac{6}{8}$.

Solution $\dfrac{7}{8} + \dfrac{6}{8} = \dfrac{7 + 6}{8} = \dfrac{13}{8}$, or $1\dfrac{5}{8}$ ■

Observe that the denominator $b \cdot d$ found in part 2 of rule 3–6, is not necessarily the LCD. For example, using part 2,

$$\dfrac{3}{8} + \dfrac{1}{4} = \dfrac{12 + 8}{32} = \dfrac{20}{32}$$

However, the fraction $\dfrac{20}{32}$ can be reduced to $\dfrac{5}{8}$, which involves the LCD of 8 and 4. In general, we continue to use the method discussed in the last section for finding LCM to find the LCD.

Subtraction can be defined in terms of addition by noting that

$$\dfrac{a}{b} - \dfrac{c}{d} = \dfrac{a}{b} + \dfrac{-c}{d} = \dfrac{a \cdot d + (-bc)}{b \cdot d}$$
$$= \dfrac{a \cdot d - b \cdot c}{b \cdot d}$$

Rule 3–7

Subtracting Fractions

$$\dfrac{a}{b} - \dfrac{c}{d} = \dfrac{ad - bc}{bd} \qquad b, d \neq 0$$

Example 2   Subtract: $\dfrac{x}{x^2 - 16} - \dfrac{4}{x^2 - 16}$.

Solution
$$\dfrac{x}{x^2 - 16} - \dfrac{4}{x^2 - 16} = \dfrac{x - 4}{x^2 - 16}$$
$$= \dfrac{x - 4}{(x - 4)(x + 4)}$$
$$= \dfrac{\cancel{(x - 4)} \cdot 1}{\cancel{(x - 4)}(x + 4)}$$
$$= \dfrac{1}{x + 4} \quad ■$$

Example 3   Subtract: $\dfrac{9}{50} - \dfrac{7}{15}$.

Solution

Find the LCM for 50 and 15.
$$50 = 2 \cdot 5^2$$
$$15 = 3 \cdot 5$$
$$\text{LCM} = \text{LCD} = 2 \cdot 3 \cdot 5^2 = 150$$
$$\frac{9}{50} - \frac{7}{15} = \frac{9 \cdot 3}{50 \cdot 3} - \frac{7 \cdot 10}{15 \cdot 10} = \frac{27}{150} - \frac{70}{150}$$
$$= \frac{27 - 70}{150} = \frac{-43}{150}$$

Example 4 Add: $\frac{5}{x-3} + \frac{6}{x-3}$.

Solution $\frac{5}{x-3} + \frac{6}{x-3} = \frac{5+6}{x-3} = \frac{11}{x-3}$

Example 5 Add: $\frac{5}{x-3} + \frac{2}{x+2}$.

Solution In this problem, the denominators are not the same. We find the LCD by methods explained in the last section.
$$x - 3 = (x - 3)^1$$
$$x + 2 = (x + 2)^1$$
$$\text{LCD} = \text{LCM} = (x - 3)(x + 2).$$

The denominator of each fraction must be changed to $(x - 3)(x + 2)$.

$$\frac{5}{x-3} + \frac{2}{x+2} = \frac{5}{x-3} \cdot \frac{x+2}{x+2} + \frac{2}{x+2} \cdot \frac{x-3}{x-3}$$
$$= \frac{5(x+2)}{(x-3)(x+2)} + \frac{2(x-3)}{(x+2)(x-3)}$$
$$= \frac{5(x+2) + 2(x-3)}{(x-3)(x+2)}$$
$$= \frac{5x + 10 + 2x - 6}{(x-3)(x+2)}$$
$$= \frac{7x + 4}{(x-3)(x+2)}$$

The answer will not simplify further because there are no common factors in the numerator and denominator.

Example 6 Subtract: $\frac{6}{x+3} - \frac{7}{2x+4}$.

Rational Expressions

Solution Find the LCD. Then subtract.
$$x + 3 = (x + 3)^2$$
$$2x + 4 = 2(x + 2)$$
$$\text{LCD} = \text{LCM} = 2(x + 2)(x + 3)$$

$$\frac{6}{x+3} - \frac{7}{2x+4} = \frac{6}{x+3} - \frac{7}{2(x+2)}$$
$$= \frac{6}{x+3} \cdot \frac{2(x+2)}{2(x+2)} - \frac{7}{2(x+2)} \cdot \frac{x+3}{x+3}$$
$$= \frac{6(2)(x+2)}{2(x+3)(x+2)} - \frac{7(x+3)}{2(x+2)(x+3)}$$
$$= \frac{12x + 24}{2(x+3)(x+2)} - \frac{7x + 21}{2(x+3)(x+2)}$$
$$= \frac{12x + 24 - (7x + 21)}{2(x+3)(x+2)}$$
$$= \frac{12x + 24 - 7x - 21}{2(x+3)(x+2)}$$
$$= \frac{5x + 3}{2(x+3)(x+2)}$$

The numerator of the second fraction should be enclosed within parentheses when going from the fourth line to the fifth line so that we will be reminded to change the sign of both terms in the numerator of the second fraction.

Example 7 Simplify: $\dfrac{3x}{x-4} - \dfrac{3x}{x+4} - \dfrac{96}{x^2-16}$.

Solution Since the factors of $x^2 - 16$ are $x - 4$ and $x + 4$, the LCD is $(x - 4)(x + 4)$.

$$\frac{3x}{x-4} - \frac{3x}{x+4} - \frac{96}{(x-4)(x+4)}$$
$$= \frac{3x}{x-4} \cdot \frac{x+4}{x+4} - \frac{3x}{x+4} \cdot \frac{x-4}{x-4} - \frac{96}{(x-4)(x+4)}$$
$$= \frac{3x(x+4) - 3x(x-4) - 96}{(x-4)(x+4)}$$
$$= \frac{3x^2 + 12x - 3x^2 + 12x - 96}{(x-4)(x+4)}$$
$$= \frac{24x - 96}{(x-4)(x+4)}$$
$$= \frac{24(x-4)}{(x-4)(x+4)}$$
$$= \frac{24\cancel{(x-4)}}{\cancel{(x-4)}(x+4)}$$
$$= \frac{24}{x+4}$$

Example 8 Add: $\dfrac{x^{3a} + 2}{x^3} + \dfrac{x^{a-1} - 2}{x}$.

Solution The LCD for x^3 and x is x^3.

$$\dfrac{x^{3a} + 2}{x^3} \cdot \dfrac{1}{1} + \dfrac{x^{a-1} - 2}{x} \cdot \dfrac{x^2}{x^2} = \dfrac{x^{3a} + 2 + x^2(x^{a-1} - 2)}{x^3}$$
$$= \dfrac{x^{3a} + 2 + x^{a+1} - 2x^2}{x^3} \blacksquare$$

3–5 Exercises

In exercises 1–43, add or subtract. All answers must be completely reduced. *(See examples 1–2 for exercises 1–13.)*

1. $\dfrac{3}{5} + \dfrac{1}{5}$
2. $\dfrac{3}{8b} + \dfrac{1}{8b}$
3. $\dfrac{3}{2a} + \dfrac{1}{2a}$
4. $\dfrac{x}{x+4} + \dfrac{4}{x+4}$
5. $\dfrac{x+3}{x-5} - \dfrac{8}{x-5}$
6. $\dfrac{x^2}{x-3} - \dfrac{9}{x-3}$
7. $\dfrac{x^2}{x-5} - \dfrac{25}{x-5}$
8. $\dfrac{x}{16-x^2} + \dfrac{4}{16-x^2}$
9. $\dfrac{y}{36-y^2} - \dfrac{6}{36-y^2}$
10. $\dfrac{2x}{x^2-9} - \dfrac{6}{x^2-9}$
11. $\dfrac{3x^2}{x^2-3x} - \dfrac{9x}{x^2-3x}$
12. $\dfrac{a^2}{a^2+5a} - \dfrac{25}{a^2+5a}$
13. $\dfrac{5c}{5c+5d} + \dfrac{5d}{5c+5d}$

See examples 3–7 for exercises 14–38.

14. $\dfrac{5}{8} + \dfrac{5}{12}$
15. $\dfrac{3}{5} - \dfrac{7}{10}$
16. $\dfrac{5}{2a} + \dfrac{6}{7a}$
17. $\dfrac{u}{3} - \dfrac{3}{2}$
18. $\dfrac{5}{2x} + \dfrac{3}{4x} - \dfrac{2}{x^2}$
19. $\dfrac{3}{z-4} + \dfrac{1}{z}$
20. $\dfrac{5}{x-3} + \dfrac{2}{3-x}$ $\left(\textit{Hint:} \text{ First multiply } \dfrac{2}{3-x} \text{ by } \dfrac{-1}{-1}.\right)$
21. $\dfrac{7}{2m-4} + \dfrac{8}{m-2}$
22. $\dfrac{6x}{5x+10} + \dfrac{3}{x}$
23. $\dfrac{5}{x^2+x-6} - \dfrac{3}{x^2+x-12}$
24. $\dfrac{2a}{2a^2-a-3} + \dfrac{a}{a^2-2a-3}$
25. $\dfrac{3}{x^2-4} + \dfrac{5}{2x^2-x-6}$
26. $\dfrac{a}{3x-9} - \dfrac{2a}{x^2+4x-21}$
27. $\dfrac{3}{p-q} - \dfrac{5}{p^2-2pq+q^2}$
28. $\dfrac{5}{n^2-16} - \dfrac{4}{n+4}$

Rational Expressions

29. $\dfrac{8}{x^3 - 8} + \dfrac{5}{x^2 + 2x + 4}$

30. $\dfrac{7}{x^3 - 27} - \dfrac{5}{x^2 + 3x + 9}$

31. $\dfrac{c}{c - 3} - \dfrac{3}{c + 3} - \dfrac{18}{c^2 - 9}$

32. $\dfrac{x}{x - 5} - \dfrac{5}{x + 5} - \dfrac{50}{x^2 - 25}$

33. $\dfrac{3x}{x - 3} - \dfrac{2}{x + 3} + \dfrac{-7x - 21}{x^2 - 9}$

34. $\dfrac{-1}{-x^2 - y^2} - \dfrac{2y^2}{y^4 - x^4} - \dfrac{1}{x^2 - y^2}$

35. $\dfrac{a}{b^2 - ab} - \dfrac{-b}{a^2 - ab}$

36. $m + \dfrac{8m}{m^2 - 4} - \dfrac{4}{m - 2}$

37. $\dfrac{-18}{x^2 - 9} + \dfrac{x}{x + 3} + \dfrac{-x}{x - 3}$

38. $\dfrac{a^2}{a - b} + \dfrac{4b^2}{b - a} + \dfrac{3ab}{a - b}$

See example 8 for exercises 39–43.

39. $\dfrac{x^n + 1}{x^2} + \dfrac{x^y - 1}{x^4}$

40. $\dfrac{x^{2n-3} + 5}{x} - \dfrac{x^{2y} - 1}{x^3}$

41. $\dfrac{x^{2c-1} - 3}{x^5} - \dfrac{x^{3c+1} + 1}{x} + \dfrac{x^c - 3}{x^2}$

42. $\dfrac{x^{4a-2} + 2}{x^5} - \dfrac{x^{-a+3} + 1}{x}$

43. $\dfrac{4^{3a} + 2}{4^2} + \dfrac{4^{-2a} - 3}{4^3}$

44. When three resistors in an electrical circuit are connected in parallel, the equivalent resistance is given by

$$\dfrac{1}{R} = \dfrac{1}{R_1} + \dfrac{1}{R_2} + \dfrac{1}{R_3}$$

Find the sum of the terms on the right side of this equation.

45. The closing price for a stock on the New York Stock Exchange on Monday was $37\tfrac{3}{8}$. On Tuesday it gained $\tfrac{3}{4}$ of a point. What was the closing price of the stock on Tuesday?

46. A calculus student was given the task of finding the derivative of the equation

$$y = \left(\dfrac{1}{x - 1} + \dfrac{1}{x + 1}\right)(x - 1)(x + 1)$$

The student simplified the right side of the equation first and then found the derivative. What was the simplified answer?

3–6 Complex Fractions

Recall from section 3–3 that a complex fraction is defined to be any fraction with a fraction in the numerator, denominator, or both. Thus

$$\frac{\frac{3}{4}}{\frac{5}{}}, \quad \frac{x+\frac{1}{x}}{\frac{3}{x}}, \quad \text{and} \quad \frac{\frac{1}{x}-\frac{1}{y}}{x^2-y^2}$$

are examples of complex fractions.

The first method of simplifying complex fractions involves using the fundamental theorem of fractions to multiply the numerator and denominator by the LCM of all the individual denominators in the given fraction. For example, $\frac{3/4}{5/9}$ is a complex fraction in which $\frac{3}{4}$ and $\frac{5}{9}$ are the individual fractions. Because 36 is the LCM for 4 and 9, we multiply the complex fraction by $\frac{36}{36}$. We get

$$\frac{\frac{3}{4}}{\frac{5}{9}} = \frac{\frac{3}{4}}{\frac{5}{9}} \cdot \frac{36}{36} = \frac{\frac{3}{4} \cdot \overset{9}{\cancel{36}}}{\frac{5}{\cancel{9}} \cdot \cancel{36}^{4}} = \frac{3 \cdot 9}{5 \cdot 4} = \frac{27}{20}$$

The use of this method to simplify complex fractions does not require the invert-and-multiply technique.

Example 1 Simplify: $\dfrac{\frac{3}{7}}{\frac{2}{3}}$

Solution The LCM for 7 and 3 is 21. Thus

$$\frac{\frac{3}{7}}{\frac{2}{3}} = \frac{\frac{3}{\cancel{7}_1} \cdot \frac{\overset{3}{\cancel{21}}}{1}}{\frac{2}{\cancel{3}_1} \cdot \frac{\cancel{21}^{7}}{1}} = \frac{3 \cdot 3}{2 \cdot 7} = \frac{9}{14}$$

Example 2 Simplify: $\dfrac{\frac{5}{6x}}{\frac{7}{5}}$

Solution The LCM for $6x$ and 5 is $30x$.

$$\frac{\frac{5}{6x}}{\frac{7}{5}} = \frac{\frac{5}{\cancel{6x}_1} \cdot \frac{\overset{5}{\cancel{30x}}}{1}}{\frac{7}{\cancel{5}_1} \cdot \frac{\cancel{30x}^{6x}}{1}} = \frac{5 \cdot 5}{7 \cdot 6x} = \frac{25}{42x}$$

Rational Expressions

Example 3 Simplify: $\dfrac{\dfrac{5}{x} + \dfrac{x}{3}}{\dfrac{6}{x^2} - \dfrac{2x}{6}}$

Solution The LCM for x, 3, x^2, and 6 is $6x^2$. (We write it as $6x^2/1$).

$$\dfrac{\dfrac{5}{x} + \dfrac{x}{3}}{\dfrac{6}{x^2} - \dfrac{2x}{6}} = \dfrac{\dfrac{6x^2}{1}\left(\dfrac{5}{x} + \dfrac{x}{3}\right)}{\dfrac{6x^2}{1}\left(\dfrac{6}{x^2} - \dfrac{2x}{6}\right)}$$ Multiply numerator and denominator by $6x^2$.

$$= \dfrac{\dfrac{6x^2}{1} \cdot \dfrac{5}{x} + \dfrac{6x^2}{1} \cdot \dfrac{x}{3}}{\dfrac{6x^2}{1} \cdot \dfrac{6}{x^2} - \dfrac{6x^2}{1} \cdot \dfrac{2x}{6}}$$ Use the distributive axiom and divide out common factors in each term.

$$= \dfrac{6x \cdot 5 + 2x^2 \cdot x}{6 \cdot 6 - x^2 \cdot 2x}$$

$$= \dfrac{30x + 2x^3}{36 - 2x^3}$$

$$= \dfrac{2(15x + x^3)}{2(18 - x^3)}$$ Factor.

$$= \dfrac{15x + x^3}{18 - x^3}$$ ■

A second method for simplifying complex fractions makes use of the fact that a fraction bar is another symbol for division. This essentially is the invert-and-multiply technique. For example,

$$\dfrac{\dfrac{6}{11}}{\dfrac{9}{16}} = \dfrac{6}{11} \div \dfrac{9}{16} = \dfrac{\overset{2}{\cancel{6}}}{11} \cdot \dfrac{16}{\underset{3}{\cancel{9}}} = \dfrac{2}{11} \cdot \dfrac{16}{3} = \dfrac{32}{33}$$

Example 4 $\dfrac{\dfrac{9x}{4}}{\dfrac{4}{x}} = \dfrac{9x}{4} \div \dfrac{4}{x} = \dfrac{9x}{4} \cdot \dfrac{x}{4} = \dfrac{9x^2}{16}$ ■

Example 5 Simplify: $\dfrac{\dfrac{x}{3} + 5}{3 + \dfrac{2}{x}}$

Solution

In this example, to use the invert-and-multiply technique, we must first find the LCD for the two terms in the numerator and the LCD for the two terms in the denominator. The steps are as follows.

$$\frac{\frac{x}{3}+5}{3+\frac{2}{x}} = \frac{\frac{x}{3}+\frac{15}{3}}{\frac{3x}{x}+\frac{2}{x}} = \frac{\frac{x+15}{3}}{\frac{3x+2}{x}}$$

$$= \frac{x+15}{3} \div \frac{3x+2}{x}$$

$$= \frac{x+15}{3} \cdot \frac{x}{3x+2} \qquad \text{Invert and multiply by the divisor.}$$

$$= \frac{x(x+15)}{3(3x+2)} \qquad \text{Multiply the numerators and denominators.}$$

In the invert-and-multiply method, the individual terms of the numerator and denominator must be combined before you invert and multiply.

3–6 Exercises

In exercises 1–28, simplify, using either method of simplification. Leave all answers in reduced form. *(See examples 1–5.)*

1. $\dfrac{\frac{3}{10}}{\frac{2}{5}}$

2. $\dfrac{\frac{7}{11}}{\frac{3}{22}}$

3. $\dfrac{\frac{3}{2x}}{\frac{5}{6}}$

4. $\dfrac{\frac{3}{x^2}}{\frac{5}{x^3}}$

5. $\dfrac{3+\frac{2}{b}}{9-\frac{3}{b}}$

6. $\dfrac{\frac{1}{x}-4}{\frac{2}{x}+5}$

7. $\dfrac{2-\frac{x}{y}}{6}$

8. $\dfrac{3c}{\frac{1}{c}+\frac{1}{4}}$

9. $\dfrac{\frac{3}{x}}{\frac{1}{3}-\frac{1}{x}}$

10. $\dfrac{\frac{1}{5}-\frac{1}{x}}{\frac{1}{25}-\frac{1}{x^2}}$

11. $\dfrac{\frac{1}{4}-\frac{1}{x}}{\frac{1}{16}-\frac{1}{x^2}}$

12. $\dfrac{1-\frac{25}{h^2}}{1+\frac{5}{h}}$

13. $\dfrac{9-\frac{16}{s^2}}{3+\frac{4}{s}}$

14. $\dfrac{3+\frac{4}{x}}{3+\frac{1}{x}-\frac{4}{x^2}}$

15. $\dfrac{\frac{81}{y^2}-4}{\frac{9}{y}+2}$

16. $\dfrac{2-\frac{1}{x}}{2-\frac{7}{x}+\frac{3}{x^2}}$

17. $\dfrac{1+\frac{4}{n}-\frac{21}{n^2}}{1+\frac{12}{n}+\frac{35}{n^2}}$

18. $\dfrac{5+\frac{4}{x}}{5+\frac{19}{x}+\frac{12}{x^2}}$

Rational Expressions

19. $\dfrac{\dfrac{1}{x} - \dfrac{2}{x+3}}{\dfrac{1}{x} + \dfrac{3}{x+3}}$

20. $\dfrac{\dfrac{2}{x-1} - \dfrac{1}{x}}{\dfrac{1}{x-1} + \dfrac{3}{x}}$

21. $\dfrac{1 + \dfrac{1}{1 + \dfrac{x}{y}}}{1 - \dfrac{1}{1 - \dfrac{x}{y}}}$

22. $\dfrac{3 + \dfrac{4}{\dfrac{3}{x} + \dfrac{y}{x}}}{5 + \dfrac{4}{\dfrac{2}{y} - \dfrac{x}{y}}}$

23. $\dfrac{\dfrac{1}{x} + \dfrac{2x}{x-3}}{\dfrac{x-1}{x}}$

24. $\dfrac{\dfrac{x+2}{x^2+5x+6}}{\dfrac{5}{x^2-9} + 1}$

25. $\dfrac{5x + \dfrac{x}{x-5}}{5x - \dfrac{x}{x-5}}$

26. $\dfrac{\dfrac{3}{y^2z^2} + \dfrac{3}{x^2y^2} - \dfrac{3}{x^2z^2}}{\dfrac{3z}{xy} + \dfrac{3x}{yz} - \dfrac{3y}{xz}}$

27. $\dfrac{\dfrac{a}{b} + 5 - \dfrac{6b}{a}}{\dfrac{a}{b} - \dfrac{b}{a}}$

28. $\dfrac{2x + \dfrac{6xy}{2x-3y}}{\dfrac{9y^2}{9y^2-4x^2} - 1}$

29. One of Einstein's equations in his special theory of relativity is

$$V_R = \dfrac{V + v}{1 + \dfrac{Vv}{c^2}}$$

The equation gives the correct way (according to Einstein) to add two velocities, V and v, to obtain the resultant velocity, V_R. The letter c represents the speed of light. Simplify the complex fraction represented by the right side of the equation.

30. If an object moves a distance s and has a beginning velocity of v_0 and a final velocity of v_f, then its acceleration, a, is given by the equation

$$a = \dfrac{1 - \dfrac{v_0^2}{v_f^2}}{\dfrac{2s}{v_f^2}}$$

(a) Simplify the right side of the equation.
(b) From the simplified equation in part a, find the acceleration if $v_f = 50$, $v_0 = 40$, and $s = 50$.

3–7 Division of Polynomials

Polynomials may be divided by an algorithm similar to the algorithm for long division of numbers. An **algorithm** is a rule or procedure for solving a problem.

We divide 1081 by 23 as follows.

$$\begin{array}{r} 47 \\ 23\overline{)1081} \\ \underline{92} \\ 161 \\ \underline{161} \end{array}$$

This illustrates the long-division algorithm. This same technique can be used to divide polynomials.

Example 1 Divide: $(x^2 + 6x^3 - 7x - 60) \div (x - 3)$.

Solution A step-by-step solution is given. Begin by writing the given problem as shown.

$$x - 3 \overline{\smash{)}\begin{array}{l} x^2 \\ x^3 + 6x^2 - 7x - 60 \end{array}}$$

Divide the x in $x - 3$ into x^3 to get x^2.

$$x - 3 \overline{\smash{)}\begin{array}{l} x^2 \\ x^3 + 6x^2 - 7x - 60 \\ \underline{x^3 - 3x^2} \end{array}}$$

Multiply x^2 by $x - 3$ and write the result under the first two terms of the dividend.

$$x - 3 \overline{\smash{)}\begin{array}{l} x^2 \\ x^3 + 6x^2 - 7x - 60 \\ \underline{x^3 - 3x^2} \\ 9x^2 - 7x \end{array}}$$

Subtract $x^3 - 3x^2$ from $x^3 + 6x^2$ and bring down $-7x$.

$$x - 3 \overline{\smash{)}\begin{array}{l} x^2 + 9x \\ x^3 + 6x^2 - 7x - 60 \\ \underline{x^3 - 3x^2} \\ 9x^2 - 7x \\ \underline{9x^2 - 27x} \end{array}}$$

Divide the x in $x - 3$ into $9x^2$ to get $9x$, and multiply $9x$ by $x - 3$. Place the results under the terms $9x^2 - 7x$.

$$x - 3 \overline{\smash{)}\begin{array}{l} x^2 + 9x \\ x^3 + 6x^2 - 7x - 60 \\ \underline{x^3 - 3x^2} \\ 9x^2 - 7x \\ \underline{9x^2 - 27x} \\ 20x - 60 \end{array}}$$

Subtract $9x^2 - 27x$ from $9x^2 - 7x$ and bring down -60.

$$x - 3 \overline{\smash{)}\begin{array}{l} x^2 + 9x + 20 \\ x^3 + 6x^2 - 7x - 60 \\ \underline{x^3 - 3x^2} \\ 9x^2 - 7x \\ \underline{9x^2 - 27x} \\ 20x - 60 \\ \underline{20x - 60} \\ 0 \end{array}}$$

Divide the x in $x - 3$ into $20x$ to get 20. Multiply 20 by $x - 3$ to finish the problem.

Rational Expressions

Not all problems will have a **remainder** of zero. Also, you must leave a space for any missing terms, including the constant term, as shown in the next example. All terms in both the dividend and divisor should be arranged in descending powers of the variable.

Example 2 Divide $x^3 + 2x$ by $-3 + x$.

Solution

$$\begin{array}{r} x^2 + 3x + 11 \\ x - 3 \overline{\smash{)}\ x^3 + 0x^2 + 2x + 0} \\ \underline{x^3 - 3x^2} \\ 3x^2 + 2x \\ \underline{3x^2 - 9x} \\ 11x + 0 \\ \underline{11x - 33} \\ 33 \end{array}$$

This problem has a remainder of 33. Write the quotient as

$$x^2 + 3x + 11 + \frac{33}{x - 3}$$

The problem then looks like this:

$$\begin{array}{r} x^2 + 3x + 11 + \dfrac{33}{x-3} \\ x - 3 \overline{\smash{)}\ x^3 + 0x^2 + 2x + 0} \\ \underline{x^3 - 3x^2} \\ 3x^2 + 2x \\ \underline{3x^2 - 9x} \\ 11x + 0 \\ \underline{11x - 33} \\ 33 \end{array}$$

3–7 Exercises

In exercises 1–20, divide the polynomials. Use the algorithm shown in this section. (See examples 1 and 2.)

1. $(x^2 + 7x + 12) \div (x + 3)$
2. $(2x^2 + 7x - 15) \div (x + 5)$
3. $(-3x - 20 + 2x^2) \div (2x + 5)$
4. $(2x^2 + 9x + 9) \div (x + 3)$
5. $(x^3 + 4x^2 - 2x - 3) \div (-1 + x)$
6. $(4x^3 - 7x + 3) \div (2x - 1)$ (Write the first expression as $4x^3 + 0x^2 - 7x + 3$ because the second degree term is missing).
7. $(2x^4 + x^3 + 4x) \div (2x + 1)$ (Write $0x^2$ and 0 for the missing terms.)
8. $(x^3 - 8) \div (x - 2)$ (Write $x^3 - 8$ as $x^3 + 0x^2 + 0x - 8$.)
9. $(27x^3 + 8) \div (3x + 2)$ (Write $27x^3 + 8$ as $27x^3 + 0x^2 + 0x + 8$.)

10. $(8x^3 - 1) \div (2x - 1)$
11. $(x^3 + 8) \div (x^2 - 2x + 4)$
12. $(x^4 + x^3 + 5x^2 + 5x + 12) \div (x^2 + 2x + 3)$
13. $(a^3 + 5a^2 + 3a - 9) \div (a^2 + 2a - 3)$
14. $(x^4 + 5x^3 + 6x - 2) \div (x - 1)$
15. $(x^4 - 6x^3 + 2x^2 - 4) \div (x + 2)$
16. $z^4 \div (z - 3)$
17. $x^3 \div (x + 1)$
18. $(x^4 + 1) \div (x + 1)$
19. $(v^4 - 3) \div (v + 1)$
20. $(x^5 - 1) \div (x + 1)$

In exercises 21–28, make the indicated divisions by:
  (a) factoring the numerator and dividing out common factors;
  (b) long division.

21. $\dfrac{3x^2 - 20x + 12}{3x - 2}$

22. $\dfrac{5x - 7x - 6}{x - 2}$

23. $\dfrac{x^4 + 4x^3 - 3x - 12}{x^3 - 3}$ (*Hint:* Factor the numerator by grouping.)

24. $\dfrac{2x^4 - 2x^3 + 3x - 3}{2x^3 + 3}$

25. $\dfrac{x^6 - x^5 - 3x^2 + 3x}{x^5 - 3x}$

26. $\dfrac{x^7 - 3x^3 - 7x^5 + 21x}{x^4 - 3}$

27. $\dfrac{15a^5 - 18a^4}{3a^2}$

28. $\dfrac{27x^2y^3 - 36x^3y}{9xy}$

3–8 Synthetic Division

Let us look again at the division of two polynomials and make a few observations that will lead to a shorter method of making the division.

Example 1

$$\begin{array}{r}
x^2 + 7x + 20 \\
x - 2 \overline{\smash{\big)}\, x^3 + 5x^2 + 6x - 3}\\
\underline{x^3 - 2x^2 }\\
7x^2 + 6x \\
\underline{7x^2 - 14x }\\
20x - 3\\
\underline{20x - 40}\\
37
\end{array}$$

If you look carefully at example 1, you will see that many terms are repeated. Also, we could do the division with detached coeffi-

Rational Expressions

cients. That is, we could use only the numerical coefficients to make the division.

$$\begin{array}{r} 1 + 7 + 20 \\ 1 - 2 \overline{\smash{)}1 + 5 + 6 - 3} \\ \underline{1 - 2 } \\ 7 + 6 \\ \underline{7 - 14 } \\ 20 - 3 \\ \underline{20 - 40} \\ 37 \end{array}$$

We did this division the same way as before, except we did not write the variable. To make the subtractions, we changed the signs of the bottom terms and added to those above.

Careful analysis of the last division will show that several numbers are repetitions of those numbers directly above them. This is illustrated now with the repeated numbers shown in parentheses.

This 1 can be eliminated. It produces numbers we are leaving out.

$$\begin{array}{r} 1 + 7 + 20 \\ 1 - 2 \overline{\smash{)}1 + 5 + 6 - 3} \\ (1) - 2 \\ 7 + (6) \\ (7) - 14 \\ 20 - (3) \\ (20) - 40 \\ 37 \end{array}$$

These numbers are repetitions and can be eliminated.

These numbers are repetitions and can be eliminated.

The division now looks like this with the repetitions eliminated:

$$\begin{array}{r} 1 + 7 + 20 \\ -2 \overline{\smash{)}1 + 5 + 6 - 3} \\ \underline{- 2 } \\ 7 \\ \underline{- 14 } \\ 20 \\ \underline{- 40} \\ 37 \end{array}$$

If the divisor in the last step is changed to +2, we can add the partial products instead of subtracting. We then have

The numbers along this dashed line are repeated in the quotient. The 7 and the 20 can be eliminated.

$$\begin{array}{r} 1 + 7 + 20 \\ 2 \overline{\smash{)}1 + 5 + 6 - 3} \\ 2 \\ 7 \\ 14 \\ 20 \\ 40 \\ 37 \end{array}$$

Now we pull the numbers left up closer to $1 + 5 + 6 - 3$ to get

$$
\begin{array}{r}
1 + 7 + 20 \\
2\overline{)1 + 5 + 6 - 3} \\
21440 \\
\hline
37
\end{array}
$$

Finally, we write the quotient numbers along the bottom line with the remainder.

$$
\begin{array}{r|rrrr}
\underline{2}| & 1 + 5 + & 6 - & 3 \\
& + 2 + 14 + & 40 \\
\hline
& 1 + 7 \,|\, 20 + & \underline{|37}
\end{array}
$$
The quotient numbers have been written below this line, along with the remainder, shown as $\underline{|37}$.

The quotient numbers below the line in the last step represent the answer. The divisor was $x - 2$, a first degree polynomial. The dividend was $x^3 + 5x^2 + 6x - 3$, a third degree polynomial. Dividing a third degree polynomial by a first degree polynomial gives a quotient that will be a second degree polynomial. In general, if the divisor is a first degree polynomial, the degree of the quotient will be one less than the degree of the dividend. The following diagram illustrates how the quotient numbers are used to write the answer to the problem.

$$
\begin{array}{cccc}
1 & + 7 & + 20 & + 37 \\
\downarrow & \downarrow & \downarrow & \downarrow \\
x^2 & + 7x & + 20 & + \dfrac{37}{x - 2}
\end{array}
$$

Thus, we see $(x^3 + 5x^2 + 6x - 3) \div (x - 2) = x^2 + 7x + 20 + \dfrac{37}{x - 2}$.

This method of dividing is called synthetic division. It is simple and fast once you catch on. Example 2 shows the synthetic division process for this same example with step-by-step instructions.

Example 2 Divide $(x^3 + 5x^2 + 6x - 3)$ by $(x - 2)$.

Solution

$1 + 5 + 6 - 3$ — Write the coefficients of the dividend.

$\underline{2}|\,\, 1 + 5 + 6 - 3$ — The number 2 is the divisor. Place it by the coefficients and draw a line.

$$
\begin{array}{r}
\underline{2}|\,\, 1 + 5 + 6 - 3 \\
+ 2 \\
\hline
1 + 7
\end{array}
$$
Bring down the 1 and multiply by the divisor, 2. Place this product under 5 and add.

$$
\begin{array}{r}
\underline{2}|\,\, 1 + 5 + 6 - 3 \\
+ 2 + 14 \\
\hline
1 + 7 + 20
\end{array}
$$
Multiply 7 by the divisor, 2. Place this product under the 6 and add.

$$
\begin{array}{r}
\underline{2}|\,\, 1 + 5 + 6 - 3 \\
+ 2 + 14 + 40 \\
\hline
1 + 7 + 20 + \underline{|37}
\end{array}
$$
Multiply 20 by the divisor, 2. Place this product under the -3 and add.

Rational Expressions

The numbers across the bottom represent the quotient and remainder. Dividing by $x - 2$ will give a quotient whose degree is one less than the degree of the dividend. Thus,

$$(x^3 + 5x^2 + 6x - 3) \div (x - 2) = x^2 + 7x + 20 + \frac{37}{x - 2}$$

Example 3 Divide $x^5 - 4x^2 + 5$ by $x - 2$ using synthetic division.

Solution First write zeros to hold the place for any missing terms. Then divide.

$$\underline{2|} \; 1 + 0 + 0 - 4 + 0 + \; 5$$
$$\phantom{\underline{2|} \; 1} \; \; 2 + 4 + 8 + 8 + 16$$
$$\overline{\phantom{\underline{2|}} \; 1 + 2 + 4 + 4 + 8 + \underline{|21}}$$

When the divisor has the form $x - a$, the degree of the quotient is one less than the degree of the dividend. The quotient is $x^4 + 2x^3 + 4x^2 + 4x + 8$ and the remainder is 21. Thus

$$\frac{x^5 - 4x^2 + 5}{x - 2} = x^4 + 2x^3 + 4x^2 + 4x + 8 + \frac{21}{x - 2}$$

If the remainder is zero in a division, then the divisor is a factor of the dividend. For example, $15 \div 3 = 5$. The remainder is zero, so 3 is a factor or 15. This idea can be used to determine if one polynomial is a factor of a second polynomial, as shown in example 4.

Example 4 Use synthetic division to see if $x - 4$ is a factor of $x^4 - 4x^3 + 5x^2 - 17x - 12$.

Solution
$$\underline{4|} \; 1 - 4 + 5 - 17 - 12$$
$$\phantom{\underline{4|} \; 1} \; \; 4 + 0 + 20 + 12$$
$$\overline{\phantom{\underline{4|}} \; 1 + 0 + 5 + \; 3 + \underline{|0}}$$

The last number in the bottom line in the synthetic division process is zero. This means the remainder is zero and $x - 4$ is a factor of $x^4 - 4x^3 + 5x^2 - 17x - 12$. (The quotient, $x^3 + 5x + 3$, is also a factor of the given polynomial.)

3–8 Exercises

In exercises 1–12, divide using synthetic division. *(See examples 2–3.)*

1. $\dfrac{x^2 + x - 6}{x - 2}$

2. $\dfrac{x^2 + 7x + 12}{x + 3}$

3. $\dfrac{x^3 + 4x^2 + x - 6}{x + 2}$

4. $\dfrac{3x^3 - 16x^2 + 7x - 10}{x - 5}$

5. $(2x^4 + 8x^3 - 3x^2 - 12x) \div (x + 4)$

6. $(r^5 + 5r^4 + 5r^2 + 22r - 15) \div (r + 5)$
7. $(y^3 - 8) \div (y - 2)$
8. $(5t^4 - 3t^2 + 1) \div (t - 4)$
9. $(3x^6 - 5x^3 - 2x) \div (x + 2)$
10. $(2p^5 - 3) \div (p - 1)$
11. $(3x^2 - 2x + 3 + x^3) \div (2 + x)$
12. $(a^3 - 14 + a^2 - a^4 + 2a^5) \div (-3 + a)$

In exercises 13–20, use synthetic division to find the remainder.

13. $(5x^3 - 4x^2 + 3x - 1) \div (x + 3)$
14. $(x^4 - 3x + 2) \div (x + 3)$
15. $(2x^5 - 4x^3 + x - 5) \div (x + 2)$
16. $(x^3 + x - 4) \div (x - 2)$
17. $\dfrac{3x^5 + 10x^4 - 5x^2}{x - 2}$
18. $\dfrac{x^4 - 6x^3 + 5x}{x + 4}$
19. $(2x^4 - 5) \div (x - 3)$
20. $(9x^4 + 3x) \div (x + 1)$

In exercises 21–26, use synthetic division to determine if the given binomial is a factor of $x^4 + 4x^3 - x^2 - 16x - 12$. *(See example 4.)*

21. $x + 1$ 22. $x - 4$ 23. $x - 3$ 24. $x + 3$ 25. $x + 2$
26. $x - 2$

27. A calculus student had to find an antiderivative for the function

$$f(x) = \frac{x^5 + 1}{x + 1} \quad x \neq -1$$

Before the student could find an antiderivative, she needed to simplify the right side of the equation by making the indicated division. Divide the right side of the equation for her, using synthetic division.

28. A student in analytic geometry had to draw the graph of

$$y = \frac{x^3 + 8}{x + 2} \quad x \neq -2$$

He divided the right side of the equation before drawing the graph and got $y = x^2 + 2x + 4$. Was he right? Divide the right side of the equation using synthetic division and check his work.

Chapter 3   Summary

Vocabulary

algorithm
complex fraction
composite number
least common denominator (LCD)
least common multiple (LCM)
prime factor

prime factorization
prime number
remainder
standard form of a fraction
synthetic division
term of a fraction

Definitions

3-1 Multiplying Rational Expressions

If A/B and C/D are rational expressions and if $B, D \neq 0$, then

$$\frac{A}{B} \cdot \frac{C}{D} = \frac{AC}{BD}$$

3-2 Dividing Rational Expressions

If A/B and C/D are rational expressions, then

$$\frac{A}{B} \div \frac{C}{D} = \frac{A}{B} \cdot \frac{D}{C} = \frac{AD}{BC}, \quad (B, C, D \neq 0)$$

3-3 Rewriting Rational Expressions

For any rational expressions A and B, $B \neq 0$,

$$\frac{A}{B} = A \cdot \frac{1}{B}$$

Rules

3-1 Equality of Fractions

$\dfrac{a}{b} = \dfrac{c}{d}$ if and only if $a \cdot d = b \cdot c \quad b, d, \neq 0$

3-2 Fundamental Theorem of Fractions

If a, b, and c are real numbers and $c \neq 0$, then

$$\frac{a}{b} = \frac{a \cdot c}{b \cdot c} \quad b \neq 0$$

3-3 A Fraction Containing One Minus Sign

If a and b are real numbers and $b \neq 0$,

$$\frac{-a}{b} = \frac{a}{-b} = -\frac{a}{b}$$

3-4 A Fraction Containing Two Minus Signs

If a and b are real numbers and $b \neq 0$,

$$\frac{-a}{-b} = \frac{a}{b} \text{ and } -\frac{-a}{b} = \frac{a}{b} \text{ and } -\frac{a}{-b} = \frac{a}{b}$$

3-5 A Fraction Containing Three Minus Signs

If a and b are real numbers and $b \neq 0$,

$$-\frac{-a}{-b} = \frac{-a}{b}$$

3–6 Adding Fractions

1. $\dfrac{x}{y} + \dfrac{z}{y} = \dfrac{x+z}{y}$ $y \neq 0$

2. $\dfrac{a}{b} + \dfrac{c}{d} = \dfrac{ad+bc}{bd}$ $b, d, \neq 0$

3–7 Subtracting Fractions

$\dfrac{a}{b} - \dfrac{c}{d} = \dfrac{ad-bc}{bd}$ $b, d, \neq 0$

Procedures

3–1 Finding the LCD

To find the LCM of two or more numbers, write a product that contains each factor raised to the largest power found on that factor in any of the prime factorizations for the given numbers.

Review Problems

Section 3–1

Are the following fractions equal?

1. $\dfrac{3}{17}$ and $\dfrac{2}{9}$
2. $\dfrac{5x}{6y}$ and $\dfrac{15x}{18y}$
3. $\dfrac{x+y}{5}$ and $\dfrac{3x+2y}{15}$

Raise the given fractions to the indicated terms.

4. $\dfrac{5}{6} = \dfrac{?}{18}$
5. $\dfrac{9x}{2y^2} = \dfrac{?}{2xy^2 + 4y^2}$
6. $\dfrac{3}{x-4} = \dfrac{?}{3x^2 - 19x + 28}$

Simplify the signs on the fraction.

7. $-\dfrac{-7}{-9}$
8. $-\dfrac{-(a-3)}{5}$
9. $-\dfrac{-6}{-(-m+7)}$

Section 3–2

Reduce the fractions to lowest terms.

10. $\dfrac{5x^3 y}{15xy^4}$
11. $\dfrac{r-3}{r^3 - 27}$
12. $\dfrac{x^3 - 27}{x^2 - 9}$
13. $\dfrac{5x^2 - 28x - 12}{5x^2 + 17x + 6}$

Rational Expressions

Section 3–3

Multiply or divide.

14. $\dfrac{8xy}{7z} \cdot \dfrac{21xz^3}{12xz^4}$

15. $\dfrac{t^2 - 16}{5t} \cdot \dfrac{25t^2}{t - 4}$

16. $\dfrac{9x^2 + 8x - 1}{x^2 - x - 2} \cdot \dfrac{x^2 - 2x}{9x^2 + 26x - 3}$

17. $\dfrac{7xy^2}{2x} \div \dfrac{14x^2}{10x^2 y}$

18. $\dfrac{2s^2 - 3s}{9s^2 - 25} \div \dfrac{2s^3 - 3s^2}{3s^2 + 11s + 10}$

Section 3–4

Find the LCM.

19. 36, 42, and 20

20. $4x^2 - 9$ and $12x - 18$

Section 3–5

Add or subtract.

21. $\dfrac{5}{2c} + \dfrac{1}{6}$

22. $\dfrac{6}{x^2 - 2x} + \dfrac{7}{x^2 - 5x + 6}$

23. $\dfrac{9w}{w^2 + 7w + 12} - \dfrac{w + 2}{w + 3}$

24. $\dfrac{2}{8x^3 - 1} - \dfrac{3x}{4x^2 + 2x + 1}$

Section 3–6

Simplify the following complex fractions.

25. $\dfrac{\tfrac{9}{16}}{\tfrac{3}{4}}$

26. $\dfrac{1 + \tfrac{7}{x} + \tfrac{10}{x^2}}{3 + \tfrac{5}{x} - \tfrac{2}{x^2}}$

27. $\dfrac{\tfrac{1}{x} - \tfrac{1}{4}}{\tfrac{1}{x^2} - \tfrac{1}{16}}$

Section 3–7

28. Find the quotient $(x^4 - 5x^3 + 2x - 3) \div (x - 2)$ by long division.

Section 3–8

29. Find the quotient $(x^5 - 4x^3 + 2x - 3) \div (x - 1)$ by using synthetic division.

30. Use synthetic division to determine if $x - 2$ is a factor of $x^4 + 2x^2 - 3x - 4$.

31. In other mathematics courses (such as calculus) it is sometimes necessary to divide a rational expression before other operations (such as integration) can be performed. For example, the function $f(x) = (x^4 - 8)/(x - 2)$ can be integrated if the right side of the equation is first divided. Use synthetic division to divide the right side of the equation. (The word *function* used in this exercise will be defined in Chapter 7).

32. What will be the degree of the quotient if $2x^5 - 4x^2 + 3$ is divided by $x + 5$?

Cumulative Test for Chapters 1–3

Chapter 1

For problems 1–3, use $A = \{1, 2, 3, 4, 5, 6\}$ and $B = \{3, 5, 7, 8\}$.

1. Find $A \cap B$.
2. Find $A \cup B$.
3. Find $A \cap A$.
4. If $a \neq b$, then either $a > b$ or $a < b$. This statement illustrates which axiom of inequality?
5. Simplify: $[12 - (-2)]^2 - [3 - (4 - 2) + 5]$.
6. Simplify: $\dfrac{-8^2 - 4^2 - [-2(-10 + 2)]}{8 - (-5)^2 - 3(-5)}$.

Chapter 2

7. Simplify: $\dfrac{(9x^0)(-2x^3)^3}{(6x^6)^2}$.
8. If $f(x) = 3x^3 - 4x$, find $f(-3)$.
9. The location of an object from the origin on a number line is given by $d = 2t^3 - 3t^2$. Find d if $t = 5$.
10. Subtract $9x^2 - 6x + 5$ from $-2x^2 + 7x - 4$.
11. Factor: $9x^2y^3 - 12x^5y^2 - 3x^2y^2$.
12. Multiply and simplify: $(5x - 3)(4x + 3)$.
13. Factor: $a^2 + 2ac - 3ab - 6bc$.
14. Factor: $2x^2 - 18$.

Chapter 3

15. Reduce: $\dfrac{5x^2 + x}{15x^2 - 2x - 1}$.
16. Multiply and simplify: $\dfrac{2x^2 - x}{6x + 1} \cdot \dfrac{6x^2 + 19x + 3}{5x^2 + 15x}$.
17. Subtract: $\dfrac{x}{y^2 - xy} - \dfrac{y}{xy - x^2}$.
18. Simplify: $\dfrac{\dfrac{1}{6} - \dfrac{1}{x}}{\dfrac{1}{x^2} - \dfrac{1}{36}}$.
19. Divide, using synthetic division: $(3x^3 - 4x + 2) \div (x - 2)$.
20. The difference in the volumes of two right circular cylinders with the same heights is $V = \pi r_2^2 h - \pi r_1^2 h$, where r_2 and r_1 are the radii of the two cylinders and h is the height of each cylinder. Factor the right side of the equation completely.

CHAPTER 4
First-Degree Equations and Inequalities

Objectives of Chapter 4

To be able to:

- ☐ Define a declarative statement. (4–1)
- ☐ Define open sentence. (4–1)
- ☐ Solve a first-degree equation. (4–1)
- ☐ Translate a word problem into an algebraic expression or equation. (4–1)
- ☐ Solve fractional equations. (4–2)
- ☐ Work word problems. (4–3)
- ☐ Solve first-degree inequalities. (4–4)
- ☐ Solve absolute value equations. (4–5)
- ☐ Solve absolute value inequalities. (4–6)
- ☐ Solve for a specified letter within an equation. (4–7)
- ☐ Evaluate formulas for specific values of the variables. (4–7)

A great deal of what we do in algebra involves solving equations. In this chapter we examine first-degree equations and develop means of solving them. First degree equations are then used to solve word problems. We also investigate first-degree inequalities and their uses.

We begin our study of these concepts with some basic definitions that relate to equations.

4–1 First-Degree Equations

In algebra we work with expressions and statements. An **expression** is the sum or difference of a number of terms. A **statement** is a declarative sentence. For example,

$$x + 5 = 8, \quad 4 = 4, \quad 2x + 3 < 7, \quad \text{and} \quad 9 = 8$$

are all statements. Clearly, $4 = 4$ is true, while $9 = 8$ is false. The statements $x + 5 = 8$ and $2x + 3 < 7$ are called **open sentences** because we do not know if they are true or false unless a value for the variable is known. If x is 3, we can declare that $x + 5 = 8$ is true.

Definition 4–1

Equation

An **equation** is a statement indicating that two quantities are equal.

The numbers we use to replace the variable in an equation come from a set of numbers called the **replacement set** for the variable. For example, if $3x = 12$ and if $x \in \{1, 2, 3, 4, 5\}$, then $\{1, 2, 3, 4, 5\}$ is the replacement set for the variable. However, if we are to have a true statement, then x must be 4. The set of numbers from the replacement set which makes an equation true is called the **solution set** for the equation.

Definition 4–2

Solution Set for an Equation

The **solution set** for an equation consists of those elements from the replacement set of the variable that makes the equation true. If there is no such number in the replacement set, then the solution set is empty.

Example 1 If $x + 5 = 12$ and $x \in \{3, 5, 9\}$, then the solution set is $\varnothing$. ■

First-Degree Equations and Inequalities

In our study of algebra, we encounter two types of equations. A **conditional equation** is an equation that is true for only certain replacements of the variable from its replacement set. For example, $5x = 15$ is a conditional equation because it is true for the condition that $x = 3$ and false for all other values of x. On the other hand, the equation $x + 1 = x + 1$ is true for every value of x in the replacement set of x—there is no condition placed upon what values may be assigned to x. Such an equation is called an **identity**.

Definition 4–3

Conditional Equation

A **conditional equation** is an equation that is true only for certain replacements of the variable from the replacement set of the variable.

Definition 4–4

Identity

An **identity** is an equation that is true for all replacements of the variable from the replacement set of the variable.

An equation is said to be *solved* when its solution set has been determined. This is accomplished by finding a series of equivalent equations. Equations are equivalent if they have the same solution set. For example, the equations

$$2x + 3 = -5$$
$$2x = -8$$
$$x = -4$$

are all equivalent because the solution set for each is $\{-4\}$.

Equivalent equations are obtained from a given equation by applying the equality addition theorem and equality multiplication theorems given in chapter 1. The basic idea in applying these theorems is to isolate the variable. When the variable has been isolated, one side of the equation will consist solely of the variable. The equation is then solved for the variable.

Recall from your first course in algebra that a first-degree equation in one variable is an equation which, when simplified, contains the variable raised only to the first power.

Example 1 Solve $5x + 3 = 13$ for x.

Solution We begin by adding -3, the additive inverse of 3, to both sides of $5x + 3 = 13$. We get

$$5x + 3 + (-3) = 13 + (-3)$$
$$5x + 0 = 10$$
$$5x = 10$$

Now we divide both sides of $5x = 10$ by 5.

$$\frac{5x}{5} = \frac{10}{5}$$

$$x = 2$$

When the equation has been solved, the value obtained for the variable may be indicated in two ways. In the case of example 1, we may say:

1. The solution is 2, or
2. The solution set is {2}.

Either method of indicating the answer is acceptable. To check that our solution is correct, we replace x with 2 in the original equation.

Check: $5(2) + 3 = 13$
$10 + 3 = 13$
$13 = 13$ ■

We do not show the checks for every equation in the following examples. However, you are encouraged to perform the checks, especially if you have reason to doubt the correctness of the solution.

Rules for operating on equations are condensed in rule 4–1.

Rule 4–1

Operating on an Equation

Any operation (except multiplying or dividing both sides by zero) applied to one side of an equation must be applied to the other side also.

Example 2   Solve $2x - 3 = 15$ for x.

Solution
$2x - 3 = 15$	Given equation.
$2x - 3 + 3 = 15 + 3$	Add 3 to both sides.
$2x = 18$	Simplify.
$x = 9$ ■	Divide both sides by 2.

Example 3   Solve $6x - 4 = x + 26$ for x.

Solution
$6x - 4 = x + 26$	Given equation.
$6x - x - 4 = x - x + 26$	Subtract x from both sides.
$5x - 4 = 26$	Simplify.
$5x - 4 + 4 = 26 + 4$	Add 4 to both sides.
$5x = 30$	Simplify.
$x = 6$ ■	Divide both sides by 5.

First-Degree Equations and Inequalities   **131**

Example 4   Solve $2(x - 3) + 5 = 3(x + 4) + 1$ for x.

Solution
$$
\begin{aligned}
2(x - 3) + 5 &= 3(x + 4) + 1 &&\text{Given equation.} \\
2x - 6 + 5 &= 3x + 12 + 1 &&\text{Distributive axiom.} \\
2x - 1 &= 3x + 13 &&\text{Simplify.} \\
2x - 3x - 1 &= 3x - 3x + 13 &&\text{Subtract } 3x \text{ from both sides.} \\
-1x - 1 &= 13 &&\text{Simplify.} \\
-x - 1 + 1 &= 13 + 1 &&\text{Add 1 to both sides.} \\
-x &= 14 &&\text{Simplify.} \\
(-1)(-x) &= (-1)(14) &&\text{Multiply both sides by } -1. \\
x &= -14 &&\text{Simplify.}
\end{aligned}
$$

Example 5   Solve $3m = m + 2m + 3$ for m.

Solution
$$
\begin{aligned}
3m &= m + 2m + 3 &&\text{Given equation.} \\
3m &= 3m + 3 &&\text{Simplify the right side.} \\
3m - 3m &= 3m - 3m + 3 &&\text{Subtract } 3m \text{ from each side.} \\
0 &= 0 + 3 &&\text{Simplify each side.} \\
0 &= 3 &&\text{Additive identity axiom.}
\end{aligned}
$$

The last line in example 5 is false. A false conclusion indicates no value of m will make $3m = m + 2m + 3$ true. The solution set is $\emptyset$.

Example 6   Solve $3(x + 2) = x + 2x + 6$ for x.

Solution
$$
\begin{aligned}
3(x + 2) &= x + 2x + 6 &&\text{Given equation.} \\
3x + 6 &= 3x + 6 &&\text{Remove parentheses and collect like terms.}
\end{aligned}
$$

We see the equation is an identity because both sides of the equation contain exactly the same terms. The solution set for this equation is the entire set of real numbers.

You can express some verbal expressions or verbal statements as algebraic expressions or equations. To be successful in doing this, you must look for key words and phrases that translate into mathematical quantities. For example, a number three times as large as x is represented by $3x$. A number c decreased by y is represented by $c - y$. The key word here is *decreased*, which means to subtract. In the first example, the key word is *times*, which means to multiply. The verbal statement "six x is equal to x increased by ten" translates as

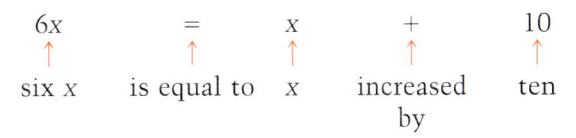

$$
\underset{\text{six } x}{6x} \quad \underset{\text{is equal to}}{=} \quad \underset{x}{x} \quad \underset{\substack{\text{increased} \\ \text{by}}}{+} \quad \underset{\text{ten}}{10}
$$

This equation can be solved by the methods just given to get $x = 2$.

Example 7 Five times a number, increased by four, is the same as the number increased by 16. Find the number.

Solution Let x represent the number. The equation is
$$5x + 4 = x + 16$$
If we solve the equation for x, we get
$$5x + 4 = x + 16$$
$$4x + 4 = 16$$
$$4x = 12$$
$$x = 3$$

Check: $5(3) + 4 = 3 + 16$
$15 + 4 = 19$
$19 = 19$

4–1 Exercises

In exercises 1–42, solve for the variable. *(See examples 1–6.)*

1. $a + 8 = 7$
2. $x - 3 = 4$
3. $t - 8 = -9$
4. $x - 10 = -12$
5. $2p - 3 = 11$
6. $2r + 8 = 14$
7. $5t + 4 = 19$
8. $7x + 5 = 26$
9. $2y + 3 = y + 8$
10. $5x + 4 = 4x - 2$
11. $6k - 4 = 2k + 12$
12. $12x - 8 = 2x + 12$
13. $2n - 4 = 5n - 19$
14. $5s - 8 = 7s + 16$
15. $7x - 3x + 4 = x + 13$
16. $2p - 5p + 3 = p - 2p + 17$
17. $6x - 9x + 5 = 3x - x + 25$
18. $2q - 7q + 3 = 10q + q + 35$
19. $t^2 + 3t - 4 = t^2 + t + 6$
20. $3x - 8 + x^2 = x + 6 + x^2$
21. $3z - 6z^2 + 5 = 2z - 6z^2 + 7$
22. $x^3 - 2x - 3 = x^3 + 3x + 7$
23. $5(y - 3) + 2 = 2(y + 5) + 1$
24. $-6(h - 3) - 4 = 3(h + 5) + 26$
25. $-2(x - 3) + 8 = -9(x + 4) + 1$
26. $-3(-x - 4) + x = 2(x - 3) - x$
27. $-6(-k + 3) - 9k = 2(k - 4) - 3k$
28. $5(x - 2) - 6x = -2(x - 4) + 9x - 2$
29. $3(r + 2) = 2r + r - 5$
30. $3x + x - 4 = 2(2x - 8)$
31. $5(b + 2) + 1 = b + 4b + 11$
32. $-2(c - 3) + 2 = -c - c + 8$
33. $(x - 5)(x + 2) = (x - 8)(x - 2) + 2$

First-Degree Equations and Inequalities

34. $3(u + 1) + (u - 3)(u + 4) = u^2 + 6u - 3$

35. $2(x - 1) + (x - 4)(x + 5) - 6 = (x - 1)(x + 2) - 6$

36. $(v - 4)(v + 3) - 6 = (v - 4)(v + 5) + 8$

37. $5 - 3[2(x - 3) - 4(x - 1)] = 2x - (x - 1)$

38. $-2[x - 3(2x + 5)] - (x + 1) = 3(x - 2) - 1$

39. $3(x - 2) + 7 = (x - 4) - 2(-x + 3)$

40. $2(-x - 8) + 3(x - 1) = 2[3(x - 1)] - 2(x + 3)$

41. $9 - \{3 - 2[x + 4(3 - x)]\} = 3x - 4(x - 1)$

42. $10 + 3[2x - 3(x - 4) + 5] = 2 - [3(x - 1)]$

In exercises 43–60, write an algebraic expression for each. Represent the unknown number with a variable.

43. Five more than an unknown number.

44. Two less than an unknown number.

45. The number x diminished by 3.

46. y less than 12.

47. x added to y.

48. The product of x and y increased by 6.

49. The sum of x and y subtracted from z.

50. John's age if he is 3 years older than Tina.

51. Mike's age if he is 3 years less than twice as old as Nikki.

52. The number of gallons of water in a tank holding x gallons if the tank is half full.

53. Susan's age 5 years from now.

54. The number of hours you have been walking if you have been walking 1 hour less than the number of hours Evan has been walking.

55. A number five times as large as y.

56. A number $\frac{1}{2}$ as large as t, decreased by 3.

57. $\frac{5}{8}$ of a number z increased by x^2.

58. The sum of two numbers is 30. If the larger is represented by m, represent the smaller in terms of m.

59. The sum of two numbers is z. If the smaller is represented by t, represent the larger in terms of t and z.

60. The difference of two numbers is d. If the larger number is represented by n, represent the smaller in terms of d and n.

In exercises 61–64, write an equation and solve for the variable. (See example 7.)

61. Nine times a number, increased by 2, is equal to twenty.

62. Twelve times a number, increased by 3, is the same as the number increased by 36.

63. Sixteen times a number, diminished by 2, is 22 less than six times the number.

64. If 18 is added to twice a number, the result is 2 more than three times the number.

65. In calculus, the equation $y' = 2(x - 3) + 2(5x - 3)(5)$ is the derivative of some function. Frequently, the right side of the equation must be set equal to zero and the resulting equation solved for x. If this were done, what value of x would be obtained?

66. The equation $s = 3t^2 + 7t + 3$ gives the distance s (in feet) that an object will travel in t seconds. The derivative of this equation is $s' = 6t + 7$; it gives the velocity (in feet per second) of the object at any time t. That is,

$$v = 6t + 7$$

where s' is replaced by v. At what time t will the velocity of the object be 25 feet per second?

67. Solve: $5.16(x - 0.14) = 2.24x + 12.84$. Round the answer to the nearest hundredth.

68. Solve: $9.14(x + 2.60) = -5.18 + 3(x - 5.10)$. Round the answer to the nearest hundredth.

4–2 Fractional Equations

A **fractional equation** is any equation in which one or more terms of the equation is a fraction. Thus $x + \frac{3}{4} = 5$; $x/2 + 5 = 8$, and $(x + 3)/(x + 1) = 6$ are examples of fractional equations.

We use the equality multiplication theorem to solve fractional equations when we multiply both sides of the equation by the LCM of all the denominators of the fractions in the equation. This gives a new equation that contains no fractions.

Example 1 Solve $\dfrac{x}{3} + \dfrac{2}{3} = 7$ for x.

Solution The LCM is 3. Multiply both sides of the equation by 3.

$$3\left(\dfrac{x}{3} + \dfrac{2}{3}\right) = 3 \cdot 7$$

$$3 \cdot \dfrac{x}{3} + 3 \cdot \dfrac{2}{3} = 21 \qquad \text{Remove the parentheses.}$$

$$\dfrac{\cancel{3} \cdot x}{\cancel{3}} + \dfrac{\cancel{3} \cdot 2}{\cancel{3}} = 21$$

$$x + 2 = 21 \qquad \text{Simplify each term on the left side.}$$

$$x = 19 \qquad \text{Solve for } x.$$

First-Degree Equations and Inequalities

You should check that the solution set is {19}.

Example 2 Solve $\dfrac{3}{2x} - \dfrac{7}{10} + \dfrac{2}{5x} = \dfrac{-2}{10x}$ for x.

Solution The LCM for $2x$, 10, $5x$, and $10x$ is $10x$. Multiply both sides of the equation by $10x$.

$$10x\left(\dfrac{3}{2x} - \dfrac{7}{10} + \dfrac{2}{5x}\right) = 10x \cdot \dfrac{-2}{10x}$$

$$\dfrac{\overset{5}{\cancel{10x}} \cdot 3}{\cancel{2x}} - \dfrac{\cancel{10}x \cdot 7}{\cancel{10}} + \dfrac{\overset{2}{\cancel{10x}} \cdot 2}{\cancel{5x}} = \dfrac{\cancel{10x} \cdot -2}{\cancel{10x}}$$

$$15 - 7x + 4 = -2$$
$$-7x + 19 = -2$$
$$-7x = -21$$
$$x = 3$$

The check for the equation to example 2 is not shown, but any proposed solution for a fractional equation in which the variable appears in one or more denominators should be placed back into the original equation to make sure no denominator of zero results. (See example 4, for instance.) In this example, 3 is the solution because this value of x does not give any zero denominators.

Example 3 Solve: $\dfrac{6}{x-3} - \dfrac{5}{3-x} = \dfrac{2}{3}$.

Solution At first glance we might say the LCM for all the denominators is $3(x-3)(3-x)$. But notice that the denominator of $5/(3-x)$ is the opposite of the denominator of $6/(x-3)$. By multiplying $5/(3-x)$ by $-1/(-1)$, we can make the denominator of $5/(3-x)$ the same as the denominator of $6/(x-3)$. We show this as the next step.

$$\dfrac{6}{x-3} - \dfrac{(-1)5}{(-1)(3-x)} = \dfrac{2}{3}$$

$$\dfrac{6}{x-3} - \dfrac{-5}{x-3} = \dfrac{2}{3}$$

The two minus signs in the term $-(-5)/(x-3)$ can be combined to give a plus sign. Our problem now is

$$\dfrac{6}{x-3} + \dfrac{5}{x-3} = \dfrac{2}{3}.$$

Now multiply each side by the LCM and solve for x. We show this without giving reasons for each step.

$$3(x-3)\left[\frac{6}{x-3} + \frac{5}{x-3}\right] = 3(x-3)\frac{2}{3}$$

$$\frac{3(x-3)6}{x-3} + \frac{3(x-3)5}{x-3} = \frac{3(x-3)2}{3}$$

$$18 + 15 = 2(x-3)$$
$$33 = 2x - 6$$
$$39 = 2x$$
$$x = \frac{39}{2}$$

See the paragraph following example 2.

Example 4

Solve $\dfrac{2}{x+4} - \dfrac{5}{x-4} = \dfrac{-40}{x^2-16}$ for x.

Solution

The LCM is $(x+4)(x-4)$. If each term of the equation is multiplied by $(x+4)(x-4)$, we get

$$\frac{(x+4)(x-4)2}{x+4} - \frac{(x+4)(x-4)5}{x-4} = \frac{-40(x+4)(x-4)}{(x+4)(x-4)}$$

$$2(x-4) - 5(x+4) = -40$$
$$2x - 8 - 5x - 20 = -40 \qquad \text{Remove parentheses.}$$
$$-3x - 28 = -40 \qquad \text{Collect like terms and solve for } x.$$
$$-3x = -12$$
$$x = 4$$

While $x = 4$ may appear to be a solution, it is not. If 4 is substituted for x the original equation, then

$$\frac{2}{8} - \frac{5}{0} = \frac{-40}{0}$$

We cannot divide by zero. The solution set is $\varnothing$. This example clearly shows that we should always check the solution in the original equation if the equation has a variable in at least one of the denominators. The number 4 is called an *extraneous solution*. An **extraneous solution** is a proposed solution that does not make the original equation true. Extraneous solutions may arise when both sides of the fractional equation are multiplied by an expression containing a variable.

4–2 Exercises

Find the solution for each equation and check for extraneous solutions. (See examples 1–4.)

1. $\dfrac{x}{5} = 6$
2. $\dfrac{x}{3} = -4$
3. $\dfrac{2y}{3} = 4$
4. $\dfrac{5z}{3} = 15$
5. $5x + \dfrac{x}{2} = 11$
6. $x + \dfrac{3}{4} = 5$
7. $\dfrac{1}{2m} + \dfrac{1}{8} = \dfrac{2}{m}$
8. $\dfrac{6n}{5} - \dfrac{2n}{3} = 8$
9. $\dfrac{2x+3}{5} + \dfrac{x-3}{2} = 5$
10. $\dfrac{3x-1}{2} + \dfrac{5x-1}{7} = x + \dfrac{4}{7}$
11. $\dfrac{2a+1}{4} - \dfrac{a-3}{4} = a - \dfrac{1}{2}$
12. $\dfrac{b-1}{3} + \dfrac{2b+5}{5} = b - \dfrac{2}{3}$
13. $\dfrac{5x}{x-1} + \dfrac{2x}{x+1} = 7$
14. $\dfrac{3x}{x+2} - \dfrac{4x}{x+3} = -1$
15. $\dfrac{3c}{c+2} - \dfrac{9c}{c-3} = -6$
16. $\dfrac{d}{d-1} + \dfrac{2d}{d-3} = 3$
17. $\dfrac{5}{x+4} - \dfrac{6}{x-4} = \dfrac{-40}{x^2-16}$

(See example 4.)

18. $\dfrac{7}{x+3} + \dfrac{2}{x-3} = \dfrac{12}{x^2-9}$
19. $\dfrac{6}{p-3} + \dfrac{2}{3-p} = 4$
20. $\dfrac{2}{q-5} - \dfrac{7}{5-q} = 3$
21. $\dfrac{9}{x-2} - \dfrac{3}{2-x} = 4$
22. $\dfrac{3}{4-x} - \dfrac{5}{x-4} = -2$
23. $\dfrac{6}{r} + \dfrac{5}{r-3} = \dfrac{15}{r^2-3r}$
24. $\dfrac{3s-2}{s^2+5s+6} + \dfrac{6}{s+3} = \dfrac{8}{s+2}$
25. $\dfrac{x-3}{2x^2-x-15} - \dfrac{5}{x-3} = \dfrac{2}{2x+5}$
26. $\dfrac{-2x}{x^2-9} + \dfrac{2}{x-3} = \dfrac{-3}{x+3}$
27. $\dfrac{6}{u^2+u-6} = \dfrac{7}{u^2+2u-8}$
28. $\dfrac{2}{2v^2+7v-15} = \dfrac{-3}{2v^2-7v+6}$

In exercises 29–32, write the statement as a fractional equation and solve for the variable.

29. The sum of one-half of a number and one-third of a number exceeds two-thirds of the number by 10. What is the number?

30. The difference between the third and fourth parts of a number is 1. What is the number?

31. A business woman uses $\frac{1}{3}$ of her yearly capital for miscellaneous expenses and $\frac{1}{9}$ of it for employee benefits. If these two expenses amount to $20,000 per year, how much is her yearly capital?

32. A banquet room is $\frac{1}{4}$ as wide as it is long. If the length were $\frac{1}{2}$ as much and the width were three times as great, the perimeter would be 500 feet. What are the dimensions of the room? (*Hint:* In working this problem, let L represent the length of the room and assume the room is a rectangle. The perimeter of a rectangle is $P = 2L + 2W$.)

4–3 Applications

In this section, we take an in-depth look at several types of word problems. Every word problem requires a translation from English into mathematical symbols. This translation is made by looking for critical words and phrases that can be written as mathematical symbols. We give some of the more common words and phrases and their mathematical equivalents.

English Word or Phrase	*Mathematical Symbol*
Is, equals, gives, is the same as, was, results in.	=
Add, added to, increased by, more than, plus, sum, the sum of.	+
Subtract, difference, diminished by, less than, decreased by, minus, subtracted from.	−
Quotient, divided by, divide.	÷
What is the number, what is the selling price, how fast did it travel, what is her age, and so on.	a variable

No one set of rules can be applied equally well to all word problems to get you started and to give you a solution for each and every problem. However, we can give a few general ideas and suggestions. Remember, though, you will become an expert at solving word problems only by working word problems. No one has yet found a substitute for practice—and then more practice.

Suggestions for Solving Word Problems

1. Read the entire problem to get a general overview of it.
2. Determine what you are trying to find.
3. Represent the quantity or quantities you are trying to find by some variable.
4. Draw a sketch, if one applies.
5. Reread the problem and begin to represent the phrases and unknowns in the problem by variables and numbers.

6. Form an equation by setting two expressions that represent the same number equal to each other.
7. Solve the equation for the variable.
8. State the answer to the problem.
9. Check the correctness of the answer according to the conditions given in the problem.

Several different kinds of examples now follow. Observe how the suggestions have been applied to each example.

Geometry Problems

Example 1 The length of a rectangle is 11 feet more than its width. If the perimeter of the rectangle is 106 feet, find the length and width of the rectangle.

Solution Select a variable to represent the width.
Let W represent the width of the rectangle.
Then $W + 11$ will represent the length because it is 11 feet more than the width.
A sketch helps to show the relationship between the length and width.

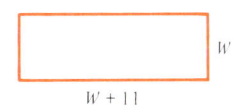

The perimeter of a rectangle is given by the equation $P = 2L + 2W$, where L and W represent the length and width, respectively.

$P = 2L + 2W$
$106 = 2(W + 11) + 2W$ Write the expressions for the length and width in the equation for the perimeter.
$106 = 2W + 22 + 2W$ Remove the parentheses.
$106 = 4W + 22$ Simplify the right side.
$84 = 4W$ Subtract 22 from each side.
$21 = W$ Divide each side by 4.

The width is 21 feet. Because $W + 11$ represents the length, $L = 21 + 11 = 32$ feet. The rectangle measures 32 feet by 21 feet.

Check: $P = 2L + 2W$
$106 = 2(32) + 2(21)$
$106 = 106$ ∎

Work Problems

The solution to a work problem is based on the fact that if a job can be completed in h hours, then the part of the job that can be completed in 1 hour is $1/h$. For example, if it takes Jeff 7 hours to paint a fence, then he can finish $\frac{1}{7}$ of the job in one hour. If he works for 5 hours, then he completes $5(\frac{1}{7}) = \frac{5}{7}$ of the job. Also, if a pipe can fill a tank in 8 hours, then in 1 hour it fills $\frac{1}{8}$ of the tank. In x hours, it would fill $x(\frac{1}{8}) = x/8$ of the tank.

Example 2 Steve can paint the front of his house in 9 hours, while it takes Ross 12 hours to do the same job. If they work together, how long will it take them to paint the front of the house?

Solution Let x represent the number of hours it takes both boys working together to paint the front of the house.
  Then $x(\frac{1}{9}) = x/9$ is the part of the job that Steve will do, and
  $x(\frac{1}{12}) = x/12$ is the part of the job that Ross will do.
The sum of these two parts must equal 1 because the entire job is completed when these two parts are added.

$$\frac{x}{9} + \frac{x}{12} = 1$$
$$4x + 3x = 36 \qquad \text{Multiply both sides by 36.}$$
$$7x = 36$$
$$x = \frac{36}{7}, \quad \text{or} \quad 5\frac{1}{7} \text{ hours}$$

Thus, the two boys can paint the house in $5\frac{1}{7}$ hours if they start at the same time and work together until the job is finished.

Check: To check the problem, multiply the number of hours each works times that part of the job each completes in one hour, and add these results to see if the sum is one.

$$\frac{36}{7} \cdot \frac{1}{9} + \frac{36}{7} \cdot \frac{1}{12} = 1$$
$$\frac{4}{7} + \frac{3}{7} = 1$$
$$\frac{7}{7} = 1$$
$$1 = 1 \quad ■$$

First-Degree Equations and Inequalities   141

Example 3 A pipe, which takes 80 minutes to fill a tank, is left running for 20 minutes and then shut off. A second pipe is opened and finishes filling the tank in 40 minutes. How long would it take for the second pipe alone to fill the tank?

Solution Let x be the number of minutes required for the second pipe alone to fill the tank. We organize the information in a table.

	That part of the tank filled in one minute	×	The number of minutes the pipe runs	=	That part of the tank filled
First pipe	$\dfrac{1}{80}$	·	20	=	$\dfrac{1}{4}$
Second pipe	$\dfrac{1}{x}$	·	40	=	$\dfrac{40}{x}$

The equation is

$$\frac{1}{4} + \frac{40}{x} = 1$$
$$x + 160 = 4x \qquad \text{Multiply both sides by } 4x.$$
$$3x = 160$$
$$x = \frac{160}{3}, \text{ or } 53\frac{1}{3} \text{ minutes}$$

The second pipe can fill the tank in $53\frac{1}{3}$ minutes. The check is not shown because it is similar to example 2.

Mixture Problems

The name *mixture problem* is frequently given to the type of problem in which two or more items are combined to form a mixture or solution. A slight variation in this type of problem occurs when the sum of the individual costs of two or more items mixed is the same as the cost of the mixture.

Example 4 How many grams of a food supplement containing 25% protein must a dietician mix with a food supplement containing 60% protein to obtain 70 grams of a supplement that contains 40% protein?

Solution

The equation is based on the fact that the amount of pure protein in the 25% supplement plus the amount of pure protein in the 60% supplement will be the same as the amount of pure protein in the 40% supplement. The amount of pure protein is found by multiplying the amount of the supplement by the percent of protein it contains.

Let g represent the number of grams of the 25% supplement to be used.

Then $70 - g$ represents the number of grams of the 60% supplement to be used.

The equation is

$$0.25g + 0.60(70 - g) = 0.40(70)$$
$$\text{pure protein} + \text{pure protein} = \text{pure protein}$$

The decimals can be eliminated if both sides of the equation are multiplied by 100.

$$25g + 60(70 - g) = 40(70)$$
$$25g + 4{,}200 - 60g = 2{,}800$$
$$-35g = -1400$$
$$g = 40$$

The dietician should use 40 grams of the 25% supplement and $70 - g = 70 - 40 = 30$ grams of the 60% supplement.

Check: To check the problem, see if the number of grams of pure protein in the two parts mixed is the same as the number of grams of pure protein in the mixture.

$$0.25(40) + 0.60(30) = 0.40(70)$$
$$10 + 18 = 28$$
$$28 = 28 \quad ■$$

Example 5

An automobile radiator has a capacity of 12 liters. If it contains a solution of water and antifreeze that is 80% antifreeze, how much of the solution should be drained and replaced with water if the new solution is to be 60% antifreeze?

Solution

Let x represent the number of liters of the 80% solution that is drained and replaced with water. Then $12 - x$ liters of the 80% solution remains in the radiator. No antifreeze is added to the radiator when it is refilled with water, so the amount of pure antifreeze in the $12 - x$ liters of the 80% solution is the same as the amount of pure antifreeze in the 12 liters of the 60% solution. A table is again helpful.

First-Degree Equations and Inequalities

	Number of Liters of the Solution	Percent Pure Antifreeze	Number of Liters of Pure Antifreeze
Original 80% solution	$12 - x$	80%	$0.80(12 - x)$
Water	x	0%	0
New 60% solution	12	60%	$0.60(12)$

The number of liters of antifreeze in the $12 - x$ liters of the 80% solution is the same as the number of liters of antifreeze in the 60% solution. The last parts of the table give the equation.

$$0.80(12 - x) = 0.60(12)$$
$$80(12 - x) = 60(12) \quad \text{Multiply by 100.}$$
$$960 - 80x = 720$$
$$-80x = -240$$
$$x = 3$$

Thus 3 liters of the solution should be drained and replaced with pure water. To check, observe that the 9 remaining liters contains $9(0.80) = 7.2$ liters of antifreeze. Also, $(12 \text{ liters})(0.60) = 7.2$ liters of antifreeze. ■

Consecutive Integer Problems

Integers are numbers belonging to the set $\{\cdots, -2, -1, 0, 1, 2, \cdots\}$. Any given integer is one more or one less than an adjacent integer. *Consecutive integers* are integers separated by one unit. If x represents any integer, then

$$x, \quad x + 1, \quad \text{and} \quad x + 2$$

represent three consecutive integers.

An integer is *even* if it is divisible by 2. An integer is even if it is a member of the set $\{\cdots, -4, -2, 0, 2, 4, \cdots\}$. *Consecutive even integers* are integers separated by two units. If x represents any even integer, then

$$x, \quad x + 2, \quad \text{and} \quad x + 4$$

represent three consecutive even integers.

An integer is *odd* if it is not even. An odd integer belongs to the set $\{\cdots, -3, -1, 1, 3, \cdots\}$. *Consecutive odd* integers are separated by two units. If x represents an odd integer, then

$$x, \quad x + 2, \quad \text{and} \quad x + 4$$

represent three consecutive odd integers.

Example 6 Four times the middle integer of three consecutive odd integers is 5 more than three times the largest integer. Find the three integers.

Solution Let x represent the smallest odd integer. Then $x + 2$ represents the middle integer and $x + 4$ represents the largest integer.

4	·	$(x + 2)$	=	5	+	3	·	$(x + 4)$
↑	↑	↑	↑	↑	↑	↑	↑	↑
Four	times	the middle integer	is	five	more than	three	times	the largest integer

Now pull the terms and factors of the equation close together.

$$4(x + 2) = 5 + 3(x + 4)$$
$$4x + 8 = 5 + 3x + 12 \quad \text{Remove parentheses.}$$
$$4x - 3x = 17 - 8$$
$$x = 9$$

We know that 9 represents the smallest of the three consecutive odd integers. Thus the three integers are

$$9, \quad 11, \quad \text{and} \quad 13$$

Check: To check the problem, see if four times the middle integer is the same as five more than three times the largest.

$$4(11) = 5 + 3(13)$$
$$44 = 5 + 39$$
$$44 = 44 \quad ■$$

Interest Problems

If $100 dollars is invested for 1 year at 7% simple interest, then the interest earned will be $100(0.07) = 7.00. The formula for simple interest is $I = prt$, where I is the interest earned, p is the amount of money invested (called the *principal*), r is the rate of interest, and t is the time in years. If $t = 1$, then the formula reduces to $I = pr$. In the examples and exercises that follow in this section, we assume the interest is simple interest.

Example 7 David has $1,000 more invested at 5% than he has invested at 6%. If the annual interest on the 5% investment is $30 less than the annual interest on the 6% investment, how much has he invested at each rate?

Solution   Let x be the amount of money invested at 6%. Then $x + 1000$ is the amount of money invested at 5%.

$$\text{(interest earned at 5\%)} = \text{(interest earned at 6\%)} - \$30$$
$$0.05(x + 1,000) = 0.06(x) - 30 \quad \text{Multiply both sides of the equation by 100 to eliminate decimals.}$$

$$5(x + 1,000) = 6x - 3,000$$
$$5x + 5,000 = 6x - 3,000$$
$$-x = -8,000$$
$$x = 8,000$$

He has $8,000 invested at 6% and $9,000 invested at 5%.

Check: The interest earned on the 5% investment is $0.05(\$9,000) = \$450$. The interest earned on the 6% investment is $0.06(\$8,000) = \$480$. We see that the interest earned from the 5% investment is $30 less than the interest earned from the 6% investment. ■

Uniform-Motion Problems

The distance traveled at a uniform rate for a certain period of time is given by $d = rt$, where d is the distance traveled, r is the uniform rate of travel, and t is the time. For example, if you travel at an average speed of 45 miles per hour for 3 hours, you will have traveled $3(45) = 135$ miles. If these figures are substituted into the equation $d = rt$, we get

$$d = rt$$
$$d = 45(3) = 135 \text{ miles}$$

Example 8   Terri can ski 3 miles down the Sacramento River in the same time she can ski 2 miles upstream. If the Sacramento River flows with a current of 5 miles per hour, what is the speed of the ski boat in still water?

Solution   We want to know how fast the boat pulling Terri would be going if there were no current in the river. Select a variable to represent this.

Let x represent the speed of the boat in still water. We also need to represent the speed of the boat going upstream and downstream. The speed of the boat is either increased or decreased by the speed of the current of the river, depending on the direction of the boat. Thus

$x + 5$ represents the speed of the boat downstream and
$x - 5$ represents the speed of the boat upstream.

A table helps to relate all the facts.

	Distance	Rate of Travel	Time $= \dfrac{d}{r}$
Upstream	2	$x - 5$	$\dfrac{2}{x - 5}$
Downstream	3	$x + 5$	$\dfrac{3}{x + 5}$

The time going upstream is the same as the time going downstream. This fact, combined with the expressions in the third column of the table, gives the equation.

$$\dfrac{3}{x + 5} = \dfrac{2}{x - 5}$$
$3(x - 5) = 2(x + 5)$ Multiply by the LCM, $(x + 5)(x - 5)$.
$3x - 15 = 2x + 10$ Remove parentheses and solve for x.
$x = 25$

The boat (and Terri) would move at a rate of 25 miles per hour if the river had no current. It does have a current, however, so the boat moves at a rate of 30 miles per hour downstream and 20 miles per hour upstream.

Check: We know:

$$\text{time upstream} = \text{time downstream}$$
$$\dfrac{d}{r} \text{ upstream} = \dfrac{d}{r} \text{ downstream}$$
$$\dfrac{2}{20} = \dfrac{3}{30}$$
$$\dfrac{1}{10} = \dfrac{1}{10}$$ ■

Example 9 One runner ran a race in 3 minutes, 5 seconds. A second runner took 3 minutes, 20 seconds to run the same race. The rate of the first runner was $1\frac{1}{2}$ feet per second faster than the second runner. What was the distance the runners ran?

Solution Let x be the rate of the second runner in feet per second. Then $x + 1\frac{1}{2}$ is the rate of the first runner in feet per second. The time must also be expressed in seconds because the rates are in feet per second.

First-Degree Equations and Inequalities

	Time (in Seconds)	Rate (in Feet per Second)	Distance = Time × Rate
First runner	185	$x + \dfrac{3}{2}$	$185\left(x + \dfrac{3}{2}\right)$
Second runner	200	x	$200x$

The distances for the two runners are the same. The equation is

$$200x = 185\left(x + \frac{3}{2}\right)$$

$$200x = 185x + \frac{3(185)}{2}$$

$$15x = \frac{555}{2}$$

$$x = \frac{555}{2 \cdot 15}$$

$$x = 18.5$$

The second runner is running at a rate of 18.5 feet per second. The rate of the first runner is $18\frac{1}{2} + 1\frac{1}{2} = 20$ feet per second. The distance that each ran is shown in the check.

Check: (18.5 feet per second)(200 seconds) = 3,700 feet
(20 feet per second)(185 seconds) = 3,700 feet

Miscellaneous Problems

Example 10 Suppose you buy some stock on Wednesday and find, much to your delight, that its value has increased 5% on Thursday. On Friday, its value decreases by 10% from the value it had on Thursday. Panic stricken, you sell the stock on Friday for $189. What was the purchase price you paid on Wednesday for the stock?

Solution Let x represent the purchase price paid on Wednesday. Then $x + 0.05x$ represents the value of the stock on Thursday. On Friday, the value of the stock was down by 10% from the value on Thursday. This is represented by $(x + 0.05x) - 0.10(x + 0.05x)$

$$\underset{\substack{\uparrow \\ \text{value on} \\ \text{Thursday}}}{(x + 0.05x)} \underset{\substack{\uparrow \\ \text{minus}}}{-} \underset{\substack{\uparrow \\ 10\% \text{ of the} \\ \text{value on} \\ \text{Thursday}}}{0.10(x + 0.05x)}$$

The expression just given represents the selling price on Friday. Thus

$$(x + 0.05x) - 0.10(x + 0.05x) = 189$$
$$1.05x - 0.10(1.05x) = 189$$
$$1.05x - 0.105x = 189$$
$$0.945x = 189$$
$$x = 200$$

The price paid for the stock on Wednesday was $200. ∎

4–3 Exercises

Set up an equation for each exercise and solve the equation.

Geometry Problems (See example 1.)

1. The length of a rectangle is 3 meters more than twice the width. If the perimeter of the rectangle is 144 meters, what are the dimensions of the rectangle?

2. The number that represents the perimeter of a rectangle is 10 times greater than the number that represents the width. If the length of the rectangle is 32 centimeters, what are the width and perimeter?

3. The length of a rectangle is twice its width. If the length is increased by 1 foot and the width decreased by 1 foot, the decrease in the area of the rectangle is 6 square feet. What were the dimensions of the rectangle before any changes were made in its dimensions?

4. The perimeter of a triangle is 45 inches. If the second side is 1 inch more than the first side and the third side is 4 inches more than the second side, what are the lengths of the three sides of the triangle?

5. A rectangle whose width is 5 feet must have an area the same as a square whose sides are 10.2 feet. What is the length of the rectangle?

6. The two equal sides of an isosceles triangle are 3 feet more than three times the length of the third side, called the *base* of the triangle. If the perimeter is 1 foot more than eight times the length of the base, find the lengths of the sides of the triangle.

7. A rectangular yard has a length that is 10 feet more than its width. Surrounding the yard is a uniform border 2 feet wide. If the border contains 296 square feet, what are the dimensions of the yard?

8. The area for a trapezoid is given by

$$A = \frac{h}{2}(B + b)$$

First-Degree Equations and Inequalities **149**

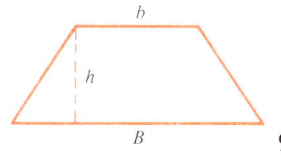

where B and b are the two bases and h is the altitude between the two bases. (See the figure). The area of the trapezoid is 5 square meters, and its altitude is 4 meters. If the lower base is four times as long as the upper base, find the dimensions of the bases.

9. The length of one side of a triangle is two-fifths the length of the perimeter. The second side is 8 feet, and the third side is one-third the length of the perimeter. Find the perimeter of the triangle.

Work Problems (See examples 2 and 3.)

10. The water tank in a certain town can be filled in 20 hours by a large inlet pipe. A smaller inlet pipe, used only in emergencies, can fill the tank in 30 hours. If the tank is emptied for cleaning, how long will it take to fill the tank if both pipes are used?

11. From Example 2, we know Steve can paint the front of his house in 9 hours, while it takes Ross 12 hours to do the same job. Suppose Steve works by himself for 3 hours and then Ross joins him in finishing the job. (Ross wants Steve to go bowling with him.) How many hours will it take the two boys to finish painting the front of the house?

12. A pile of logs at a sawmill in northern California can be cut into lumber in 4 days. At a smaller mill located nearby, the same pile can be cut into lumber in 10 days. If they both work together sawing the logs, how long will it take to finish the job?

13. It takes a pipe 50 minutes to fill a tank. The pipe is left running for 30 minutes and then shut off. A second pipe is opened and finishes filling the tank in 30 minutes. How long would it take for the second pipe alone to fill the entire tank?

14. John needs to pick the apples off his apple tree. He can pick all the apples in 5 hours. His friend, Sue, can pick all the apples in 6 hours. Sam, another friend of John's, can pick them in 3 hours. If John and Sue start picking the apples at 8 A.M. and if Sam joins them at 8:30 A.M., what time can they finish picking all the apples?

15. Mary can assemble a ten-speed bicycle three times as fast as Tom can. If both work together, starting at the same time, they can put the bicycle together in 2 hours. How long would it take Tom to do the job by himself?

16. One pipe can fill a tank in 6 hours and a second pipe can fill it in 8 hours. A third pipe can drain the tank in 12 hours. The two fill pipes are both turned on for 2 hours. At that time, someone accidently opens the drain pipe. How long will it take to finish filling the tank with both fill pipes and the drain pipe open?

Mixture Problems (See examples 4 and 5.)

17. A car radiator has a capacity of 20 liters. How many liters of a solution that is 30% antifreeze should be mixed with a solution that is 70% antifreeze to obtain 20 liters that is 60% antifreeze?

18. Julie is on a diet, but she hates skimmed milk. How many cups of milk that is 4% cream should she mix with some skimmed milk (0% cream) to obtain 18 cups of milk that is 3% cream?

19. How many liters of an 8% salt solution and how many liters of a 13% salt solution should be mixed to obtain 10 liters of a 10% salt solution?

20. How many grams of a silver alloy that is 70% silver should be melted with 20 grams of an alloy that is 10% silver to obtain an alloy that is 40% silver?

21. Pure sulfuric acid is added to 40 quarts of a 30% sulfuric acid solution to obtain a solution that is 60% sulfuric acid. How many quarts of pure sulfuric acid must be used?

22. A 50-pound mixture of walnuts and almonds costs $42.50. If the walnuts costs 75¢ per pound and the almonds cost $1.00 per pound, how many pounds of each kind of nut are in the mixture?

23. Some tea worth $8.00 per pound is to be mixed with tea worth $6.00 per pound to obtain a mixture of 96 pounds worth $6.50 per pound. How many pounds of each kind of tea should be used?

24. Four types of candy, A, B, C, and D, are to be mixed to obtain a mixture selling for $3.75 per pound. Twenty pounds of A, worth $3 per pound, 10 pounds of B, worth $4 per pound, and 15 pounds of C, worth $3.50 per pound are to be used. How many pounds of type D, worth $4.50 per pound, should be used?

25. An alloy that is one part lead and three parts zinc is added to another alloy that is two parts lead and three parts zinc. How much of each alloy must be used to make 50 pounds of an alloy that will be 34% lead?

Consecutive Integer Problems (See example 6.)

26. Find three consecutive integers with a sum of 66.

27. Find three consecutive integers such that twice the first increased by the sum of the other two has a value of 83.

28. Find three consecutive odd integers such that three times the sum of the first two diminished by 137 is equal to the third.

29. Find three consecutive even integers such that twice the second, increased by the first, is the same as the third diminished by 36.

30. Find three consecutive integers such that their sum is zero.

31. If the sum of the first and second of three consecutive integers is subtracted from the sum of the second and third, then the result is 2. Find the three integers. (What general statement can be made about the three integers?)

32. Find three consecutive integers with a sum of 25. (What conclusion must you draw about this problem?)

33. Find three consecutive even integers so that the product of the first and -2 is the same as the sum of the other two diminished by -66.

34. The square of the smallest of three consecutive integers increased by the largest integer is 20 less than the square of the middle integer. What are the three integers?

35. A long trip is to be divided into three segments that correspond to three consecutive multiples of 10. (Multiples of 10 are 10, 20, 30, ...) Find the length of the entire trip if three times the length of the shortest segment exceeds the largest segment by 440 miles.

36. An inheritance is to be divided into three parts corresponding to three consecutive multiples of 50. (Multiples of 50 are 50, 100, 150, ...) Find the amount of the inheritance if twice the largest part exceeds the smallest part by $8,700.

37. Prove that the sum of five consecutive integers is the same as five times the middle integer.

Interest Problems (See example 7.)

38. Chris has a certain amount of money invested at 6% interest. He has a second amount of money, which is $2,000 more than the first amount, invested at 8% interest. If the annual income from these two investments is $1,000, how much does he have invested at each rate?

39. Grace has $12,000 divided between two investments. The first investment earns 8% interest and the second earns 10% interest. If the second investment earns $300 more per year than the first investment, how much is invested at each rate?

40. A man has a certain amount of money invested in stocks at 7% and half that amount invested in bonds at 8%. If the bonds earn $189 a year more than the stocks, how much does he have invested in each?

41. A man has $10,000 to invest. He invests $2,000 at 10%. Part of the remaining money is invested at 12% and the other part at 9%. If his annual return on all three investments is $1,070, how much does he have invested at 12% and 9%?

42. A certain amount of money is invested for one year at 8%. If the interest rate were to be increased to 10%, the interest earned would be $160 more than the interest earned at 8%. How much money was invested at 8%?

43. Suppose you have $11,500 to invest. If you invest $2,000 at 8% and $4,500 at 5½%, at what rate must you invest the rest to have an annual return of $832.50 from all three investments?

44. A jeweler invests $12,000 in earrings and necklaces. The jeweler made a 40% profit on the necklaces and a 10% profit on the earrings. These two profits combined were the same as a 30% return on the original investment. How much did the jeweler invest in earrings and necklaces?

45. Is a single discount of 25% from the price P of a coat the same as two successive discounts of 10% and 15%? If they are not the same, which method gives the greater discount?

Uniform Motion Problems (See examples 8 and 9.)

46. Joy has 10 hours to use in training for the cross-country marathon. How far can she ride a bicycle at an average rate of 9 miles per hour and be able to return at the end of 10 hours if she jogs back at an average rate of 6 miles per hour?

47. Two cars leave Salt Lake City at the same time, traveling in opposite directions. The average rate of travel of one car is 5 miles per hour faster than the average rate of travel for the other car. If the two cars are 420 miles apart at the end of 4 hours, what is the rate of each?

48. Two cars leave Memphis, traveling south. The first car leaves at 11 A.M. Two hours later the second car leaves, traveling at an average rate that is $11\frac{2}{3}$ miles per hour more than the first car. If the second car overtakes the first one at 7 P.M., what is the average rate of both cars?

49. Brooke decides to go for a bike ride and invites Becky to go with her. However, Becky will not be ready to leave for another 30 minutes. Brooke leaves without her at 8 A.M. and averages 16 miles per hour. Becky leaves at 8:30 A.M. and averages 20 miles per hour. When does she overtake Brooke?

50. A highway patrol officer noticed a car go by a certain point at 80 miles per hour. One minute later she gave chase and overtook the car after 5 minutes. What was her average rate of speed?

51. A faster runner ran a distance of 1,065 yards in the same amount of time that a slower runner ran a distance of 994 yards. The average speed of the faster runner was $\frac{1}{2}$ yard per second faster than the average speed of the other runner. What was the average speed of each runner, and how much time did each take to run the race?

52. A faster train can cover a distance of 594 miles in the same amount of time that a slower train goes a distance of 451 miles. If the average speed of the faster train is 13 miles per hour faster than the average speed of the slower train, how fast was each train going?

53. Carolyn flew her plane with a 30 mile per hour tailwind for $1\frac{3}{4}$ hours; she then landed at the Napa airport and had lunch. On the return trip, with the same wind now as a headwind, she flew for 2 hours and was still 75 miles from where she started the trip. What was the speed of the plane in still air, and what was the one-way distance to Napa?

54. A plane has a cruising speed of 200 miles per hour. The pilot wants to fly due south against a headwind of 40 miles per hour and then return. If he must make the round trip in 2 hours, at what time should he turn back if he leaves at 8 A.M.? Assume he has a 40-mile-per-hour tailwind on the return trip.

Miscellaneous Problems (See example 10.)

55. A 6% sales tax on a new automobile was $336. What was the price of the automobile before the tax was added? How much did the customer pay for the automobile, including the sales tax?

56. A house and lot are worth $90,000. If the value of the house is $6\frac{1}{2}$ times the value of the lot, what is the value of each?

57. A coat was discounted 20% from the original price. It did not sell and was discounted another 20% from the first discount price. If the coat sold for $51.20 after the two discounts, what was the original selling price?

58. A company offers a 2% discount to its customers if they pay their bills

within 30 days. If a customer saved $120 by paying within the 30-day time limit, how much was the bill before it was discounted the 2%?

59. One-third of a woman's gross monthly check was withheld for federal income taxes and $\frac{1}{10}$ of it was withheld for state income taxes. If her take-home pay was $1,020, how much was her gross pay?

60. A shopper can buy one can of string beans for a certain price. If he buys the cans by the case, the price per can is $\frac{1}{11}$ less than the price of a single can. If he paid $7.20 for a case of 24 cans, how much was the price of a single can?

4—4 First-Degree Inequalities

In mathematics, we often encounter quantities that are unequal rather than equal. Recall from chapter 1 that a statement that two quantities are unequal is called an *inequality*. Our study here will be limited to first-degree inequalities.

The symbol for inequality is $\neq$. which is read "is not equal to." If $a \neq b$, then by the trichotomy axiom given in section 1–5, $a > b$ or $a < b$.

Inequalities are solved in very much the same way as equations, with one exception that we will mention shortly. All substitutions from the replacement set for the variable that make the inequality true will be called the *solution set* of the inequality.

An equation (conditional) has a finite solution set, but an inequality often has an infinite solution set. For example, $x < 5$ has an infinite solution set if x is a real number.

Several rules (or theorems) regulate how we work with inequalities. One rule states that the same number may be added to or subtracted from both sides of an inequality without changing the direction or sense of the inequality. For example,

$$7 < 12$$

is a true statement. Suppose 5 is added to or subtracted from each side of the inequality.

Add the same number to each side.	Subtract the same number from each side.
$7 < 12$	$7 < 12$
$7 + 5 \; ? \; 12 + 5$	$7 - 5 \; ? \; 12 - 5$
$12 < 17$	$2 < 7$

The **sense,** or direction, of the inequality remains the same for both operations.

Rule 4–2

Addition and Subtraction Properties of Inequalities

If a, b, and c are real numbers and $a < b$, then

1. $a + c < b + c$
2. $a - c < b - c$

We now take a look at what happens to the sense of an inequality if each side of it is multiplied or divided by the same positive number or by the same negative number.

Suppose we start with $6 < 12$ and first multiply each side by 3 and then divide each side by 3.

Multiply each side by the same positive number.	Divide each side by the same positive number.
$6 < 12$	$6 < 12$
$6(3) \;?\; 12(3)$	$\dfrac{6}{3} \;?\; \dfrac{12}{3}$
$18 < 36$	$2 < 4$

Notice the sense of the inequality did not change if both sides were either multiplied or divided by the same positive number.

Now begin with $6 < 12$; first multiply both sides by -2 and then divide both sides by -2.

Multiply each side by the same negative number.	Divide each side by the same negative number.
$6 < 12$	$6 < 12$
$-2(6) \;?\; -2(12)$	$\dfrac{6}{-2} \;?\; \dfrac{12}{-2}$
$-12 > -24$	$-3 > -6$

The sense of the inequality is reversed when both sides of the inequality are either multiplied or divided by the same negative number.

Rule 4–3 summarizes the ideas for multiplying or dividing an inequality by some nonzero number.

Rule 4–3

Multiplication and Division Properties of Inequalities

1. If a and b are real numbers, c is any positive real number, and $a < b$, then

$$ac < bc \quad \text{and} \quad \frac{a}{c} < \frac{b}{c}$$

First-Degree Equations and Inequalities   155

2. If a and b are real numbers, c is any negative real number, and $a < b$, then

$$ac > bc \quad \text{and} \quad \frac{a}{c} > \frac{b}{c}$$

Rules 4–2 and 4–3 are also true if the symbol $<$ in the statement $a < b$ is replaced with any of the symbols $>$, $\geq$, or $\leq$.

The operations applied to solve an equation are applied in the same way to solve an inequality, with the one exception that the sense of the inequality is reversed if both sides of the inequality are multiplied or divided by the same negative number.

Example 1 Solve $x + 5 < 11$ for x.

Solution
$$x + 5 < 11 \qquad \text{Given inequality.}$$
$$x + 5 + (-5) < 11 + (-5) \qquad \text{Add } -5 \text{ to both sides.}$$
$$x < 6 \qquad \text{Simplify.}$$

The solution set is $\{x|x < 6\}$.
The solution set consists of every real number less than 6. Figure 4–1 shows the graph of the solution set on a number line.

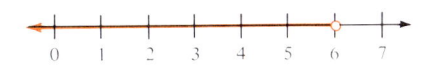

Figure 4–1

The open circle at 6 indicates that 6 is not a member of the solution set. The arrow on the left end of the graph indicates the graph continues in that direction indefinitely.

Example 2 Solve $2r - 6 < 4$ for r and draw the graph of the solution set.

Solution
$$2r - 6 < 4 \qquad \text{Given inequality.}$$
$$2r - 6 + 6 < 4 + 6 \qquad \text{Add 6 to both sides.}$$
$$2r < 10 \qquad \text{Simplify.}$$
$$r < 5 \qquad \text{Divide each side by 2.}$$

The solution set is $\{r|r < 5\}$. The graph of the solution set is shown in figure 4–2.

Figure 4–2

Example 3 Solve $-3x - 4 < x + 4$ for x and draw the graph of the solution set.

Solution

$-3x - 4 < x + 4$	Given inequality.
$-3x - 4 + 4 < x + 4 + 4$	Add 4 to each side.
$-3x < x + 8$	Simplify.
$-3x - x < x + 8 - x$	Subtract x from each side.
$-4x < 8$	Simplify.
$x > -2$	Divide each side by -4.
	Reverse the sense.

The solution set is $\{x \mid x > -2\}$. The graph of the solution set is shown in figure 4–3.

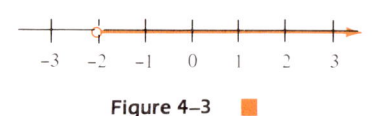

Figure 4–3

Example 4 Solve $7 < 5x - 3 < 12$ for x and draw the graph of the solution set.

Solution This is a compound statement consisting of

$$7 < 5x - 3 \quad \text{and} \quad 5x - 3 < 12.$$

The word *and* indicates we want the intersection of the two solution sets. Graphically, the solution will consist of those parts of the two graphs that overlap. Solving (you supply the reasons)

$7 < 5x - 3$	and	$5x - 3 < 12$
$10 < 5x$	and	$5x < 15$
$2 < x$	and	$x < 3$
$x > 2$	and	$x < 3$

The solution set is $\{x \mid 2 < x < 3\}$. Figure 4–4 shows how the graphs of the two parts of this example provide the graph of the solution set.

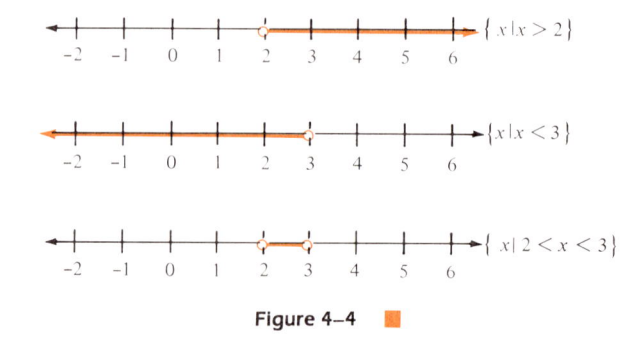

Figure 4–4

First-Degree Equations and Inequalities 157

The inequality shown in example 4 can be solved more quickly if we think of it as one continued inequality made up of three parts. We then solve it for x by applying the same operation to all three parts of the inequality. Example 5 shows how to solve example 4 by this method.

Example 5   Solve $7 < 5x - 3 < 12$ for x by isolating x between the two inequality symbols.

Solution

$$7 < 5x - 3 < 12 \quad \text{Given inequality.}$$
$$7 + 3 < 5x - 3 + 3 < 12 + 3 \quad \text{Add 3 to all three parts of the inequality.}$$
$$10 < 5x < 15 \quad \text{Simplify.}$$
$$2 < x < 3 \quad \text{Divide each part by 5.}$$

The solution is $\{x | 2 < x < 3\}$, which is the same result obtained in example 4.

Example 6   Solve for x in the compound statement $3x - 2 \leq 10$ or $-2x - 5 > -4x + 7$. Draw the graph of the solution set.

Solution   In this problem we want the union of the two solution sets because the statements are connected by the word *or*. We use rules 4–2 and 4–3 to obtain the solution sets.

$$3x - 2 \leq 10 \quad \text{or} \quad -2x - 5 > -4x + 7$$
$$3x \leq 12 \quad \text{or} \quad 2x > 12$$
$$x \leq 4 \quad \text{or} \quad x > 6$$

The solution set is $\{x | x \leq 4 \text{ or } x > 6\}$. Graphically, the solution set is represented by all the points on one or the other or both the graphs (see figure 4–5).

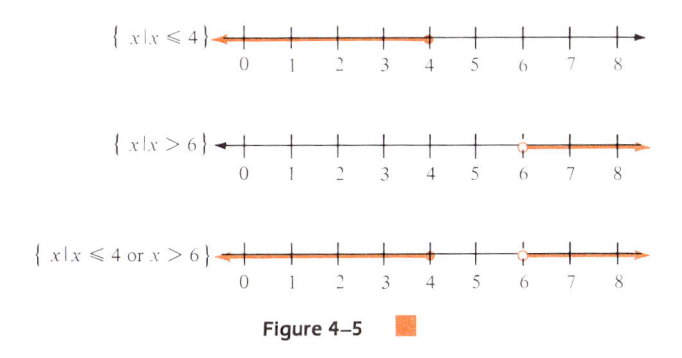

Figure 4–5

Example 7 Write the following as an inequality and solve for the variable: Five times a number, diminished by 2, is less than thirteen.

Solution Let n represent the number. Then

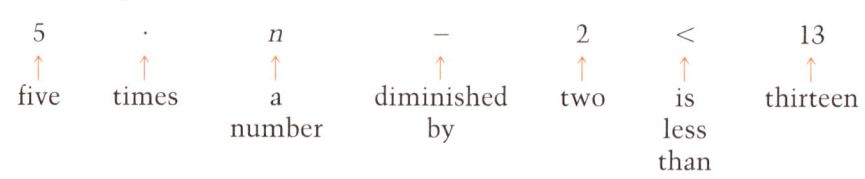

$$\underset{\text{five}}{5} \cdot \underset{\text{times}}{} \underset{\substack{\text{a} \\ \text{number}}}{n} - \underset{\substack{\text{diminished} \\ \text{by}}}{} \underset{\text{two}}{2} \underset{\substack{\text{is} \\ \text{less} \\ \text{than}}}{<} \underset{\text{thirteen}}{13}$$

Thus,

$$5n - 2 < 13$$
$$5n < 15 \quad \text{Add 2 to each side.}$$
$$n < 3 \quad \text{Divide each side by 5.} \blacksquare$$

4–4 Exercises

In exercises 1–34, solve each inequality and draw the graph of the solution set. *(See examples 1–3.)*

1. $t + 5 > 8$
2. $a + 3 < 5$
3. $x - 4 < -2$
4. $x - 3 < -5$
5. $2c + 5 < 9$
6. $-4x + 3 \geq -9$
7. $\dfrac{-m}{3} - 4 \geq -5$ (*Hint:* Multiply each term of inequality by 3 and simplify.)
8. $\dfrac{-n}{5} + \dfrac{2}{3} < \dfrac{4}{5}$ (*Hint:* Multiply each term by 15, the LCM for 3 and 5.)
9. $\dfrac{-x}{2} + \dfrac{3}{4} \leq -x + \dfrac{11}{4}$
10. $\dfrac{-8p}{3} - 1 < \dfrac{2p}{3} + \dfrac{17}{3}$
11. $3x + 4 < 2x - 5$
12. $5x - 4 \geq 3x + 8$
13. $-7x + 3 > -4x + 9$
14. $-8x + 5 < 5x - 21$
15. $8x - 3 + x \leq 2x + 4$
16. $2x - 4 \geq x + 4 - 3x$
17. $8y - 4 + 2y \leq y + 7 - 2y$
18. $6t - 3 > 0$
19. $3x - 5 + 2x < 0$
20. $-m < 2m$
21. $5m > m$
22. $2m > -\tfrac{1}{2}m$
23. $3m < -\tfrac{1}{3}m$
24. $0 - 3p < 2p - 5$
25. $2(y - 3) < 5(-y + 4) + 2$
26. $5(x - 2) + 3 \geq -2(x - 3) + 1$
27. $-3(-x - 3) - 4 \geq 4(x - 2)$
28. $2(-a + 5) < -(a + 3) - 10$
29. $5(d - 2) \leq 2(d - 5)$
30. $-7(x - 4) \leq 3(x - 6) - 4$
31. $5(2y - 3) - 2(3y - 2) < 5y$
32. $-2(s - 3) + 5s > -(s + 2)$
33. $-3p - (2p - 5) \leq 3(p - 2) - p - 3$
34. $3(x - 5) - (x - 3) < 2(x - 1)$

First-Degree Equations and Inequalities

In exercises 35–53, solve each compound inequality and draw the graph of the solution set. (See examples 4–6.)

35. $2d \geq 6$ and $-5d < -20$
36. $2z + 3 > -5$ and $5z < 15$
37. $-3 \leq x + 5 < 3$
38. $6 < 2x + 2 < 12$
39. $9 \leq 3x + 6 \leq 18$
40. $-16 \leq 4x - 4 \leq 20$
41. $25 < 5x - 5 < 35$
42. $7 < 2x - 3 < 11$
43. $9 < 5x + 4 < 19$
44. $-17 \leq 3x + 1 < -11$
45. $2x - 5 > 7$ or $x - 7 < -4$
46. $8x - 1 \leq -9$ or $-x - 5 \geq 2x + 7$
47. $9x + 2(x - 1) < 19$ or $2(x - 3) \geq x - 1$
48. $x + 3(x - 1) \leq 4(x - 4) + x$ or $2x + 1 < 7$
49. $9(a + 5) \geq 2(a + 5)$ or $5a - (a + 2) \leq 3(a - 2)$
50. $8c - 4(c - 5) < 3(c + 8) - 2$ or $4c - 5 > -(c + 10)$
51. $3y < 6$ and $4y > 12$
52. $2x - 3 \geq 7$ and $2x < 10$
53. $2(x - 5) + 3(x + 2) < 8x + 2$ and $4(x + 2) < -12$

In exercises 54–62, write an algebraic inequality and solve for the variable. (See example 7.)

54. Four times a number, diminished by 8, is greater than the number increased by 1.
55. Two diminished by five times a number is less than three times the number diminished by 14.
56. Two times a number, diminished by negative four, is greater than or equal to the square of four.
57. Negative five times y, increased by 8, is greater than the square of three increased by 14.
58. The quotient of a number and 2, increased by 5, is less than three times the number.
59. The quotient of a number and 3 is greater than fifteen.
60. A person can lease a piece of farm machinery for $750 down and $350 a month. On the other hand, the person can buy the equipment for a down payment of $2,400 and a monthly payment of $75 a month. For how many months would leasing be no more expensive than buying?
61. An electronics company has two factories in which it manufactures electronic calculators. In one factory, the labor cost is $3.00 per calculator, and in the other it is $2.50 per calculator. The total labor cost for both factories must be no more than $7,000 a week, and the company must manufacture a total of 2,600 calculators. If both factories must remain in operation, what is the maximum number of calculators that can be manufactured at the factory with the $3.00 labor cost?

62. Julie has tests scores of 83, 91, and 87 on three tests. What is the minimum score that she can make on the fourth test if her average must be 88 or more?

63. Solve: $0.02(x - 0.05) < 3(0.06)$.

64. Solve: $-0.16(x - 1.52) \geq 7.01(x - 0.01)$.

4–5 Absolute Value Equations

The symbol $|x|$ is defined in chapter 1 to be

$$|x| = \begin{cases} x & \text{if } x \geq 0 \\ -x & \text{if } x < 0 \end{cases}$$

If x is replaced by $x - c$ in this definition, we get

$$|x - c| = \begin{cases} x - c & \text{if } x - c \geq 0 \\ -(x - c) & \text{if } x - c < 0 \end{cases}$$

This thought is stated as definition 4–5.

Definition 4–5

Absolute Value of a Difference

$$|x - c| = \begin{cases} x - c & \text{if } x - c \geq 0 \\ -(x - c) & \text{if } x - c < 0 \end{cases}$$

Definition 4–5 gives us a way to find the solutions of absolute value equations.

If $|x - c| = b$, then

$$x - c = b \quad \text{or} \quad -(x - c) = b$$
$$x = b + c \quad \text{or} \quad x - c = -b$$
$$x = b + c \quad \text{or} \quad x = c - b$$

That is, x represents two numbers b units from c. One of these numbers is $c + b$ and the other is $c - b$. Figure 4–6 shows the graph of these two numbers.

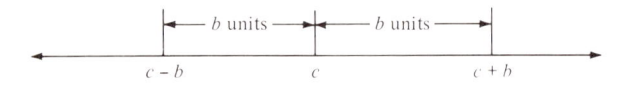

Figure 4–6

In terms of a specific example, if $|x - 2| = 5$, then x represents two numbers 5 units from 2. We know these numbers are 7 and -3 (see figure 4–7).

First-Degree Equations and Inequalities

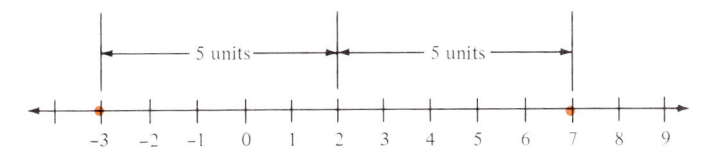

Figure 4–7

If the expression inside the absolute value bars is more complicated than this, the graphical method of solving for the variable becomes cumbersome. Let us solve $|x - 2| = 5$ by using definition 4-5. Our solution must have two steps to correspond to the two parts of the definition. In the first step, we assume $x - 2$ is nonnegative. In the second, we assume $x - 2$ is negative.

1. Assume $x - 2 \geq 0$. Then, $|x - 2| = 5$ means

$$x - 2 = 5$$
$$x = 7$$

2. Assume $x - 2 < 0$. Then, $|x - 2| = 5$ means

$$-(x - 2) = 5 \quad \text{Multiply both sides by } (-1).$$
$$x - 2 = -5 \quad \text{Solve for } x.$$
$$x = -3$$

The solution set is $\{-3, 7\}$. This is the same answer that we obtained by graphing.

Example 1

Solution

Solve $|x + 6| = 1$ for x.

1. Assume $x + 6 \geq 0$. Then,

$$x + 6 = 1$$
$$x = -5$$

2. Assume $x + 6 < 0$.

$$-(x + 6) = 1$$
$$x + 6 = -1 \quad \text{Multiply both sides by } -1.$$
$$x = -7$$

The solution set is $\{-5, -7\}$. ∎

Example 2

Solution

Solve $|2x - 3| = 5$ for x.

1. Assume $2x - 3 \geq 0$.

$$2x - 3 = 5$$
$$2x = 8$$
$$x = 4$$

2. Assume $2x - 3 < 0$.

$$-(2x - 3) = 5$$
$$2x - 3 = -5 \quad \text{Multiply both sides by } (-1).$$
$$2x = -2$$
$$x = -1$$

The solution set is $\{4, -1\}$. ■

Example 3 Solve $|p - 6| = -5$ for p.

Solution This problem has no solution. It is impossible for the left side, which is positive or zero, to equal the right side, which is negative. ■

Example 4 Solve $|2x - 6| = |x - 5|$ for x.

Solution This problem has four cases. However, these can be reduced to just two cases. All four cases are listed so that we may see what they are.

(1) $2x - 6 = x - 5$; assume $2x - 6 \geq 0$ and $x - 5 \geq 0$.
(2) $2x - 6 = -(x - 5)$; assume $2x - 6 \geq 0$ and $x - 5 < 0$.
(3) $-(2x - 6) = x - 5$; assume $2x - 6 < 0$ and $x - 5 \geq 0$.
(4) $-(2x - 6) = -(x - 5)$; assume $2x - 6 < 0$ and $x - 5 < 0$.

Notice (1) and (4) are the same, for if we multiply both sides of the equation in (4) by (-1) it becomes the equation in (1). Also, (2) and (3) are the same, for if we multiply both sides of the equation in (3) by (-1), it becomes (2). Therefore, we see $|2x - 6| = |x - 5|$ becomes

1. $2x - 6 = x - 5$
$2x = x + 1$
$x = 1$

2. $2x - 6 = -(x - 5)$
$2x - 6 = -x + 5$
$3x = 11$
$x = \frac{11}{3}$

The solution set is $\{1, \frac{11}{3}\}$. ■

Example 5 Solve $|a - 3| = |a + 2|$ for a.

Solution 1. $a - 3 = a + 2$
$0 = 5$

This case has no solution because $0 \neq 5$ for any value of a.

First-Degree Equations and Inequalities

2. $a - 3 = -(a + 2)$
$a - 3 = -a - 2$
$2a = 1$
$a = \dfrac{1}{2}$

The solution set is $\{\tfrac{1}{2}\}$. ■

4–5 Exercises

In exercises 1–32, solve each equation. (See examples 1–3.)

1. $|x| = 5$
2. $|y| = 6$
3. $|z| = -7$
4. $|x| = -3$
5. $|x - 4| = 8$
6. $|n + 5| = 9$
7. $|m - 7| = 3$
8. $\left|\dfrac{t - 3}{3}\right| = 4$
9. $\left|\dfrac{2x - 4}{3}\right| = 8$
10. $|3k - 4| = 8$
11. $|3 - 5x| = 7$
12. $|4 + 2x| = 9$
13. $|3 - 6x| = 9$
14. $|2 - 5x| = 13$
15. $|7r - 3| + 5 = 16$
16. $|2s - 4| + 3 = 11$
17. $|3x + 7| - 3 = 7$
18. $|2x + 3| - 6 = 5$
19. $-|u - 3| = -5$
20. $-|v + 5| = -6$
21. $|x - 2| + 5 = 6$
22. $|x + 7| - 6 = -4$
23. $|2y - 3| + 7 = 6$
24. $|-3y + 2| = 7$
25. $|-2y - 5| = 13$
26. $|3 - 4(x + 1)| = 11$
27. $|5c - 3(c - 2)| + 8 = 10$
28. $|3k - 7| - 5 = 0$
29. $|5s - 6| - 4 = 0$
30. $|x| = 0$
31. $2|x| = 0$
32. $-5|x - 4| = 0$

In exercises 33–40, solve each equation. (See examples 3–5.)

33. $|2x - 5| = |x + 3|$
34. $|3x - 4| = |2x + 3|$
35. $|a - 6| = |8 - a|$
36. $|5b + 4| = |8 + 3b|$
37. $|3x - 2| = |x + 6|$
38. $|4x - 1| = |x + 8|$
39. $|2x - 5| = -4$
40. $|3 + 2a| = -9$

41. Show by example that $|ab| = |a||b|$. For example, let $a = 5$ and $b = -6$. Also, choose values for a and b so that both are positive and both are negative.

42. Show by example, similar to the method explained in exercise 41, that $|a - b| = |b - a|$.

43. Show by example that $|a + b| = |a| + |b|$ is not true by finding values for a and b for which it is false.

44. Solve: $|x - 7.19| = 9.16$.
45. Solve: $|2x + 8.76| = 12.82$.
46. Solve: $|5.14x + 7.24| = 28.16$. (Give your answers to the nearest integer.)

4-6 ■ Absolute Value Inequalities

Absolute value inequalities may be solved using the definition of absolute value. For example, $|x| < 5$ can be written in two parts using the definition of absolute value.

$x < 5$ or $-x < 5$
$\qquad\qquad\quad x > -5$ Multiply each side by -1.
$\qquad\qquad -5 < x$ Write as an inequality involving less than.

The two solutions can be combined as the continued inequality $-5 < x < 5$. That is, $|x| < 5$ means $-5 < x < 5$. We say that the solution set for $|x| < 5$ is $\{x| -5 < x < 5\}$.

Rule 4-4

Absolute Value Inequalities Involving Less Than
1. $|x| < k$ means $-k < x < k$, $k > 0$.
2. $|ax - c| < k$ means $-k < ax - c < k$, $k > 0$.

Example 1

Solve $|x - 3| < 7$ and draw the graph of the solution set.

Solution

$|x - 3| < 7$ means

$\qquad -7 < x - 3 < 7$
$-7 + 3 < x - 3 + 3 < 7 + 3$ Add 3 to all three parts of the inequality.
$\qquad -4 < x < 10$ Simplify.

The solution set is $\{x| -4 < x < 10\}$. The graph of the solution set is shown in figure 4-8.

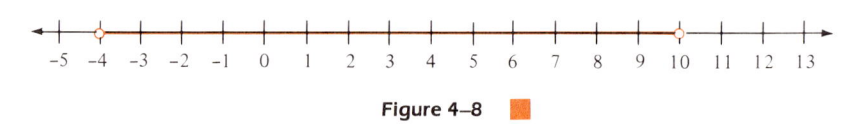

Figure 4-8

Example 2

Solve $|3x - 4| \leq 8$.

Solution

$|3x - 4| \leq 8$ means

$\qquad -8 \leq 3x - 4 \leq 8$
$-8 + 4 \leq 3x - 4 + 4 \leq 8 + 4$ Add 4 to all three parts of the inequality.
$\qquad -4 \leq 3x \leq 12$ Simplify.
$\qquad -\dfrac{4}{3} \leq x \leq 4$ Divide each part of the inequality by 3.

First-Degree Equations and Inequalities

The solution set is $\{x| -\frac{4}{3} \leq x \leq 4\}$. The graph of the solution set is shown in figure 4–9.

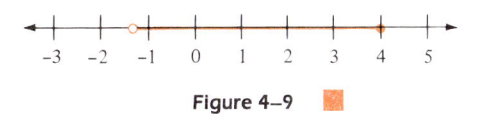

Figure 4–9

Now consider an absolute value inequality of the form $|x| > k$. If we solve using the definition of absolute value, we have

$$x > k \quad \text{or} \quad -x > k$$
$$x < -k \quad \text{Multiply both sides by } -1.$$

These two inequalities cannot be combined into one continued inequality. Thus we see $|x| > k$ means $x < -k$ or $x > k$. This is true in general for inequalities of the form $|ax - c| > k$, which means $ax - c < -k$ or $ax - c > k$.

Rule 4–5

Absolute Value Inequalities Involving Greater Than

1. $|x| > k$ means $x < -k$ or $x > k$, $k > 0$.
2. $|ax - c| > k$ means $ax - c < -k$ or $ax - c > k$, $k > 0$.

Because the inequalities in rule 4–5 are connected by the word *or*, the solution set is the union of the solution sets of the two inequalities.

Example 3 Solve $|x - 3| > 5$ using rule 4–5.

Solution $|x - 3| > 5$ means

$$x - 3 > 5 \quad \text{or} \quad x - 3 < -5$$
$$x > 8 \qquad\qquad x < -2$$

The solution set is $\{x|x > 8 \text{ or } x < -2\}$. Do not make the mistake of writing these two inequalities as $8 < x < -2$. The transitive property for inequalities applied to $8 < x < -2$ gives $8 < -2$, which is false.

Example 4 Solve $|x - 3| > -5$.

Solution This equation is true for all values of x. This fact becomes evident when we realize the left side represents a nonnegative number, which is always greater than -5. The solution set is the entire set of real numbers.

4–6 Exercises

In exercises 1–36, solve each inequality using rules 4–4 and 4–5. (See examples 1–4.)

1. $|x| < 3$
2. $|x| > 5$
3. $|a| > 8$
4. $|b| > 9$
5. $|x| < 10$
6. $|x| < 8$
7. $|s - 3| < 10$
8. $|t + 5| \geq 8$
9. $\left|\dfrac{x}{3} + 4\right| \geq 5$
10. $\left|x - \dfrac{6}{5}\right| \leq 5$
11. $|2u - 3| < 9$
12. $|3v - 1| > 5$
13. $|5x - 1| \leq 9$
14. $|12x - 5| < 19$
15. $\left|\dfrac{7z}{2} + \dfrac{3}{4}\right| > 17$
16. $\left|\dfrac{6q}{5} + \dfrac{1}{2}\right| \leq 13$
17. $|-2x + 5| + 6 \leq 15$
18. $|-3x - 1| + 3 \geq 14$
19. $|-5a - 3| + 6 > 18$
20. $|-6b - 1| - 4 < 7$
21. $7|5x - 4| < 42$
22. $4|3x - 2| < 16$
23. $5|2c - 3| \geq 15$
24. $3|2c + 5| \leq 45$
25. $|2x - 3| > -7$ (See example 4.)
26. $|5x - 3| > -18$
27. $|2(x - 3) + 4| \geq 12$
28. $|-3(x - 4) + 5| \leq 2$
29. $|2(x + 7) - (x + 3)| > 6$
30. $|-4(k - 2) - 7| - 6 > 3$
31. $|7(t + 3) - 3t| < 1$
32. $-2|3x - 7| \leq -10$
33. $-3|5r - 6| \geq -12$
34. $|3 - 2x + 5| < 16$
35. $|9 - 3x - 2| > 19$
36. $|-x - 3| \leq 24$
37. Show that if $|f(x) - L| < \epsilon$, then $L - \epsilon < f(x) < L + \epsilon$.
38. The average length of a rod of steel is 45.3 inches. The tolerance, or error, from the average length can be no more than 0.01 inches. Use absolute value notation to show that a particular length L does not vary more than 0.01 inches from the average length.
39. Megan's height, M, and Jaymie's height, J, differ by no more than 0.5 inches. Use absolute value notation to express this fact.
40. Show by example that $|a + b| \leq |a| + |b|$. Choose values of a and b such that both are positive, both are negative, and one is positive and the other negative.
41. Write the continued inequality $-5 < x < 11$ as an absolute value inequality.
42. Write the continued inequality $-17 \leq x \leq 7$ as an absolute value inequality.
43. Solve: $|5x - 0.1| < 0.65$.
44. Solve: $|2x + 0.08| \geq 7.66$.

First-Degree Equations and Inequalities 167

4–7 Changing the Subject of an Equation

On many occasions in algebra, physics, chemistry, and numerous engineering courses, we must know the value of a particular variable other than the one for which the equation is solved. An equation is solved for a particular variable when that variable is isolated on one side of the equation. For example, the formula for the area of a rectangle is $A = LW$, where A is the area, L is the length, and W is the width. For the rectangle shown in figure 4–10, the area

$$A = LW$$
$$= 8 \cdot 2$$
$$= 16 \text{ square centimeters}$$

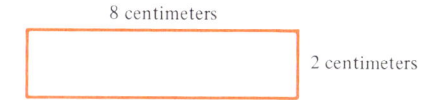

8 centimeters

2 centimeters

Figure 4–10

Now suppose we know the area of a rectangle is 42 square meters and its width is 3 meters (figure 4–11). What is the length?

$L = ?$

42 square meters   3 meters

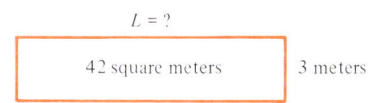

Figure 4–11

We use the equation $A = LW$, solving for the variable whose value we seek. To solve for L, we must divide both sides of $A = LW$ by W.

$$A = LW$$
$$\frac{A}{W} = \frac{LW}{W} \quad \text{Divide both sides by } W.$$
$$\frac{A}{W} = L \quad \text{Simplify.}$$
$$L = \frac{A}{W} \quad \text{Symmetric axiom.}$$

If we now substitute the given values in $L = A/W$, we get

$$L = \frac{42}{3} = 14 \text{ meters}$$

Example 1 Solve $f = ma$ for a.

Solution
$$f = ma$$
$$\frac{f}{m} = \frac{ma}{m} \quad \text{Divide both sides by } m.$$
$$\frac{f}{m} = \frac{\cancel{m}a}{\cancel{m}} \quad \text{Simplify.}$$
$$a = \frac{f}{m}$$

Example 2 Solve $3ax + 2 = 2x + y$ for x.

Solution If the variable for which we are solving appears in two or more terms, we must rearrange the equation, if necessary, so that those terms appear on one side of the equation and the others appear on the other side of the equation.

$$3ax + 2 = 2x + y$$
$$3ax - 2x = y - 2$$
$$x(3a - 2) = y - 2 \quad \text{Distributive axiom.}$$
$$x = \frac{y - 2}{3a - 2} \quad \text{Divide both sides by } 3a - 2.$$

Example 3 If $s = 64$ and $t = 2$, find the value of a by solving $s = \frac{1}{2}at^2$ for a and substituting in the given values.

Solution
$$s = \frac{1}{2}at^2$$
$$2s = at^2$$
$$\frac{2s}{t^2} = \frac{at^2}{t^2} \quad \text{Divide both sides by } t^2 \text{ and simplify.}$$
$$a = \frac{2s}{t^2}$$
$$a = \frac{2(64)}{2^2} = \frac{128}{4} = 32 \quad \text{Substitute 64 and 2 for } s \text{ and } t, \text{ respectively.}$$

4–7 Exercises

The following equations (or formulas) are taken from either physics, chemistry, mathematics, or other related courses. Solve for the indicated variable or constant. (See example 1.)

1. $A = \pi r^2$; π
2. $A = \frac{h(B + b)}{2}$; B
3. $d = rt$; t
4. $c^2 = a^2 + b^2$; a^2

First-Degree Equations and Inequalities

5. $E = \frac{1}{2}mv^2$; m **6.** $V = \frac{KT}{p}$; p

7. $y^2 = 4px$; x **8.** $b^2x^2 + a^2y^2 = a^2b^2$; x^2

In exercises 9–20, solve for x. (See example 2.)

9. $bx + 3 = -ax$ **10.** $ax + bx = a^2 - b^2$

11. $2ax + 3 = 4ax + c$ **12.** $(x - 3)(a + 4) = c$

13. $\frac{2}{x} + \frac{3}{y} = \frac{1}{4}$ **14.** $\frac{6 - 3x}{5} = 2a$

15. $c(c - x) = d(d - x)$ **16.** $8 = \frac{1}{a}x - c$

17. $\frac{5}{3} + \frac{2}{x} = \frac{a}{5x}$ **18.** $ax = bx + a^3 - b^3$

19. $\frac{c - 4x}{b} = d$ **20.** $C = \frac{1}{r} - \frac{R^2x}{h^2}$

In exercises 21–26, first solve for the variable indicated. Then find the value of that variable for the given values of the other variables. (See example 3.)

21. $s = rx$ for x; $s = 50, r = 5$

22. $F = \frac{W}{g}a$ for a; $W = 100, F = 5, g = 32$

23. $F = \frac{mv^2}{r}$ for r; $F = 20{,}000, m = 120{,}000, v = 10$

24. $S = \frac{P}{1 - dt}$ for P; $S = 1000, d = 0.06, t = \frac{1}{2}$

25. $P = \frac{S}{1 + rt}$ for r; $P = 100, r = 0.06; t = 1, s = 106$

26. $P = 2L + 2W$ for L; $P = 60, W = 10$

27. $S = \frac{180(n - 2)}{n}$ for n; $S = 108$

28. $\frac{1}{R} = \frac{1}{R_1} + \frac{1}{R_2}$ for R_1; $R = \frac{20}{3}, R_2 = 20$

Chapter 4   Summary

Vocabulary

conditional equation   identity
equation   open sentence
equivalent equations   replacement set
expression   sense of an inequality
extraneous solution   solution set
fractional equation   statement

Definitions

4–1 Equation

An equation is a statement indicating that two quantities are equal.

4–2 Solution Set

The solution set for an equation consists of those elements from the replacement set of the variable which makes the equation true. If there is no such number in the replacement set, then the solution set is empty.

4–3 Conditional Equation

A conditional equation is an equation that is true only for certain replacements of the variable from the replacement set of the variable.

4–4 Identity

An identity is an equation that is true for all replacements of the variable from the replacement set of the variable.

4–5 Absolute Value of a Difference

$$|x - c| = \begin{cases} x - c & \text{if } x - c \geq 0 \\ -(x - c) & \text{if } x - c < 0 \end{cases}$$

Rules

4–1 Operating on an Equation

Any operation (except multiplying or dividing both sides by zero) applied to one side of an equation must be applied to the other side also.

4–2 Addition and Subtraction Properties of Inequalities

If a, b, and c are real numbers and $a < b$, then:

1. $a + c < b + c$
2. $a - c < b - c$

4–3 Multiplication and Division Properties of Inequalities

1. If a and b are real numbers, c is any positive real number, and $a < b$, then

$$ac < bc \quad \text{and} \quad \frac{a}{c} < \frac{b}{c}$$

2. If a and b are real numbers, c is any negative real number, and $a < b$, then

$$ac > bc \quad \text{and} \quad \frac{a}{c} > \frac{b}{c}$$

4-4 Absolute Value Inequalities Involving Less Than
1. $|x| < k$ means $-k < x < k$, $k > 0$.
2. $|ax - c| < k$ means $-k < ax - c < k$, $k > 0$.

4-5 Absolute Value Inequalities Involving Greater Than
1. $|x| > k$ means $x < -k$ or $x > k$, $k > 0$.
2. $|ax - c| > k$ means $ax - c < -k$ or $ax - c > k$, $k > 0$.

Review Problems

Section 4-1

In exercises 1-4, solve the equations for the variable.

1. $x - 6 = 4$
2. $5x - 3 = 17$
3. $t(t - 4) + t = (t - 1)(t + 2) - 2$
4. $5s - 3 + s = 6(s + 5) - s$
5. Twice a number, increased by 4, is 40. Find the number.

Section 4-2

In exercises 6-9, solve the fractional equations for the variable.

6. $\dfrac{m}{5} + \dfrac{m}{2} = 7$
7. $\dfrac{6n}{3} - \dfrac{3 - 2n}{2} = n + \dfrac{1}{2}$
8. $\dfrac{x + 4}{2} + x = \dfrac{3 + x}{4} + \dfrac{3x - 2}{5} + \dfrac{1}{10}$
9. $\dfrac{x + 3}{x} - \dfrac{x + 1}{x - 1} = \dfrac{2x - 3}{x^2 - x}$

10. John spends half of his weekly salary for food and one third of his weekly salary for gasoline. If these two expenses are $100 per week, what is his weekly salary?

Section 4-3

11. A woman has $2,000 more invested at 9% than she does at 10%. If the annual interest on the 9% investment is $100 more than the annual interest on the 10% investment, how much does she have invested at 9%?

12. One train can travel a distance of 294 miles in the same time another train can travel a distance of 210 miles. If the faster train was traveling at an average rate of 14 miles per hour faster than the slower train, how fast was each train going?

Section 4-4

In exercises 13-16, solve the inequalities for the variable.

13. $u - 3 < 2$
14. $3v - 6 - 8v \leq 9$
15. $4(x - 3) - 3(x + 2) > 2x - 4$

16. $(3x - 1)(x - 2) \geq (3x + 1)(x - 3)$

17. Five times a number, increased by 5, must be at least 70. What are the values the number may have?

Section 4–5

In exercises 18–21, solve the absolute value equations for the variable.

18. $|p + 6| = 5$ **19.** $|3q - 2| = 5$

20. $|5x - 4| = |2x + 11|$ **21.** $|3x + 2| - 8 = 6$

22. If the absolute value of a number is decreased by 2, the result is zero. What values may the number have?

Section 4–6

In exercises 23–26, solve the absolute value inequalities for the variable.

23. $|c + 3| \leq -4$ **24.** $|-3d + 4| > 25$

25. $7|2x + 5| - 14 < 7$ **26.** $|2x + 3| > x + 7$

Section 4–7

In exercises 27–30, solve the equations for the indicated variable or constant.

27. $V = \frac{4}{3}\pi r^3$; π **28.** $\frac{x^2}{a^2} + \frac{y^2}{b^2} = 1$; a^2

29. $m = \frac{y_2 - y_1}{x_2 - x_1}$; x_1 **30.** $\frac{1}{a} = \frac{1}{b} + \frac{1}{c}$; a

Cumulative Test for Chapters 1–4

Chapter 1 For problems 1–3, answer true or false.

1. Every whole number is a rational number.
2. If the intersection of two sets is the empty set, then the two sets have no common elements.
3. A repeating decimal is an irrational number.
4. Simplify: $\dfrac{3 - 6^2 + 11(3)}{[9 - (-2)]^2}$.

Chapter 2
5. Simplify: $\dfrac{-(-3)^2 x^5 y^2}{27 x^4 y^3}$.
6. Simplify: $t^{2n+3} \, t^{5n-4}$.
7. If $P(x) = x^3 + 5x - 8$, find $P(2y)$.
8. Factor: $(7x - 2)(2x) - (7x - 2)(5)$.
9. Factor completely: $2x^2 + 22x + 60$.

Chapter 3
10. Multiply and simplify: $\dfrac{5x^2 + 19x + 12}{x^2 - 9} \cdot \dfrac{x^2 - 2x - 3}{2x^2 + 5x - 3}$.
11. Divide and simplify: $\dfrac{7x^2 - 22x + 3}{2x^2 + 11x - 6} \div \dfrac{7x^2 + 34x - 5}{x^2 + 3x - 18}$.
12. Subtract: $\dfrac{9}{x^2 - 16} - \dfrac{2x}{x^2 + 3x - 10}$.
13. Simplify: $\dfrac{\dfrac{1}{9} - \dfrac{1}{a}}{\dfrac{1}{81} - \dfrac{1}{a^2}}$.
14. Divide: $x - 3 \,\overline{\smash{\big)}\, x^3 - 3x^2 - 5x + 15}$.

Chapter 4
15. Solve for k: $3(k + 6) = 5k + 6$.
16. Solve for x: $(x - 1)(x + 2) - 3 = (x + 2)(x - 4) - 7$.
17. Find the solution and draw the graph of the solution set for $8n - 7 \geq 4n + 5$.
18. Find the solution for $|t - 6| = 7$.
19. Find the solution for $3|2x - 5| \geq 21$.
20. The sum of one-third of a number and one-fourth of the number exceeds one-half of the number by 3. Find the number.
21. The square of the smallest of three consecutive odd integers, increased by the largest integer, is 51 less than the square of the middle integer. What are the three integers?
22. An automobile radiator has a capacity of 15 liters. It contains a water-and-antifreeze solution that is 80% antifreeze. How much of the solution should be drained and replaced with water if the new solution is to be 50% antifreeze?

CHAPTER 5 | Exponents, Roots, and Radicals

Objectives of Chapter 5

To be able to:

☐ Extend the rules for exponents to include rational exponents. (5–1)

☐ Write a number in scientific notation. (5–1)

☐ Use scientific notation in computations. (5–1)

☐ Write a number with a rational exponent as a radical. (5–2)

☐ Give the restrictions on the base when rational exponents are used. (5–2)

☐ Find the principal root of a radical expression. (5–2)

☐ Simplify a radical expression. (5–3)

☐ Add or subtract radical expressions. (5–4)

☐ Multiply radical expressions. (5–4)

☐ Rationalize the denominator of a fraction containing radical expressions. (5–4)

☐ Identify an extraneous root. (5–5)

☐ Solve radical equations. (5–5)

Exponents, Roots, and Radicals

In this chapter, we extend the concept of an exponent to rational numbers and consider exponents such as one-half, one-third, and five-halves, as well as integer exponents. We also examine the relationship between rational exponents and radicals. We briefly review the rules for whole number exponents, which are covered in chapter 2.

5–1 Integral Exponents

In chapter 2, we defined a^n to be $a \cdot a \cdot a \cdots a$ for n factors of a, n a natural number. Also, we have the following rules for whole number exponents.

1. $a^m a^n = a^{m+n}$ for natural numbers m and n.
2. $\dfrac{a^m}{a^n} = \begin{cases} a^{m-n} & \text{for natural numbers } m > n \text{ and } a \neq 0. \\ \dfrac{1}{a^{n-m}} & \text{for natural numbers } n > m \text{ and } a \neq 0. \end{cases}$
3. $a^0 = 1$ for $a \neq 0$.
4. $(a^m)^n = a^{mn}$ for natural numbers m and n.
5. $(a^m b^n)^p = a^{mp} b^{np}$ for natural numbers m, n, and p.
6. $\left(\dfrac{a}{b}\right)^m = \dfrac{a^m}{b^m}$ for any natural number m and $b \neq 0$.

The next rule for exponents was not given in chapter 2.

7. $a^{-n} = \dfrac{1}{a^n}$ for any natural number n and $a \neq 0$

Rule 7 deals with the case in which the exponent is a negative integer. To see the meaning of an expression such as 3^{-2}, assume for the moment that rule 1 is true if m and n are integers rather than natural numbers. Then

$$\begin{aligned} 3^2 \cdot 3^{-2} &= 3^{2+(-2)} \\ &= 3^0 \\ &= 1 \qquad \text{rule 3} \end{aligned}$$

We know $3^2 = 9$. Therefore, 3^{-2} must be $\tfrac{1}{9}$ because the only way the product of two numbers can be 1 is if they are reciprocals. We must conclude then that

$$3^{-2} = \dfrac{1}{9} = \dfrac{1}{3^2}$$

That is, $3^{-2} = 1/3^2$, and in general, $a^{-n} = 1/a^n$ if n is a natural number and $a \neq 0$.

Rules 1–6 are true for whole number exponents. We shall also assume rules 1–7 are valid when the exponents represented by m, n, and p are integers. The next few examples illustrate how the rules are applied when the exponents are integers. In each example, we simplify and leave all exponents positive.

Example 1 Simplify: $h^3 \cdot h^{-4}$

Solution
$$h^3 \cdot h^{-4} = h^{3+(-4)} = h^{-1} = \frac{1}{h}$$

Example 2 Simplify: $(m^{-2}n^4)^{-2}$

Solution
$$(m^{-2}n^4)^{-2} = m^{(-2)(-2)} \cdot n^{4(-2)}$$
$$= m^4 \cdot n^{-8}$$
$$= m^4 \cdot \frac{1}{n^8} = \frac{m^4}{n^8}$$

Example 3 Simplify: $\left(\dfrac{x^{-4}}{y^2}\right)^{-2}$

Solution
$$\left(\frac{x^{-4}}{y^2}\right)^{-2} = \frac{x^{(-4)(-2)}}{y^{2(-2)}}$$
$$= \frac{x^8}{y^{-4}}$$
$$= \frac{x^8}{\frac{1}{y^4}}$$
$$= x^8 \cdot \frac{y^4}{1} = x^8 y^4$$

Example 4 Simplify: $\dfrac{x^{-2}y^4}{z^{-3}}$

Solution
$$\frac{x^{-2}y^4}{z^{-3}} = \frac{\frac{1}{x^2} \cdot y^4}{\frac{1}{z^3}}$$
$$= \frac{\frac{y^4}{x^2}}{\frac{1}{z^3}}$$
$$= \frac{y^4}{x^2} \cdot \frac{z^3}{1} = \frac{y^4 z^3}{x^2}$$

Notice in example 4 that those factors with negative exponents simply "moved" to the other part of the fraction and the exponent became positive in that new position. That is, x^{-2} in the numerator became x^2 in the denominator and z^{-3} in the denominator became z^3 in the numerator. This observation provides us with a shortcut that works if the numerator and denominator consist only of factors. Example 5 uses this shortcut.

Example 5 Simplify: $\dfrac{u^{-6}y^4}{z^2}$

Solution $\dfrac{u^{-6}y^4}{z^2} = \dfrac{y^4}{u^6 z^2}$; u^{-6} moves to the denominator and becomes u^6.

Example 6 Simplify $\left(\dfrac{3^{-4}x^{-2}y^2}{z^{-4}}\right)^{-2}$ using the shortcut.

Solution
$\left(\dfrac{3^{-4}x^{-2}y^2}{z^{-4}}\right)^{-2} = \dfrac{3^8 x^4 y^{-4}}{z^8}$ Raise each factor to the -2 power.

$= \dfrac{3^8 x^4}{y^4 z^8}$ y^{-4} moves to the denominator and becomes y^4.

Example 7 Simplify $\dfrac{x^{-m}}{(x^n)^{-2}}$ using the shortcut.

Solution $\dfrac{x^{-m}}{(x^n)^{-2}} = \dfrac{x^{-m}}{x^{-2n}} = \dfrac{x^{2n}}{x^m} = x^{2n-m}$, or $\dfrac{1}{x^{m-2n}}$

In some areas of study, very large or very small numbers are frequently encountered. For example, the electrical charge on an electron is 0.00000000000000000016 coulombs. The weight of the earth is 6,000,000,000,000,000,000,000,000,000 grams. Both these numbers have too many zeros to be entered in the usual way on a calculator. The calculator must have a scientific notation capability to be able to accomodate numbers with this many digits.

Scientific notation is a way of writing numbers using exponents. A number is written in **scientific notation** if it is written as a product of a number greater than or equal to 1 and less than 10 and some integral power of 10. For example, in scientific notation, 6,300 is 6.3×10^3.

In scientific notation, $0.00000000000000000016 = 1.6 \times 10^{-19}$ and $6,000,000,000,000,000,000,000,000,000 = 6 \times 10^{27}$. The -19 in 1.6×10^{-19} tells us the decimal point must be moved 19 places to the left to return it to its original position.

178   Intermediate Algebra

Example 8 Write the following in scientific notation.
(a) 7,200   (b) 0.000132

Solution
(a) $7,200 = 7.2 \times 10^3$
(b) $0.000132 = 1.32 \times 10^{-4}$ ■

Example 9 Use scientific notation to simplify: $\dfrac{6,400 \times 0.00012}{8,000 \times 40}$

Solution
$$\dfrac{6,400 \times 0.00012}{8,000 \times 40} = \dfrac{6.4 \times 10^3 \times 1.2 \times 10^{-4}}{8 \times 10^3 \times 4 \times 10^1}$$
$$= \dfrac{\overset{0.8}{\cancel{(6.4)}}\overset{0.3}{\cancel{(1.2)}}}{\cancel{(8)}\cancel{(4)}} \times \dfrac{10^{-1}}{10^4}$$
$$= (0.8)(0.3) \times 10^{-5}$$
$$= 0.24 \times 10^{-5}$$
$$= 0.0000024 \; ■$$

5–1 Exercises

In exercises 1–12, write the expressions as a single number or a fraction in lowest terms. (See examples 1–7.)

1. 5^{-3}
2. $3 \cdot 2^{-4}$
3. $8 \cdot 2^{-4}$
4. $\dfrac{5}{2^{-3}}$
5. $\dfrac{6}{(-3)^{-1}}$
6. $\dfrac{4^{-2}}{2^{-1}}$
7. $3^{-1} + 4^{-1}$
8. $\dfrac{2^0}{2^{-3}}$
9. $(2^2)^{-2}$
10. $(4^{-2})^2$
11. $\dfrac{6^{-2}}{6^{-1}} + \dfrac{3^{-4}}{3^{-3}}$
12. $\dfrac{8^{-3}}{8^{-2}} + \dfrac{4^{-3}}{4^{-2}}$

Example
(a) $\dfrac{x^{-4}}{x^2 y^{-2}}$
(b) $\left(\dfrac{x^{-2} y^{-4}}{x^{-3}}\right)^{-2}$

Solution
(a) $\dfrac{x^{-4}}{x^2 y^{-2}} = \dfrac{y^2}{x^2 x^4} = \dfrac{y^2}{x^6}$
(b) $\left(\dfrac{x^{-2} y^{-4}}{x^{-3}}\right)^{-2} = \dfrac{x^4 y^8}{x^6} = \dfrac{y^8}{x^2}$ ■

13. $\dfrac{5^{-3} x^{-4}}{y^{-3}}$
14. $\dfrac{2x^{-7}}{y^4}$
15. $w^{-8} \cdot w^4$
16. $\dfrac{b^{-4} c^5}{b^3}$
17. $(x^{-3} y^4)^2$
18. $(x^{-2} y^{-3})^{-4}$
19. $\dfrac{h^{-4} h^6}{k^{-5} k^4}$
20. $\dfrac{a^{-8} b^{-2}}{c^{-4}}$
21. $\dfrac{x^{-8} y^{-4} z^2}{x^{-2} y^4 z^2}$
22. $\dfrac{x^{-2} y^4 z^{-6}}{xy^{-2}}$
23. $\dfrac{(s^{-3} t^0 z)^{-2}}{(st^4)^3}$
24. $\dfrac{(t^2 u^{-2} v^{-3})^2}{(tu^3)^{-3}}$

Exponents, Roots, and Radicals

25. $\dfrac{3^{-1}x^{-4}y^2}{(3xy^{-3})^{-2}}$ 26. $\left(\dfrac{2r}{3s}\right)^{-1}\left(\dfrac{r^{-3}}{s^2}\right)^{-2}$ 27. $\left(\dfrac{a^{-1}b^{-2}c^3}{a^2bc^2}\right)^{-1}$

28. $\left(\dfrac{a^2c^3b^{-2}}{a^{-3}b^4}\right)^2$ 29. $\left[\left(\dfrac{a^{-2}b^{-1}c^0}{ab^{-2}c^3}\right)^{-1}\right]^2$

30. $\left[\left(\dfrac{ab^{-3}c^4}{a^{-1}b^{-2}c^{-3}}\right)^{-2}\right]^{-3}$

In exercises 31–38, write each as an expression in which each variable appears only once and all exponents are negative.

Example (a) $\dfrac{5^{-2}x^{-3}}{y^4}$ (b) $\left(\dfrac{3a^{-2}b^2}{2ab^{-3}}\right)^{-2}$

Solution (a) $\dfrac{5^{-2}x^{-3}}{y^4} = 5^{-2}y^{-4}x^{-4}$ (b) $\left(\dfrac{3a^{-2}b^2}{2ab^{-3}}\right)^{-2} = \dfrac{3^{-2}a^4b^{-4}}{2^{-2}a^{-2}b^6}$
$= \dfrac{3^{-2}b^{-10}}{2^{-2}a^{-6}}$ ∎

31. $\dfrac{5^2a^{-3}}{b}$ 32. $\dfrac{x^{-3}y^4}{xy^6}$ 33. $(3x^{-2}y^4)^{-2}$

34. $\dfrac{x^{-3}yz^{-4}}{(xyz^5)^{-1}}$ 35. $\left(\dfrac{x^{-3}yz^{-1}}{2x^{-1}z}\right)^{-2}$ 36. $\dfrac{(8a^{-3}b^{-1}c^0)^{-3}}{(5ab^{-2}c^2)^2}$

37. $\dfrac{[(a+b)^{-2}]^{-1}}{(2x^{-3})^{-1}}$ 38. $\dfrac{(c-d)^2}{[(c-d)^3]^{-1}}$

In exercises 39–44, write each so that each variable appears only once and no denominator (except 1) is left in the problem. Exponents may be left with whatever sign is necessary to accomplish this task.

Example $\dfrac{(xy)^{-2}}{x^{-3}}$

Solution $\dfrac{(xy)^{-2}}{x^{-3}} = \dfrac{x^{-2}y^{-2}}{x^{-3}} = xy^{-2}$ ∎

39. $\dfrac{m^{-3}t^2}{m^2t^{-3}}$ 40. $\dfrac{(sv)^{-1}}{s^{-3}v^4}$ 41. $\dfrac{(x^{-2}y^3)^{-2}}{(x^{-4}y^{-2})^2}$

42. $\left(\dfrac{x^3yz^0}{x^{-1}y^4z^5}\right)^3$ 43. $\left[\left(\dfrac{a^3b^{-1}c}{a^{-1}bc^2}\right)^{-1}\right]^{-2}$ 44. $\left[\left(\dfrac{a^{-1}b^2c^{-3}}{abc}\right)^2\right]^{-1}$

In exercises 45–54, write each as a single fraction with positive exponents.

Example $\quad x^{-1} + y^{-2}$

Solution $\quad x^{-1} + y^{-2} = \dfrac{1}{x} + \dfrac{1}{y^2} \quad$ The LCD is xy^2; we rewrite each fraction with this new denominator.

$\qquad\qquad\qquad = \dfrac{y^2}{xy^2} + \dfrac{x}{xy^2}$

$\qquad\qquad\qquad = \dfrac{y^2 + x}{xy^2}$ ■

45. $u^{-2} + v^{-1}$
46. $p^{-1} + q^{-3}$
47. $\dfrac{x^{-1}}{y^{-2}} + \dfrac{y}{x}$
48. $x^{-1} + y$
49. $c^{-2} + cd$
50. $\dfrac{h^{-1} + k^{-1}}{h}$
51. $\dfrac{x^{-3} + y^{-4}}{2xy}$
52. $\dfrac{x^{-2} + y}{2x^{-1}}$
53. $(a^{-2} + b^{-2})^{-1}$
54. $(m + n^{-3})^{-2}$

In exercises 55–60, write each in scientific notation. (See example 8.)

55. 630
56. 9,430
57. 73,000,000
58. 0.00015
59. 0.0000613
60. 973.4

In exercises 61–66, simplify using scientific notation. (See example 9.)

61. $(0.000012)(76,000)$
62. $(80,000)(12,000)$
63. $\dfrac{0.00012 \times 6,300}{4,000 \times 0.00021}$
64. $\dfrac{95 \times 0.00084}{0.00019 \times 4,200}$
65. $\dfrac{7,500 \times 0.000026 \times 18,000}{0.0013 \times 250 \times 90}$
66. $\dfrac{30,000 \times 0.000016 \times 84}{8,000 \times 0.05 \times 0.0084}$

67. The weight of the earth is about 6×10^{27} grams.
    (a) If 1 pound equals 454 grams, what is the weight of the earth in pounds? Write the answer in scientific notation. Round off that part of the answer between 1 and 10 to the nearest hundredth.
    (b) If a kilogram is 10^3 grams, what is the weight of the earth in kilograms?

5–2 Rational Exponents

If N is a number, then b is a **square root** of N if

$$b^2 = N$$

For example, 4 is a square root of 16 because
$$4^2 = 16$$
Also, -4 is a square root of 16 because
$$(-4)^2 = 16$$

If N is a positive real number, then the number represented by N has two square roots, which are opposites of each other.

The symbol $\sqrt{}$ is called a **radical sign.** The number or expression under the radical sign is called the **radicand.** The small number written outside the radical sign is called the **index.** For a square root radical, the index is understood to be 2. For the cube root of N, we have the following.

$$\text{radical sign} \to \overset{\text{index}}{\sqrt[3]{N}} \leftarrow \text{radicand}$$

The positive square root of the positive number N is denoted by $\sqrt{N}$. The negative square root of N is written as $-\sqrt{N}$. The symbol $\sqrt{N}$ is called the **principal square root** of N. For example,
$$\sqrt{25} = 5$$
and 5 is the principal square root of 25. The negative square root of 25 is written as
$$-\sqrt{25} = -5$$
When the negative square root of a positive number is required, a minus sign must be written in front of the radical sign.

A negative real number does not have a square root that is a real number. For example, $\sqrt{-16}$ does not exist as a real number. There is no real number b such that $b^2 = -16$.

A cube root of a number N is indicated by
$$\sqrt[3]{N}$$
In general, c is a **cube root** of N if
$$c^3 = N$$
For example, $\sqrt[3]{8} = 2$ because $2^3 = 8$. Also, $\sqrt[3]{-8} = -2$ because $(-2)^3 = -8$. From this we see that if a real number is positive, its real number cube root is positive. If the real number is negative, its real number cube root is negative.

Let us assume now that rule 1 for exponents (sec. 5–1) holds for rational numbers of the form $1/n$, where n is a natural number. Assuming this rule to be valid, we can calculate
$$9^{1/2} \cdot 9^{1/2} = 9^{(1/2)+(1/2)} = 9^{2/2} = 9^1 = 9$$

Be sure you understand the impact of the line of numbers just given. The number $9^{1/2}$ used twice as a factor is equal to 9. But, according to our discussion at the beginning of this section, the only number with this property is the square root of 9. Therefore, we must conclude $9^{1/2} = \sqrt{9}$, or that $9^{1/2} = 3$, because $\sqrt{9}$ and 3 are the same number. In general, if $a > 0$, then $a^{1/2} = \sqrt{a}$.

Definition 5–1

Square Root

$a^{1/2} = \sqrt{a}$ if $a > 0$.

In a similar manner, $8^{1/3} \cdot 8^{1/3} \cdot 8^{1/3} = 8^{3/3} = 8$. But we also know that $2 \times 2 \times 2 = 8$. From this, we conclude that $8^{1/3} = \sqrt[3]{8} = 2$. In general, if a is any real number, $a^{1/3} = \sqrt[3]{a}$.

Definition 5–2

Cube Root

$a^{1/3} = \sqrt[3]{a}$, where a is any real number.

Definitions 5–1 and 5–2 are rather specific. In general, we will define $a^{1/n} = \sqrt[n]{a}$. This definition seems reasonable if rule 4 (sec. 5–1) is to hold for rational exponents. Assuming rule 4 is valid, we can write $(a^{1/n})^n = a^{n/n} = a$. Thus we see that $a^{1/n}$ is one of n equal factors of a, and we call $a^{1/n}$ the **nth root of a**.

There are some restrictions on the values that a may take in $\sqrt[n]{a}$. If $a < 0$, then $\sqrt[n]{a}$ does not exist as a real number if n is even. For example, $\sqrt{-4}$ and $\sqrt[4]{-16}$ are not real numbers. If n is odd, then $\sqrt[n]{a}$ exists for any real value of a and $n > 2$. For example, $\sqrt[3]{8} = 2$ and $\sqrt[5]{-32} = -2$.

Definition 5–3

nth Root

$a^{1/n} = \sqrt[n]{a}$. If $a > 0$ and n is any even natural number, then two real nth roots of a exist. One is denoted as $\sqrt[n]{a}$ and is called the **principal nth root of a**. The other root is written as $-\sqrt[n]{a}$. If a is any real number and n is an odd natural number greater than 2, then one real nth root of a exists. This root is negative if a is negative and positive if a is positive.

Example 1

Find the indicated root.

(a) $27^{1/3}$ (b) $(-32)^{1/5}$ (c) $\dfrac{9^{1/2}}{-(36)^{1/2}}$

Solution

(a) $27^{1/3} = \sqrt[3]{27} = 3$

Exponents, Roots, and Radicals

(b) $(-32)^{1/5} = \sqrt[5]{-32} = -2$

(c) $\dfrac{9^{1/2}}{-(36)^{1/2}} = \dfrac{\sqrt{9}}{-\sqrt{36}} = \dfrac{3}{-6} = -\dfrac{1}{2}$ ■

The next case for fractional exponents involves fractions of the form m/n, where m/n is any rational number and $n \ne 0$. First, let's take a specific example. Suppose we want to know the value of $8^{2/3}$. If rule 4 (sec. 5–1) is to be valid for fractional exponents, $8^{2/3}$ can be written as $8^{2/3} = (8^{1/3})^2$. Thus $8^{2/3} = (8^{1/3})^2 = (\sqrt[3]{8})^2 = (2)^2 = 4$. We can also work this problem as $8^{2/3} = (8^2)^{1/3} = \sqrt[3]{8^2} = \sqrt[3]{64} = 4$. The two answers are the same. However, the first method gives a smaller radicand (8) than the second method (64). In general, it is easier to use the method that keeps the radicand as small as possible. This discussion is formalized as definition 5–4.

Definition 5–4

Rational Exponent, Nonnegative Base

If m and n are natural numbers and a is any nonnegative real number, then $a^{m/n} = (a^{1/n})^m = (a^m)^{1/n}$.

Example 2   Find $27^{2/3}$.

Solution
$27^{2/3} = (27^{1/3})^2$
$= (\sqrt[3]{27})^2$
$= 3^2$
$= 9$ ■

Example 3   Find $-9^{3/2}$.

Solution
$-9^{3/2} = -(9^{1/2})^3$ The minus sign is not part of the base for the exponent.
$= -(\sqrt{9})^3$
$= -(3)^3$
$= -27$ ■

If a is negative and m and n are both even natural numbers, then what is the meaning of $a^{m/n}$? For example, $(-9)^{2/2} = [(-9)^2]^{1/2} = 81^{1/2} = 9$. If we reduce the exponent in $(-9)^{2/2}$ to $(-9)^1$, we get $(-9)^{2/2} = (-9)^1 = -9$. Surely both answers cannot be correct. We agree under these conditions to write $(-9)^{2/2} = |-9|^{2/2} = |-9|^1 = 9$, which is the principal root.

183

Definition 5–5

Rational Exponent, Negative Base

If a is any negative real number and if m and n are both even natural numbers, then $a^{m/n} = (a^m)^{1/n} = |a|^{m/n}$.

Example 4 Simplify $(-9)^{4/2}$.

Solution By definition 5–5,

$$(-9)^{4/2} = |-9|^{4/2}$$
$$= |-9|^2$$
$$= 9^2$$
$$= 81$$

Rational exponents may be negative, and it seems natural to define them as we do negative integral exponents. Recall that 3^{-2} means $1/3^2 = \frac{1}{9}$. If m and n are natural numbers, then expressions such as $m/(-n)$, $-m/n$, or $-(m/n)$ represent negative numbers. For example,

$$8^{-2/3} = (8^{1/3})^{-2} = \frac{1}{(8^{1/3})^2} = \frac{1}{2^2} = \frac{1}{4}$$

Definition 5–6

Negative Rational Exponents

If $a \neq 0$ and m and n are natural numbers,

$$a^{-m/n} = \frac{1}{a^{m/n}}$$

Example 5 Simplify $(-8)^{-2/3}$.

Solution
$$(-8)^{-2/3} = \frac{1}{(-8)^{2/3}}$$
$$= \frac{1}{(\sqrt[3]{-8})^2}$$
$$= \frac{1}{(-2)^2}$$
$$= \frac{1}{4}$$

Although we shall not do so, we could prove that rules 1, 2, 4, 5, 6, and 7 (sec. 5–1) are true for all real number exponents.

Exponents, Roots, and Radicals

Example 6   Simplify $\dfrac{x^{2/3}}{x^{1/6}}$, $x > 0$.

Solution $\dfrac{x^{2/3}}{x^{1/6}} = x^{(2/3)-(1/6)} = x^{(4/6)-(1/6)} = x^{3/6} = x^{1/2} = \sqrt{x}$

Example 7   Simplify $\left(\dfrac{x^{-1/3}y^{1/4}}{x^2}\right)^{-1/3}$ if $x, y > 0$.

Solution
$$\left(\dfrac{x^{-1/3}y^{1/4}}{x^2}\right)^{-1/3} = \dfrac{x^{1/9}y^{-1/12}}{x^{-2/3}}$$
$$= \dfrac{x^{1/9} \cdot x^{2/3}}{y^{1/12}}$$
$$= \dfrac{x^{1/9} \cdot x^{6/9}}{y^{1/12}}$$
$$= \dfrac{x^{7/9}}{y^{1/12}} = \dfrac{(\sqrt[9]{x})^7}{\sqrt[12]{y}}$$

5–2 Exercises

In exercises 1–24, simplify. (See examples 1–3.)

1. $25^{1/2}$
2. $8^{1/3}$
3. $-36^{1/2}$ (See example 3.)
4. $-81^{1/2}$ (See example 3.)
5. $25^{3/2}$
6. $16^{3/4}$
7. $16^{3/2}$
8. $9^{-1/2}$
9. $16^{-1/2}$
10. $-4^{-1/2}$
11. $\left(\dfrac{1}{4}\right)^{1/2}$
12. $\left(\dfrac{1}{4}\right)^{-1/2}$
13. $-\left(\dfrac{1}{9}\right)^{-1/2}$
14. $\left(\dfrac{4}{9}\right)^{1/2}$
15. $\left(\dfrac{16}{25}\right)^{1/2}$
16. $\left(\dfrac{9}{16}\right)^{-1/2}$
17. $\left(\dfrac{8}{27}\right)^{-2/3}$
18. $8^{2/3} \cdot 8^{-1/3}$
19. $27^{2/3} \cdot 27^{-1/3}$
20. $16^{1/4} \cdot 16^{-3/4}$
21. $-8^{1/2} \cdot 8^{-1/6}$
22. $-4^{5/8} \cdot 4^{7/8}$
23. $9^{9/6} \cdot 9^{1/2}$
24. $8^{2/3} \cdot 8^{-4/3}$

In exercises 25–38, simplify so that each variable appears only once in the problem and with a positive exponent. Write your answer as a radical if the final exponent is not an integer. Assume all variables represent positive real numbers.

Example $\dfrac{(x^{1/5}y^{-1/3})^{1/2}}{(x^{-1/2})^{1/10}}$

Solution

$$\frac{(x^{1/5}y^{-1/3})^{1/2}}{(x^{-1/2})^{1/10}} = \frac{x^{1/10}y^{-1/6}}{x^{-1/20}}$$

$$= \frac{x^{1/10} \cdot x^{1/20}}{y^{1/6}}$$

$$= \frac{x^{(2/20)+(1/20)}}{y^{1/6}}$$

$$= \frac{x^{3/20}}{y^{1/6}}$$

$$= \frac{(\sqrt[20]{x})^3}{\sqrt[6]{y}} \ \blacksquare$$

25. $\dfrac{u^{2/3}}{u^{-1/3}}$

26. $\dfrac{v^{1/5}}{v^{-1/3}}$

27. $\dfrac{x^{1/4}y^{2/3}}{x^{-1/2}y^{1/2}}$

28. $\dfrac{x^{-1/2}y^{-1/3}}{x^{1/4}y^{1/2}}$

29. $\left(\dfrac{a^{1/2}}{b^{1/5}}\right)^{-4}$

30. $\left(\dfrac{h^{-1/5}}{k^{-1/3}}\right)^2$

31. $\dfrac{(x^{1/2}y^{-1/3})^{1/2}}{(x^3y^{1/4})^{1/3}}$

32. $\dfrac{x^{2/3}y^{1/4}}{(x^{-1/3}y^{1/2})^{-1/2}}$

33. $(a^{1/2}b^{-1/3}c^{1/4})^2$

34. $(a^{1/3}b^{-1/2}c^2)^{1/2}$

35. $(a^{1/6}b^{-1/3}c^4)^{-2}$

36. $(a^{-1/2}b^{1/3}c^{-3})^{-1/3}$

37. $\dfrac{(16x^{-1/2}y^{1/4})^{1/2}}{(25x^{1/3}y^{-1/4})^{1/2}}$

38. $\dfrac{(8x^{1/3}y^{1/4})^{1/3}}{(27x^{-1/3}y^{1/2})^{-1/3}}$

In exercises 39–46, use the distributive axiom to write each product as a sum or difference. Leave your answer in exponential form. Assume the variable represents a positive real number.

Example

$x^{1/2}(x^{3/8} + x^{1/2})$

Solution

$x^{1/2}(x^{3/8} + x^{1/2}) = x^{1/2} \cdot x^{3/8} + x^{1/2} \cdot x^{1/2}$
$= x^{(4/8)+(3/8)} + x^{(1/2)+(1/2)}$
$= x^{7/8} + x \ \blacksquare$

39. $a^{1/4}(a^{1/2} + a)$

40. $b^{1/2}(b^{1/3} + b^2)$

41. $x^{-1/2}(x^{-1/3} - x^{1/3})$

42. $x^{4/5}(x^{1/3} - x^{-1/5})$

43. $h^{-1/3}(h^{-2/5} - h^{-1/4})$

44. $m^{1/2}(m^{-4/5} + m^{-3/5})$

45. $x^3(x^{3/2} - x^{-5/2})$

46. $x^{-2}(x^{-5/2} + x^{-7/2})$

In exercises 47–62, find each square. Assume the variables represent positive real numbers. Leave all answers in exponential form. Reduce fractional exponents if possible.

Example

$(x^{1/2} - x^{-1/3})^2$

Exponents, Roots, and Radicals

Solution

$$(x^{1/2} - x^{-1/3})^2 = (x^{1/2})^2 - 2x^{1/2} \cdot x^{-1/3} + (-x^{-1/3})^2$$
$$= x^{2/2} - 2x^{3/6} \cdot x^{-2/6} + x^{-2/3}$$
$$= x - 2x^{1/6} + x^{-2/3} \blacksquare$$

47. $(t^{1/2} - t^{-1/3})^2$
48. $(r^{1/2} + r^{1/3})^2$
49. $(x^{1/3} + x^{1/2})^2$
50. $(x^{1/5} - x^{1/4})^2$
51. $(s^{1/2} - s^{1/4})^2$
52. $(v^3 - v^{1/2})^2$
53. $(x^{1/3} - y^{2/3})^2$
54. $(x^{2/5} - 3y^{1/2})^2$
55. $(2x^{1/7} - 3y^{2/5})^2$
56. $(2s - 3t^{1/2})^2$
57. $(4a^{-1} + 3b^{-2})^2$ (Leave the answer with negative exponents.)
58. $(-2a^{-1/5} + 2b^{-1})^2$ (See exercise 57.)
59. $(2r^{1/4} - 3s^{2/5})^2$
60. $(-7a^{1/3} - 2b^{1/6})^2$
61. $(8a^{1/3}b^2 + 5c^{1/5}d^3)^2$
62. $(5r^{1/5}s^{1/3} - 2u^{1/2}v^{2/7})^2$

In exercises 63–70, simplify. All variable exponents represent positive real numbers, and x, y, and z represent positive real numbers. Leave all exponents positive.

Example

$$\frac{(x^n)^{2/5}}{x^{n/2}}$$

Solution

$$\frac{(x^n)^{2/5}}{x^{n/2}} = \frac{x^{2n/5}}{x^{n/2}} = x^{(2n/5)-(n/2)}$$
$$= x^{(4n/10)-(5n/10)}$$
$$= x^{-n/10}$$
$$= \frac{1}{x^{n/10}} \blacksquare$$

63. $\dfrac{(x^n)^{3/5}}{x^{n/3}}$
64. $\dfrac{(x^{2n})^{1/5}}{(x^n)^2}$
65. $\dfrac{(x^{3/4})^n}{(x^{-1/4})^n}$
66. $\dfrac{(x^{-n})^{1/5}}{(x^{-2n})^{1/3}}$
67. $\left(\dfrac{x^n}{y}\right)^{1/2} \cdot \left(\dfrac{x^{2n}}{y^{1/2}}\right)^2$
68. $\dfrac{(x^2)^{n/3}}{(x^{n+2})^2}$
69. $\left(\dfrac{x^a y^{3a}}{z^{-2a}}\right)^{1/a}$
70. $\left(\dfrac{x^{2a} y^{-4a}}{z^{-8a}}\right)^{1/(2a)}$

5–3 Simplifying Radicals

The previous two sections should have convinced you that radicals and exponents are inseparable. In fact, any number of the form $a^{m/n}$ may be written as $\sqrt[n]{a^m}$.

At times, it is more convenient to work with radicals than it is with rational exponents. We need three rules for the purpose of simplifying radicals.

Rule 5–1

The nth Root of a Product

$\sqrt[n]{ab} = \sqrt[n]{a}\,\sqrt[n]{b}$ for $a, b > 0$ and n a natural number greater than 1.

Proof

Statements	Reasons
1. $\sqrt[n]{ab} = (ab)^{1/n}$	1. Definition 5–3
2. $\quad = a^{1/n}b^{1/n}$	2. Rule 5 (sec. 5–1) extended for rational exponents
3. $\quad = \sqrt[n]{a}\sqrt[n]{b}$	3. Definition 5–3

Example 1 Use rule 5–1 to simplify $\sqrt{50}$.

Solution
$\sqrt{50} = \sqrt{25(2)}$
$\quad\;\; = \sqrt{25}\sqrt{2}$; rule 5–1
$\quad\;\; = 5\sqrt{2}$ ■

Rule 5–2

The nth Root of a Quotient

$\sqrt[n]{\dfrac{a}{b}} = \dfrac{\sqrt[n]{a}}{\sqrt[n]{b}}$ for $a, b > 0$ and n a natural number greater than 1.

Rule 5–2 can be proved in a manner similar to rule 5–1. The proof is left as an exercise.

Example 2 Use rule 5–2 to simplify $\sqrt{\dfrac{15}{9}}$

Solution
$\sqrt{\dfrac{15}{9}} = \dfrac{\sqrt{15}}{\sqrt{9}}$; rule 5–2
$\quad\;\; = \dfrac{\sqrt{15}}{3}$ ■

Exponents, Roots, and Radicals

Rule 5–3

Special Radicals

1. $\sqrt[n]{a^n} = |a|$ if n is even and $a < 0$; $\sqrt[n]{a^n} = a$ if n is even and $a > 0$.
2. $\sqrt[n]{a^n} = a$ if n is odd.

A word of explanation is needed for part 1 of rule 5–3, which states that $\sqrt[n]{a^n} = |a|$ if n is even and $a < 0$. The absolute value bars are needed to insure that we always obtain the principal nth root. If we omitted them, we might write $\sqrt[4]{(-4)^4}$ as -4, which would be wrong because -4 is not the principal root of $\sqrt[4]{(-4)^4}$.

We must proceed cautiously when reducing a rational exponent if we do not know whether the base for the exponent is positive or negative. For example, is $x^{2/4} = x^{1/2}$? If $x > 0$, then $x^{2/4} = x^{1/2}$ is a true statement. For example, if $x = 16$, then

$$x^{2/4} = 16^{2/4} = (\sqrt[4]{16})^2 = 2^2 = 4$$

and

$$x^{1/2} = 16^{1/2} = \sqrt{16} = 4$$

We conclude $x^{2/4} = x^{1/2}$ if $x > 0$. However, if x is negative, we can see $x^{2/4} \neq x^{1/2}$. Let $x = -16$. Then,

$$x^{2/4} = (-16)^{2/4} = \sqrt[4]{(-16)^2} = \sqrt[4]{256} = 4$$

and

$$x^{1/2} = (-16)^{1/2} = \sqrt{-16}$$

Because $\sqrt{-16}$ is not a real number, we conclude $x^{2/4} \neq x^{1/2}$ if $x < 0$.

If $x < 0$, then is $x^{3/9} = x^{1/3}$? Let $x = -512$. Then,

$$x^{3/9} = (-512)^{3/9} = (\sqrt[9]{-512})^3 = (-2)^3 = -8$$
$$x^{1/3} = (-512)^{1/3} = \sqrt[3]{-512} = -8.$$

In this example, we see $x^{3/9} = x^{1/3}$ is true if $x < 0$ because odd roots of negative numbers are real numbers.

In general, an expression such as $a^{cm/cn}$ can always be reduced to $a^{m/n}$ if $a > 0$. However, if $a < 0$, then we must examine the expression $a^{cm/cn}$ for each case because $a^{cm/cn} = a^{m/n}$ may be true for some cases and false for others.

A radical expression is considered to be written in its simplest form if it meets all the conditions given in rule 5–4.

Rule 5–4

Simplifying a Radical

To write a radical in simplified form, write the radical so that:

1. No fraction appears under the radical.
2. No radical appears in the denominator of a fraction.
3. The exponent of each factor in the radicand is less than the index.
4. The index of the radical has no factors common with powers of the factors in the radicand.

Example 3 Simplify $\sqrt{75}$.

Solution $\sqrt{75} = \sqrt{(25)(3)} = \sqrt{25}\sqrt{3} = 5\sqrt{3}$ ■

Example 4 Simplify $\sqrt[3]{54x^4}$.

Solution $\sqrt[3]{54x^4} = \sqrt[3]{27(2)x^3 \cdot x} = \sqrt[3]{27x^3}\sqrt[3]{2x} = 3x\sqrt[3]{2x}$ ■

Example 5 Simplify $\sqrt[4]{16x^4}$.

Solution $\sqrt[4]{16x^4} = \sqrt[4]{16}\sqrt[4]{x^4} = 2|x|$

We must write $\sqrt[4]{x^4}$ as $|x|$ because we do not know if x is positive or negative. ■

Example 6 Simplify $\sqrt{\dfrac{3}{8}}$.

Solution We must eliminate all radicals from the denominator of the given problem. This is called **rationalizing the denominator** of the fraction. Multiply both the numerator and denominator by $\sqrt{2}$ so the denominator can be expressed as the square root of a perfect square.

$$\sqrt{\dfrac{3}{8}} = \dfrac{\sqrt{3}}{\sqrt{8}} = \dfrac{\sqrt{3}}{\sqrt{8}} \cdot \dfrac{\sqrt{2}}{\sqrt{2}} = \dfrac{\sqrt{6}}{\sqrt{16}} = \dfrac{\sqrt{6}}{4}$$ ■

Example 7 Rationalize $\sqrt[3]{\dfrac{3}{x}}$.

Solution In this example, the denominator must be written as the cube root of a perfect cube.

$$\sqrt[3]{\dfrac{3}{x}} = \dfrac{\sqrt[3]{3}}{\sqrt[3]{x}} = \dfrac{\sqrt[3]{3}}{\sqrt[3]{x}} \cdot \dfrac{\sqrt[3]{x^2}}{\sqrt[3]{x^2}} = \dfrac{\sqrt[3]{3x^2}}{\sqrt[3]{x^3}} = \dfrac{\sqrt[3]{3x^2}}{x}$$ ■

Exponents, Roots, and Radicals

Example 8 Simplify $\sqrt{50x^6y^4}$. Assume x and y represent positive numbers.

Solution $\sqrt{50x^6y^4} = \sqrt{25 \cdot 2x^6y^4} = \sqrt{25x^6y^4}\sqrt{2} = 5x^3y^2\sqrt{2}$ ■

Example 9 Use rational number exponents and write $\sqrt{2}\sqrt[3]{3}$ as an expression containing one radical.

Solution
$$\begin{aligned}\sqrt{2}\sqrt[3]{3} &= 2^{1/2} \cdot 3^{1/3} &&\text{Write the given problem in exponential form.}\\ &= 2^{3/6} \cdot 3^{2/6} &&\text{Write the exponents with the LCD of 6.}\\ &= (2^3 \cdot 3^2)^{1/6} &&\text{Rule 5 (sec. 5--1) extended for rational exponents.}\\ &= \sqrt[6]{2^3 \cdot 3^2} &&\text{Write the problem in radical form.}\\ &= \sqrt[6]{72} &&\text{Simplify the radicand.}\end{aligned}$$ ■

Example 10 Write $x^{1/5}y^{3/4}z^{1/2}$ as a single radical expression.

Solution
$$\begin{aligned}x^{1/5}y^{3/4}z^{1/2} &= x^{4/20}y^{15/20}z^{10/20}\\ &= (x^4y^{15}z^{10})^{1/20}\\ &= \sqrt[20]{x^4y^{15}z^{10}}\end{aligned}$$ ■

NOTE: Combining two or more radicals with different indices does not necessarily give a radical in its simplest form. In examples 9 and 10, part (4) of rule 5–4 does not hold.

5–3 Exercises

In exercises 1–14, simplify. *(See examples 1–5.)*

1. $\sqrt{27}$
2. $\sqrt{32}$
3. $\sqrt{72}$
4. $\sqrt[3]{16}$
5. $\sqrt[3]{54}$
6. $\sqrt[4]{32}$
7. $\sqrt{8x^2}$ (See example 5.)
8. $\sqrt{18x^2}$ (See example 5.)
9. $\sqrt{12x^4}$
10. $\sqrt{20x^6}$
11. $\sqrt[3]{-16}$
12. $\sqrt[3]{-54}$
13. $\sqrt[3]{-16p^3}$
14. $\sqrt[3]{8r^4}$

In exercises 15–30, simplify. Assume all variables represent positive real numbers. *(See example 8.)*

15. $\sqrt{16x^3y^5}$
16. $\sqrt{20a^5b^7}$
17. $-\sqrt{32c^4d^5}$
18. $-\sqrt{16x^6y^3}$
19. $-\sqrt[3]{-8x^4y}$
20. $\sqrt[4]{32h^5k^6}$
21. $\sqrt{\dfrac{12}{16}}$
22. $\sqrt{\dfrac{5x^5y^2}{4}}$
23. $\sqrt{\dfrac{4}{9}}\sqrt{9u^5v}$
24. $\sqrt{\dfrac{16}{25}}\sqrt{18w^4z^5}$
25. $\sqrt[3]{x^4y^2}\sqrt[3]{x^5y^2}$
26. $\dfrac{\sqrt{x^3y^2}\sqrt{9x^4y}}{\sqrt{x^4}}$

27. $\dfrac{\sqrt{6a^5b^2}\sqrt{9a^5b^4}}{\sqrt{9a^6}}$

28. $\dfrac{\sqrt[3]{t^3}\sqrt[3]{t^4}}{\sqrt{t^5}}$

29. $\dfrac{\sqrt[3]{x^2y^3}\sqrt[3]{x^4y^5}}{\sqrt[3]{8x^6}}$

30. $\dfrac{\sqrt[4]{16x^3}\sqrt[4]{3x^5}}{\sqrt[4]{x^6}}$

In exercises 31–52, rationalize each denominator. Assume all variables in the radicands represent positive real numbers. (See examples 6 and 7.)

31. $\dfrac{1}{\sqrt{5}}$

32. $\dfrac{5}{\sqrt{2}}$

33. $\dfrac{\sqrt{2}}{\sqrt{6}}$

34. $\dfrac{\sqrt{5}}{\sqrt{3}}$

35. $\dfrac{\sqrt{12}}{\sqrt{3}}$

36. $\dfrac{\sqrt{x}}{\sqrt{y}}$

37. $\dfrac{\sqrt{3p}}{\sqrt{2q}}$

38. $\sqrt{\dfrac{3r}{s}}$

39. $\sqrt{\dfrac{8x}{y^3}}$

40. $\sqrt{\dfrac{3}{2t}}$

41. $\dfrac{\sqrt[4]{3u}}{\sqrt[4]{2u^3}}$

42. $\dfrac{\sqrt[3]{x}}{\sqrt[3]{y}}$

43. $\dfrac{\sqrt{50}}{\sqrt{z^3}}$

44. $\dfrac{5}{4\sqrt{200}}$

45. $\dfrac{\sqrt{9x^2}}{3\sqrt{50}}$

46. $\dfrac{\sqrt[5]{x}\sqrt[5]{xy^2}}{\sqrt[5]{x^3}}$

47. $\dfrac{\sqrt{(a+b)^2}}{\sqrt{a+b}}$

48. $\dfrac{\sqrt{m-n}}{\sqrt{m+n}}$

49. $\dfrac{\sqrt{r+s}}{\sqrt{(r+s)^3}}$

50. $\sqrt[3]{\left(\dfrac{x}{y}\right)^4}$

51. $\sqrt[3]{\left(\dfrac{2x}{3y}\right)^2}$

52. $\sqrt{\left(\dfrac{x^2}{y}\right)^3}$

In exercises 53–60, reduce the index. Assume all variables in the radicands represent positive real numbers.

Example   (a) $\sqrt[6]{x^4}$ (b) $\sqrt[6]{16x^4}$

Solution   (a) $\sqrt[6]{x^4} = x^{4/6}$
$= x^{2/3}$
$= \sqrt[3]{x^2}$

(b) $\sqrt[6]{16x^4} = \sqrt[6]{2^4x^4}$
$= (2^4x^4)^{1/6}$
$= 2^{4/6}x^{4/6}$
$= 2^{2/3}x^{2/3}$
$= (2^2x^2)^{1/3}$
$= \sqrt[3]{4x^2}$

53. $\sqrt[6]{81b^4}$

54. $\sqrt[8]{64c^6}$

55. $\sqrt[10]{x^5}$

56. $\sqrt[8]{x^6y^6}$

57. $\sqrt[4]{h^2k^2}$

58. $\sqrt[4]{16n^2t^2}$

59. $\sqrt[12]{9x^2}$

60. $\sqrt[6]{8x^3}$

In exercises 61–74, write each as a single radical expression. (See examples 9 and 10.)

61. $\sqrt{3}\sqrt[3]{4}$

62. $\sqrt{2}\sqrt[5]{3}$

63. $\sqrt{5}\sqrt[3]{4}$

64. $\sqrt[3]{3}\sqrt{5}$

65. $\sqrt[5]{5}\sqrt{3}$

66. $\sqrt[3]{5}\sqrt{3}\sqrt[6]{2}$

67. $\sqrt[3]{2h}\sqrt{3h}$

68. $\sqrt{x}\sqrt[5]{3x^4}$

69. $a^{3/4}b^{1/2}c^{2/3}$

Exponents, Roots, and Radicals

70. $p^{2/5}q^{3/10}r^{1/2}$ 71. $x^{3/4}y^{1/6}z^{-1/3}$ 72. $a^{1/2}b^{1/3}c^{-1/6}$

73. $\dfrac{\sqrt{3}}{\sqrt[3]{4}}$ 74. $\dfrac{\sqrt[3]{7}}{\sqrt[4]{27}}$

75. According to Einstein's theory of relativity, the mass m of an object moving with a velocity of v is given by

$$m = \frac{m_0}{\sqrt{1 - v^2/c^2}}$$

where m_0 is the mass of the object at rest and c represents the velocity of light. Find m in terms of m_0 if the velocity of the object is $\tfrac{1}{3}$ the speed of light. That is, $v = \tfrac{1}{3}c$.

76. The period T of a simple pendulum is given by

$$T = 2\pi\sqrt{\frac{L}{g}}$$

where L is the length of the pendulum and g is a constant which gives the acceleration due to gravity. Find T if $L = 490$ and $g = 980$.

77. Evaluate and simplify the expression $\sqrt{x + 4}/\sqrt{x^2 + 3}$ for $x = 5$.

78. Evaluate the expression $\sqrt{(x_2 - x_1)^2 + (y_2 - y_1)^2}$ for $x_1 = -3$, $x_2 = 3$, $y_1 = 2$ and $y_2 = -2$.

79. Prove rule 5–2.

5–4 Operations with Radicals

Expressions that contain radicals can often be added or subtracted by using the distributive axiom. However, our work will usually go faster if we simply think of the process as collecting like radicals. Before we define what is meant by radicals being alike, let's take a look at a few examples in which we use the distributive axiom to explain the process.

Example 1   Add: $2\sqrt{3} + 5\sqrt{3}$.

Solution $2\sqrt{3} + 5\sqrt{3} = \sqrt{3}(2 + 5) = \sqrt{3}(7) = 7\sqrt{3}$ ■

Example 2   Add: $5\sqrt{18} + 4\sqrt{50}$.

Solution
$$\begin{aligned}
5\sqrt{18} + 4\sqrt{50} &= 5\sqrt{9}\sqrt{2} + 4\sqrt{25}\sqrt{2} \\
&= 5 \cdot 3\sqrt{2} + 4 \cdot 5\sqrt{2} \\
&= 15\sqrt{2} + 20\sqrt{2} \\
&= \sqrt{2}(15 + 20) \\
&= \sqrt{2}(35) = 35\sqrt{2}
\end{aligned}$$ ■

Example 3   Add: $3\sqrt[3]{5} + 2\sqrt{5}$.

Solution   These two radicals cannot be combined as one term because we cannot apply the distributive axiom to remove a common factor. The difficulty arises because the index of the first radical is 3 and the index of the second radical is 2.

Example 4   Add: $3\sqrt{7} + 2\sqrt{5}$.

Solution   These radicals cannot combine as one term because, as in example 3, we cannot apply the distributive axiom to remove any common factor. The difficulty arises here because the radicands are not the same.

Using examples 3 and 4 as a guide, we can now state under what conditions two or more radicals may be added (or subtracted).

Rule 5–5

Adding or Subtracting Radicals

Two radicals may be added or subtracted to obtain a single term if:

1. The radicands are the same.
2. The indices of both radicals are the same.

The radicals are combined using the distributive axiom.

Example 5   Add: $6x\sqrt{18x^3} + \sqrt{72x^5}$. Assume $x > 0$.

Solution   We first simplify the radicals.

$$\begin{aligned}
6x\sqrt{18x^3} + \sqrt{72x^5} &= 6x\sqrt{9 \cdot 2 \cdot x^2 \cdot x} + \sqrt{36 \cdot 2x^4 x} \\
&= 6x\sqrt{9x^2}\sqrt{2x} + \sqrt{36x^4}\sqrt{2x} \\
&= 6x \cdot 3x \cdot \sqrt{2x} + 6x^2\sqrt{2x} \\
&= 18x^2\sqrt{2x} + 6x^2\sqrt{2x} \\
&= (18 + 6)x^2\sqrt{2x} \\
&= 24x^2\sqrt{2x}
\end{aligned}$$

Example 6   Add: $3\sqrt[3]{54} + 2\sqrt[3]{16}$.

Solution

$$\begin{aligned}
3\sqrt[3]{54} + 2\sqrt[3]{16} &= 3\sqrt[3]{27 \cdot 2} + 2\sqrt[3]{8 \cdot 2} \\
&= 3 \cdot 3\sqrt[3]{2} + 2 \cdot 2\sqrt[3]{2} \\
&= 9\sqrt[3]{2} + 4\sqrt[3]{2} \\
&= (9 + 4)\sqrt[3]{2} = 13\sqrt[3]{2}
\end{aligned}$$

Exponents, Roots, and Radicals

Two radicals are **like radicals** if the radicands and indices of each are the same. Examples 1, 2, 5, and 6 indicate that like radicals can be added or subtracted by adding or subtracting the coefficients of the radicals and then multiplying this result by the common radical factor in each term. For example,

$$3\sqrt{2} + 5\sqrt{2} = (3 + 5)\sqrt{2} = 8\sqrt{2}$$

where $(3+5)$ are the coefficients of the radicals and $\sqrt{2}$ is the common radical factor.

Radical expressions may also be multiplied by recalling $\sqrt[n]{a}\sqrt[n]{b} = \sqrt[n]{ab}$ from rule 5–1. For example,

$$\sqrt{5}(\sqrt{2} + \sqrt{5}) = \sqrt{5}\sqrt{2} + \sqrt{5}\sqrt{5}$$
$$= \sqrt{10} + \sqrt{25}$$
$$= \sqrt{10} + 5$$

Example 7   Multiply: $\sqrt{6}(\sqrt{3} + 5)$.

Solution
$$\sqrt{6}(\sqrt{3} + 5) = \sqrt{6}\sqrt{3} + \sqrt{6} \cdot 5$$
$$= \sqrt{18} + 5\sqrt{6}$$
$$= \sqrt{9 \cdot 2} + 5\sqrt{6}$$
$$= 3\sqrt{2} + 5\sqrt{6}$$

This example does not simplify any further because we cannot combine terms whose radicands are not alike (See rule 5–5). ■

Example 8   Multiply and simplify: $(\sqrt{6} + \sqrt{3})^2$.

Solution
$$(\sqrt{6} + \sqrt{3})^2 = (\sqrt{6})^2 + 2\sqrt{6}\sqrt{3} + (\sqrt{3})^2$$
$$= 6 + 2\sqrt{18} + 9$$
$$= 15 + 2\sqrt{9 \cdot 2}$$
$$= 15 + 2 \cdot 3\sqrt{2}$$
$$= 15 + 6\sqrt{2}$$ ■

Example 9   Multiply and simplify: $(\sqrt{5} - \sqrt{x})^2$. Assume $x > 0$.

Solution
$$(\sqrt{5} - \sqrt{x})^2 = (\sqrt{5})^2 - 2\sqrt{5}\sqrt{x} + (-\sqrt{x})^2$$
$$= 5 - 2\sqrt{5x} + x$$ ■

Example 10   Multiply and simplify: $(\sqrt[3]{x^2} - \sqrt[3]{y^2})(\sqrt[3]{x} + \sqrt[3]{y})$.

Solution

$$(\sqrt[3]{x^2} - \sqrt[3]{y^2})(\sqrt[3]{x} + \sqrt[3]{y}) = \sqrt[3]{x^2} \cdot \sqrt[3]{x} + \sqrt[3]{x^2} \cdot \sqrt[3]{y} - \sqrt[3]{y^2}\sqrt[3]{x} - \sqrt[3]{y^2}\sqrt[3]{y}$$
$$= \sqrt[3]{x^3} + \sqrt[3]{x^2 y} - \sqrt[3]{xy^2} - \sqrt[3]{y^3}$$
$$= x + \sqrt[3]{x^2 y} - \sqrt[3]{xy^2} - y \quad ■$$

In expressions such as $5/(\sqrt{7} - \sqrt{2})$, we sometimes need to write an equivalent expression that contains no radicals in the denominator. Recall that such a process is called *rationalizing the denominator*. To accomplish this, recall expressions such as $a - b$ and $a + b$ produce no middle terms when multiplied. That is,

$$(a - b)(a + b) = a^2 + ab - ab - b^2 = a^2 - b^2$$

The middle terms cancel. Also, the terms which are left are perfect squares. Expressions such as $(a - b)$ and $(a + b)$ are called **conjugates**; $\sqrt{5} + 2$ and $\sqrt{5} - 2$ are specific examples of conjugates. Conjugates can be used to rationalize the denominator of expressions such as $5/(\sqrt{7} - \sqrt{2})$.

Example 11 Rationalize the denominator of $\dfrac{5}{\sqrt{7} - \sqrt{2}}$.

Solution We multiply the numerator and denominator of the given problem by $(\sqrt{7} + \sqrt{2})$, the conjugate of the denominator.

$$\frac{5}{\sqrt{7} - \sqrt{2}} \cdot \frac{\sqrt{7} + \sqrt{2}}{\sqrt{7} + \sqrt{2}} = \frac{5(\sqrt{7} + \sqrt{2})}{\sqrt{7}\sqrt{7} - \sqrt{2}\sqrt{2}}$$
$$= \frac{5(\sqrt{7} + \sqrt{2})}{\sqrt{49} - \sqrt{4}}$$
$$= \frac{5(\sqrt{7} + \sqrt{2})}{7 - 2}$$
$$= \frac{5(\sqrt{7} + \sqrt{2})}{5}$$
$$= \sqrt{7} + \sqrt{2} \quad ■$$

Exercises 5–4

In exercises 1–32, add or subtract the radicals. Simplify all answers. Assume the variables appearing in the radicands represent positive real numbers. (See examples 1–6.)

1. $3\sqrt{5} + 4\sqrt{5}$
2. $8\sqrt{3} - 2\sqrt{3}$
3. $9\sqrt{7} + 6\sqrt{7} - 2\sqrt{7}$
4. $18\sqrt{6} - 2\sqrt{6} - 9\sqrt{6}$
5. $-9\sqrt{3} - 8\sqrt{3} + 2\sqrt{3}$
6. $9\sqrt{9} - 2\sqrt{3} - 4\sqrt{3}$
7. $8\sqrt{16} - 3\sqrt{7} + 4\sqrt{7}$
8. $3\sqrt{4} + 5\sqrt{9} - 2\sqrt{7}$
9. $3\sqrt{18} - 5\sqrt{2}$
10. $16\sqrt{8} + 8\sqrt{2}$
11. $9\sqrt{18} - 7\sqrt{32} + 4\sqrt{50}$
12. $3\sqrt{40} - 2\sqrt{90}$

13. $2\sqrt{5} - 3\sqrt{45} + 4\sqrt{80}$
14. $5\sqrt{20} - 8\sqrt{45} - \sqrt{5}$
15. $4\sqrt[3]{54} - 3\sqrt[3]{16} + \sqrt[3]{2}$
16. $7\sqrt[3]{128} - \sqrt[3]{16}$
17. $8\sqrt[4]{32} - 10\sqrt[4]{162}$
18. $\sqrt[4]{48} - 5\sqrt[4]{3}$
19. $3\sqrt[5]{64} - 2\sqrt[5]{2}$
20. $\sqrt[5]{2x} - 9\sqrt[5]{64x}$
21. $9\sqrt{4x^3} + 5x\sqrt{x}$
22. $3\sqrt{300c^2d} - c\sqrt{27d}$
23. $3\sqrt{m^4n^3} - 5n\sqrt{m^4n}$
24. $7\sqrt{x^7y} - 8x^3\sqrt{xy}$
25. $8\sqrt[3]{27x^7} - 9x^2\sqrt[3]{x}$
26. $5\sqrt[3]{t} - 4\sqrt[3]{t} - 2\sqrt{t}$
27. $9\sqrt[4]{z^3} - 2\sqrt[4]{z^3} - 5\sqrt{z}$
28. $6\sqrt[5]{x^2} - 8\sqrt[5]{x} + 7\sqrt[5]{x^2}$
29. $\sqrt{\dfrac{3}{4}} + \sqrt{\dfrac{1}{12}}$
30. $\sqrt{\dfrac{5}{9}} + \sqrt{\dfrac{1}{20}}$
31. $\sqrt{\dfrac{7}{16}} - \sqrt{\dfrac{1}{63}}$
32. $\sqrt{\dfrac{3}{16}} - \sqrt{\dfrac{1}{27}}$

In exercises 33–56, multiply and simplify where possible. Assume all variables in the radicands represent positive real numbers. (See examples 7–10.)

33. $5(\sqrt{2} - 3)$
34. $9(\sqrt{5} - 4)$
35. $2(\sqrt{3} + \sqrt{7})$
36. $8(\sqrt{5} + \sqrt{7})$
37. $\sqrt{2}(\sqrt{3} + \sqrt{5})$
38. $\sqrt{5}(\sqrt{3} - \sqrt{7})$
39. $\sqrt{2}(\sqrt{6} + \sqrt{10})$
40. $\sqrt{3}(\sqrt{6} - \sqrt{7})$
41. $\sqrt[3]{3}(\sqrt[3]{9} + \sqrt[3]{5})$
42. $\sqrt[3]{4}(\sqrt[3]{2} - \sqrt[3]{16})$
43. $(\sqrt{3} - \sqrt{5})(\sqrt{3} + \sqrt{5})$
44. $(\sqrt{3} + \sqrt{7})(\sqrt{3} - \sqrt{7})$
45. $(5\sqrt{3} - \sqrt{7})(5\sqrt{3} + \sqrt{7})$
46. $(3 - 2\sqrt{6})(3 + 2\sqrt{6})$
47. $(x - \sqrt{y})(x + \sqrt{y})$
48. $(2\sqrt{x} - 3\sqrt{y})(2\sqrt{x} + 3\sqrt{y})$
49. $(\sqrt{3} - 2)^2$
50. $(\sqrt{7} + 5)^2$
51. $(\sqrt{7} + \sqrt{2})^2$
52. $(\sqrt{2} - \sqrt{7})^2$
53. $(2 - \sqrt[4]{5^2})(2 + \sqrt[4]{5^2})$
54. $(3 + \sqrt[6]{a^3})(3 - \sqrt[6]{a^3})$
55. $(3\sqrt[3]{b^2} - 2\sqrt[3]{c^2})(\sqrt[3]{b} + \sqrt[3]{c})$
56. $(\sqrt[3]{x} - \sqrt[3]{y})(\sqrt[3]{x^2} + \sqrt[3]{y^2})$

In exercises 57–62, reduce to lowest terms. Assume variables in the radicand represent positive real numbers.

Example

(a) $\dfrac{6 + 3\sqrt{5}}{3}$

(b) $\dfrac{2x - \sqrt{x^3}}{3x}$

Solution

(a) $\dfrac{6 + 3\sqrt{5}}{3} = \dfrac{3(2 + \sqrt{5})}{3}$
$= 2 + \sqrt{5}$

(b) $\dfrac{2x - \sqrt{x^3}}{3x} = \dfrac{2x - x\sqrt{x}}{3x}$
$= \dfrac{x(2 - \sqrt{x})}{3x}$
$= \dfrac{2 - \sqrt{x}}{3}$

57. $\dfrac{8 - 2\sqrt{5}}{4}$ 58. $\dfrac{9 + 3\sqrt{5}}{3}$ 59. $\dfrac{2u - 3\sqrt{u^3}}{3u}$

60. $\dfrac{2x^2y - 2\sqrt{x^5y}}{2x^2}$ 61. $\dfrac{3x^2y - \sqrt[3]{x^7y^3}}{x^2y}$ 62. $\dfrac{\sqrt{v} - 2\sqrt{v^3}}{3\sqrt{v}}$

In exercises 63–71, rationalize the denominator. Assume variables in the radicands represent positive real numbers. (See example 11.)

63. $\dfrac{3}{\sqrt{2} - 3}$ 64. $\dfrac{9}{\sqrt{8} - 2}$ 65. $\dfrac{\sqrt{3}}{\sqrt{2} - 4}$

66. $\dfrac{\sqrt{5}}{\sqrt{7} - 5}$ 67. $\dfrac{\sqrt{3} - \sqrt{2}}{\sqrt{5} + \sqrt{6}}$ 68. $\dfrac{\sqrt{13} + \sqrt{3}}{\sqrt{5} - \sqrt{2}}$

69. $\dfrac{5}{2 - \sqrt{y}}$ 70. $\dfrac{9}{\sqrt{x} - 3}$ 71. $\dfrac{\sqrt{2} - \sqrt{3}}{\sqrt{y} - \sqrt{x}}$

72. The escape velocity, v_e, for a satellite orbiting the earth is $v_e = \sqrt{2}v_0$, where v_0 is the orbital velocity. What is the escape velocity for an earth satellite if its orbital velocity is 17,500 miles per hour? Use 1.414 for $\sqrt{2}$.

5–5 Introduction to Radical Equations.

Radical equations are equations that contain a variable expression under one or more radical signs. The general procedure, if the radicals in the equation are of order 2, is to square both sides of the equation. If the radical equation contains radicals of order 3, then we cube both sides of the equation, and so forth.

When both sides of the equation are raised to some power for the purpose of eliminating the radicals, then every solution of the original equation is a solution of the new equation. But we may get more than we asked for. Not only will the solution set of the new equation contain the solution set of the original equation, but it may contain numbers that are not in the solution set of the original equation. Recall that such numbers are called *extraneous roots*. A necessary part of solving any radical equation, then, consists of checking each solution in the original equation. Any extraneous solution must be discarded.

Example 1 Solve $\sqrt{x - 3} = 4$ for x.

Solution Squaring both sides we get,

$$(\sqrt{x - 3})^2 = 4^2$$
$$x - 3 = 16$$
$$x = 19$$

Exponents, Roots, and Radicals

The solution set of the squared equation is {19}. Check this in the original equation.

Check: $\sqrt{19 - 3} = 4$

$\sqrt{16} = 4$

$4 = 4$

The solution set of the original equation is {19}.

Example 2 Solve $\sqrt{3x + 4} = 5$ for x.

Solution
$(\sqrt{3x + 4})^2 = 5^2$

$3x + 4 = 25$

$3x = 21$

$x = 7$

We must now check to see that 7 is the solution of the original equation.

Check: $\sqrt{3 \cdot 7 + 4} = 5$

$\sqrt{21 + 4} = 5$

$\sqrt{25} = 5$

$5 = 5$

The solution set for the original equation is {7}.

Example 3 Solve $\sqrt{3x - 1} + \sqrt{2x + 3} = 0$ for x.

Solution First, isolate one of the radicals on one side of the equation.

$\sqrt{3x - 1} = -\sqrt{2x + 3}$

$(\sqrt{3x - 1})^2 = (-\sqrt{2x + 3})^2$

$3x - 1 = 2x + 3$

$x = 4$

The solution set is {4} for the squared equation.

Check: $\sqrt{3 \cdot 4 - 1} + \sqrt{2 \cdot 4 + 3} = 0$

$\sqrt{11} + \sqrt{11} = 0$

$\sqrt{11} + \sqrt{11} \neq 0.$

Hence 4 is not a solution of the original equation. The solution set is $\emptyset$.

Example 4 Solve $\sqrt{x + 3} = 1 + \sqrt{x - 1}$ for x.

Solution

$$(\sqrt{x+3})^2 = (1 + \sqrt{x-1})^2$$
$$x + 3 = 1 + 2\sqrt{x-1} + (x-1)$$
$$x + 3 = x + 2\sqrt{x-1}$$
$$3 = 2\sqrt{x-1}$$
$$9 = 4(x-1) \quad \text{Square again.}$$
$$9 = 4x - 4$$
$$13 = 4x$$
$$x = \frac{13}{4}$$

The solution set is $\{\frac{13}{4}\}$ for the squared equation.

Check: $\sqrt{\frac{13}{4} + 3} = 1 + \sqrt{\frac{13}{4} - 1}$

$$\sqrt{\frac{13}{4} + \frac{12}{4}} = 1 + \sqrt{\frac{13}{4} - \frac{4}{4}}$$

$$\sqrt{\frac{25}{4}} = 1 + \sqrt{\frac{9}{4}}$$

$$\frac{5}{2} = 1 + \frac{3}{2}$$

$$\frac{5}{2} = \frac{5}{2}$$

The solution set is $\{\frac{13}{4}\}$.

Example 5 Solve $\sqrt[3]{x-3} = 4$ for x.

Solution
$$\sqrt[3]{x-3} = 4$$
$$(\sqrt[3]{x-3})^3 = 4^3 \quad \text{Cube both sides of the equation.}$$
$$x - 3 = 64$$
$$x = 67$$

The solution set is $\{67\}$ for the cubed equation. The usual check shows that 67 is a solution to the original equation.

5–5 Exercises

In exercises 1–40, solve for the variable. In every case, check your solution in the original equation. *(See examples 1–5.)*

1. $\sqrt{x - 4} = 5$
2. $\sqrt{x + 5} = 4$
3. $\sqrt{2m - 3} = 7$
4. $\sqrt{3n + 2} = 6$

Exponents, Roots, and Radicals

5. $\sqrt{3x-4} - 2 = 0$ (*Hint:* Isolate the radical on the left side of the equation.)

6. $\sqrt{5x-3} - 4 = 0$
7. $5\sqrt{x-3} = 10$
8. $8\sqrt{2x+3} = 24$
9. $3\sqrt{3x-4} = 9$
10. $2\sqrt{x-3} = 5$
11. $-2\sqrt{x+3} = 4$
12. $-\sqrt{2x+7} = 3$
13. $\sqrt{3a+1} + 2 = 5$
14. $\sqrt{7c-1} + 2 = 9$
15. $3 + \sqrt{6x-5} = 7$
16. $\sqrt[3]{a-1} = 4$
17. $\sqrt[3]{5k-2} = -3$
18. $\sqrt[4]{3t+4} = 2$
19. $\sqrt[4]{2x-2} = 2$
20. $\sqrt[5]{-a-8} = 2$
21. $\sqrt{2a+4} = -4$
22. $\sqrt{3b-3} = -3$
23. $\sqrt{2x+2} = \sqrt{3x}$
24. $\sqrt{9x-3} = \sqrt{x+5}$
25. $\sqrt{3c-3} - \sqrt{2c+5} = 0$
26. $\sqrt{5d+4} - \sqrt{2d+7} = 0$
27. $\sqrt{2x-3} + \sqrt{x+1} = 0$
28. $\sqrt{1-x} + \sqrt{3x-3} = 0$
29. $h + 3 = \sqrt{h^2 + 2h - 3}$
30. $k - 2 = \sqrt{k^2 - 2k + 12}$
31. $2x - 1 = \sqrt{4x^2 - 2x + 3}$
32. $2x = 3 + \sqrt{4x^2 + 5x - 8}$
33. $\sqrt[3]{p-3} = 2$
34. $\sqrt[4]{q+5} = 2$
35. $\sqrt[5]{x-3} = -2$
36. $\sqrt[3]{2x+2} = -2$
37. $\sqrt{t+1} = 1 - \sqrt{t+1}$
38. $\sqrt{r+2} = 3 - \sqrt{r-1}$
39. $\sqrt{x+1} = 2 - \sqrt{x+1}$
40. $\sqrt{x+2} + 1 = \sqrt{x-3}$

41. In Exercise 76 of section 5–3, the equation for the period (T) of a simple pendulum was given as $T = 2\pi\sqrt{L/g}$. Find the length (L) in feet that will give a period of 2 seconds. Use 32 as the value for g and 3.14 as the value of π. Give the answer correct to the nearest hundredth.

42. The velocity of a wave in shallow water is given by $V = \sqrt{gh}$, where h is the depth of the water and g is the acceleration due to gravity. Solve this equation for h.

43. What is the velocity (in feet per second) of a water wave if g = 32 and h = 3. Use the equation given in exercise 42. Give the answer correct to the nearest tenth.

44. The equation $V = \sqrt{Y/d}$ gives the velocity of a longitudinal compression wave along certain materials, such as a steel rod or aluminum rod. In the equation, Y represents a number known as Young's modulus and d represents the density of the material. Find V (in centimeters per second) if $Y = 1.9 \times 10^{12}$ and d = 7.5. Leave the answer in scientific notation. Give that part of the answer that occurs between 1 and 10 to the nearest whole number.

Chapter 5 Summary

Vocabulary

conjugate
cube root
index
like radicals
nth root
principal nth root

radical sign
radical equation
rationalizing a denominator
radicand
scientific notation
square root

Definitions

5–1 Square Root

$a^{1/2} = \sqrt{a}$ if $a > 0$.

5–2 Cube Root

$a^{1/3} = \sqrt[3]{a}$, where a is any real number.

5–3 nth Root

$a^{1/n} = \sqrt[n]{a}$. If $a > 0$ and n is any even natural number, then two real nth roots of a exist. One is denoted as $\sqrt[n]{a}$ and is called the **principal nth root of a.** The other root is written as $-\sqrt[n]{a}$. If a is any real number and n is an odd natural number, then one real nth root of a exists. This root is negative if a is negative and positive if a is positive.

5–4 Rational Exponent, Nonnegative Base

If m and n are natural numbers and a is any nonnegative real number, then $a^{m/n} = (a^{1/n})^m = (a^m)^{1/n}$.

5–5 Rational Exponent, Negative Base

If a is any negative real number and if m and n are both even natural numbers, then $a^{m/n} = (a^m)^{1/n} = |a|^{m/n}$.

5–6 Negative Rational Exponents

If $a \neq 0$ and m and n are natural numbers,

$$a^{-m/n} = \frac{1}{a^{m/n}}$$

Scientific Notation

A number is written in scientific notation if it is written as a product of a number greater than or equal to 1 and less than 10 and some integral power of 10.

Rules

Rules for Rational Exponents

1. $a^m a^n = a^{m+n}$ for rational numbers m and n.
2. $\dfrac{a^m}{a^n} = a^{m-n}$ for rational numbers m and n ($a \neq 0$).
3. $a^0 = 1$ for $a \neq 0$.
4. $(a^m)^n = a^{mn}$ for rational numbers m and n.
5. $(a^m b^n)^p = a^{mp} b^{np}$ for rational numbers m, n, and p.
6. $\left(\dfrac{a}{b}\right)^m = \dfrac{a^m}{b^m}$ for rational number m and $b \neq 0$.
7. $a^{-n} = \dfrac{1}{a^n}$ for rational number n and $a \neq 0$.

Rules for Radicals

5–1 The nth Root of a Product

$\sqrt[n]{ab} = \sqrt[n]{a}\sqrt[n]{b}$ for $a, b > 0$ and n a natural number greater than 1.

5–2 The nth Root of a Quotient

$\sqrt[n]{\dfrac{a}{b}} = \dfrac{\sqrt[n]{a}}{\sqrt[n]{b}}$ for $a, b > 0$ and n a natural number greater than 1.

5–3 Special Radicals

1. $\sqrt[n]{a^n} = |a|$ if n is even and $a < 0$; $\sqrt[n]{a^n} = a$ if n is even and $a > 0$.
2. $\sqrt[n]{a^n} = a$ if n is odd.

5–4 Simplifying a Radical

To write a radical in simplified form, write the radical so that:

1. No fraction appears under the radical.
2. No radical appears in the denominator of a fraction.
3. The exponent of each factor in the radicand is less than the index.
4. The index of the radical has no factors common with the powers of the factors in the radicand.

5–5 Adding or Subtracting Radicals

Two radicals may be added or subtracted to obtain a single term if:

1. The radicands are the same.
2. The indices of both radicals are the same.

The radicals are combined using the distributive axiom.

Review Problems

Section 5–1

Simplify.

1. 3^{-3}
2. $8 \cdot 3^{-2}$
3. $\dfrac{3}{4^{-2}}$
4. $\dfrac{5^{-1}}{2^{-4}}$

Simplify by writing each exercise as an expression in which each variable appears only once and all exponents are positive.

5. $\dfrac{m^{-4}n^2}{m^{-3}}$
6. $(p^{-3}q^2)^4$
7. $\dfrac{x^{-8}y^2}{z^{-2}}$
8. $\left(\dfrac{x^2y^2z^{-3}}{x^{-4}y^{-3}z^0}\right)^{-2}$

Write each as an expression in which each variable appears only once and all exponents are negative.

9. $\dfrac{5^3 r^{-3}}{s^2}$
10. $\left(\dfrac{3a^{-2}y^{-4}z^{-2}}{a^{-1}y^2 z^3}\right)^3$

Write each as a single fraction with positive exponents.

11. $x^{-3} + y^{-2}$
12. $\dfrac{x^4 + y^{-2}}{x}$

Use scientific notation to simplify.

13. $\dfrac{0.00016 \times 36{,}000}{0.00012 \times 4{,}000}$

Section 5–2

Simplify the following.

14. $36^{1/2}$
15. $49^{-1/2}$
16. $\left(\dfrac{1}{16}\right)^{-1/2}$

Simplify so that each variable appears only once in the problem and with a positive exponent. Write your answer as a radical if the final exponent is not an integer. All variables represent positive real numbers.

17. $\dfrac{t^{5/8}}{t^{-1/4}}$
18. $\left(\dfrac{w^{1/2}}{z^{2/3}}\right)^{-1/2}$
19. $\left(\dfrac{4x^{-2/3}y^{1/2}z^{1/5}}{9x^{-1/2}y^{2/3}z^{1/10}}\right)^{-1/2}$

Use the distributive axiom to write each product as a sum or difference. Leave your answer in exponential form. All variables represent positive real numbers.

20. $x^{-1/2}(x^{3/4} + x^{1/2})$
21. $k^{2/3}(k^{-1/2} - k^{3/7})$

Exponents, Roots, and Radicals

Simplify the following. All variable exponents represent positive real numbers. Assume x represents a positive real number. Leave all exponents positive.

22. $\dfrac{(x^{2n})^{3/4}}{x^{n/2}}$

23. $\dfrac{(x^{-3n})^{1/2}}{(x^{-n})^{1/3}}$

Section 5–3

Simplify each radical. All variables in the radicands represent positive real numbers.

24. $\sqrt{20}$

25. $\sqrt{16n^5 t^3}$

26. $\sqrt[3]{54p^4 s^5}$

27. $\dfrac{\sqrt[4]{16x^5}\,\sqrt[4]{3x^7}}{\sqrt[4]{x^8}}$

28. $\dfrac{\sqrt{5}}{\sqrt{2}}$

29. $\dfrac{\sqrt{5h}}{\sqrt{2k}}$

30. $\sqrt[3]{\dfrac{3}{2p}}$

Reduce the index. Assume all variables in the radicands represent positive real numbers.

31. $\sqrt[8]{x^4}$

32. $\sqrt[4]{x^2 y^2}$

33. $\sqrt[10]{9a^2}$

Section 5–4

Add or subtract. All answers must be simplified. Assume all variables appearing in the radicands represent positive real numbers.

34. $3\sqrt{18} + 2\sqrt{50}$

35. $5\sqrt{27} - 2\sqrt{75}$

36. $3x\sqrt{27y} - \sqrt{3x^2 y}$

Multiply and simplify.

37. $\sqrt{3}(\sqrt{6} - \sqrt{3})$

38. $(\sqrt{3} - \sqrt{6})^2$

39. $(\sqrt{5} - 4)^2$

Rationalize the denominators. All variables in the radicands represent positive real numbers.

40. $\dfrac{5}{\sqrt{7} - \sqrt{2}}$

41. $\dfrac{\sqrt{3} - \sqrt{2}}{\sqrt{7} + \sqrt{2}}$

42. $\dfrac{\sqrt{3} - \sqrt{x}}{\sqrt{x} + \sqrt{y}}$

Multiply the radicals. Assume all variables in the radicands represent positive real numbers.

43. $\sqrt[3]{2}\,\sqrt[6]{3}$

44. $\sqrt{2x}\,\sqrt[3]{xy}$

45. $\sqrt[3]{b^2 c}\,\sqrt[4]{b^3 c^2}$

Section 5–5

Solve for x and check for extraneous roots.

46. $\sqrt{z - 8} = 25$

47. $\sqrt[3]{2x - 4} = 6$

48. $\sqrt{2x - 3} = \sqrt{x + 2}$

49. $\sqrt{r + 5} + \sqrt{2r + 1} = 0$

50. $\sqrt{t + 1} - 1 = \sqrt{t + 1}$

51. The equation $c = \sqrt{E/m}$ is a well-known equation found in any physics course. Solve this equation for E to see it in its more popular form.

52. Solve the equation $x = \sqrt{y/z}$ for z and then find the value of z if $y = 75 \times 10^{12}$ and $x = 5 \times 10^3$.

Cumulative Test for Chapters 1–5

Chapter 1   *Answer true or false for problems 1–4.*

1. Some rational numbers are whole numbers.
2. The set of real numbers consists of the set of rational numbers and the set of irrational numbers.
3. The fraction $\frac{5}{3}$ is not a real number.
4. When simplifying an expression containing both addition and multiplication (and no grouping symbols), the addition must be performed first.
5. Which axiom of the real number system is illustrated by the following: $ab + c = ba + c$.
6. Which of the following (if any) represents a negative number?

    (a) -3^2 (b) $(-3)^2$ (c) $-(-3)^2$

Chapter 2

7. Simplify: $(8x^3)(-2x)^4(-3x^2y)(-y^4)$.
8. Simplify: $\dfrac{(9x^3)^0(6x^4)}{(-2x)^2}$.
9. Subtract and simplify: $(5x^n - 2x + 3) - (-4x^n + 2x - 8)$.
10. Factor and simplify: $x^3 - (a + 3)^3$.

Chapter 3

11. What expression should replace the question mark in the following?

    $$\frac{3c}{c-3} = \frac{?}{c^2 - 9}$$

12. Is the fraction $\dfrac{2x}{3}$ equal to the fraction $\dfrac{6x}{9}$? Why or why not?
13. Simplify the signs on $-\dfrac{-(x-6)}{5}$.
14. Simplify: $n + \dfrac{32n}{n^2 - 16} - \dfrac{16}{n-4}$.
15. Add: $\dfrac{x^n + 3}{x^2} + \dfrac{x^m - 1}{x}$.
16. Divide: $2x - 3 \overline{\smash{\big)}\, 2x^3 + 9x^2 - 24x + 9}$

Chapter 4

17. Solve for x: $5x - 4 = 3(x - 8) + 2x$.
18. Find the solution for the compound statement

    $$5x - 3 < x + 9 \quad \text{or} \quad -6(x - 5) \leq -4x + 18$$

    Draw the graph of the solution set.
19. Solve for m: $|2m - 6| = |m - 2|$.
20. Solve for x: $5ax - 4 = 2c + 3x$.
21. One side of a triangle is $\frac{1}{4}$ the perimeter, the second side is $\frac{1}{3}$ the perimeter, and the third side is 20 inches. Find the perimeter of the triangle.

22. Find the number such that when it is increased by 5, the result is 10 more than 6 times the number.

23. A coat that sold for $120 was discounted by 20%. It did not sell and it was discounted another 10%. What was the final selling price of the coat?

Chapter 5

24. Simplify: $\dfrac{5^{-2}}{2^{-2}}$.

25. Add: $2^{-3} + 3^{-2}$.

26. Simplify $(x^{-4}y^2z^{-2})^{-2}$. Give the final answer with positive exponents.

27. Simplify $\left(\dfrac{x^{-3}y^{-2}z^2}{x^{-4}y^2z^{-3}}\right)^2$. Write the final answer with positive exponents.

28. Write 0.0000528 in scientific notation.

29. Simplify $\sqrt{72x^3y^7}$. Assume all variables represent positive numbers.

30. Add $7\sqrt{20} + 2\sqrt{45}$.

31. Rationalize the denominator of $\dfrac{3}{\sqrt{11} - \sqrt{5}}$.

32. Solve for x. Check your solution.
$$\sqrt{3x + 4} - \sqrt{2x + 8} = 0.$$

CHAPTER 6
Second-Degree Equations and Inequalities

Objectives of Chapter 6

To be able to:

☐ Give the definition of a quadratic equation. (6–1)

☐ Write a quadratic equation in standard form. (6–1)

☐ Solve quadratic equations by factoring. (6–1)

☐ Write a quadratic equation if its rational roots are given. (6–1)

☐ Form perfect square trinomials (6–2)

☐ Solve a quadratic equation by completing the square. (6–2)

☐ Give the definition of the imaginary unit. (6–3)

☐ Give the definition of an imaginary (or complex) number. (6–3)

☐ Simplify powers of i. (6–3)

☐ Add and subtract complex numbers. (6–3)

☐ Multiply and divide complex numbers. (6–3)

☐ Use the definition of equal complex numbers to solve equations. (6–3)

☐ Solve equations having complex solutions. (6–3)

☐ Simplify square roots containing negative radicands. (6–3)

☐ Solve a quadratic equation by using the quadratic formula. (6–4)

☐ Determine the nature of the roots of a quadratic equation. (6–4)

☐ Solve radical equations. (6–5)

☐ Solve equations that are quadratic in form. (6–6)

☐ Solve quadratic inequalities. (6–7)

☐ Solve word problems that yield quadratic equations. (6–8)

Second-Degree Equations and Inequalities 209

We have seen in the previous chapters how to find the solution sets for first-degree equations. In this chapter, we learn how to solve quadratic equations and inequalities by factoring, by completing the square, and by using the quadratic formula.

6–1 Finding Solution Sets for Quadratic Equations by Factoring

The type of **quadratic equation** with which we shall be working is of the form $ax^2 + bx + c = 0$, where $a \neq 0$. This is the **standard form of a quadratic equation.**

Definition 6–1

Standard Form of a Quadratic Equation

The **standard form of a quadratic equation** is $ax^2 + bx + c = 0$, where $a \neq 0$. The coefficients a, b, and c represent real numbers.

In the definition just given, $a \neq 0$. If a is zero, then the equation is not quadratic. If either b or c is equal to zero, then the quadratic equation is called an **incomplete quadratic equation.**

The quadratic equations of this section may be solved by factoring. This is done by using the principle that if $p \cdot q = 0$, then either $p = 0$ or $q = 0$ (or both). You should also note that if either $p = 0$ or $q = 0$, then the product $p \cdot q = 0$. We use this latter idea to write a quadratic equation given its **roots,** or solutions.

Rule 6–1

Zero Factor Rule

For real numbers p and q, $pq = 0$ if and only if $p = 0$ or $q = 0$ (or both).

Example 1 Find the solution set for $(x - 1)(x + 1) = 0$ by using rule 6–1.

Solution Because $(x - 1)(x + 1) = 0$, then either

$$x - 1 = 0 \quad \text{or} \quad x + 1 = 0$$

Solving for x gives $x = 1$ or $x = -1$. The solution set is $\{-1, 1\}$. ■

Example 2 Find the solution set for $3r^2 + r = 0$.

Solution This is an incomplete quadratic equation. Because r is a common factor, we write
$$3r^2 + r = 0$$
$$r(3r + 1) = 0$$
Therefore, $r = 0$ or $3r + 1 = 0$. Solving for r gives $r = 0$ or $r = -\frac{1}{3}$.

Check: For $r = 0$:
$$3r^2 + r = 0$$
$$3(0)^2 + 0 = 0$$
$$0 + 0 = 0$$
$$0 = 0$$

For $r = -\frac{1}{3}$:
$$3r^2 + r = 0$$
$$3\left(-\frac{1}{3}\right)^2 + \left(-\frac{1}{3}\right) = 0$$
$$3\left(\frac{1}{9}\right) + \left(-\frac{1}{3}\right) = 0$$
$$\frac{1}{3} + \left(-\frac{1}{3}\right) = 0$$
$$0 = 0$$

The solution set is $\{-\frac{1}{3}, 0\}$. ■

Example 3 Solve $3x^2 = 27$ for x.

Solution
$$3x^2 = 27$$
$$3x^2 - 27 = 0 \quad \text{Write the equation in standard form.}$$
$$3(x^2 - 9) = 0 \quad \text{Remove the common factor.}$$
$$3(x - 3)(x + 3) = 0 \quad \text{Completely factor the left side.}$$
$$x - 3 = 0 \text{ or } x + 3 = 0 \quad \text{Rule 6-1 (we cannot set 3 equal to 0).}$$
$$x = 3 \text{ or } x = -3 \quad \text{Solve for } x.$$

Check: For $x = 3$:
$$3x^2 = 27$$
$$3(3)^2 = 27$$
$$3(9) = 27$$
$$27 = 27$$

For $x = -3$:
$$3x^2 = 27$$
$$3(-3)^2 = 27$$
$$3(9) = 27$$
$$27 = 27$$

The solution set is $\{-3, 3\}$. ■

The method of checking solutions is shown for examples 2 and 3 so that you may see how to check your proposed solutions. We do not show a check for every example. You are encouraged to check your proposed solutions from time to time just to make sure everything is being done correctly.

Second-Degree Equations and Inequalities 211

Example 4 Solve $2p^2 + p - 21 = 0$ for p.

Solution
$2p^2 + p - 21 = 0$
$(2p + 7)(p - 3) = 0$ Factor the left side.
$2p + 7 = 0$ or $p - 3 = 0$ Set each factor equal to zero.
$p = -\frac{7}{2}$ or $p = 3$ Solve for p.

The solution set is $\{-\frac{7}{2}, 3\}$. ■

Example 5 Solve $x^2 + 6x + 9 = 0$ for x.

Solution
$x^2 + 6x + 9 = 0$
$(x + 3)(x + 3) = 0$
$x + 3 = 0$ or $x + 3 = 0$
$x = -3$ or $x = -3$

The solution set is $\{-3\}$. ■

Example 5 illustrates that both solutions (or roots) of a quadratic equation may be the same. It can be shown that every quadratic equation has two roots. When both roots are the same, they are said to occur with a **multiplicity** of two.

Example 6 Solve $\frac{x}{2} = \frac{x^2 + 4}{3x + 2} - \frac{1}{2}$ for x.

Solution First clear the equation of fractions by multiplying by $2(3x + 2)$, the LCM for the denominators of all the fractions.

$2(3x + 2) \cdot \frac{x}{2} = 2(3x + 2) \cdot \frac{x^2 + 4}{3x + 2} - 2(3x + 2) \cdot \frac{1}{2}$
$x(3x + 2) = 2(x^2 + 4) - (3x + 2)$ Divide out the denominators.
$3x^2 + 2x = 2x^2 + 8 - 3x - 2$ Remove parentheses.
$3x^2 + 2x = 2x^2 - 3x + 6$ Collect like terms.
$x^2 + 5x - 6 = 0$ Write the equation in standard form.
$(x + 6)(x - 1) = 0$ Factor.
$x + 6 = 0$ or $x - 1 = 0$ Set each factor equal to zero.
$x = -6$ or $x = 1$ Solve for x.

Neither value of x causes any denominator in the original equation to be zero and both values check, so the solution set is $\{-6, 1\}$. ■

Example 7 Write a quadratic equation whose roots are 3 and $\frac{1}{2}$.

Solution We know $x = 3$ or $x = \frac{1}{2}$.

$x - 3 = 0$ or $x - \frac{1}{2} = 0$ Set each equation equal to zero.

$$(x - 3)\left(x - \frac{1}{2}\right) = 0 \quad \text{Rule 6–1.}$$

$$x^2 - \left(\frac{1}{2}\right)x - 3x + \frac{3}{2} = 0 \quad \text{Multiply.}$$

$$2x^2 - x - 6x + 3 = 0 \quad \text{Multiply the equation by 2.}$$

$$2x^2 - 7x + 3 = 0 \quad \text{Collect like terms.}$$

An equation whose roots are 3 and $\frac{1}{2}$ is $2x^2 - 7x + 3 = 0$.

Example 8 Write a quadratic equation whose roots are $-\frac{1}{3}$ is and $\frac{2}{5}$.

Solution If the roots are fractions, it is easier to clear the fractions early in the work rather than later.

$x = -\frac{1}{3}$ or $x = \frac{2}{5}$

$3x = -1$ or $5x = 2$ Clear the equations of fractions.

$3x + 1 = 0$ or $5x - 2 = 0$ Set each equation equal to zero.

$(3x + 1)(5x - 2) = 0$ Rule 6–1.

$15x^2 - x - 2 = 0$ Multiply and collect like terms.

An equation whose roots are $-\frac{1}{3}$ and $\frac{2}{5}$ is $15x^2 - x - 2 = 0$.

6–1 Exercises

In exercises 1–42, find the solution sets for each equation. *(See examples 1–6.)*

1. $(x - 2)(x + 3) = 0$
2. $(x + 5)(x - 3) = 0$
3. $u^2 - u - 6 = 6$
4. $(y - 3)(y + 1) = 12$
5. $(2x - 3)(x + 2) = 4$
6. $4(p - 4)\left(p - \frac{1}{3}\right) = 0$
7. $5\left(t - \frac{3}{4}\right)\left(t + \frac{1}{2}\right) = 0$
8. $\left(2x - \frac{3}{4}\right)\left(5x - \frac{1}{2}\right) = 0$
9. $\left(3x - \frac{1}{4}\right)\left(2x - \frac{3}{8}\right) = 0$
10. $\left(2r - \frac{5}{9}\right)(r - 6) = 0$

Second-Degree Equations and Inequalities

11. $(s - 4)(s + 5)(s - 6) = 0$
12. $(x + 7)(x - 9)(x + 4) = 0$
13. $(2x - 5)(3x + 2)(5x + 4) = 0$
14. $(9m - 1)(6m + 5)(m - 3) = 0$
15. $(n^2 + 7n + 12)(n - 7) = 0$ (*Hint:* Factor $n^2 + 7n + 12$ first.)
16. $(2x^2 - x - 3)(x - 7) = 0$ (*Hint:* Factor $2x^2 - x - 3$ first.)
17. $(3x^2 + 7x + 2)(x - 5) = 0$
18. $(2t^2 - 7t + 3)(2t + 1) = 0$
19. $3r^2 + 5r = 0$
20. $9x^2 - 6x = 0$
21. $x^2 = 25$
22. $9s^2 = 16$
23. $\frac{9}{4}v^2 = \frac{25}{36}$ (See example 6.)
24. $\frac{4}{25}x^2 = \frac{49}{64}$ (See example 6.)
25. $x^2 - x = 6$
26. $p^2 = -p + 20$
27. $2q^2 + 5q - 12 = 0$
28. $3x^2 - 11x = 4$
29. $x(2x + 11) = 6$
30. $y(12y + 1) = 1$
31. $a(4a - 7) = -2(a^2 + 1)$
32. $2x(2x + 3) - 3 = -4x(x + 1)$
33. $3x - \frac{8}{x} - 2 = 0$
34. $10 + \frac{1}{u} - \frac{2}{u^2} = 0$
35. $-t = \frac{t^2 + 12}{t - 11}$
36. $x = \frac{2(3 - 5x)}{10x + 3} - \frac{3}{10x + 3}$
37. $x = \sqrt{5x - 6}$
38. $c = \sqrt{c + 12}$
39. $h = \sqrt{\frac{-9h + 5}{2}}$
40. $x = \sqrt{\frac{x + 4}{3}}$
41. $9x^2 - 12x + 4 = 0$
42. $25m^2 = -40m - 16$

In exercises 43–50, write a quadratic equation with the indicated roots. (See examples 7 and 8.)

43. 3 and 4
44. 5 and -1
45. -2 and -8
46. 10 and -2
47. $\frac{3}{4}$ and -3
48. $-\frac{5}{6}$ and 3
49. $\frac{1}{2}$ is the only solution.
50. $\frac{3}{4}$ is the only solution.

In exercises 51–54, solve for x in terms of the other letters.

Examples (a) $x^2 - 9c^2 = 0$ (b) $2x^2 + ax - a^2 = 0$

Solution (a) $\quad x^2 - 9c^2 = 0$
$(x - 3c)(x + 3c) = 0$
$x - 3c = 0 \quad$ or $\quad x + 3c = 0$
$x = 3c \quad$ or $\quad x = -3c$

The solution set is $\{3c, -3c\}$.

(b) $2x^2 + ax - a^2 = 0$
$(2x - a)(x + a) = 0$
$2x - a = 0$ or $x + a = 0$
$x = \dfrac{a}{2}$ or $x = -a$

The solution set is $\{a/2, -a\}$. ■

51. $4x^2 - 9a^2 = 0$
52. $25x^2 = 16c^2$
53. $2x^2 - ax - a^2 = 0$
54. $4x^2 + 4ax - 3a^2 = 0$

6–2 Completing the Square

If a quadratic equation can be simplified to the form

$$x^2 = a, \quad a > 0$$

then the equation may be solved by taking the square root of both sides of the equation. For example, the equation

$$x^2 = 25$$

has two solutions, 5 and -5. These solutions can be found by taking the square root of both sides of the given equation. Both the positive square root and the negative square root of 25 must be taken if two solutions are to be obtained. That is, $x^2 = 25$ gives

$$x = 5 \quad \text{or} \quad x = -5$$

as the two solutions. In general, if $x^2 = a$, then

$$x = \sqrt{a} \quad \text{or} \quad x = -\sqrt{a}$$

for any real number $a > 0$.

The principle of taking the square root of both sides of a quadratic equation may also be applied to those quadratic equations of the form

$$(x - a)^2 = b, \quad b > 0$$

to obtain

$$x - a = b \quad \text{or} \quad x - a = -b$$
$$x = a + b \quad \text{or} \quad x = a - b$$

(See example 1).

Example 1 Find the solution set for $(m - 3)^2 = 16$.

Second-Degree Equations and Inequalities

Solution

$(m - 3)^2 = 16$
$m - 3 = 4 \quad \text{or} \quad m - 3 = -4$ Take the square root of each side.
$\quad m = 7 \quad \text{or} \qquad m = -1$ Solve for m.

The solution set is $\{7, -1\}$. ◼

Example 2   Find the solution set for $\dfrac{2x^2}{5} = 8$.

Solution   First solve the equation for x^2.

$$\frac{2x^2}{5} = 8$$
$$2x^2 = 40 \quad \text{Multiply each side by 5.}$$
$$x^2 = 20 \quad \text{Divide each side by 2.}$$

At this point, we have an equation similar to the one in example 1. By taking square roots, $x^2 = 20$ implies,

$x = \sqrt{20} \quad \text{or} \quad x = -\sqrt{20}$
$x = 2\sqrt{5} \quad \text{or} \quad x = -2\sqrt{5}$ Simplify the radicals.

The solution set is $\{2\sqrt{5}, -2\sqrt{5}\}$. ◼

We now turn our attention to finding a more general way of solving quadratic equations. The equations in section 6–1 and the examples just given are all illustrations of quadratic equations that are factorable or already essentially have the form $x^2 = a$. If the equation is not already of the form $x^2 = a$, it can be written in that form by a process known as **completing the square.** This method makes one side of the equation into a perfect-square trinomial, which can then be factored into the form $(x - a)^2 = b$. Then, methods already discussed in this section can be applied to find the solution set. The procedure is explained by solving an example, showing the steps.

Example 3   Solve $2x^2 + x - 6 = 0$ by completing the square.

Solution   A set of instructions will be given and numbered to correspond to the step to which it applies.

Step 1   Arrange the given equation in the form $ax^2 + bx = -c$. $2x^2 + x = 6$

Step 2   Divide both sides of the equation by the numerical coefficient of the second-degree term. (In this case, divide by 2). $x^2 + \dfrac{x}{2} = 3$

Step 3 Complete the square on the left side of the equation by taking one-half of the numerical coefficient of the first-degree term, squaring this number, and adding it to both sides of the equation. The left side is now a perfect square trinomial.

$$x^2 + \frac{x}{2} + \frac{1}{16} = 3 + \frac{1}{16}$$

Step 4 Factor the left side of the equation and simplify the right side of the equation.

$$\left(x + \frac{1}{4}\right)^2 = \frac{49}{16}$$

Step 5 Take the square root of both sides of the equation, form two (first-degree) equations, and solve for x.

$$x + \frac{1}{4} = \frac{7}{4} \quad \text{or} \quad x + \frac{1}{4} = -\frac{7}{4}$$

$$x = \frac{7}{4} - \frac{1}{4} \quad \text{or} \quad x = -\frac{7}{4} - \frac{1}{4}$$

$$x = \frac{3}{2} \quad \text{or} \quad x = -2$$

The solution set is $\{\frac{3}{2}, -2\}$. ■

From this example, we see the steps to be followed in solving an equation by completing the square are as follows.

Procedure 6-1

Solving a Quadratic Equation by Completing the Square:

Step 1 Arrange the equation in the form $ax^2 + bx = -c$.

Step 2 If $a \neq 1$, divide both sides of the equation by a, where a is the coefficient of the second-degree term.

Step 3 Complete the square on the left side by taking one-half of the coefficient of the first-degree term, squaring this number, and adding it to both sides of the equation. The left side is now a perfect-square trinomial.

Step 4 Express the left side of the equation as the square of a binomial and simplify the right side of the equation.

Step 5 Take the square root of both sides of the equation, form two (first-degree) equations, and solve for x.

Example 4 Solve $5x^2 + 2x - 4 = 0$ for x by completing the square.

Solution The steps are numbered to correspond to the general instructions just listed. Compare each numbered step to the corresponding step in procedure 6-1.

Step 1 $5x^2 + 2x = 4$

Step 2 $x^2 + \dfrac{2}{5}x = \dfrac{4}{5}$

Step 3 $x^2 + \dfrac{2}{5}x + \dfrac{1}{25} = \dfrac{4}{5} + \dfrac{1}{25}$; $\dfrac{1}{2}$ of $\dfrac{2}{5} = \dfrac{1}{5}$ and $\left(\dfrac{1}{5}\right)^2 = \dfrac{1}{25}$.

Step 4 $\left(x + \dfrac{1}{5}\right)^2 = \dfrac{21}{25}$

Step 5 $x + \dfrac{1}{5} = \dfrac{\sqrt{21}}{5}$ or $x + \dfrac{1}{5} = \dfrac{-\sqrt{21}}{5}$

$x = -\dfrac{1}{5} + \dfrac{\sqrt{21}}{5}$ or $x = -\dfrac{1}{5} - \dfrac{\sqrt{21}}{5}$

$x = \dfrac{-1 + \sqrt{21}}{5}$ or $x = \dfrac{-1 - \sqrt{21}}{5}$

The solution set is

$$\left\{\dfrac{-1 + \sqrt{21}}{5}, \dfrac{-1 - \sqrt{21}}{5}\right\}.$$

The two solutions can also be combined and written as

$$\left\{\dfrac{-1 \pm \sqrt{21}}{5}\right\}$$ ■

6–2 Exercises

In exercises 1–24, solve for the variable by writing each equation in the form $x^2 = a$ or $(ax - b)^2 = c$. Then, take the square root of both sides of the equation. (See examples 1 and 2.)

1. $x^2 - 9 = 0$ **2.** $x^2 = 16$ **3.** $m^2 = \dfrac{25}{16}$ **4.** $n^2 = \dfrac{49}{4}$

5. $3x^2 = 27$ **6.** $4x^2 = 12$ **7.** $\dfrac{p^2}{2} = 8$ **8.** $\dfrac{t^2}{4} = 6$

9. $\dfrac{2x^2}{3} = 8$ **10.** $\dfrac{4x^2}{3} = 3$

11. $(y - 3)^2 = 4$ **12.** $(r + 5)^2 = 16$

13. $(x + 2)^2 = 8$ **14.** $(x - 3)^2 = 12$

15. $(2a - 3)^2 = 16$ **16.** $(3b - 1)^2 = 4$

17. $(3x + 1)^2 = 8$ **18.** $(2x + 5)^2 = 7$

19. $(3h + 2)^2 - 18 = 0$ **20.** $(5k + 4)^2 - 20 = 0$

21. $(2x + 5)^2 - 24 = 1$ **22.** $\left(x - \dfrac{3}{4}\right)^2 - \dfrac{24}{16} = \dfrac{1}{16}$

23. $\left(u - \dfrac{2}{3}\right)^2 - \dfrac{15}{9} = \dfrac{1}{9}$ **24.** $\left(r - \dfrac{1}{5}\right)^2 - \dfrac{3}{25} = \dfrac{22}{25}$

In exercises 25–34, determine what number must be added to each expression to form a perfect-square trinomial. Once the perfect-square trinomial is formed, factor it as the square of a binomial.

Example Form a perfect-square trinomial from $x^2 + 6x$.

Solution To form the third term, we must add the square of one-half the numerical coefficient of x. Thus,

$$\left[\frac{1}{2}(6)\right]^2 = 3^2 = 9$$

The trinomial is $x^2 + 6x + 9$. In factored form,

$$x^2 + 6x + 9 = (x + 3)^2$$

25. $x^2 + 4x$
26. $x^2 - 6x$
27. $x^2 - 8x$
28. $x^2 + 12x$
29. $x^2 - 7x$
30. $x^2 + 5x$
31. $x^2 + \frac{2}{3}x$
32. $x^2 + \frac{3}{4}x$
33. $x^2 - \frac{5}{3}x$
34. $x^2 + \frac{5}{2}x$

In exercises 35–50, solve by completing the square. (See examples 3 and 4.)

35. $x^2 - 12x + 27 = 0$
36. $x^2 + 9x + 20 = 0$
37. $m^2 + m - 20 = 0$
38. $n^2 - 4n + 3 = 0$
39. $2x^2 + 3x = 2$
40. $3x^2 = 7x - 2$
41. $4p^2 + 8p = 12$
42. $3y^2 = y + 4$
43. $x^2 + 5x - 2 = 0$
44. $x^2 + 6x - 1 = 0$
45. $2a^2 - a = 5$
46. $3r^2 + 6r = -1$
47. $8x^2 = -12x + 3$
48. $7x^2 - 2x = \frac{48}{7}$
49. $5t^2 - 3t = \frac{7}{20}$
50. $3x^2 + 2x = \frac{14}{3}$

6–3 Complex Numbers

The set of real numbers, consisting of the set of rational numbers and the set of irrational numbers, would appear to be sufficient for all our needs. However, the quadratic equation $x^2 = -4$ has no solution in the set of real numbers, for the solution consists of numbers whose square is -4. Because every real number has a square that is greater

Second-Degree Equations and Inequalities 219

than or equal to zero, we suddenly find ourselves in the uncomfortable position of not being able to solve this simple quadratic equation (over the set of real numbers).

If we solve $x^2 = -4$ by the methods already discussed, we get

$$x = \sqrt{-4} \quad \text{or} \quad x = -\sqrt{-4}$$

The question we now face is that of finding an expression for $\sqrt{-4}$. We see $\sqrt{-4} \neq -2$, since $(-2)(-2) = 4$, not -4. Also, $\sqrt{-4} \neq 2$, since $(2)(2) = 4$, not -4.

To find a way out of this, recall $\sqrt{ab} = \sqrt{a}\sqrt{b}$ if a and b are both positive. Let us assume this rule also holds if one of the two numbers a and b is positive and the other negative. Then we can write

$$\sqrt{-4} = \sqrt{4(-1)}$$
$$= \sqrt{4}\sqrt{-1}$$
$$= 2\sqrt{-1}$$

Let us now agree to represent $\sqrt{-1}$ with the letter i. We then have

$$\sqrt{-4} = 2\sqrt{-1} = 2i$$

We now define the **imaginary unit**, $\sqrt{-1}$, as follows.

Definition 6–2

The Imaginary Unit
$$i = \sqrt{-1} \quad \text{and} \quad i^2 = -1$$

From the discussion prior to definition 6–2, we see $\sqrt{-4} = 2i$. We can check the accuracy of this result by noting

$$(2i)^2 = (2i)(2i) = 4i^2$$
$$= 4(-1) \quad \text{definition 6–2}$$
$$= -4$$

Also, $-\sqrt{-4} = -2i$ because,

$$(-2i)^2 = (-2i)(-2i)$$
$$= 4i^2$$
$$= 4(-1) \quad \text{definition 6–2}$$
$$= -4$$

Thus -4 has two square roots, $2i$ and $-2i$. The quadratic equation $x^2 = -4$ has the solution set $\{2i, -2i\}$.

In general, if a is any positive real number, the meaning of $\sqrt{-a}$ is as given in definition 6–3.

Definition 6–3

Principal Square Root of $-a$

If a is any positive real number, $\sqrt{-a} = i\sqrt{a}$. The symbol $\sqrt{-a}$ is called the *principal square root* of $-a$.

Any number such as $\sqrt{-a}$, where a is any positive real number, is called a **pure imaginary number**.

We now define a **complex number**.

Definition 6–4

Complex Number

If a and b are real numbers, then $a + bi$ is called a **complex number**.

In the complex number $a + bi$, a is the *real part* of the complex number and bi is the *imaginary part*.

The set of real numbers is a subset of the complex numbers. We may see this by observing that any real number a may be written in the form $a + 0i = a + 0 = a$.

We have already seen that $i = \sqrt{-1}$ and $i^2 = -1$. Also $i^3 = i^2 \cdot i = (-1)i = -i$ and $i^4 = i^2 \cdot i^2 = (-1)(-1) = 1$. You should memorize rule 6–2.

Rule 6–2

Powers of i

$$i = i$$
$$i^2 = -1$$
$$i^3 = -i$$
$$i^4 = 1$$

Any power of i greater than 4 repeats one of the values given in rule 6–2. For example, $i^{10} = i^4 \cdot i^4 \cdot i^2 = (i^4)^2 \cdot i^2 = (1)^2 \cdot i^2 = i^2 = -1$. If n is zero or a positive integer, then

$$i^{4n+1} = i$$
$$i^{4n+2} = -1$$
$$i^{4n+3} = -i$$
$$i^{4n+4} = 1$$

For example, $i^{19} = i^{4 \cdot 4 + 3} = -i$.

The meaning of equality between two complex numbers is given in definition 6–5.

Definition 6–5

Equality of Complex Numbers

For any two complex numbers $a + bi$ and $c + di$, $a + bi = c + di$ if and only if $a = c$ and $b = d$.

Example 1

Use definition 6–5 to find the values of the real numbers x and y in the equation $2x + yi = 6 - 4i$.

Solution

By definition 6–5, $2x + yi = 6 - 4i$ only if

$$2x = 6 \quad \text{and} \quad y = -4$$
$$x = 3 \quad \text{and} \quad y = -4$$

Complex numbers may be added, subtracted, multiplied or divided according to the following definitions.

Definition 6–6

Sum of Complex Numbers

The sum of two complex numbers $a + bi$ and $c + di$ is

$$(a + bi) + (c + di) = (a + c) + (b + d)i$$

Definition 6–7

Difference of Complex Numbers

The difference of two complex numbers $a + bi$ and $c + di$ is

$$(a + bi) - (c + di) = (a + bi) + (-c - di)$$
$$= (a - c) + (b - d)i$$

Example 2

Find the sum of $3 + 4i$ and $2 - 7i$.

Solution

$$(3 + 4i) + (2 - 7i) = (3 + 2) + (4 - 7)i$$
$$= 5 + (-3)i$$
$$= 5 - 3i$$

Example 3

Subtract $8 - 6i$ from $5 - 4i$.

Solution

$$(5 - 4i) - (8 - 6i) = (5 - 8) + [-4 - (-6)]i$$
$$= (5 - 8) + (-4 + 6)i$$
$$= -3 + 2i$$

The quick, easy way to add or subtract complex numbers is to think of them as if they were polynomials in i and collect like terms. For example,

$$(3 + 2i) + (8 - 3i) - (2 - 3i)$$
$$= 3 + 2i + 8 - 3i - 2 + 3i \qquad \text{Remove parentheses.}$$
$$= 9 + 2i \qquad \text{Collect like terms.}$$

Definition 6-8

> **Product of Complex Numbers**
>
> The product of two complex numbers $a + bi$ and $c + di$ is
>
> $$(a + bi)(c + di) = (ac - bd) + (ad + bc)i$$

Fortunately, we do not need to memorize definition 6–8. Multiply two complex numbers just as you would multiply any two binomials. Then simplify powers of i and collect like terms. Study the next example carefully.

Example 4

Multiply and simplify: $(3 + 2i)(4 - 3i)$.

Solution

$$\begin{aligned}(3 + 2i)(4 - 3i) &= 3(4) - 3(3i) + (2i)(4) - (2i)(3i) \\ &= 12 - 9i + 8i - 6i^2 \\ &= 12 - i - 6(-1) \quad \text{Recall } i^2 = -1. \\ &= 12 - i + 6 \\ &= 18 - i \quad \blacksquare\end{aligned}$$

We define division of two complex numbers in terms of multiplying the numerator and denominator by the conjugate of the denominator. The *conjugate* of $a + bi$ is $a - bi$. Specifically, the conjugate of $3 + 4i$ is $3 - 4i$. Also, $-2 - 5i$ and $-2 + 5i$ are conjugates.

Definition 6-9

> **Quotient of Complex Numbers**
>
> The quotient of two complex numbers $a + bi$ and $c + di$ is
>
> $$\begin{aligned}\frac{a + bi}{c + di} &= \frac{a + bi}{c + di} \cdot \frac{c - di}{c - di} \\ &= \frac{(ac + bd) + (bc - ad)i}{c^2 + d^2}\end{aligned}$$

This definition is rather cumbersome, and as in the case for the definition of multiplication of complex numbers, the work on an actual problem is merely a matter of multiplying, simplifying, and collecting similar terms.

Example 5

Divide and simplify: $\dfrac{3 + 4i}{2 - 3i}$.

Solution

Multiply the numerator and denominator by $2 + 3i$, the conjugate of $2 - 3i$.

Second-Degree Equations and Inequalities

$$\frac{3+4i}{2-3i} = \frac{3+4i}{2-3i} \cdot \frac{2+3i}{2+3i} = \frac{(3+4i)(2+3i)}{(2-3i)(2+3i)}$$
$$= \frac{6+9i+8i+12i^2}{4+6i-6i-9i^2} \quad \text{Perform the indicated multiplications.}$$
$$= \frac{6+17i+12(-1)}{4-9(-1)} \quad \text{Collect like terms and simplify } i^2.$$
$$= \frac{-6+17i}{4+9} \quad \text{Simplify and collect like terms.}$$
$$= \frac{-6}{13} + \frac{17i}{13} \quad \blacksquare$$

The imaginary unit i may be written as $0 + i$. As such, its conjugate is $0 - i = -i$. That is, the conjugate of i is $-i$.

Example 6 Divide and simplify: $\frac{3+2i}{i}$.

Solution
$$\frac{3+2i}{i} = \frac{3+2i}{i} \cdot \frac{-i}{-i}$$
$$= \frac{-i(3+2i)}{-i^2}$$
$$= \frac{-3i - 2i^2}{-(-1)}$$
$$= \frac{-3i - 2(-1)}{1}$$
$$= 2 - 3i \quad \blacksquare$$

While we do not do so, it can be shown that the properties developed in chapter 1 for real numbers also hold for complex numbers. The one exception to this statement involves the inequality axioms. While we may order the real numbers, we lose this important ability with the complex numbers. That is, we cannot say which is the larger of $3 + 2i$ and $2 + 3i$. In general, if $a + bi$ and $c + di$ are complex numbers, we cannot say which is larger, although we can tell if they are equal by definition 6–5.

Example 7 Simplify $\sqrt{-6}\sqrt{-3}$.

Solution Use definition 6–3 to write $\sqrt{-6}$ as $i\sqrt{6}$ and $\sqrt{-3}$ as $i\sqrt{3}$. Then
$$\sqrt{-6}\sqrt{-3} = i\sqrt{6}i\sqrt{3}$$
$$= i^2\sqrt{6}\sqrt{3}$$
$$= i^2\sqrt{18}$$
$$= (-1)\sqrt{9 \cdot 2}$$
$$= -1\sqrt{9}\sqrt{2}$$
$$= -3\sqrt{2}$$

You are cautioned *not* to write

$$\sqrt{-6}\sqrt{-3} = \sqrt{(-6)(-3)}$$
$$= \sqrt{18}$$
$$= 3\sqrt{2}$$

This is *incorrect* because we are assuming $\sqrt{a}\sqrt{b} = \sqrt{ab}$ to be true if both a and b are negative. This assumption is *not* correct, and to avoid making this mistake, always write numbers of the form $\sqrt{-a}$ ($a > 0$) as $\sqrt{-a} = i\sqrt{a}$ before carrying out the indicated operations.

6–3 Exercises

In exercises 1–8, express each square root as expressions using i. Simplify the radicand if possible.

Example

$\sqrt{-8} = i\sqrt{8} = i\sqrt{4 \cdot 2} = i\sqrt{4}\sqrt{2} = 2i\sqrt{2}$

1. $\sqrt{-9}$
2. $\sqrt{-16}$
3. $\sqrt{-12}$
4. $\sqrt{-18}$
5. $-\sqrt{-4}$
6. $-\sqrt{-16}$
7. $-\sqrt{-24}$
8. $-\sqrt{-32}$

In exercises 9–14, multiply or divide. Simplify all powers of i and reduce radicands when possible.

Example

$$\sqrt{-8}\sqrt{-6} = (i\sqrt{8})(i\sqrt{6})$$
$$= i^2\sqrt{48}$$
$$= -1\sqrt{16 \cdot 3}$$
$$= -4\sqrt{3}$$

9. $\sqrt{-9}\sqrt{-4}$
10. $\sqrt{-3}\sqrt{-4}$
11. $\sqrt{-3}\sqrt{-6}$
12. $\sqrt{-9}\sqrt{-5}$
13. $\dfrac{\sqrt{-28}\sqrt{-24}}{\sqrt{-6}}$
14. $\dfrac{\sqrt{-32}\sqrt{-5}}{\sqrt{-10}\sqrt{-16}}$

In exercises 15–22, simplify, using rule 6–2.

15. i^6
16. i^7
17. i^{14}
18. i^{23}
19. i^{37}
20. i^{45}
21. i^{78}
22. i^{103}

In exercises 23–30, add or subtract.

Example

$$\sqrt{-24} + 3\sqrt{-6} = i\sqrt{24} + 3i\sqrt{6}$$
$$= i\sqrt{4 \cdot 6} + 3i\sqrt{6}$$
$$= 2i\sqrt{6} + 3i\sqrt{6}$$
$$= 5i\sqrt{6}$$

23. $\sqrt{-32} + 3\sqrt{-50}$
24. $\sqrt{-27} - 4\sqrt{-3}$
25. $3\sqrt{-32} - 5\sqrt{18} - 4\sqrt{72}$
26. $3\sqrt{-8} - 4\sqrt{-12} + 5\sqrt{-48}$
27. $5 + \sqrt{-4} - 9 - 5\sqrt{-16}$
28. $9 + 3\sqrt{-9} - 6\sqrt{-16} + 7$
29. $8 - 7\sqrt{25} - 4\sqrt{-16} + 3\sqrt{9}$
30. $8 - 2\sqrt{49} + 8\sqrt{-25} - 6\sqrt{4}$

In exercises 31–60, perform the indicated operations and leave your answer in the form a + bi. (This may mean some answers will take the form a + 0i or 0 + bi.)

31. $(5 + 2i) + (6 - 3i)$
32. $(6 - 2i) + (-9 - 4i)$
33. $(7 - 3i) + (4 + 2i)$
34. $(8 - 12i) + (9 - 7i)$
35. $(23i - 6) + (7 - 4i)$
36. $(8i - 3) + (5 - 4i)$
37. $(9 - 8i) - (5 - 2i)$
38. $(6 - 5i) - (7 - 4i)$
39. $(7 - 4i) - (8 - 4i)$
40. $(8 - 6i) + (9 + 6i)$
41. $(9 + 5i) - (9 - 3i)$
42. $(8 + 9i) - (8 + 9i)$
43. $(3 + 2i)(9 - 2i)$
44. $(6 - 2i)(5 - i)$
45. $(6 - 3i)(2 - 3i)$
46. $(7 - i)(8 + 2i)$
47. $(15 - i)(3 + 2i)$
48. $(9 + 3i)(2i - 3)$
49. $(3 + 2i)^2$
50. $(3 - 4i)^2$
51. $(6 - 4i)^2$
52. $(7 + 2i)^2$
53. $\dfrac{4 + 3i}{2 - 3i}$
54. $\dfrac{9 + 2i}{3 - 2i}$
55. $\dfrac{2 - 4i}{3 + 2i}$
56. $\dfrac{2 - 3i}{5 + 2i}$
57. $\dfrac{4 - 3i}{i}$
58. $\dfrac{3 - 2i}{2i}$
59. $\dfrac{3 + i}{i^3}$
60. $\dfrac{5 + 3i}{i^5}$

In exercises 61–68, solve for x and y. (See example 1.)

61. $3x - 2yi = 9 + 8i$
62. $5x + 3yi = 10 - 6i$
63. $6x + 5yi = -12 - 20i$
64. $3x + 2yi = -12 + 4i$
65. $3x^2 + 2yi = 27 + 6i$
66. $2x^2 - 3yi = 32 - 27i$
67. $5x + 2yi = (3 + 2i)(8 - 2i)$
68. $3x - 2yi = \dfrac{5 + 3i}{4 - 2i}$

6–4 The Quadratic Formula

In the previous sections, we solved quadratic equations either by factoring or by completing the square. A third method of solving quadratic equations involves using the quadratic formula. This method gives a means of solving any quadratic equation of the form $ax^2 + bx + c = 0$ $(a \neq 0)$. This method is a favorite of students, and it is a general method for solving all quadratic equations, including those that can be solved by factoring.

We derive the quadratic formula by completing the square on $ax^2 + bx + c = 0$ $(a \neq 0)$. We number the steps to correspond to the steps given in procedure 6–1.

Step 1 Arrange the equation in the form $ax^2 + bx = -c$.

$$ax^2 + bx = -c$$

Step 2 Divide both sides of the equation by a.

$$x^2 + \frac{b}{a}x = -\frac{c}{a}$$

Step 3 Take one-half of $\frac{b}{a}$, square it, and add this number to each side of the equation.

$$x^2 + \frac{b}{a}x + \frac{b^2}{4a^2} = \frac{b^2}{4a^2} - \frac{c}{a}$$

Step 4 Write the left side of the equation as the square of a binomial and simplify the right side.

$$\left(x + \frac{b}{2a}\right)^2 = \frac{b^2 - 4ac}{4a^2}$$

Step 5 Take the square root of both sides of the equation. Form two (first-degree) equations and solve for x. (Here, we combine the terms on the right side over the common denominator, $2a$.)

$$x + \frac{b}{2a} = \pm\frac{\sqrt{b^2 - 4ac}}{2a}$$

$$x = \frac{-b \pm \sqrt{b^2 - 4ac}}{2a}$$

The last equation in step 5 is known as the **quadratic formula**; it should be memorized.

Rule 6–3

The Quadratic Formula

The solutions of any quadratic equation having the form $ax^2 + bx + c = 0$, $a \neq 0$, are given by

$$x = \frac{-b \pm \sqrt{b^2 - 4ac}}{2a}$$

Example 1 Solve $3x^2 + 4x = 15$ for x by using the quadratic formula.

Solution The standard form of the given equation is $3x^2 + 4x - 15 = 0$. This gives $a = 3$, $b = 4$, and $c = -15$. Substitute these values into the quadratic formula.

$$x = \frac{-b \pm \sqrt{b^2 - 4ac}}{2a}$$

$$x = \frac{-4 \pm \sqrt{4^2 - 4(3)(-15)}}{2 \cdot 3}$$

$$x = \frac{-4 \pm \sqrt{16 + 180}}{6}$$

$$x = \frac{-4 \pm \sqrt{196}}{6}$$

$$x = \frac{-4 \pm 14}{6}$$

The last equation gives

$$x = \frac{-4 + 14}{6} \quad \text{or} \quad x = \frac{-4 - 14}{6}$$

$$x = \frac{5}{3} \quad \text{or} \quad x = -3$$

The solution set is $\left\{\frac{5}{3}, -3\right\}$. ■

Example 2 Solve $x^2 = -x - \frac{5}{2}$ for x by using the quadratic formula.

Solution The standard form of the equation is $2x^2 + 2x + 5 = 0$. Substitute the values $a = 2$, $b = 2$, and $c = 5$ into the formula.

$$x = \frac{-b \pm \sqrt{b^2 - 4ac}}{2a}$$

$$x = \frac{-2 \pm \sqrt{(2)^2 - 4(2)(5)}}{2(2)}$$

$$x = \frac{-2 \pm \sqrt{4 - 40}}{4}$$

$$x = \frac{-2 \pm \sqrt{-36}}{4}$$

$$x = \frac{-2 \pm i\sqrt{36}}{4} \qquad \text{Apply definition 6–3.}$$

$$x = \frac{-2 \pm 6i}{4}$$

$$x = \frac{2(-1 \pm 3i)}{4} \qquad \text{Factor the numerator.}$$

$$x = \frac{-1 \pm 3i}{2} \qquad \text{Divide out the common factor of 2.}$$

The solution set is $\left\{\dfrac{-1+3i}{2}, \dfrac{-1-3i}{2}\right\}$. ■

Example 3 Solve $\dfrac{x^2-3}{2}+x=\dfrac{3}{4}$ for x by using the quadratic formula.

Solution Write the equation in standard form. First multiply both sides by 4.

$2(x^2-3)+4x=3$
$2x^2-6+4x=3$ Multiply, and collect like terms.
$2x^2+4x-9=0$ Write the equation in standard form.

Now that we have the equation in standard form, $a=2$, $b=4$, and $c=-9$. Therefore,

$$x=\dfrac{-4\pm\sqrt{4^2-4(2)(-9)}}{2(2)}$$

$$x=\dfrac{-4\pm\sqrt{16+72}}{4}$$

$$x=\dfrac{-4\pm\sqrt{88}}{4}$$

$$x=\dfrac{-4\pm\sqrt{4(22)}}{4}$$

$$x=\dfrac{-4\pm 2\sqrt{22}}{4}$$

$$x=\dfrac{2(-2\pm\sqrt{22})}{4}$$

$$x=\dfrac{-2\pm\sqrt{22}}{2}$$

The solution set is $\left\{\dfrac{-2+\sqrt{22}}{2}, \dfrac{-2-\sqrt{22}}{2}\right\}$. ■

The kind of numbers (or the nature of the roots) in the solution set for a quadratic equation is determined by b^2-4ac, which is known as the **discriminant** of the equation. There are four cases.

Discriminant	Nature of the Roots
1. $b^2-4ac>0$ and a perfect square	The roots are real, rational, and unequal.
2. $b^2-4ac>0$ and not a perfect square	The roots are real, irrational, and unequal.

Second-Degree Equations and Inequalities

3. $b^2 - 4ac = 0$ The roots are real, rational, and equal.
4. $b^2 - 4ac < 0$ The roots are complex conjugates.

Example 4   Without solving for x, determine the nature of the roots for $5x^2 - 4x + 3 = 0$

Solution   For $a = 5$, $b = -4$ and $c = 3$,
$$b^2 - 4ac = (-4)^2 - 4(5)(3)$$
$$= 16 - 60$$
$$= -44$$

Because $-44 < 0$, the roots are complex conjugates. ■

Example 5   Determine the values of k so that the solution of $x^2 - kx + 4 = 0$ has one solution.

Solution   The equation $x^2 - kx + 4 = 0$ has one solution if $b^2 - 4ac = 0$. Letting $a = 1$, $b = -k$ and $c = 4$,
$$b^2 - 4ac = (-k)^2 - 4(1)(4) = 0$$
$$k^2 - 16 = 0$$
$$k^2 = 16$$
$$k = 4 \quad \text{or} \quad k = -4$$

Thus $x^2 - kx + 4 = 0$ has one solution if k is 4 or -4. ■

6–4 Exercises

In exercises 1–26, solve each quadratic equation by using the quadratic formula. (See examples 1–3.)

1. $x^2 - x + 20 = 0$
2. $x^2 - 9x + 18 = 0$
3. $2m^2 + 9m - 5 = 0$
4. $3n^2 + 5n - 2 = 0$
5. $2x^2 + 5x - 12 = 0$
6. $5x^2 - 6x - 8 = 0$
7. $2a^2 = 7a + 4$
8. $15d^2 = -7d + 2$
9. $x^2 - 4x + 13 = 0$
10. $x^2 - 10x + 29 = 0$
11. $t^2 - 6t + 58 = 0$
12. $s^2 + 12s + 52 = 0$
13. $2x^2 - 4x + 5 = 0$
14. $4x^2 - 2x - 3 = 0$
15. $5p^2 - 4p - 3 = 0$
16. $3r^2 - 4r - 2 = 0$
17. $x^2 - \frac{2}{3}x = \frac{1}{2}$
18. $x^2 - \frac{1}{3}x = \frac{3}{2}$
19. $\frac{c^2 - 4}{2} + c = \frac{3}{4}$
20. $\frac{h^2 + 3}{3} + \frac{h}{3} = 4$
21. $\frac{2x^2 - 3}{2} + x = \frac{1}{2}$

22. $\dfrac{3x^2 - 1}{2} + \dfrac{x}{3} = 1$ 23. $4u^2 + 9 = 0$ 24. $3z^2 + 7 = 0$

25. $5x^2 + 2x = 0$ 26. $6x^2 + 5x = 0$

In exercises 27–36, determine the nature of the roots by using the discriminant of the quadratic formula. Do not solve for x. (See example 4.)

27. $x^2 + 7x + 12 = 0$ 28. $x^2 - 5x + 6 = 0$ 29. $4x^2 + 12x = -9$

30. $16x^2 - 8x = -1$ 31. $x^2 = 6x - 13$ 32. $x^2 = 18x - 85$

33. $x^2 + \dfrac{x}{2} - \dfrac{5}{2} = 0$ 34. $\dfrac{3}{2}x^2 + x - \dfrac{5}{2} = 0$ 35. $\dfrac{2}{7}x^2 = -\dfrac{2}{7}x - \dfrac{3}{7}$

36. $\dfrac{1}{2}x^2 = -\dfrac{1}{5} - \dfrac{3}{10}x$

37. Determine the values of k so that $x^2 + kx + 3 = 0$ will have one solution. (See example 5.)

38. Determine the values of k so that $x^2 + 3x + k - 1 = 0$ will have imaginary roots.

39. Determine the values of k so that $kx^2 - 3x + 2 = 0$ will have real roots.

40. Determine the values of k so that $3x^2 + 5x + k + 2 = 0$ will have imaginary roots.

6–5 Radical Equations

Radical equations were introduced in section 5–5. The general procedure given in that section for solving equations containing radicals is used in this section. Generally, as you recall, the procedure is to isolate a radical term with a variable in the radicand. If the radical in the isolated term is of order 2, then both sides of the equation are squared. If it is of order 3, then both sides of the equation are cubed, and so forth. This procedure is repeated as many times as necessary to eliminate all radicals in the equation. As usual, it is necessary to check the proposed solutions in the original equation for extraneous roots.

Example 1 Solve $\sqrt{x + 1} = x - 1$ for x.

Solution Square both sides of the equation.

$$(\sqrt{x + 1})^2 = (x - 1)^2$$
$$x + 1 = x^2 - 2x + 1$$
$$x^2 - 3x = 0 \qquad \text{Write the equation in standard form.}$$
$$x(x - 3) = 0 \qquad \text{Factor the left side.}$$
$$x = 0 \quad \text{or} \quad x = 3 \qquad \text{Solve for } x.$$

Second-Degree Equations and Inequalities 231

Check: We now check for extraneous roots.

For $x = 0$:   For $x = 3$:
$\sqrt{0 + 1} = 0 - 1$ $\sqrt{3 + 1} = 3 - 1$
$1 = -1$ $\sqrt{4} = 2$
$1 \neq -1$ $2 = 2$
0 is extraneous.   3 checks as a solution.

The solution set for $\sqrt{x + 1} = x - 1$ is $\{3\}$. ■

Example 2   Solve $\sqrt{3 - x} + \sqrt{x + 2} = 3$ for x.

Solution   First, arrange the equation so that one radical term is on each side of the equation.

$\sqrt{3 - x} = 3 - \sqrt{x + 2}$
$(\sqrt{3 - x})^2 = (3 - \sqrt{x + 2})^2$ Square both sides of the equation.
$3 - x = 9 - 6\sqrt{x + 2} + x + 2$
$-2x - 8 = -6\sqrt{x + 2}$ Collect all terms, except the radical term, on the left side of the equation.
$x + 4 = 3\sqrt{x + 2}$ Divide the equation by (-2).
$(x + 4)^2 = (3\sqrt{x + 2})^2$ Square both sides again.
$x^2 + 8x + 16 = 9(x + 2)$
$x^2 + 8x + 16 = 9x + 18$ Simplify the right side.
$x^2 - x - 2 = 0$ Write the equation in standard form.
$(x - 2)(x + 1) = 0$ Factor the left side.
$x - 2 = 0$ or $x + 1 = 0$ Solve for x.
$x = 2$ or $x = -1$

Check: For $x = 2$:   For $x = -1$:
$\sqrt{3 - 2} + \sqrt{2 + 2} = 3$ $\sqrt{3 - (-1)} + \sqrt{-1 + 2} = 3$
$\sqrt{1} + \sqrt{4} = 3$ $\sqrt{4} + \sqrt{1} = 3$
$1 + 2 = 3$ $2 + 1 = 3$
$3 = 3$ $3 = 3$
2 is a solution. -1 is a solution.

The solution set is $\{2, -1\}$. ■

Example 3   Solve $\sqrt[3]{x^2 - 2x} = 2$ for x.

Solution   Cube both sides of the equation.

$$(\sqrt[3]{x^2 - 2x})^3 = (2)^3$$
$$x^2 - 2x = 8$$
$$x^2 - 2x - 8 = 0 \quad \text{Write the equation in standard form.}$$
$$(x - 4)(x + 2) = 0 \quad \text{Factor.}$$
$$x = 4 \quad \text{or} \quad x = -2$$

Check: For $x = 4$:

$$\sqrt[3]{4^2 - 2(4)} = 2$$
$$\sqrt[3]{16 - 8} = 2$$
$$\sqrt[3]{8} = 2$$
$$2 = 2$$

For $x = -2$:

$$\sqrt[3]{(-2)^2 - 2(-2)} = 2$$
$$\sqrt[3]{4 + 4} = 2$$
$$\sqrt[3]{8} = 2$$
$$2 = 2$$

The solution set is $\{4, -2\}$. ■

6–5 Exercises

In exercises 1–26, solve each equation. Check for extraneous roots. (See examples 1–3.)

1. $\sqrt{x^2 + 9} = 5$
2. $\sqrt{x^2 + 36} = 10$
3. $\sqrt{m^2 + 6m} = 4$
4. $\sqrt{2n^2 + 3n} = 3$
5. $x = \sqrt{14x - 48}$
6. $-x = \sqrt{7x - 12}$
7. $\sqrt{2p^2 + 12} = \sqrt{11p}$
8. $\sqrt{3t^2 + 13t} = \sqrt{10}$
9. $3x - 2 = \sqrt{7(-3x + 2)}$
10. $2x - 1 = \sqrt{2(2x + 11)}$
11. $4 - \sqrt{s - 2} = s$ *(See example 2.)*
12. $3 - \sqrt{u + 3} = u$
13. $\sqrt{x + 2} - \sqrt{2x + 2} = -1$
14. $\sqrt{x + 5} + 1 = \sqrt{2x + 11}$
15. $\sqrt{a + 6} - \sqrt{3a + 7} - 1 = 0$
16. $\sqrt{2r + 27} - \sqrt{r + 10} - 2 = 0$
17. $\sqrt{x + 1} + \sqrt{x + 2} = \sqrt{2x + 3}$
18. $\sqrt{x + 3} + \sqrt{x} = \sqrt{x + 8}$
19. $\sqrt{v + 2} - \sqrt{2v + 5} = -\sqrt{v - 1}$
20. $\sqrt{u + 4} + \sqrt{u + 9} = \sqrt{u + 25}$
21. $2\sqrt{x + 3} - \sqrt{x + 10} = \dfrac{6}{\sqrt{x + 3}}; \quad x \neq -3$
22. $3\sqrt{x + 1} - \sqrt{x + 6} = \dfrac{6}{\sqrt{x + 1}}; \quad x \neq -1$
23. $\sqrt[3]{p^2 - 12p} = -3$ *(See example 3.)*
24. $\sqrt[3]{n^2 - 6n} = -2$

Second-Degree Equations and Inequalities

25. $\sqrt[4]{x^2 + 6x} = 2$

26. $\sqrt[4]{2x^2 - 9x} = 3$

27. The resonant frequency, f, for a tuned circuit is

$$f = \frac{1}{2\pi\sqrt{LC}}$$

Solve this equation for L.

28. Use the equation given in exercise 27 and find the frequency (in hertz) for a tuned circuit in which $L = 250 \times 10^{-6}$ and $C = 100 \times 10^{-12}$. Use $\pi = 3.14$.

29. A calculus student differentiated a function and obtained the derivative, y', to be

$$y' = \frac{2x}{\sqrt{2x^2 + 4}}$$

For what value (or values) of x will y' have a value of 1? (All proposed solutions must be checked.)

30. Find the value of y' in exercise 29 if $x = -2$.

6–6 Equations that are Quadratic in Form

Many times an equation that is not quadratic can be made to take the form of a quadratic equation by an appropriate substitution. In the equation $x^4 - 13x^2 + 36 = 0$, if y is substituted for x^2, the equation becomes $y^2 - 13y + 36 = 0$, which is now quadratic and may be solved by the methods already discussed for solving quadratic equations. An equation such as $(x^2 - 4)^2 + 5(x^2 - 4) + 4 = 0$ becomes quadratic if $y = x^2 - 4$ is substituted into the equation to give $y^2 + 5y + 4 = 0$. When the equation for y has been solved, the solution to the original equation can be found by substituting the expression represented by y into the y-equation. A few examples should make the procedure clear. All proposed solutions should be checked in the original equation to determine if any are extraneous.

Example 1   Solve $x^4 - 13x^2 + 36 = 0$ for x.

Solution   Substitute y for x^2 and y^2 for x^4 in the equation to make it quadratic.

$y^2 - 13y + 36 = 0$	Substitute.
$(y - 4)(y - 9) = 0$	Factor.
$y = 4 \quad$ or $\quad y = 9$	Solve for y.
$x^2 = 4 \quad$ or $\quad x^2 = 9$	Substitute x^2 for y.
$x = \pm 2 \quad$ or $\quad x = \pm 3$.	Solve for x.

Check: For $x = \pm 2$: For $x = \pm 3$:
$$(\pm 2)^4 - 13(\pm 2)^2 + 36 = 0 \qquad (\pm 3)^4 - 13(\pm 3)^2 + 36 = 0$$
$$16 - 52 + 36 = 0 \qquad 81 - 117 + 36 = 0$$
$$52 - 52 = 0 \qquad 117 - 117 = 0$$
$$0 = 0 \qquad 0 = 0$$

In the checks, both 2 and -2 (as well as 3 and -3) are checked simultaneously, since all powers in $x^4 - 13x^2 + 36 = 0$ are even powers.

The solution set is $\{2, -2, 3, -3\}$. ■

Example 2 Solve $(x^2 - 4)^2 + 5(x^2 - 4) + 4 = 0$ for x.

Solution Let $y = x^2 - 4$ and let $y^2 = (x^2 - 4)^2$.

$$y^2 + 5y + 4 = 0 \qquad \text{Substitute.}$$
$$(y + 4)(y + 1) = 0 \qquad \text{Factor.}$$
$$y = -4 \quad \text{or} \quad y = -1 \qquad \text{Solve for } y.$$
$$x^2 - 4 = -4 \quad \text{or} \quad x^2 - 4 = -1 \qquad \text{Substitute } x^2 - 4 \text{ for } y.$$
$$x^2 = 0 \quad \text{or} \quad x^2 = 3$$
$$x = 0 \quad \text{or} \quad x = \pm\sqrt{3}. \qquad \text{Solve for } x.$$

Check: For $x = 0$:
$$(0^2 - 4)^2 + 5(0^2 - 4) + 4 = 0$$
$$16 + (-20) + 4 = 0$$
$$0 = 0$$

For $x = \pm\sqrt{3}$:
$$[(\pm\sqrt{3})^2 - 4]^2 + 5[(\pm\sqrt{3})^2 - 4] + 4 = 0$$
$$[3 - 4]^2 + 5[3 - 4] + 4 = 0$$
$$(-1)^2 + 5(-1) + 4 = 0$$
$$1 - 5 + 4 = 0$$
$$0 = 0$$

The solution set is $\{-\sqrt{3}, 0, \sqrt{3}\}$. ■

Example 3 Solve $x^{2/3} - 4x^{1/3} + 3 = 0$ for x.

Solution Let $y = x^{1/3}$ and let $y^2 = x^{2/3}$.

$$y^2 - 4y + 3 = 0 \qquad \text{Substitute.}$$
$$(y - 3)(y - 1) = 0 \qquad \text{Factor.}$$
$$y = 3 \quad \text{or} \quad y = 1 \qquad \text{Solve for } y.$$
$$x^{1/3} = 3 \quad \text{or} \quad x^{1/3} = 1 \qquad \text{Substitute } x^{1/3} \text{ for } y.$$
$$x = 27 \quad \text{or} \quad x = 1 \qquad \text{Cube both sides.}$$

Second-Degree Equations and Inequalities 235

Check: For $x = 27$:
$$(27)^{2/3} - 4(27)^{1/3} + 3 = 0$$
$$(3)^2 - 4(3) + 3 = 0$$
$$9 - 12 + 3 = 0$$
$$0 = 0$$

For $x = 1$:
$$1^{2/3} - 4(1)^{2/3} + 3 = 0$$
$$1 - 4 + 3 = 0$$
$$0 = 0$$

The solution set is $\{1, 27\}$.

Example 4 Solve for x: $x^4 - x^2 - 20 = 0$.

Solution Let $y = x^2$ and let $y^2 = x^4$.

$$y^2 - y - 20 = 0 \quad \text{Substitute.}$$
$$(y - 5)(y + 4) = 0 \quad \text{Factor.}$$
$$y = 5 \quad \text{or} \quad y = -4 \quad \text{Solve for } y.$$
$$x^2 = 5 \quad \text{or} \quad x^2 = -4 \quad \text{Substitute } x^2 \text{ for } y.$$
$$x = \pm\sqrt{5} \quad \text{or} \quad x = \pm 2i \quad \text{Solve for } x.$$

Check: For $x = \pm\sqrt{5}$:
$$(\pm\sqrt{5})^4 - (\pm\sqrt{5})^2 - 20 = 0$$
$$25 - 5 - 20 = 0$$
$$25 - 25 = 0$$
$$0 = 0$$

For $x = \pm 2i$:
$$(\pm 2i)^4 - (\pm 2i)^2 - 20 = 0$$
$$16i^4 - 4i^2 - 20 = 0$$
$$16 - 4(-1) - 20 = 0$$
$$16 + 4 - 20 = 0$$
$$20 - 20 = 0$$
$$0 = 0$$

The solution set is $\{\sqrt{5}, -\sqrt{5}, 2i, -2i\}$.

6-6 Exercises

In exercises 1–24, find the solution set for each equation. Check for extraneous solutions. Some solution sets may contain complex numbers. *(See examples 1–4.)*

1. $x^4 - 5x^2 + 4 = 0$
2. $x^4 - 10x^2 + 9 = 0$
3. $y^4 - 13y^2 + 36 = 0$
4. $z^4 + z^2 - 20 = 0$
5. $x^4 + 6x^2 - 27 = 0$
6. $x^4 + 3x^2 - 4 = 0$
7. $t^6 + 26t^3 - 27 = 0$
8. $u^6 - 7u^3 - 8 = 0$
9. $x^{2/3} + 2x^{1/3} - 3 = 0$ *(See example 3.)*
10. $x^{2/3} + 7x^{1/3} + 10 = 0$
11. $s^{4/3} - 6s^{2/3} + 8 = 0$
12. $t^{3/2} - 3t^{3/4} + 2 = 0$
13. $x + \sqrt{x} - 6 = 0$
14. $x - 2\sqrt{x} - 3 = 0$
15. $v + 2\sqrt{v} - 24 = 0$
16. $p - 6\sqrt{p} - 27 = 0$
17. $(x^2 - 1)^2 + 2(x^2 - 1) - 3 = 0$

18. $(2x^2 + 2)^2 + 7(2x^2 + 2) + 6 = 0$
19. $(x^2 + 2)^2 - 5(x^2 + 2) + 6 = 0$
20. $(x^2 + 1)^2 - 2(x^2 + 1) + 1 = 0$
21. $m^{-2} + 5m^{-1} + 6 = 0$
22. $10n^{-2} + 7n^{-1} + 1 = 0$
23. $20x^{-2} + 9x^{-1} + 1 = 0$
24. $7x^{-2} + 8x^{-1} + 1 = 0$
25. Solve the equation $x^2 + 4xy + 4y^2 - 2 = 0$ for x in terms of y. Think of the equation as $1 \cdot x^2 + (4y)x + (4y^2 - 2) = 0$. Then use the quadratic formula, letting $a = 1$, $b = 4y$, and $c = (4y^2 - 2)$.
26. Solve the equation $2x^2 + 6xy + 2y^2 - 6 = 0$ for x in terms of y. See exercise 25.

6–7 ■ Quadratic Inequalities

First-degree inequalities are discussed in section 4–4. Rules relating to linear inequalities are also discussed there. We now turn our attention to the type of quadratic inequalities that may be solved by factoring. Quadratic inequalities may be defined as follows.

Definition 6–10

Quadratic Inequalities

Inequalities of the form $ax^2 + bx + c > 0$, $ax^2 + bx + c < 0$, $ax^2 + bx + c \geq 0$, or $ax^2 + bx + c \leq 0$ where a, b, and c are constants and $a \neq 0$, are called **quadratic inequalities.**

Consider the quadratic inequality $x^2 + x - 12 > 0$. The left side may be factored as $(x + 4)(x - 3) > 0$. The left side is now the product of two factors. The process of finding the solution set of $x^2 + x - 12 > 0$ is started by taking a look at these two factors, $x + 4$ and $x - 3$. We note that $x + 4$ has a value of zero if $x = -4$ and $x - 3$ has a value of zero if $x = 3$. That is,

$$x + 4 = 0 \quad \text{if} \quad x = -4$$

and

$$x - 3 = 0 \quad \text{if} \quad x = 3$$

The values of x that cause the factors in $(x + 4)(x - 3) > 0$ to have a value of zero are called **critical numbers.** These two numbers, -4 and 3, separate a real number line into three regions. The solution set of the inequality will be contained in one or more of these three regions. The three regions are shown and labeled as regions A, B, and C in figure 6–1.

Second-Degree Equations and Inequalities   237

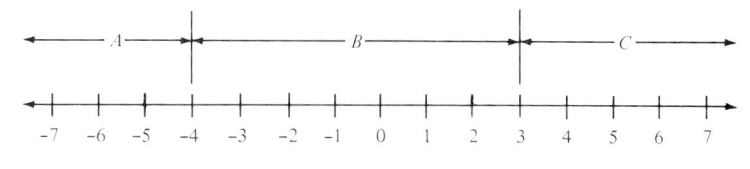

Figure 6–1

To determine which regions, or intervals, on the number line contain numbers in the solution set of the inequality, pick an arbitrary number from each interval and test it in the factored form of the inequality. We pick -5, 0, and 4 from regions A, B, and C, respectively.

$(x + 4)(x - 3) > 0$	$(x + 4)(x - 3) > 0$	$(x + 4)(x - 3) > 0$
If $x = -5$, then:	If $x = 0$, then:	If $x = 4$, then:
$(-5 + 4)(-5 - 3) > 0$	$(0 + 4)(0 - 3) > 0$	$(4 + 4)(4 - 3) > 0$
$(-1)(-8) > 0$	$4(-3) > 0$	$8(1) > 0$
$8 > 0$	$-12 > 0$	$8 > 0$
True	False	True

We get true results for numbers contained in regions A and C. Since region A is the set of all real numbers x such that $x < -4$ and region C is the set of all real numbers x such that $x > 3$, the solution set is $\{x|x < -4 \text{ or } x > 3\}$. The graph of the solution set is shown in figure 6–2.

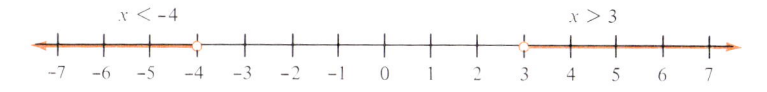

Figure 6–2

The endpoints of the graph are shown as open circles to indicate that -4 and 3 are not members of the solution set.

Procedure 6–2

Solving Favorable Quadratic Inequalities

1. Find the critical numbers by equating the factors of the inequality to zero and solving for the variable.
2. Plot the critical numbers on a number line.
3. Choose an arbitrary number in each region determined by the critical numbers and test it in the factored form of the inequality. Draw the graph of the solution set over the region or regions containing arbitrary numbers that made the inequality a true statement.

Example 1

Find the solution set for $2x^2 - 5x - 12 \leq 0$, and draw the graph for the solution set.

Solution The given inequality factors as $(2x + 3)(x - 4) \leq 0$. The critical numbers are obtained by setting the factors equal to zero.

$$2x + 3 = 0 \qquad x - 4 = 0$$
$$x = -\frac{3}{2} \qquad x = 4$$

Plot these on a number line, as shown in figure 6–3.

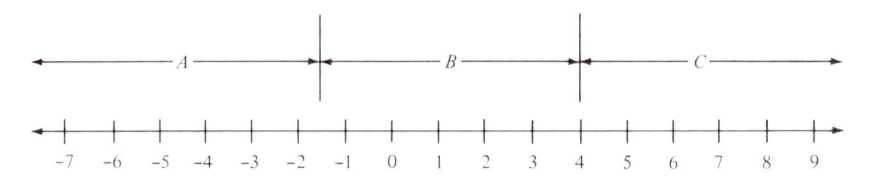

Figure 6–3

Pick -2, 0, and 5 from regions A, B, and C, respectively, as test numbers to be used in $(2x + 3)(x - 4) \leq 0$.

If $x = -2$, then:
$$[2(-2) + 3)][-2 - 4] \leq 0$$
$$(-4 + 3)(-6) \leq 0$$
$$(-1)(-6) \leq 0$$
$$6 \leq 0$$
False

If $x = 0$, then:
$$(2 \cdot 0 + 3)(0 - 4) \leq 0$$
$$3(-4) \leq 0$$
$$-12 \leq 0$$
True

If $x = 5$, then:
$$(2 \cdot 5 + 3)(5 - 4) \leq 0$$
$$(10 + 3)(1) \leq 0$$
$$13(1) \leq 0$$
$$13 \leq 0$$
False

The numbers that make the inequality true lie in the interval on the number line between the two critical numbers. The critical numbers also make the inequality true. Thus the solution set is $\{x| -\frac{3}{2} \leq x \leq 4\}$. The graph of the solution set is shown in figure 6–4.

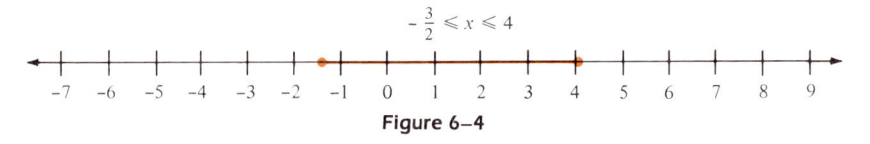

Figure 6–4

The endpoints of the graph are shown as closed dots to indicate that $-\frac{3}{2}$ and 4 are members of the solution set.

Second-Degree Equations and Inequalities   239

Example 2 Find the solution set for $\dfrac{x-2}{x-3} \leq 2$. Also draw the graph of the solution set.

Solution You may be tempted to multiply both sides of the inequality by $x - 3$ to clear it of fractions. However, $x - 3$ contains a variable and we do not know if $x - 3$ represents a positive or negative number. Therefore, if we did multiply both sides by $x - 3$, we would first need to assume it is a positive quantity and then assume it is a negative quantity. This is a tedious procedure and can be avoided if a single fraction is formed from

$$\frac{x-2}{x-3} \leq 2$$

The critical numbers obtained from the fraction are found by determining which numbers cause the numerator and denominator to be zero. (Of course, the one that causes the denominator to be zero will not be part of the solution set, but we may still use it as a critical number.) Let us continue; we write the inequality as a single fraction.

$$\frac{x-2}{x-3} \leq 2 \qquad \text{Given inequality.}$$

$$\frac{x-2}{x-3} - 2 \leq 0 \qquad \text{Subtract 2 from each side of the inequality.}$$

$$\frac{x-2}{x-3} - \frac{2(x-3)}{x-3} \leq 0 \qquad \text{Write each of the terms on the left side with a common denominator of } x - 3.$$

$$\frac{x-2-2x+6}{x-3} \leq 0 \qquad \text{Combine both terms on the left side over the common denominator.}$$

$$\frac{-x+4}{x-3} \leq 0 \qquad \text{Simplify the numerator.}$$

The numerator is zero when $x = 4$. The denominator is zero when $x = 3$. Thus the critical numbers are 3 and 4. Plot these (see figure 6–5).

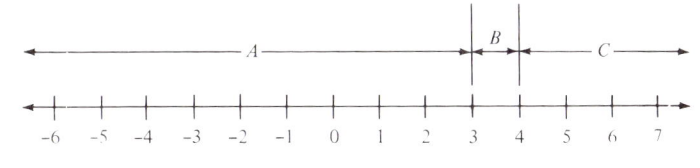

Figure 6–5

Pick numbers from regions A, B, and C. We choose 0, 3.5, and 5. Substitute these into $\frac{-x + 4}{x - 3} \leq 0$.

If $x = 0$, then:
$$\frac{0 + 4}{0 - 3} \leq 0$$
$$\frac{4}{-3} \leq 0$$
True

If $x = 3.5$, then:
$$\frac{-3.5 + 4}{3.5 - 3} \leq 0$$
$$\frac{0.5}{0.5} \leq 0$$
$$1 \leq 0$$
False

If $x = 5$, then:
$$\frac{-5 + 4}{5 - 3} \leq 0$$
$$\frac{-1}{2} \leq 0$$
True

Regions A and C of the number line contain numbers that give true results. Thus we draw the graph over these regions, as in figure 6–6.

Figure 6–6

The number 3 is not in the solution set because it would cause the denominator to be zero. The solution set is $\{x | x < 3 \text{ or } x \geq 4\}$.

6–7 Exercises

In exercises 1–18, find the solution set of each inequality. Draw the graph of the solution set. (See example 1.)

1. $x^2 + 2x - 15 > 0$
2. $x^2 + 7x + 10 > 0$
3. $x^2 - 8x + 12 < 0$
4. $x^2 + x - 6 < 0$
5. $x^2 \leq -x + 20$
6. $x^2 \leq -7x - 6$
7. $2x^2 + 5x < 3$
8. $2x^2 - 5x < -2$
9. $3x^2 \geq -14x + 5$
10. $2x^2 \geq -x + 3$
11. $4x^2 > -x + 5$
12. $5x^2 + 14x > 3$
13. $\frac{2x - 3}{x + 2} < 3$ (See example 2.)
14. $\frac{x + 1}{x - 3} < 1$ (This inequality has only one critical number.)
15. $\frac{x + 1}{x + 4} \geq 1$
16. $-4 < \frac{x + 5}{x - 5}$
17. $\frac{x^2 + 5x}{6} > -1$
18. $\frac{x^2 + 6x}{8} > -1$

In exercises 19–22, plot each critical number and find the solution set by examining each region. Draw the graph of the solution set.

19. $(x + 2)(x - 3)(x + 5) < 0$ (Hint: Plot the three critical numbers and check the inequality with a number from each interval formed by the three critical numbers.)

Second-Degree Equations and Inequalities

20. $(x + 3)(x - 1)(x - 4) < 0$
21. $(2x + 1)(x - 5)(x + 4) > 0$
22. $(3x + 4)(x - 3)(x - 1) > 0$

In exercises 23–26, finish factoring the left side of each inequality. Then find the solution set of each and draw the graph of the solution set.

23. $(x - 1)(x^2 + 4x + 3) < 0$
24. $(x + 1)(x^2 - x - 6) \leq 0$
25. $(x + 2)(2x^2 - 9x + 4) \geq 0$
26. $(x + 2)(3x^2 - 7x + 4) > 0$

In exercises 27–30, the inequalities are not factorable over the set of rational numbers. However, the critical numbers can be found by using the quadratic formula. Find the critical numbers, plot them, and draw the graph of the solution set.

27. $x^2 + 4x - 3 < 0$
28. $x^2 - 2x - 1 > 0$
29. $x^2 + x - 1 \geq 0$
30. $5x^2 - 2x + 1 \leq 0$

In exercises 31–32, find the solution set. (See example 2.)

31. $\dfrac{x^2 + 2x - 15}{x^2 + 2x - 3} \geq 0$
32. $\dfrac{x^2 - 9}{x^2 + 10x + 25} < 0$

6–8 Applications

Many problems of a practical nature can be solved by finding the solution set of a quadratic equation. Because every quadratic equation has two solutions, it is necessary to check both solutions in the context of the original problem. Sometimes, one (or maybe both) of the solutions will not fit the situation stated in the problem. For example, distance, rate of travel, height, and time are considered as positive quantities. If a quadratic equation has a negative number as one of its solutions in situations such as these, we would need to discard it as not being a practical solution. Consider the following examples.

Example 1 Find two consecutive, odd, positive integers whose product is 35.

Solution In section 4–3, we saw how to represent two consecutive odd integers as x and $x + 2$. Using these two representations for the integers, we write:

$$\underbrace{x \cdot (x + 2)}_{\text{Find two consecutive, odd, positive integers whose product}} = \underbrace{35}_{\text{is 35}}$$

Solve the equation.

$$x(x + 2) = 35$$
$$x^2 + 2x = 35$$
$$x^2 + 2x - 35 = 0 \qquad \text{Write the equation in standard form.}$$
$$(x + 7)(x - 5) = 0 \qquad \text{Factor.}$$
$$x + 7 = 0 \quad \text{or} \quad x - 5 = 0 \qquad \text{Set each factor equal to zero.}$$
$$x = -7 \quad \text{or} \quad x = 5$$

If $x = 5$, then $x + 2 = 7$ and the numbers are 5 and 7. We cannot use -7, since the problem stated the two numbers had to be positive. ■

Example 2 The difference of a whole number and twice its reciprocal is $\frac{47}{7}$. What is the value of the whole number?

Solution The reciprocal of any nonzero whole number n is $1/n$. Using these expressions to represent the numbers, we write

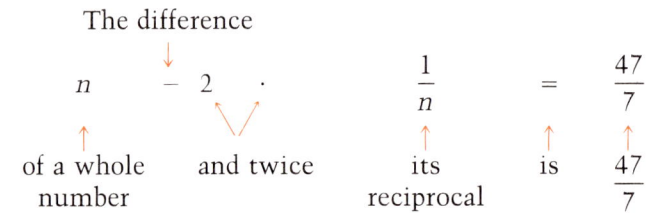

$$n - 2 \cdot \frac{1}{n} = \frac{47}{7}$$

(The difference of a whole number and twice its reciprocal is $\frac{47}{7}$)

Now solve the equation for n.

$$n - 2\left(\frac{1}{n}\right) = \frac{47}{7}$$
$$7n^2 - 14 = 47n \qquad \text{Multiply both sides by } 7n.$$
$$7n^2 - 47n - 14 = 0 \qquad \text{Write the equation in standard form.}$$
$$(n - 7)(7n + 2) = 0 \qquad \text{Factor.}$$
$$n - 7 = 0 \quad \text{or} \quad 7n + 2 = 0 \qquad \text{Set each factor equal to zero.}$$
$$n = 7 \quad \text{or} \quad n = -\frac{2}{7} \qquad \text{Solve for } n.$$

We discard $-\frac{2}{7}$, since it is not a whole number. The number is 7. As a check, we note that $7 - \frac{2}{7} = \frac{47}{7}$ is a true statement. ■

Example 3 The length of a rectangle is 2 centimeters more than twice its width. If the area of the rectangle is 60 square centimeters, what are its dimensions?

Solution Refer to figure 6–7.

Second-Degree Equations and Inequalities   243

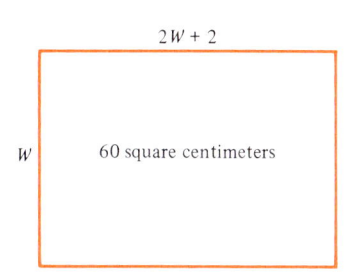

Figure 6–7

The area of a rectangle is $A = LW$. If we let W represent the width of the rectangle, then $2W + 2$ will represent the length. If we substitute $A = 60$ and $L = 2W + 2$ into $A = LW$, we have the following.

$$60 = (2W + 2)(W)$$
$$60 = 2W^2 + 2W \quad \text{Simplify the right side.}$$
$$W^2 + W - 30 = 0 \quad \text{Divide both sides by 2 and write in standard form.}$$
$$(W + 6)(W - 5) = 0 \quad \text{Factor.}$$
$$W + 6 = 0 \quad \text{or} \quad W - 5 = 0 \quad \text{Solve for } W.$$
$$W = -6 \quad \text{or} \quad W = 5$$

Length is a positive quantity, so -6 must be discarded. The dimensions are 5 centimeters wide and 12 centimeters long. ■

Example 4   A box with an open top is to be formed by cutting a 1-inch square from each corner of a rectangular piece of metal and bending up the sides. If the length of the piece of metal is 3 inches more than its width, find the dimensions of the box if its volume is to be 70 cubic inches.

Solution   For problems of this type, a sketch is helpful. See figure 6–8.

(a)   (b)

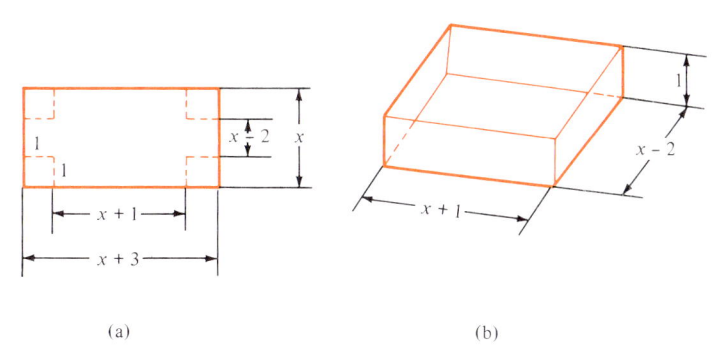

Figure 6–8

Let x and $x + 3$ represent the width and length, respectively, of the metal before the squares are removed. Once the squares are cut from the corners, the width and length of the box will be $x - 2$ and $x + 1$, respectively, and its height will be the same as the dimension of one side of the squares cut from each corner. The volume of a rectangular box is $V = LWH$, where L is the length, W is the width, and H is the height. Use figure 6–8(b).

$(x + 1)(x - 2)(1) = 70$	$LWH = V.$
$x^2 - x - 2 = 70$	
$x^2 - x - 72 = 0$	Write in standard form.
$(x - 9)(x + 8) = 0$	Factor.
$x - 9 = 0$ or $x + 8 = 0$	Set each factor equal to zero.
$x = 9$ or $x = -8$	Solve for x.

If $x = 9$, the length, $x + 1$, is 10. The width, $x - 2$, is 7. The box will have dimensions 10 by 7 by 1 inches. The solution -8 is not used because length is a positive quantity. ■

Example 5

Pete's boat can travel 15 miles per hour in still water. It takes him 2 hours more to travel 50 miles upstream than it does to travel 60 miles downstream. Find the speed of the current in the river.

Solution

Let s be the speed of the current. Going upstream, the current opposes the boat and the speed of the boat is $15 - s$. The speed of the boat going downstream $15 + s$. A table is helpful in organizing the information. We fill in spaces within the box corresponding to distance, rate, and time for the upstream and downstream trips. We have numbers and expressions for the distances and rates. These are entered first. We get,

	Upstream	Downstream
d	50	60
r	$15 - s$	$15 + s$
t		

Once two lines are filled in within the table, the third line (for time, in this case) is obtained from the two completed lines. already filled in. If we solve the equation $d = rt$ for t, we get $t = d/r$. The time spaces in the third line can now be filled in by dividing the rates into the distances immediately above them. Doing this, we get the following.

Second-Degree Equations and Inequalities

	Upstream	Downstream
d	50	60
r	$15 - s$	$15 + s$
t	$\dfrac{50}{15 - s}$	$\dfrac{60}{15 + s}$

The equation needed for solving the problem is written from the expressions contained in the last line filled in, which was the third line. We know

(Time upstream) − (time downstream) = 2 hours

Thus

$$\frac{50}{15 - s} - \frac{60}{15 + s} = 2$$

$50(15 + s) - 60(15 - s) = 2(15 - s)(15 + s)$ Multiply both sides by $(15 - s)(15 + s)$.

$750 + 50s - 900 + 60s = 450 - 2s^2$ Remove parentheses.

$ -150 + 110s = 450 - 2s^2$ Collect like terms.

$2s^2 + 110s - 600 = 0$ Write the equation in standard form.

$s^2 + 55s - 300 = 0$ Divide both sides by 2.

$(s - 5)(s + 60) = 0$ Factor the left side.

$s = 5 \text{ or } s = -60$ Set each factor equal to zero and solve for s.

The current is moving at 5 miles per hour. The value −60 has no meaning in this problem.

Check: The boat goes $15 - 5 = 10$ miles per hour upstream. It goes $15 + 5 = 20$ miles per hour downstream. To go 50 miles upstream requires $\frac{50}{10} = 5$ hours. To go 60 miles downstream requires $\frac{60}{20} = 3$ hours, and we see it takes 2 hours longer to go 50 miles upstream than it does to go 60 miles downstream.

6–8 Exercises

In exercises 1–24, translate each into an equation and solve for the variable.

1. Two positive numbers differ by 4 and their product is 45. Find the two numbers.

2. One positive number is 3 more than another positive number. If their product is 88, find the two numbers.

3. The sum of two numbers is 11, and their product is 28. Find the two numbers.
4. The sum of two numbers is 13. The sum of their squares is 89. Find the two numbers.
5. The sum of two numbers is 16. The sum of their squares is 130. Find the two numbers.
6. Find two consecutive integers whose product is 90.
7. Find two consecutive even integers whose product is 80.
8. The sum of a number and its reciprocal is $\frac{65}{8}$. What is the number?
9. A whole number minus twice its reciprocal is $5\frac{2}{3}$. Find the whole number.
10. A whole number minus four times its reciprocal is $6\frac{3}{7}$. What is the whole number?
11. The length of a rectangle is 2 centimeters more than its width. If its area is 35 square centimeters, what are its dimensions?
12. The length of a rectangle is 1 meter more than twice its width. If its area is 21 square meters, find the width and length.
13. The length of a rectangle is 3 feet more than twice its width. If the area is 65 square feet, find its dimensions.
14. The area of a triangle is $A = \frac{1}{2}bh$, where b is the base and h is the altitude. If a triangle has an altitude that is $\sqrt{3}$ times one-half its base and the area is $25\sqrt{3}$ square inches, what are the dimensions of the base and altitude?
15. A triangle has an area of $81\sqrt{3}$ square centimeters. If the altitude is $\sqrt{3}$ times one-half the base, find the dimensions of the altitude and base. (See exercise 14 for the formula for the area of a triangle.)
16. A box with an open top is to be formed by cutting squares 2 centimeters on a side from each corner of a rectangular sheet of metal and bending up the sides. If the length of the piece of metal is 2 centimeters more than its width, find the dimensions of the box if its volume is to be 48 cubic centimeters. (See example 4).
17. A box with an open top is to be formed by cutting squares from each corner of a rectangular sheet of aluminum and bending up the sides. The squares cut from each corner have dimensions equal to one-fifth the width of the aluminum sheet. If the box has a volume of 27 cubic inches and a length of 9 inches, what size square was removed from each corner of the aluminum sheet?
18. A rectangular plot of ground is 25 feet by 20 feet. A uniform concrete border is to be poured on all four sides of the plot of ground. The number of square feet in the border is to be 196. How wide should it be?
19. Mike wishes to dress up his rectangular swimming pool by making a uniform brick walkway around two sides and one end of the pool. If the pool has dimensions of 15 feet by 25 feet, how wide should he make the walkway if the walkway is to contain 183 square feet? Assume the dimension of the pool whose measure is 25 feet to be the end of the pool and the ones whose dimensions are 15 feet to the sides of the pool.

20. A rectangle has a length that is 5 feet more than its width. If the length is decreased by 2 feet and the width is increased by 2 feet, the area of the new rectangle is 506 square feet. Find the dimensions of the original rectangle.

21. The length of a rectangle is 4 centimeter more than twice its width. If the length is decreased by 2 centimeter and the width increased by 1 centimeter, the area of the new rectangle is 200 square centimeter. Find the dimensions of the original rectangle.

22. Lori recently received her private pilot's license and wanted to take Kathy on a cross-country flight. It took $\frac{5}{6}$ of an hour longer to make the 250-mile trip with a headwind of 25 miles per hour than it did to return the same distance with a tail wind of 25 miles per hour. What would be the ground speed of the plane if there were no head or tail winds?

23. Steve and Ross finished painting the front of Steve's house (see exercise 11, section 4–3), and decided to take a bike ride. They had ridden a distance of 26 miles when the bicycle has a flat on both tires. (They were both riding the same bike). They had ridden the last 10 miles of the trip at a speed that was an average of 3 miles per hour slower than the average speed of the first part of the trip. If the entire trip took 4 hours, find the average rate of speed for both parts of the trip.

24. Terri made a car trip of 180 miles and then returned over the same route. It took her $\frac{3}{5}$ of an hour longer to return than it did to go. The average rate of speed going was 10 miles per hour faster than her average return speed. Find the time that she needed to make the return trip.

For problems 25–28, see the description of a right triangle given in problem 25.

25. From geometry, the longest side of a right triangle is called the *hypotenuse* and the two sides which form the right angle are called the *legs*.

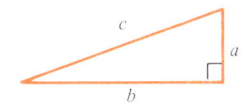

    The Pythagorean theorem states that if c is the length of the hypotenuse and a and b are the lengths of the legs, then $c^2 = a^2 + b^2$. If the hypotenuse of a right triangle is 10 centimeters long and one leg is 8 centimeters long, find the length of the other leg.

26. Find the length of the hypotenuse of a right triangle if the two legs have lengths of 5 meters and 12 meters.

27. In a right triangle, one of the legs is 1 inch longer than the other leg. If the hypotenuse has a length of 5 inches, find the length of each leg.

28. In a right triangle, one leg is 2 feet more than the other leg. The hypotenuse is 4 feet more than the shorter leg. Find the lengths of all three sides of the triangle.

29. A rectangular storage shed for tools and other equipment is to be built at the back of a factory. The shed will have an area of 700 square feet. One

wall of the shed will be the existing wall of the factory. If the sum of the lengths of the other three sides of the shed is 90 feet, find the dimensions of the shed.

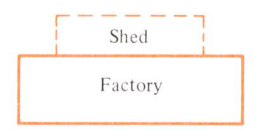

30. A rectangular solid measures 5 inches by 4 inches by 2 inches. The volume of the solid must be increased by 40%. The length will remain the same and the other two dimensions will be increased by equal amounts. What will be the new dimensions of the solid (to the nearest tenth of an inch)?

31. What are the smallest values that two consecutive, odd, positive integers can take if their product is at least 63?

32. Melody has scores of 70, 92, 83, and 91 on her history tests. A grade of B requires a test average between 80 and 90. What is the least grade she can make on her fifth test and maintain a B?

Chapter 6 Summary

Vocabulary

completing the square
complex number
critical numbers
discriminant
imaginary unit
incomplete quadratic equation
multiplicity of roots
pure imaginary number
quadratic equation
quadratic formula
quadratic inequality
standard form of a quadratic equation
roots of a quadratic equation

Definitions

6–1 The **standard form** of a quadratic equation is $ax^2 + bx + c = 0$, where $a \neq 0$. The coefficients a, b, and c represent real numbers.

6–2 The Imaginary Unit

$i = \sqrt{-1}$ and $i^2 = -1$

6-3 Principal Square Root of $-a$

If a is any positive real number, $\sqrt{-a} = i\sqrt{a}$. The symbol $\sqrt{-a}$ is called the *principal square root* of $-a$.

6-4 Complex Number

If a and b are real numbers, then $a + bi$ is called a **complex number.**

6-5 Equality of Complex Numbers

For any two complex numbers $a + bi$ and $c + di$, $a + bi = c + di$ if and only if $a = c$ and $b = d$.

6-6 Sum of Complex Numbers

The sum of two complex numbers $a + bi$ and $c + di$ is

$$(a + bi) + (c + di) = (a + c) + (b + d)i$$

6-7 Difference of Complex Numbers

The difference of two complex numbers $(a + bi)$ and $(c + di)$ is

$$(a + bi) - (c + di) = (a + bi) + (-c - di)$$
$$= (a - c) + (b - d)i$$

6-8 Product of Complex Numbers

The product of two complex numbers $a + bi$ and $c + di$ is

$$(a + bi)(c + di) = (ac - bd) + (ad + bc)i$$

6-9 Quotient of Complex Numbers

The quotient of two complex numbers $a + bi$ and $c + di$ is

$$\frac{a + bi}{c + di} = \frac{a + bi}{c + di} \cdot \frac{c - di}{c - di} = \frac{(ac + bd) + (bc - ad)i}{c^2 + d^2}$$

6-10 Quadratic Inequalities

Inequalities of the form $ax^2 + bx + c > 0$, $ax^2 + bx + c < 0$, $ax^2 + bx + c \geq 0$, or $ax^2 + bx + c \leq 0$ where a, b, and c are constants and $a \neq 0$, are called **quadratic inequalities.**

Rules

6-1 Zero Factor Rule

For real numbers p and q, $pq = 0$ if and only if $p = 0$ or $q = 0$ (or both).

6-2 Powers of i

$i = i$
$i^2 = -1$
$i^3 = -i$
$i^4 = 1$

6-3 The Quadratic Formula

The solutions of any quadratic equation having the form $ax^2 + bx + c = 0$, $a \neq 0$, are given by

$$x = \frac{-b \pm \sqrt{b^2 - 4ac}}{2a}$$

Procedures

6-1 Solving a Quadratic Equation by Completing the Square

Step 1. Arrange the equation in the form $ax^2 + bx = -c$.

Step 2. If $a \neq 1$, divide both sides of the equation by a, where a is the coefficient of the second-degree term.

Step 3. Complete the square on the left side by taking one-half of the coefficient of the first-degree term, squaring this number, and adding it to both sides of the equation. The left side is now a perfect-square trinomial.

Step 4. Express the left side of the equation as the square of a binomial and simplify the right side of the equation.

Step 5. Take the square root of both sides of the equation, form two linear (first-degree) equations, and solve for x.

6-2 Solving Factorable Quadratic Inequalities

1. Find the critical numbers by equating the factors of the inequality to zero and solving for the variable.
2. Plot the critical numbers on a number line.
3. Choose an arbitrary number in each region determined by the critical numbers and test it in the factored form of the inequality. Draw the graph of the solution set over the region or regions containing arbitrary numbers that made the inequality a true statement.

Review Problems

Section 6-1

In problems 1–3, find the solution set by factoring.

1. $(2x - 3)(x + 2) = 0$
2. $6x^2 + 7x = -2$

Second-Degree Equations and Inequalities   251

3. $m = \sqrt{\dfrac{3m + 20}{2}}$

4. Find a quadratic equation whose roots are 5 and $-\dfrac{3}{4}$.

Section 6–2 In problems 5 and 6, solve for x by completing the square.

5. $x^2 - 2x - 15 = 0$ 　　　　　　　　6. $2x^2 + x - 4 = 0$

Section 6–3
7. Simplify: $\sqrt{-32}$.
8. Simplify: $3\sqrt{-27}\sqrt{-12}$.
9. Simplify: i^{22}.
10. Add: $3\sqrt{-8} + 4\sqrt{-18}$.

In problems 11–14, perform the indicated operation and simplify. Leave your answers in the form $a + bi$.

11. $(6 + 5i) + (-2 - 4i)$ 　　　　　　12. $(9 - 7i) - (-8 + 6i)$

13. $(2 - 4i)(3 + 2i)$ 　　　　　　　　14. $\dfrac{2 + 5i}{3 - 4i}$

15. Solve for x and y: $3x + 2yi = -12 - 12i$.

Section 6–4
16. Solve $2x^2 - x - 6 = 0$ by using the quadratic formula.
17. Determine the nature of the roots of $5x^2 - 2x - 3 = 0$ by using the discriminant of the quadratic formula.
18. Determine the values of k so that $2x^2 + kx + 2 = 0$ will have two equal roots.
19. What values of k will cause $4x^2 + 5x + k = 0$ to have imaginary roots?

Section 6–5
20. Solve $\sqrt{r - 6} = r - 8$ for r and check for extraneous roots.
21. Solve $\sqrt{x + 4} + \sqrt{3x + 1} = \sqrt{9x + 4}$ for x and check for extraneous roots.

Section 6–6
22. Solve: $z^{2/3} - 7z^{1/3} - 8 = 0$

Section 6–7
23. Find the solution set for $x^2 - 2x - 15 \geq 0$ and draw the graph of the solution set.
24. Find the solution set for $\dfrac{x + 2}{x - 3} < 2$. Draw the graph of the solution set.

Section 6–8
25. One positive number is 5 more than another positive number. If their product is 24, find the two numbers.
26. The sum of two integers is 11 and the sum of their squares is 65. Find the two integers.

27. The length of the base of a triangle is 1 centimeter more than three times the length of the altitude. If the area is 40 square centimeters, how long is the base of the triangle?

28. Steve and Ross decided that bicycle riding was not for them after having a flat on both tires of the bike. (See exercise 23, section 6–8). Being adventurous, they decided to take a boat trip. They had gone 16 miles upstream when they hit a submerged rock and ruined the motor on their boat. Being unable to fix it, they had to drift with the current of the river the entire return distance. (Their parents were getting a little worried about them.) If the trip drifting back home took 2 hours longer than the trip upstream and the boat can go 12 miles per hour in still water, find the rate at which they were drifting back downstream.

Cumulative Test for Chapters 1–6

Chapter 1

In problems 1–3, answer true or false.

1. If $A = \{5, 8, 9\}$, then $5 \in A$.
2. If $B = \{3, 7, 8\}$, then B is a well-defined set.
3. If $A = \{1, 3, 7\}$ and $B = \{3, 7, 9\}$, then $A \cup B = \{3, 7\}$.
4. Supply grouping symbols to give the indicated answer: $5 \cdot 3 + 2 - 5 = 20$.
5. Which axiom of equality is illustrated by the following? (There is more than one.)

$$\text{If } 10 = \frac{a}{b} \text{ and if } \frac{a}{b} = \frac{20}{2}, \text{ then } 10 = \frac{20}{2}.$$

Chapter 2

6. What is the degree of the polynomial $7x + 3 - 6x^3 + 5x^2$?
7. What is the coefficient of $9y^2$ in the monomial $9x^5y^2z$?
8. Is the expression $3x - (2x - 3)$ a monomial, a binomial, or a trinomial (before any simplifying is done)?
9. Factor completely: $14x^3y + 49x^2y - 84xy$.
10. Square and simplify: $(9x - 2)^2$

Chapter 3

11. What expression should replace the question mark in the following: $\dfrac{5}{x - 5} = \dfrac{?}{2x^2 - 7x - 15}$.
12. Simplify the signs on the fraction $-\dfrac{-6}{-7}$.
13. Multiply and simplify: $\dfrac{2x^2 + x}{10x^2 - 8x} \cdot \dfrac{40x^2 - 32x}{4x}$.
14. Add; reduce the answer if possible: $\dfrac{2x}{x^2 - 16} + \dfrac{8}{x^2 - 16}$.
15. Subtract: $\dfrac{9}{x - 3} - \dfrac{2}{3 - x}$.
16. Simplify: $\dfrac{-8}{x^2 - 4} + \dfrac{x}{x - 2} - \dfrac{x}{x - 2}$.

Chapter 4

17. Solve for r: $r(r - 5) - r = r(r + 6) - 8$.
18. Solve for m. Check your solution.

$$\frac{m - 5}{m} - \frac{m + 1}{m - 1} = \frac{3m - 5}{m^2 - m}$$

19. Solve for x: $|3x + 2| - 14 = 12$.
20. Find the solution for $3|x + 6| - 12 < 15$.

253

21. A coin box contains nickels, dimes, and quarters. It contains four more dimes than nickels and seven more nickels than quarters. How many of each kind of coin are in the coin box if the value of all the coins is $2.25?

22. A house was purchased for $80,000. The first year after purchase the house appreciated (went up in value) by 8%. In the second year after purchase, it appreciated by 12%. What was the house worth at the end of the second year after purchase?

Chapter 5

23. Simplify; leave all exponents positive: $\left(\dfrac{x^3 y^{-2} z}{x^{-2} y z^{-3}}\right)^{-2}$.

24. Simplify: $\sqrt[3]{24 a^7 b^{11}}$.

25. Rationalize the denominator of $\dfrac{8}{\sqrt{2}}$ and simplify.

26. Multiply and simplify: $2\sqrt{3}(\sqrt{6} - 3\sqrt{12})$.

27. Add: $5x\sqrt{45} + 3\sqrt{20x^2}$; $x \geq 0$.

28. Multiply by finding a common index. Leave your answer in radical form: $\sqrt[3]{3}\sqrt[5]{2}$.

29. Solve for x and check your solution: $2\sqrt{x + 3} = -\sqrt{5x - 1}$.

Chapter 6

30. Solve for x: $(2x - 3)(x + 5) = 0$

31. Solve for p by factoring: $2p^2 - 3p = 20$.

32. Solve for x by using the quadratic formula: $2x^2 - 6x = 3$.

33. Simplify: $3\sqrt{-20} + 4\sqrt{-5}$.

34. Multiply and simplify: $(5 - 2i)(3 + 4i)$.

35. Find the values of k so that $kx^2 + 2x - 3 = 0$ will have real number solutions.

36. Solve for t and check your solutions. $\sqrt{t + 3} = t - 3$

37. Find the solution set for $x^2 - 5x - 14 < 0$. Draw the graph of the solution set.

38. A rectangular box with an open top is to be made by cutting 1-inch squares from each corner of a rectangular piece of metal and then turning up the sides. The piece of rectangular metal is twice as long as it is wide. If the volume of the box is 12 cubic inches, find the dimensions of the rectangular piece of metal.

CHAPTER 7
Relations and Functions: Part I

Objectives of Chapter 7

To be able to:

- ☐ Define an ordered pair. (7–1)
- ☐ Define a relation. (7–1)
- ☐ Find ordered pairs that are solutions to an equation or inequality. (7–1)
- ☐ Give the domain and range of a relation. (7–1)
- ☐ Use terminology associated with a rectangular coordinate system. (7–1)
- ☐ Draw the graph of a relation. (7–1)
- ☐ Determine if a relation is a function. (7–2)
- ☐ Find the domain and range of a function. (7–2)
- ☐ Use the $f(x)$ notation for a function. (7–2)
- ☐ Find the distance between two points. (7–3)
- ☐ Calculate the slope of a line. (7–3)
- ☐ Know when a line has an undefined slope or a slope of zero. (7–3)
- ☐ Recognize a linear function. (7–4)
- ☐ Draw the graph of a linear function. (7–4)
- ☐ Find the slope and y-intercept of the graph of a linear function. (7–4)
- ☐ Write an equation of a linear function if the slope and y-intercept of its graph are given. (7–4)
- ☐ Write an equation of a linear function using the point-slope form of the equation of a line. (7–4)
- ☐ Determine if two lines are parallel. (7–4)
- ☐ Determine if two lines are perpendicular. (7–4)
- ☐ Draw the graph of a linear function by using its x- and y- intercepts. (7–4)
- ☐ Draw the graph of a linear inequality. (7–5)

The idea of a function is one of the most important concepts in mathematics. We say that a distance traveled is a function of time; the pressure in an automobile tire is a function of the temperature; and the cost of electricity is a function of the number of kilowatt-hours used.

7–1 Ordered Pairs, Relations, and Solutions of Equations in Two Variables

Recall that an **ordered pair** of real numbers is represented by the notation (x, y). The number represented by x is called the **first component**, or **abscissa**, and the number represented by y is called the **second component**, or **ordinate**. The ordered pair $(3, 5)$ is different from the ordered pair $(5, 3)$, since the order in which the numbers appear in the two notations is not the same.

Any set of ordered pairs is called a **relation**. The set of all first numbers in the ordered pairs in a relation is called the **domain** of the relation. The set of all second numbers in the ordered pairs in a relation is called the **range** of the relation.

Example 1   List the domain and range for the relation

$$\{(2, 0), (5, 4), (-4, 2), (2, 6)\}$$

Solution   The domain is $\{2, 5, -4\}$. The range is $\{0, 4, 2, 6\}$.

The solution set of an equation in two variables is a set of ordered pairs. For example, the equation $x + y = 6$ has the ordered pair $(4, 2)$ as one of its solutions because $4 + 2 = 6$ is a true statement. The ordered pairs $(3, 3)$, $(5, 1)$, $(-6, 12)$ are also solutions. (We cannot list all the ordered pairs in the solution set because the solution set is infinite).

Example 2   List three ordered pairs in the solution set of $2x + y = 12$.

Solution   Ordered pairs can be found more easily if we solve the equation for one variable. We solve for y.

$$2x + y = 12$$
$$y = 12 - 2x$$

Now assign any number to x. We pick $x = -1, 0, 5$.

When $x = -1$, $y = 12 - 2(-1) = 12 + 2 = 14$.
When $x = 0$, $y = 12 - 2(0) = 12 - 0 = 12$.
When $x = 5$, $y = 12 - 2(5) = 12 - 10 = 2$.

Thus, three ordered pairs are $(-1, 14)$, $(0, 12)$, and $(5, 2)$.

A relation is frequently given by an equation. For example, the equation $2x + y = 6$ can be thought of as a rule telling us how to assign numbers from the domain and range of the relation to obtain a sum of 6. If solved for y, $2x + y = 6$ becomes $y = 6 - 2x$. If $x \in \{0, 1, -2\}$, then the relation generated by $y = 6 - 2x$ is

$$\{(0, 6), (1, 4), (-2, 10)\}$$

Example 3 If $x \in \{2, 0, -3\}$, use the equation $3x - y = 4$ to form a relation.

Solution Solving $3x - y = 4$ for y, we get $y = 3x - 4$. If we substitute the given values of x into $y = 3x - 4$, we get the relation $\{(2, 2), (0, -4), (-3, -13)\}$.

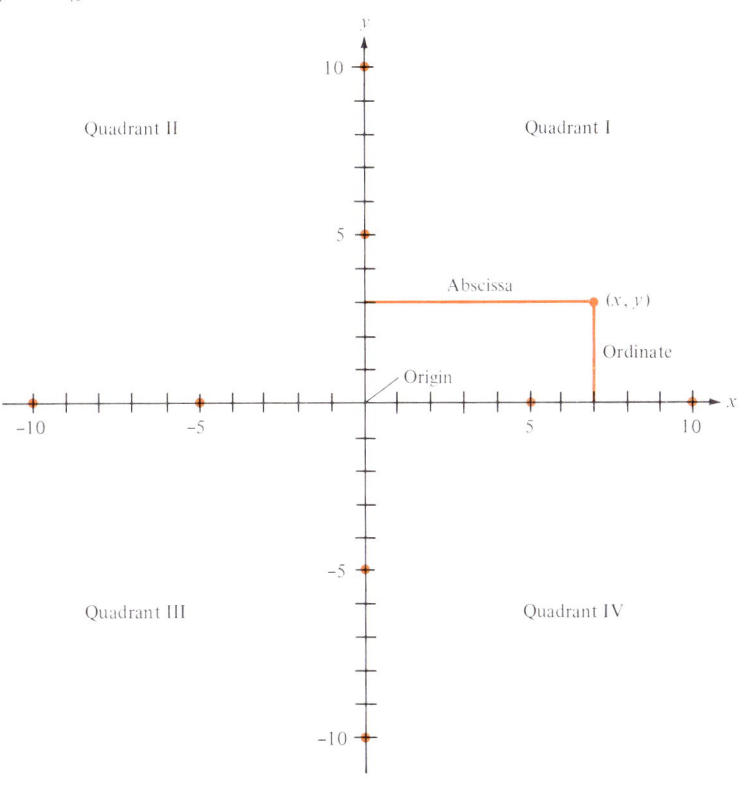

Figure 7-1

If all the ordered pairs forming a relation are plotted on a rectangular coordinate system, the resulting graph of all the ordered pairs is called the **graph of the relation.**

A **rectangular coordinate system** is illustrated in figure 7–1. It consists of two lines, called the axes, drawn perpendicular to each other and marked with some convenient scale. The horizontal line is usually labeled as the x-axis and the vertical line is usually labeled as the y-axis. The point where they intersect is called the **origin.** The two lines divide the plane into four parts, called **quadrants.** The quadrants are numbered counterclockwise from the upper right quadrant. The abscissa of an ordered pair is the distance a point is located to the right or left of the y-axis. The ordinate of an ordered pair gives the distance that a point is located above or below the x-axis. The arrows on the ends of the two lines indicate the positive ends of the two axes.

Example 4 Draw the graph of $x + y = 2$ if $x \in \{0, 1, 2\}$.

Solution Solving for y in $x + y = 2$, we get $y = 2 - x$. If the values for x are substituted into $y = 2 - x$, the relation $\{(0, 2), (1, 1), (2, 0)\}$ is formed. The graph of this relation is shown in figure 7–2.

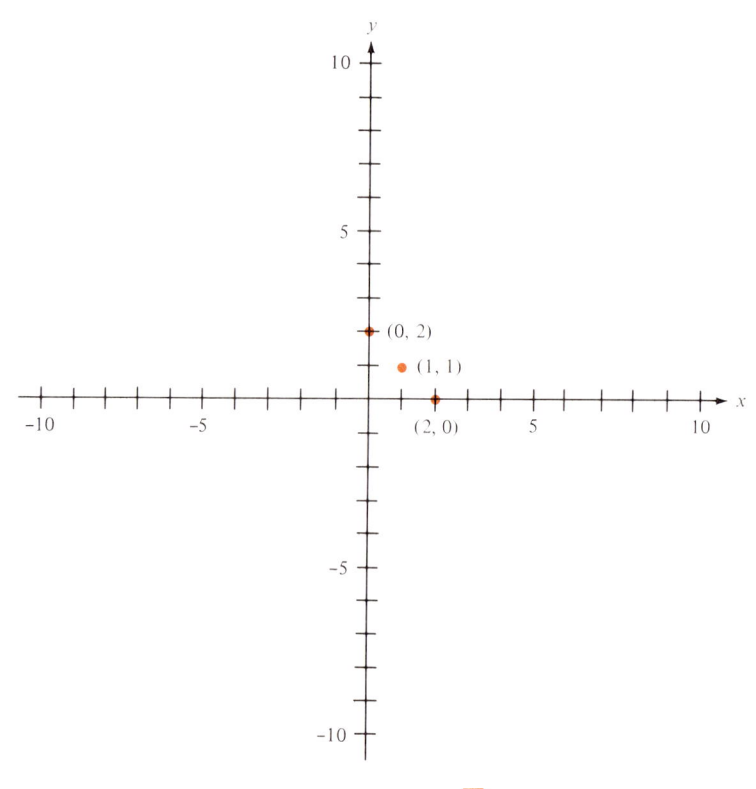

Figure 7–2

Example 5 From the set $A = \{0, 1, 2, 7, 8\}$, form the relation "y is exactly one more than x" if $x, y \in A$.

Solution We must write all ordered pairs in which the second component is one larger than the first component, with the condition that both coordinates are members of set A. Following these instructions, we obtain the relation

$$\{(0, 1), (1, 2), (7, 8)\}$$

7–1 Exercises

In exercises 1–4, either an equation or an inequality is given. For each equation or inequality, find three ordered pairs that will make it true. (See example 2.)

1. $x + y = 4$ 2. $x - y = 5$ 3. $x - y < 10$ 4. $x + y > 7$

In exercises 5–10, two ordered pairs are given for each equation. One coordinate is missing in each ordered pair. Determine what the missing component should be and finish writing each ordered pair. (See example 3.)

5. $x + y = 10$; (7,), (, 6)
6. $x - y = 8$; (3,), (, -4)
7. $2x - y = 4$; (-4,), (, -2)
8. $3x - 2y = 12$; (-2,), (, 6)
9. $\dfrac{x + y}{2} = 8$; (4,), (, -2)
10. $\dfrac{2x - y}{3} = 4$; (, -2), (0,)

In exercises 11–14, write the domain and range of each relation. (See example 1.)

11. $\{(5, 3), (-2, 4), (0, 1)\}$
12. $\{(7, 2), (-2, 3), (-4, 2)\}$
13. $\{(0, 0), (-1, 0), (0, -1)\}$
14. $\{(-2, 3), (3, -2)\}$

In exercises 15–18, find: (a) the relation indicated; (b) the domain of the relation; and (c) the range of the relation. (See example 5.)

15. From the set $A = \{-1, 0, 7\}$, form the relation that represents x is less than y if $x, y \in A$.

16. From the set $A = \{-4, 0, 4\}$, form the relation that represents x is equal to y if $x, y \in A$.

17. From the set $A = \{-10, 2, 4\}$, form the relation that represents the first component is greater than or equal to the second component, if $x, y \in A$.

18. From the set $A = \{5, 10, 15, 20\}$, form the relation that represents the first component has a value of one-fifth the second component, if $x, y \in A$.

In exercises 19–24, find the relation indicated by the given equation and the given values for x or y. Draw the graph of the relation. (See example 4.)

19. $y = 2x - 3$; $x \in \{0, 1, 2\}$
20. $y = x + 1$; $x \in \{-1, 2, 3\}$

21. $2x + y = 4$; $y \in \{-2, 0, 2\}$
22. $3x + y = 12$; $y \in \{0, 3, 6\}$
23. $x + y = 4$; $x \in \left\{-\frac{1}{2}, -\frac{1}{3}, \frac{1}{2}\right\}$
24. $x - y = 2$; $y \in \left\{-\frac{1}{2}, 0, \frac{1}{2}\right\}$

In exercises 25–32, answer true or false.

25. The abscissa is the first component in an ordered pair.
26. The horizontal axis is usually labeled the y-axis.
27. The quadrants are labeled clockwise, beginning with the upper right-hand quadrant.
28. Any ordered pair in quadrant II will have signs $(-, +)$.
29. The *second component* and the *ordinate* are different names for the same number in an ordered pair.
30. The ordered pair $(-3, 5)$ is the same as $(5, -3)$.
31. Any ordered pair lying on the x-axis will have an ordinate of zero.
32. The ordered pair $(0, 0)$ is associated with the origin.

7–2 Functions

A **function** is a special kind of relation, as given by definition 7–1.

Definition 7–1

Function

A **function** is a relation in which no two ordered pairs have the same first components and different second components.

From this definition we see that for each element in the domain of the function, there will be one and only one element in the range of the function.

Example 1   Is the relation $\{(5, 3), (-4, 0), (7, 5)\}$ a function?

Solution   Yes. Associated with each first component in every ordered pair is a unique second component.

Example 2   Is the relation $\{(-6, 0), (5, 2), (3, 1), (5, -4)\}$ a function?

Solution   No. There are two ordered pairs, $(5, 2)$ and $(5, -4)$, with 5 as the first component, but different numbers for the second components.

Relations and Functions: Part I

We can recognize a function from its graph. In any function, each value of x produces one and only one value for y. Graphically, this means any vertical line that intersects the graph of a function will intersect it in exactly one point.

Example 3 Determine if the graphs shown in figures 7–3, 7–4, 7–5, and 7–6 are graphs of functions.

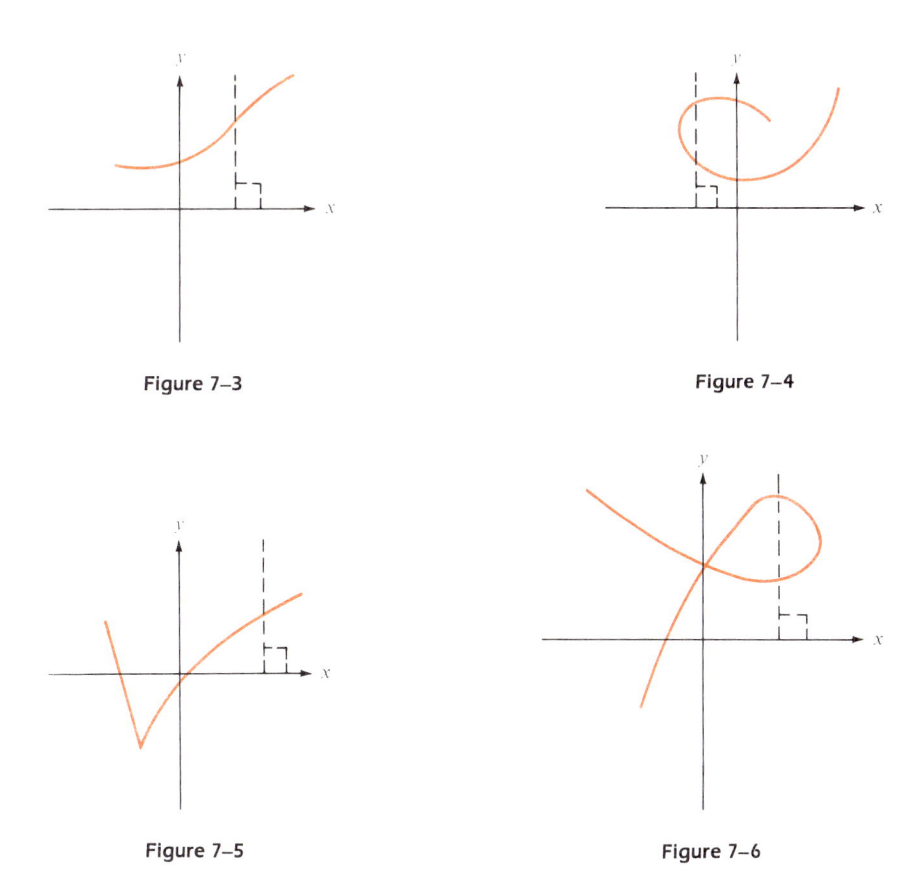

Figure 7–3

Figure 7–4

Figure 7–5

Figure 7–6

Solution Figures 7–3 and 7–5 are graphs of functions because any vertical line that intersects the graphs does so in exactly one point. Figures 7–4 and 7–6 do not represent functions. In these graphs, a vertical line can be made to intersect each graph in at least two places.

A function always consists of the three parts given in definition 7–2.

Definition 7-2

Parts of a Function
1. A set of numbers called the *domain of the function*.
2. A set of numbers called the *range of the function*.
3. A *rule of correspondence* that assigns to each member in the domain of the function one and only one member in the range of the function.

The rule of correspondence in definition 7-2 can be either an English sentence or it can be an equation.

Example 4

Find the function such that the ordinate in each ordered pair is twice the abscissa, if the domain and range of the function are both subsets of the set $U = \{1, 2, 3, 4, 5, 6\}$.

Solution

We must pick those numbers from set U in a manner that will make the second component in each ordered pair twice the first component. The ordered pairs in the following set meet the requirement.

$$\{(1, 2), (2, 4), (3, 6)\}$$

The domain of this function is $\{1, 2, 3\}$ and the range is $\{2, 4, 6\}$. ■

Example 5

Find $\{(x, y) | y = x + 1\}$ if $x, y \in U$ and $U = \{1, 2, 3, 4, 5, 6\}$.

Solution

The rule in this example is given by the equation $y = x + 1$. The ordered pairs that make $y = x + 1$ true if $x, y \in U$ are

$$\{(1, 2), (2, 3), (3, 4), (4, 5), (5, 6)\}.$$

The domain of this function is $\{1, 2, 3, 4, 5\}$ and the range is $\{2, 3, 4, 5, 6\}$. ■

In examples 4 and 5, the domain and range were limited. However, this is not usually the case, and we assume the domain consists of every real number for which the function is defined.

Example 6

What is the domain of the function $y = \dfrac{6}{x - 2}$?

Solution

The equation is a fraction, so we know the denominator must not take a value of zero. A zero denominator occurs if $x = 2$. Thus the domain for $y = 6/(x - 2)$ is the entire set of real numbers except 2. This can be stated in set notation as

The domain is $\{x | x \text{ is a real number and } x \neq 2\}$. ■

Example 7 What is the domain for $y = \sqrt{x - 3}$?

Solution If y is to represent a nonnegative real number, then the radicand must always be greater than or equal to zero. That is, $x - 3 \geq 0$ implies $x \geq 3$. For any $x \geq 3$, the radicand is nonnegative. The domain is $\{x \mid x \geq 3\}$.

Symbols such as $f(x)$ or $g(x)$ are frequently used to indicate functional values. Each of these is another name for y. For example, instead of writing $y = x + 3$, we may write $f(x) = x + 3$ or $g(x) = x + 3$. The symbol $f(x)$ is read "f of x" and means "the value of f at x." This notation has the advantage of allowing us to see both the domain and range elements being used in the function. The notation $f(3)$, for example, gives the value of a function when $x = 3$, where 3 is in the domain and $f(3)$ is in the range. (See example 8.)

Example 8 If $f(x) = 3x + 2$, find (a) $f(3)$; (b) $f(-4)$; (c) $f(a + h)$.

Solution

(a) $f(x) = 3x + 2$ Given function.
$f(3) = 3(3) + 2$ Substitute 3 for x.
$f(3) = 11$

The ordered pair formed is $(3, 11)$.

(b) $f(x) = 3x + 2$ Given function.
$f(-4) = 3(-4) + 2$ Substitute -4 for x.
$f(-4) = -10$

The ordered pair formed is $(-4, -10)$.

(c) $f(x) = 3x + 2$ Given function.
$f(a + h) = 3(a + h) + 2$ Substitute $a + h$ for x.
$f(a + h) = 3a + 3h + 2$

The ordered pair formed is $(a + h, 3a + 3h + 2)$.

7–2 Exercises

In exercises 1–6, determine if the relation is a function. Give the domain and range of the relation. (See examples 1 and 2.)

1. $\{(-4, 3), (2, 5), (0, 3), (4, 3)\}$
2. $\{(-5, 7), (8, 3), (-5, 4)\}$
3. $\{(9, 6), (7, 3), (9, 8)\}$
4. $\{(5, 6), (7, 8), (9, 8)\}$.
5. $\{(4, 9), (-4, 3), (-8, 6)\}$
6. $\{(8, 3), (9, 3), (-4, 3)\}$

In exercises 7–10, determine if the graphs represent graphs of functions. (See example 3.)

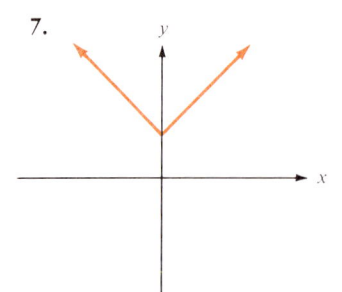

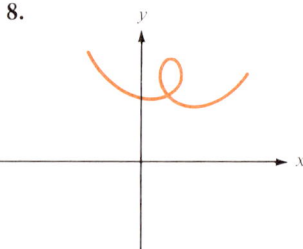

7.

8.

9.

10.

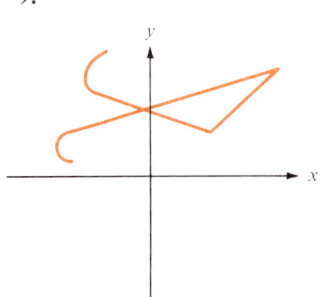

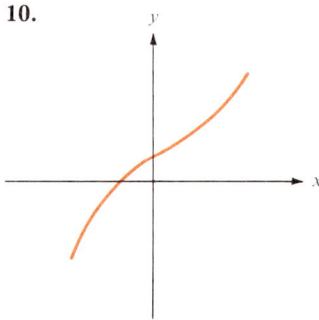

In exercises 11–18, let $U = \{-1, 0, 1, 2\}$ and let $x, y \in U$. Draw the graphs of the given relations and tell which ones are functions. (See examples 4 and 5.)

11. $\{(x, y) | x = y\}$ **12.** $\{(x, y) | y = x - 1\}$ **13.** $\{(x, y) | x > y\}$
14. $\{(x, y) | y = x + 1\}$ **15.** $\{(x, y) | y = 2x\}$ **16.** $\{(x, y) | y > x\}$
17. $\{(x, y) | y = |x|\}$ **18.** $\left\{(x, y) | y = \frac{1}{2}x\right\}$

In exercises 19–28, state the domain of each function. (See examples 6 and 7.)

19. $y = \dfrac{3x}{x - 4}$ **20.** $y = \dfrac{5}{x + 6}$ **21.** $y = \dfrac{2x - 3}{3x - 4}$

22. $y = \dfrac{5x + 4}{3x + 6}$ **23.** $y = \dfrac{x^2 + 1}{x^2 - 1}$ **24.** $y = \dfrac{9x + 1}{x^2 - 4}$

25. $y = \sqrt{x + 7}$ **26.** $y = \sqrt{x - 6}$ **27.** $y = \dfrac{5}{\sqrt{x - 4}}$

28. $y = \dfrac{2}{\sqrt{x + 6}}$

In exercises 29–40, if $f(x) = x^2 + 4$, find each value. (See example 8.)

29. $f(3)$ **30.** $f(-2)$ **31.** $f(-9)$ **32.** $f(6)$
33. $f(a)$ **34.** $f(-a)$ **35.** $f(x + h)$ **36.** $f(h^2)$

Relations and Functions: Part I

37. $f(3) - f(2)$
38. $f(5) - f(-2)$
39. $f(x + h) - f(x)$
40. $\dfrac{f(x + h) - f(x)}{h}$

41. The velocity v (in feet per second) with which an object will fall in a vacuum is related to time t (in seconds) by the equation $v = 32t$. The towers that support the Golden Gate bridge are 746 feet tall.
    (a) Find the velocity of an object dropped from the top of a tower when it hits the water below if it takes 7 seconds to reach the water.
    (b) Find the velocity at the end of 5 seconds.
    (c) Find the velocity when $t = 0$.
    (d) Plot the graph of the ordered pairs found in parts a, b, and c. Label the horizontal axis t and the vertical axis v. Does $v = 32t$ represent a function?

42. The temperature in degrees Fahrenheit is given by $F = \frac{9}{5}C + 32$, where C is the temperature measured in degrees Celsius. What is the reading in degrees Fahrenheit for each of the following?
    (a) 5° Celsius
    (b) 0° Celsius
    (c) $-40°$ Celsius
    (d) Plot the graph of the ordered pairs found in parts a, b, and c. Label the horizontal axis C and the vertical axis F. Does the relationship between F and C represent a function?

7–3 Distance Formula and Slope

We now turn our attention to the problem of finding the distance between two points on a rectangular coordinate system. We need to review the Pythagorean theorem first. In figure 7–7, triangle ABC is a right triangle. Side AB is called the *hypotenuse*. Sides AC and BC are called the *legs*. The little square mark at vertex C indicates the two legs form a right (90°) angle. In any right triangle, the Pythagorean theorem states that $c^2 = a^2 + b^2$, where c is the length of the hypotenuse and b and c are the lengths of the two legs.

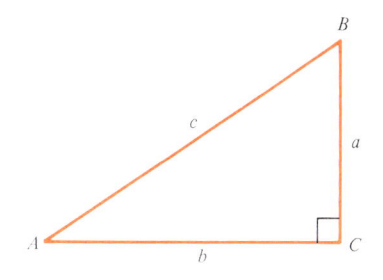

Figure 7–7

In figure 7–8, we want to find the distance between points P_1 and P_2, whose coordinates are (x_1, y_1) and (x_2, y_2), respectively. Let us call this distance d. Form a right triangle $P_1P_2P_3$, as shown in figure 7–8. The length of horizontal line segment P_1P_3 is given by $|x_2 - x_1|$. The length of vertical line segment P_2P_3 is given by $|y_2 - y_1|$. These two lengths form the legs of right triangle $P_1P_2P_3$. The Pythagorean theorem applied to triangle $P_1P_2P_3$ gives $d^2 = (x_2 - x_1)^2 + (y_2 - y_1)^2$. If we now take the square root of $d^2 = (x_2 - x_1)^2 + (y_2 - y_1)^2$, we get

$$d = \sqrt{(x_2 - x_1)^2 + (y_2 - y_1)^2}$$

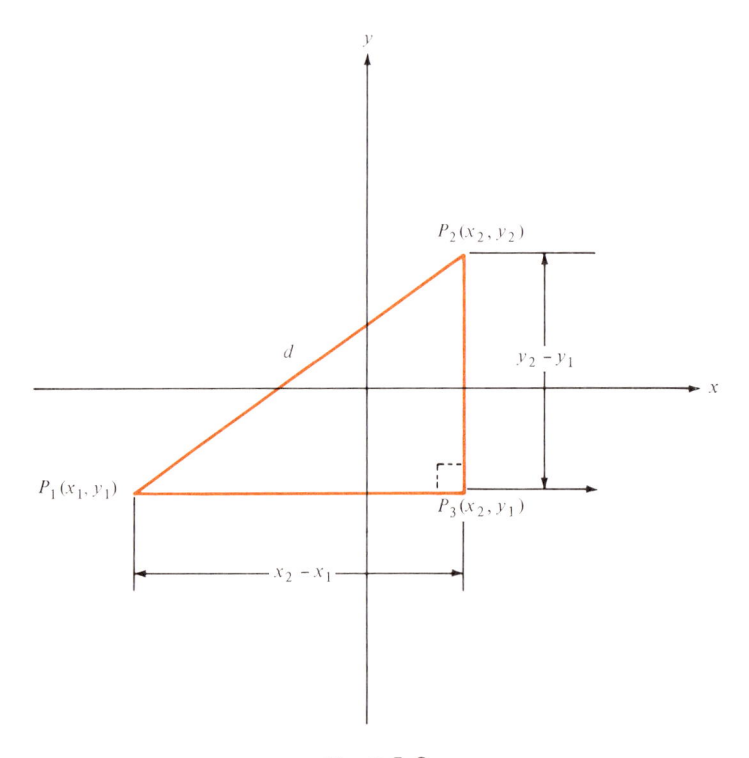

Figure 7–8

Rule 7–1

The Distance Formula

If P_1 and P_2 are any two points on a rectangular coordinate system with coordinates (x_1, y_1) and (x_2, y_2), respectively, the distance between P_1 and P_2 is given by

$$d = \sqrt{(x_2 - x_1)^2 + (y_2 - y_1)^2}$$

Example 1 Find the distance between $(5, -3)$ and $(2, 4)$.

Solution Either ordered pair can be designated as (x_1, y_1) and the other as (x_2, y_2). Putting the values given into the distance formula, we get

$$\begin{aligned} d &= \sqrt{(5-2)^2 + (-3-4)^2} \\ &= \sqrt{3^2 + (-7)^2} \\ &= \sqrt{9 + 49} \\ &= \sqrt{58} \end{aligned}$$

Example 2 Find the perimeter of the triangle whose vertices are located at $P_1(3, 2)$, $P_2(6, 6)$, and $P_3(-1, 1)$.

Solution We must find the lengths of the three line segments P_1P_2, P_1P_3, and P_2P_3 and then add these lengths to obtain the perimeter.

$$\begin{aligned} \overline{P_1P_2} &= \sqrt{(3-6)^2 + (2-6)^2} \\ &= \sqrt{(-3)^2 + (-4)^2} \\ &= \sqrt{9 + 16} \\ &= \sqrt{25} \\ &= 5 \\ \overline{P_1P_3} &= \sqrt{[3-(-1)]^2 + (2-1)^2} \\ &= \sqrt{4^2 + 1^2} \\ &= \sqrt{16 + 1} \\ &= \sqrt{17} \\ \overline{P_2P_3} &= \sqrt{[6-(-1)]^2 + (6-1)^2} \\ &= \sqrt{7^2 + 5^2} \\ &= \sqrt{49 + 25} \\ &= \sqrt{74} \end{aligned}$$

The perimeter is $5 + \sqrt{17} + \sqrt{74}$.

Another property associated with a line is its **slope,** or inclination. Roughly, we can think of the slope of a line as that property which tells us how fast the line is rising or falling. But we need a more precise definition for computational purposes. The slope of a line is the ratio of the rise to the run between two distinct points on the line. If we designate the slope by the letter m then

$$m = \frac{\text{rise}}{\text{run}}$$

This definition of slope, using P_1 and P_2, is shown in figure 7–9.

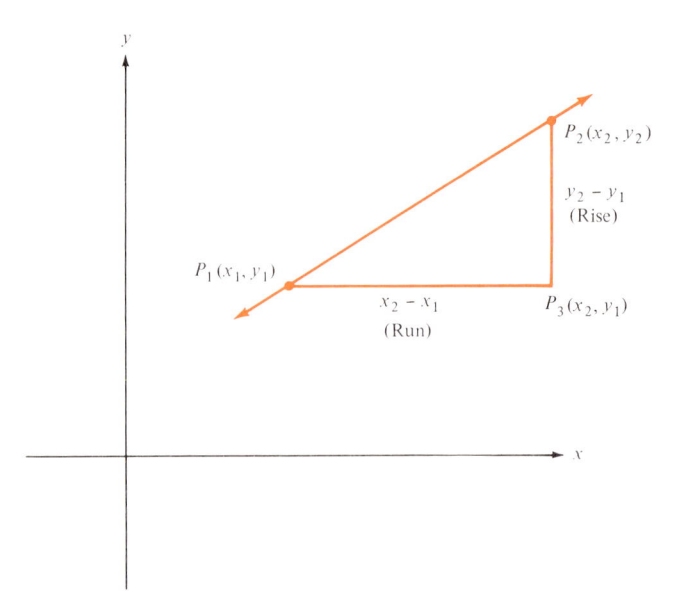

Figure 7-9

From figure 7-9,

$$m = \frac{\text{rise}}{\text{run}} = \frac{y_2 - y_1}{x_2 - x_1}$$

Definition 7-3 If P_1 and P_2 are two distinct points on a rectangular coordinate system with coordinates (x_1, y_1) and (x_2, y_2), respectively, then the slope of the line (or a line segment) joining P_1 and P_2 is

$$m = \frac{y_2 - y_1}{x_2 - x_1}$$

Example 3 Find the slope of the line going through the two points $P_1(5, 4)$ and $P_2(-2, 6)$. (See figure 7-10.)

Solution

$$m = \frac{y_2 - y_1}{x_2 - x_1}$$
$$= \frac{6 - 4}{-2 - 5}$$
$$= \frac{2}{-7} = -\frac{2}{7}$$

We can also calculate m as

$$m = \frac{4 - 6}{5 - (-2)}$$
$$= \frac{-2}{7} = -\frac{2}{7} \blacksquare$$

Relations and Functions: Part I

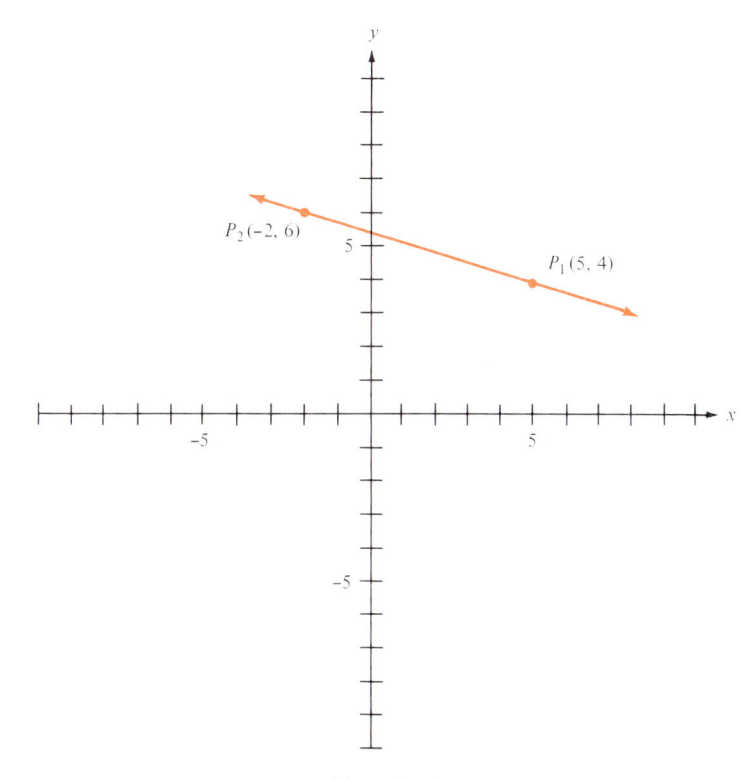

Figure 7–10

Notice from the graph in figure 7–10 that $\overleftrightarrow{P_1P_2}$ is drawn downward and to the right for increasing values of x. Lines for which this is true will always have negative slopes. Lines drawn upward and to the right for increasing values of x have positive slopes.

Two special lines whose slopes we need to examine separately are vertical and horizontal lines. Horizontal lines have the form $y = k$ and vertical lines have the form $x = k$, where k is a constant. A horizontal line segment, P_1P_2, is shown in figure 7–11(a).

(a) (b)

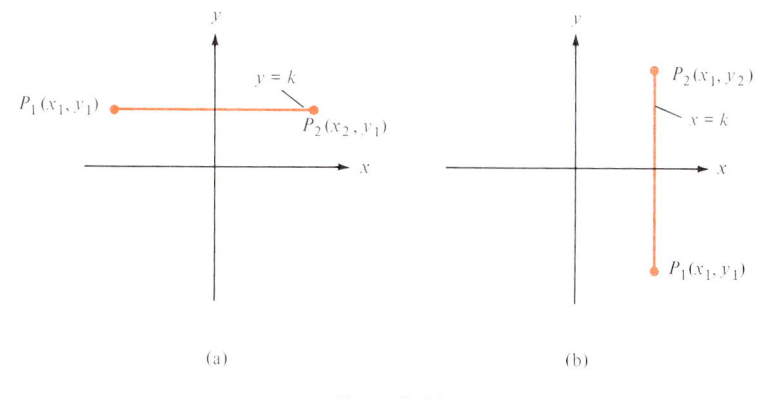

Figure 7–11

If we use definition 7–3 to calculate its slope, we get

$$m = \frac{y_2 - y_1}{x_2 - x_1}$$

$$= \frac{y_1 - y_1}{x_2 - x_1} \qquad y_1 = y_2 \text{ for any horizontal line.}$$

$$= \frac{0}{x_2 - x_1}$$

$$= 0$$

This illustrates that all horizontal lines have a slope of zero.

In figure 7–11(b), a vertical line segment, P_1P_2, is shown. The slope of P_1P_2 is

$$m = \frac{y_2 - y_1}{x_2 - x_1}$$

$$= \frac{y_2 - y_1}{x_1 - x_1} \qquad x_2 = x_1 \text{ for any vertical line.}$$

$$= \frac{y_2 - y_1}{0} \qquad \text{Undefined.}$$

We cannot divide by zero; hence any vertical line is said to have an undefined slope.

Two lines L_1 and L_2 with slopes m_1 and m_2 are *parallel* if $m_1 = m_2$. For example, any line with a slope of $\frac{3}{5}$ is parallel to any other line with a slope of $\frac{3}{5}$.

If two lines are neither horizontal nor vertical, they are perpendicular if their slopes are negative reciprocals. For example, if line L_1 has a slope of $\frac{2}{3}$, then any line perpendicular to L_1 has slope of $-\frac{3}{2}$.

Rule 7–2

Perpendicular and Parallel Lines

Two lines L_1 and L_2 with slopes m_1 and m_2, respectively, are:

1. Parallel if and only if $m_1 = m_2$.
2. Perpendicular if and only if $m_1 = -1/m_2$ ($m_1, m_2 \neq 0$).

Example 4

Determine if the line going through $P(3, 3)$ and $Q(-1, -5)$ is parallel to the line going through $R(-6, -7)$ and $S(2, 9)$.

Solution

Calculate the slopes of both lines and see if their slopes are the same.

$$m_{PQ} = \frac{3 - (-5)}{3 - (-1)} = \frac{8}{4} = 2$$

$$m_{RS} = \frac{-7 - 9}{-6 - 2} = \frac{-16}{-8} = 2$$

Each line has a slope of 2, so the two lines are parallel. ■

Example 5 Calculate the value of y if the line going through the points given by $(5, y)$ and $(4, -2)$ is to have a slope of $\frac{3}{4}$.

Solution We know $m = \frac{3}{4}$.

$$m = \frac{y_2 - y_1}{x_2 - x_1}$$

$\frac{3}{4} = \frac{y - (-2)}{5 - 4}$ Substitute the given values into the slope formula.

$\frac{3}{4} = \frac{y + 2}{1}$ Simplify the right side.

$3 = 4(y + 2)$ Clear the equation of fractions.

$3 = 4y + 8$ Remove parentheses.

$4y = -5$

$y = -\frac{5}{4}$

Example 6 The distance between the two points given by $(x, 3)$ and $(2, -3)$ is $2\sqrt{10}$. Find x.

Solution Use the distance formula.

$2\sqrt{10} = \sqrt{(x - 2)^2 + [3 - (-3)]^2}$

$4(10) = (x - 2)^2 + [3 - (-3)]^2$ Square each side.

$40 = x^2 - 4x + 4 + 36$ Square the two terms on the right side.

$40 = x^2 - 4x + 40$ Collect like terms on the right side.

$x^2 - 4x = 0$ Write the equation in standard form.

$x(x - 4) = 0$ Factor the left side.

$x = 0$ or $x - 4 = 0$ Set each factor equal to zero.

$x = 0$ or $x = 4$

Two solutions are obtained. This means the points given by $(0, 3)$ and $(4, 3)$ are both $2\sqrt{10}$ units from the point represented by $(2, -3)$.

7-3 Exercises

In exercises 1–10, find the distance between the two given points. (See example 1.)

1. $(-1, 6), (2, 2)$
2. $(9, -6), (3, 2)$
3. $(6, -1), (1, 11)$
4. $(-1, -2), (2, 2)$
5. $(8, 3), (-2, -1)$
6. $(-4, -1), (2, -3)$

7. $(4, 3), (4, -5)$ 8. $(3, -4), (5, -4)$ 9. $(5, a), (2, b)$
10. $(c, 3), (d, -4)$

In exercises 11–16, two ordered pairs are given with one of the coordinates in one of the ordered pairs unknown. The distance between the points represented by the ordered pairs is also given. Use example 6 *as a model and find the unknown coordinate.*

11. $(x, 3), (4, 5); \ d = 2\sqrt{2}$ 12. $(5, -1), (x, 2); \ d = \sqrt{13}$
13. $(2, y), (-3, 4); \ d = 5\sqrt{2}$ 14. $(-6, 3), (-2, y); \ d = 5$
15. $(0, y), (-3, -4); \ d = \sqrt{10}$ 16. $(-3, -2), (x, 5); \ d = \sqrt{65}$

In exercises 17–26, find the slope of the line going through the given points. (See example 3.)

17. $(4, 5), (-2, -4)$ 18. $(8, 5), (4, 6)$ 19. $(9, 6), (-2, 1)$
20. $(-8, 3), (5, -4)$ 21. $(7, -3), (-4, 2)$ 22. $(9, 0), (0, -4)$
23. $(8, 6), (8, -4)$ 24. $(4, -3), (8, -3)$ 25. $(5, c), (-4, d)$
26. $(a, -2), (a, 5)$.

In exercises 27–30, find the perimeter of the triangles whose vertices are located at the given points. (See example 2.)

27. $P_1(2, 2), P_2(-1, 2), P_3(-7, -6)$
28. $P_1(3, 6), P_2(-2, -6), P_3(-1, -2)$
29. $P_1(2, 3), P_2(-2, 4), P_3(0, 1)$
30. $P_1(0, 0), P_2(0, 3), P_3(4, 0)$

In exercises 31–34, determine if the three given points lie on the same line by using the distance formula. The three points will lie on the same line if the sum of the two shorter distances equals the largest distance.

31. $P_1(0, 0), P_2(1, 1), P_3(-2, -2)$ 32. $P_1(0, 1), P_2(1, 2), P_3(-1, 0)$
33. $P_1(0, -1), P_2(1, 1), P_3(2, 1)$ 34. $P_1(0, 0), P_2(1, 2), P_3(-1, -2)$

35–36. Solve Exercises 33 and 34 by using the formula for the slope of a line given in definition 7–3. (*Hint:* If all three points lie on the same line, what do you believe must be true concerning the slopes of P_1P_2, P_1P_3, and P_2P_3?)

In exercises 37–46, determine if the line going through the points given by the two ordered pairs P and Q is parallel or perpendicular (or neither) to the line going through the two points given by the two ordered pairs R and S. (See example 4.)

37. $P(0, -3), Q(1, -1); \ R(-1, -5), S(5, 7)$
38. $P(7, 11), Q(6, 9); \ R(10, 17), S(-3, -9)$
39. $P(0, 5), Q(2, 4); \ R(12, 21), S(-8, -19)$
40. $P(-6, 8), Q(4, 3); \ R(-5, -13), S(-2, -7)$

Relations and Functions: Part I

41. $P(5, 9)$, $Q(10, -1)$; $R(0, 5)$, $S(6, 0)$
42. $P(0, 2)$, $Q(3, 3)$; $R(-3, 1)$, $S(-6, 0)$
43. $P(5, 3)$, $Q(-5, -5)$; $R(0, 7)$, $S(4, 2)$
44. $P(0, 5)$, $Q(9, 5)$; $R(10, 12)$, $S(10, -4)$
45. $P(-6, 10)$, $Q(8, 10)$; $R(-3, 2)$, $S(-3, -5)$
46. $P(8, 1)$, $Q(-4, 3)$; $R(-4, 0)$, $S(-7, -2)$
47. What will be the slope of any line parallel to the line that goes through the points given by $(5, -4)$ and $(-4, -3)$?
48. What will be the slope of any line perpendicular to the line which goes through the points given by $(-6, 10)$ and $(7, -2)$?
49. What can be said about the slope of any line perpendicular to a vertical line?
50. What can be said about the slope of any line perpendicular to a horizontal line?
51. The slope of line PQ in the accompanying right triangle is $\frac{3}{4}$. If PR is 16 feet long, how long is QR?

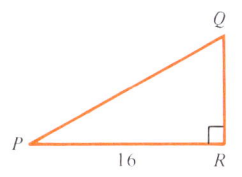

52. Line PQ is perpendicular to line RS. If line RS has a slope of $\frac{4}{5}$, find the value of y for the ordered pair labeled Q.

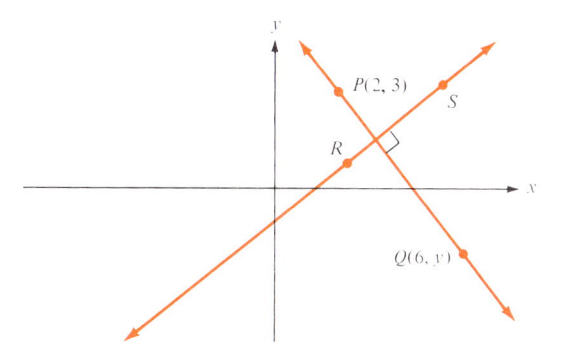

7-4 The Linear Function

Linear functions take the form $ax + by = c$, $b \neq 0$, where a, b, and c are real numbers. A linear function is said to be in *standard* *form* if it is written as $ax + by = c$. For example, $3x + 2y = 5$ and $x - 6y = 3$ are linear functions in standard form. If $b = 0$ in $ax + by = c$, the equation reduces to $ax = c$, or $x = c/a$, which is not a function. The graphs of linear functions, as the name implies, are straight lines.

Example 1 Draw the graph of $3x + 4y = 12$ by finding three ordered pairs that make $3x + 4y = 12$ a true statement. Plot the three ordered pairs and draw a straight line through the points to obtain the desired graph. Then calculate the slope of the line.

Solution If we arbitrarily let x be 0, 4, and -4 and calculate y for these values of x, we get the ordered pairs $(0, 3)$, $(4, 0)$, and $(-4, 6)$.

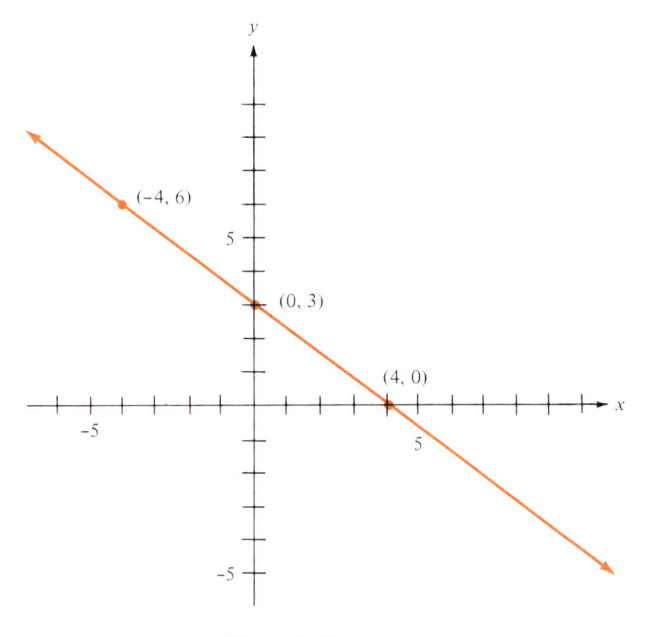

Figure 7–12

The graph is shown in figure 7–12. The slope can be found by using any two of the three ordered pairs. We use $(0, 3)$ and $(4, 0)$.

$$m = \frac{y_2 - y_1}{x_2 - x_1} = \frac{0 - 3}{4 - 0} = \frac{-3}{4}$$

One convenient form for a linear function is the *slope-intercept form* of an equation of the line. The word *intercept* used here refers to the *y*-intercept for the graph of the function. The *y*-intercept is that point on the *y*-axis where the graph intersects the *y*-axis. In figure 7–13, let $P_1(0, b)$ be any point on the *y*-axis. Let $P_2(x, y)$ be any point different from P_1 and not located on the *y*-axis.

Relations and Functions: Part I

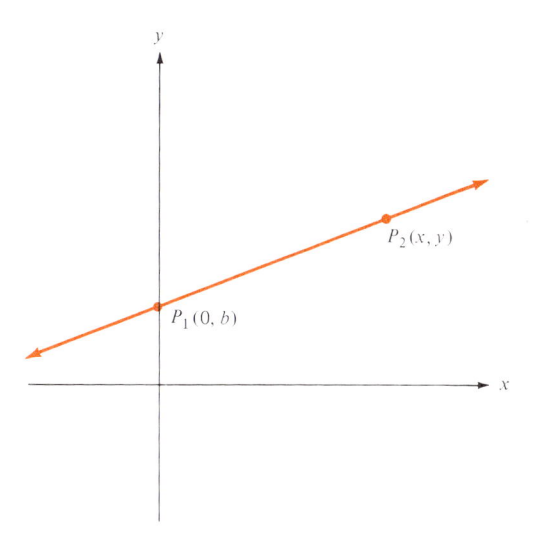

[Figure 7-13: Graph showing line through $P_1(0, b)$ and $P_2(x, y)$]

Figure 7–13

We find the slope of $\overleftrightarrow{P_1P_2}$.

$$m = \frac{y_2 - y_1}{x_2 - x_1}$$

$$= \frac{y - b}{x - 0}$$

$$= \frac{y - b}{x}$$

$$mx = y - b$$

$$y = mx + b$$

The equation $y = mx + b$ is known as the **slope-intercept form** of an equation of the line and should be memorized. The slope, m, and y-intercept, b, of the graph can be read directly from the equation.

Rule 7–3

Slope-Intercept Form of an Equation of the Line

The **slope-intercept form** for an equation of the straight line is

$$y = mx + b$$

The slope of the line is given by m and the y-intercept is given by b.

Example 2   Find the slope and y-intercept for the graph of the equation $3x + 4y = 12$ given in example 1.

Solution

First, the equation must be solved for y.
$$3x + 4y = 12$$
$$4y = 12 - 3x$$
$$y = 3 - \left(\frac{3}{4}\right)x$$
$$y = -\left(\frac{3}{4}\right)x + 3$$

From the equation $y = -\frac{3}{4}x + 3$, we see that the slope m is $-\frac{3}{4}$ and the y-intercept b is 3. ■

Example 3

Use the slope-intercept form of a linear function to write an equation of the line with slope $\frac{3}{5}$ and y-intercept 4. Leave the equation in standard form.

Solution

We know $m = \frac{3}{5}$ and $b = 4$. Substitute these into $y = mx + b$ to get
$$y = \frac{3}{5}x + 4$$

$5y = 3x + 20$ Multiply both sides by 5.
$5y - 3x = 20$ Write the equation in standard form.
$3x - 5y = -20$ ■   Multiply both sides by -1.

Example 4

Draw the graph of an equation if the graph has slope $-\frac{3}{2}$ and y-intercept -1.

Solution

Starting at the y-intercept $(0, -1)$, we "mark off" the slope. For three units of change in y in a negative direction, we must mark off two units of change in x in a positive direction. If this is done several times, the graph can be drawn through the points obtained by this process (see figure 7–14). (You should verify that the results will be the same if the slope is given as $3/(-2)$ instead of $-3/2$.)

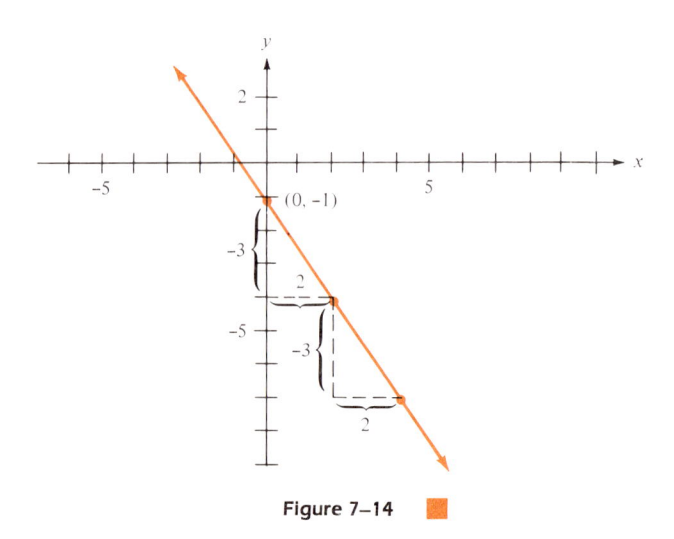

Figure 7–14 ■

Relations and Functions: Part I

Another useful form of a linear function is the **point-slope form** of an equation of the line. This equation is given by $y - y_1 = m(x - x_1)$ and should be memorized. We now derive this important equation.

Let $P_1(x_1, y_1)$ be a point whose coordinates are known. Let $P(x, y)$ be a point whose coordinates are not known, and $x \neq x_1$ (see figure 7–15).

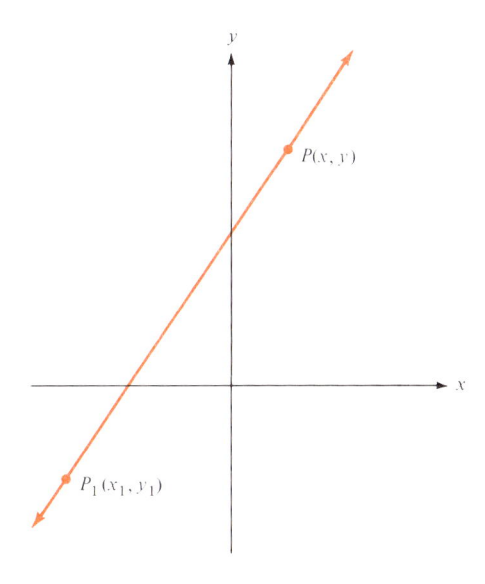

Figure 7–15

We first find the slope of the line going through $P_1(x_1, y_1)$ and $P(x, y)$.

$$m = \frac{y - y_1}{x - x_1}$$

$$m(x - x_1) = y - y_1 \qquad \text{Multiply both sides by } x - x_1.$$

$$y - y_1 = m(x - x_1)$$

Rule 7–4

Point-Slope Form for an Equation of the Straight Line

The point-slope form for an equation of the straight line is $y - y_1 = m(x - x_1)$. The slope of the line is given by m and x_1 and y_1 are the coordinates of a point through which the line passes.

Example 5

Write an equation of the line whose slope is $-\frac{1}{2}$ and that goes through $(-2, 4)$. Leave the equation in standard form.

Solution

We know $m = -\frac{1}{2}$, $x_1 = -2$, and $y_1 = 4$.

$$y - y_1 = m(x - x_1)$$
$$y - 4 = -\frac{1}{2}[x - (-2)]$$
$$2y - 8 = -1(x + 2) \quad \text{Multiply both sides by 2.}$$
$$2y - 8 = -x - 2 \quad \text{Remove the parentheses.}$$
$$x + 2y = 6 \quad \blacksquare \quad \text{Write the equation in standard form.}$$

Example 6

Write an equation of the line that goes through the points $(-3, 4)$ and $(5, -8)$. Leave the equation in standard form.

Solution

First, find the slope.

$$m = \frac{4 - (-8)}{-3 - 5} = \frac{12}{-8} = -\frac{3}{2}$$

Either ordered pair can be used as (x_1, y_1). We use $(-3, 4)$. (You should verify that $(5, -8)$ gives the same result.)

$$y - y_1 = m(x - x_1)$$
$$y - 4 = -\frac{3}{2}[x - (-3)]$$
$$2y - 8 = -3(x + 3) \quad \text{Multiply both sides by 2.}$$
$$2y - 8 = -3x - 9 \quad \text{Remove the parentheses.}$$
$$3x + 2y = -1 \quad \blacksquare \quad \text{Write the equation in standard form.}$$

Example 7

Write an equation of the line perpendicular to the graph of $2x + 3y = 5$ and passing through the point $P(7, 1)$.

Solution

The graph of the equation for $2x + 3y = 5$ and the point $P(7, 1)$ are shown in figure 7–16. The desired line is shown as a dashed line. The given equation solved for y is

$$2x + 3y = 5$$
$$y = -\frac{2}{3}x + \frac{5}{3}$$

From this last equation, we see that the graph of $2x + 3y = 5$ has slope $-\frac{2}{3}$. The line whose equation we want will have a slope of $\frac{3}{2}$. Because we know a point through which the line goes, we can use $y - y_1 = m(x - x_1)$, $m = \frac{3}{2}$, and $(7, 1)$ to obtain

Relations and Functions: Part I

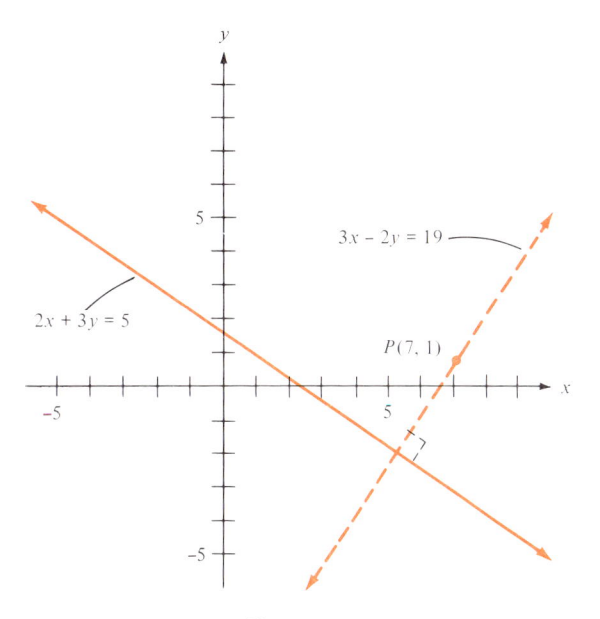

Figure 7-16

$$y - 1 = \frac{3}{2}(x - 7)$$

$2(y - 1) = 3(x - 7)$ Multiply both sides by 2.
$2y - 2 = 3x - 21$ Remove the parentheses.
$3x - 2y = 19$ Write the equation in standard form.

Thus the graph of $3x - 2y = 19$ goes through $P(7, 1)$ and is perpendicular to the graph of $2x + 3y = 5$. ■

The graph of a linear equation can be drawn by using the x- and y-intercepts. The *x-intercept* is that point on the x-axis where the line crosses the x-axis. The *y-intercept* is that point on the y-axis where the line crosses the y-axis. Of course, there is no guarantee that every line will have both x- and y-intercepts. To find the x-intercept, let $y = 0$ and solve for x; to find the y-intercept, let $x = 0$ and solve for y.

Example 8   Draw the graph of $2x - 3y = 6$ by using the x- and y-intercepts.

Solution   If $x = 0$, then $2x - 3y = 6$ becomes $-3y = 6$, which gives $y = -2$. The y-intercept is $(0, -2)$. If $y = 0$, then $2x - 3y = 6$ becomes $2x = 6$, which gives $x = 3$. The x-intercept is $(3, 0)$. Thus the line intersects the y-axis 2 units below the origin, and it intersects the x-axis 3 units to the right of the origin. The graph, including these two points, is shown in figure 7–17.

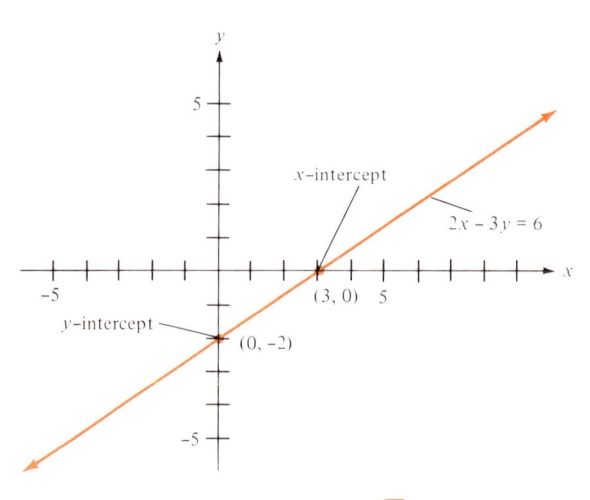

Figure 7–17

7–4 Exercises

In exercises 1–14, write each equation in the form $y = mx + b$ and give the slope and y-intercept for the graph of the equation by reading these numbers directly from the equation. *(See example 2.)*

1. $3x + 4y = 12$
2. $2x + 3y = 6$
3. $x - y = 6$
4. $2x - y = 4$
5. $5x - 3y = 6$
6. $9x - 7y = -2$
7. $-3x + 4y = -5$
8. $9x - 3y = -4$
9. $2(x - 3) + 2y = 5(x + 4)$
10. $x(x - 3) + 2y = -x(5 - x) + 8$
11. $\dfrac{x}{5} = \dfrac{3y}{10} - \dfrac{2}{5}$
12. $\dfrac{5x - 3}{4} + \dfrac{x - 3}{5} = y$
13. $\dfrac{6x - y}{2} - \dfrac{5y}{5} = 3$
14. $\dfrac{5x^2 - 3}{5} - \dfrac{2x^2 + y}{2} = 3x$

In exercises 15–20, use $y = mx + b$ to write an equation for the graph whose slope and y-intercept are given. Leave your answer in standard form. *(See example 3.)*

15. $m = 5, b = -2$
16. $m = -3, b = 4$
17. $m = \dfrac{1}{2}, b = -\dfrac{1}{3}$
18. $m = \dfrac{-4}{3}, b = \dfrac{1}{5}$
19. $m = \dfrac{3}{5}, b = 5$
20. $m = 7, b = \dfrac{5}{3}$

In exercises 21–24, find an equation of the line whose slope and y-intercept are given. Draw a sketch of the graph. *(See example 4.)*

21. $m = 2, b = 3$
22. $m = -3, b = -2$
23. $m = \dfrac{-2}{3}, b = 4$
24. $m = -\dfrac{1}{3}, b = -1$

In exercises 25–32, use the point-slope form of the equation for a straight line to write an equation of the line going through the following points. Leave your answer in standard form. (See example 6.)

25. (3, 5); (4, −1)   26. (−5, 2); (3, 6)   27. (−4, 3); (5, 3)
28. (8, 2); (−7, −4)   29. (−5, −4); (−2, 0)
30. (9, 3); (4, 5)   31. (c, d); (−8, 2)
32. (a, b); (c, d)

In exercises 33–40, determine if the lines represented by the equations are parallel, perpendicular, or neither parallel nor perpendicular. Make this determination by writing each equation in the form $y = mx + b$ and comparing the two slopes from each equation.

33. $y = \frac{2}{3}x + 3$; $y = \frac{2}{3}x - 4$ 34. $y = \frac{3}{5}x - 4$; $y = -\frac{5}{3}x + 6$

35. $3x + 5y = 20$; $5x - 3y = -9$ 36. $2x + 3y = 4$; $3y + 2x = -4$

37. $2x + 5y = 6$; $5x + 2y = -8$ 38. $5x + 2y = -3$; $5x + 2y = 8$

39. $9x + 6y = 4$; $9x + 6y = -8$ 40. $x - 2y = 6$; $2x - 3y = 4$

In exercises 41–46, draw the graph of the equation by using x- and y-intercepts. (See example 8.)

41. $2x - 3y = 12$ 42. $x - 2y = 8$ 43. $3x + 2y = 8$

44. $9x - 4y = 18$ 45. $x = \frac{3}{4}y - 2$ 46. $y = \frac{5}{3}x + 6$

47. Write an equation of the line parallel to the graph of $2x - 4y = 5$ and containing the point $P(-1, 3)$. (*Hint:* What will be true of the two slopes of the two lines? Now see example 7.)

48. Write an equation of the line perpendicular to the graph of $2x - 5y = -6$ and containing the point $P(5, -2)$. (See example 7.)

49. Write equations of the lines parallel to and perpendicular to the graph of $x - 3y = 4$ and going through the point $P(-2, 4)$.

50. Write an equation of the line going through the points $P_1(-2, 6)$ and $P_2(5, 4)$. Then, find the equation of the line perpendicular to $\overleftrightarrow{P_1P_2}$ and containing the point P_2.

51. Write an equation of the line containing the point $P(5, 8)$ and having the same y-intercept as the graph of $2x - 4y = 5$.

52. Determine the point where the graph of $2x - 3y = 18$ intersects the x-axis. Then write an equation of the line whose slope is $\frac{2}{5}$ and which goes through this point.

53. Given that x and y are linearly related and that $y = 20$ when $x = 5$ and $y = 16$ when $x = 4$.
    (a) Write an equation that expresses y in terms of x.
    (b) Use the equation from part a to find y when $x = 3.5$.

54. The Hypothetical Company makes Gadgets. The total cost for making 30 Gadgets is $90 and the total cost for making 60 Gadgets is $150. If c

is the total cost of making g Gadgets, find an equation that gives c in terms of g if c and g are linearly related. Then find the total cost for making 100 Gadgets. (*Hint:* The ordered pairs will have the form (g, c)).

55. Suppose an automobile depreciates 20% of its original purchase price each year. If it cost $10,000 when new, find an equation that gives its value V in terms of t, where t is the number of years after it was bought. Also, t must be a number between 0 and 5.

   (a) Find an equation that gives the value V of the automobile in terms of t. (*Hint:* To obtain some ordered pairs of the form (t, V), what is the value of the automobile when t = 0? When t = 1?)

   (b) What is the value of the automobile at the end of 3 years?

7–5 Linear Inequalities

Linear inequalities are relations of one of the forms $ax + by < c$ or $ax + by > c$. The symbols $\leq$ or $\geq$ are also used in linear inequalities.

Definition 7–4

Linear Inequality

A **linear inequality** is an inequality of one of the following forms:

$$ax + by < c \quad \text{or} \quad ax + by > c$$
$$ax + by \leq c \quad \text{or} \quad ax + by \geq c$$

where a, b, and c are real numbers.

A linear inequality can be graphed on a rectangular coordinate system just as a linear function can, but its graph does not consist solely of a straight line. The graph is made up of infinitely many points, which will lie on one side or the other of a straight line. The line may also be part of the graph. The line is called the *boundary line*, and it divides the plane into two regions called **half-planes**.

To find the boundary line, we temporarily change the inequality to the equation $ax + by = c$. Then we draw the graph of $ax + by = c$. This graph is the boundary line for the inequality. The graph of the boundary line is drawn as a dashed line to indicate that it is not part of the graph of the inequality if the inequality has the form $ax + by < c$ or $ax + by > c$. The graph of the boundary line is drawn as a solid line to show that it is part of the graph of the inequality if the inequality is given by $ax + by \leq c$ or $ax + by \geq c$. The rest of the

Relations and Functions: Part I

graph is indicated by shading one of the half-planes, and it is found by taking a test point in either half-plane. (The test point should never be taken from the boundary line.) If the coordinates of the test point make the inequality true, shade the half-plane containing the point. Otherwise, shade the opposite half-plane. The shaded half-plane is the graphical representation of the solution set if the inequality has the form $ax + by < c$ or $ax + by > c$. For inequalities of the form $ax + by \leq c$ or $ax + by \geq c$, the shaded half-plane and the boundary line represent the graph of the solution set. Linear inequalities do not represent functions because it is possible to obtain infinitely many values of y for a given value of x.

Example 1 Draw the graph of $2x - 3y < 6$.

Solution Use the intercept method to draw the graph of $2x - 3y = 6$.

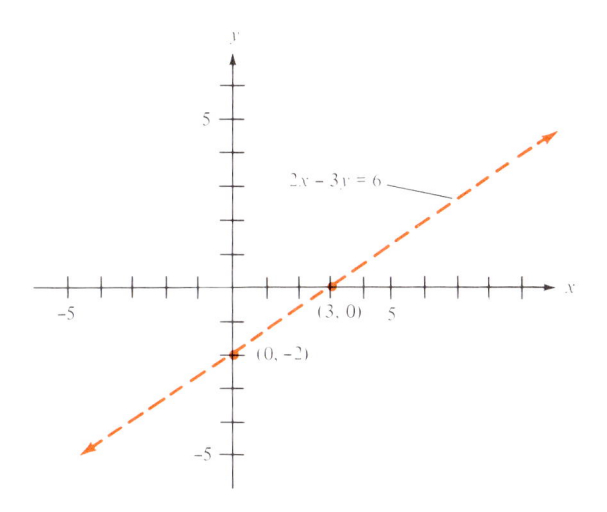

Figure 7–18

The boundary line is drawn as a dashed line in figure 7–18, since no point on the boundary line has coordinates that make $2x - 3y < 6$ a true statement. If the boundary line does not go through the origin, the point (0, 0) is the most convenient point to use as a test point. Inserting the coordinates of (0, 0) into $2x - 3y < 6$, we get

$$2(0) - 3(0) < 6$$
$$0 - 0 < 6$$
$$0 < 6$$

which is a true statement. We shade the half-plane that contains the point (0, 0). This is shown in figure 7–19.

284 Intermediate Algebra

Figure 7–19

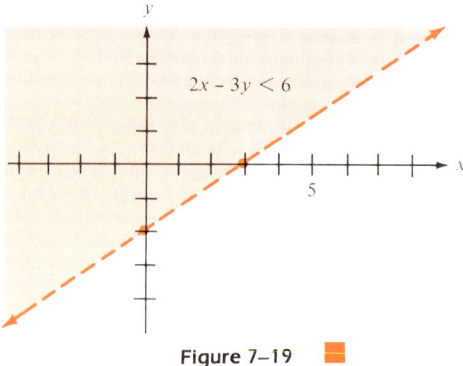

(Graph showing $2x - 3y < 6$)

A convenient way to indicate a linear inequality is with the set notation $\{(x, y) | ax + by < c\}$. This notation tells us to find every ordered pair (x, y) that will make $ax + by < c$ a true statement. Of course, the symbols $>$, $\leq$, or $\geq$ can also be used.

Example 2 Draw the graph of $\{(x, y) | x - y \leq 4\} \cap \{(x, y) | 3x + y \geq 6\}$.

Solution The intersection of these two inequalities is found by graphing each inequality separately and noting where the two graphs overlap. The double-shaded region (the overlap) is the graphical representation of the solution set. We show the steps that produce the properly shaded region in figure 7–20 and figure 7–21. The graphs shown in figure 7–20 are the graphs of each inequality drawn separately. In figure 7–21, these two graphs are both drawn on the same coordinate system. The double-shaded region in figure 7–21 is the graph of the solution set for the intersection of the two inequalities.

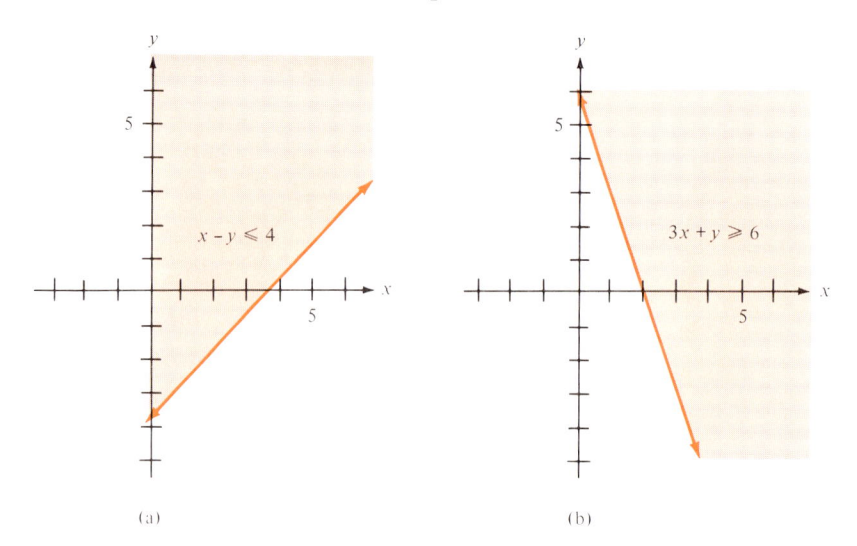

(a) Graph of $x - y \leq 4$

(b) Graph of $3x + y \geq 6$

Figure 7–20

Relations and Functions: Part I 285

[Figure 7-21: Graph showing the intersection of $\{(x,y) | x - y \leq 4\} \cap \{(x,y) | 3x + y \geq 6\}$]

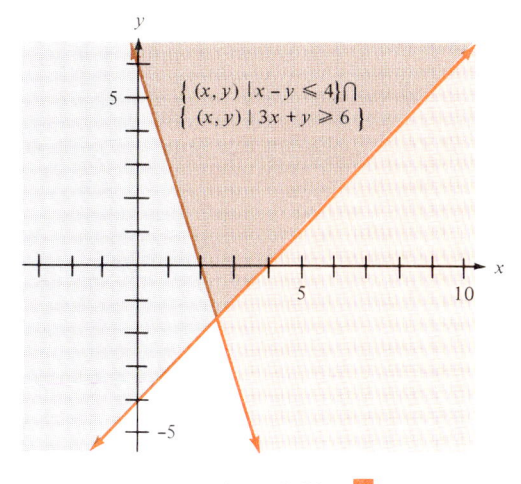

Figure 7-21

Example 3 Draw the graph of $\{(x, y) | y \geq |x - 2|\}$.

Solution In an example of this type, the boundary lines are found by considering the two cases inherent in the definition of absolute value. If $(x - 2) \geq 0$, then $y = |x - 2|$ becomes $y = x - 2$.

Case 1 $\quad x - 2 \geq 0 \quad$ and $\quad y = x - 2$
$\quad\quad\quad\quad\quad x \geq 2 \quad$ and $\quad y = x - 2$

The graph for this part of the boundary line is shown in figure 7-22. The domain for this part of $y = x - 2$ is $\{x | x \geq 2\}$.

[Figure 7-22: Graph of $y = x - 2$ and $x \geq 2$]

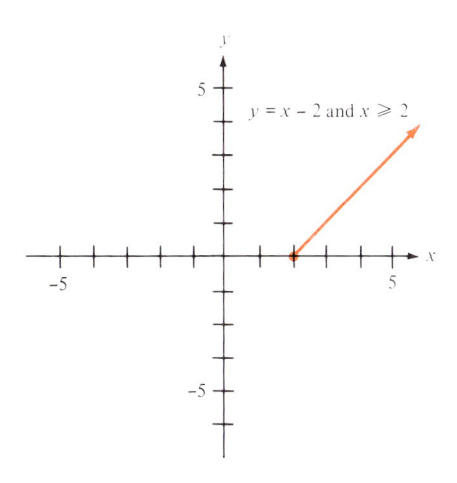

Figure 7-22

Now consider the case where $(x - 2) < 0$. Then $y = |x - 2|$ becomes $y = -(x - 2)$.

Case 2 $\quad x - 2 < 0 \quad$ and $\quad y = -(x - 2)$
$\qquad\qquad\qquad x < 2 \quad$ and $\quad y = -x + 2$

The graph for this part of the boundary line is shown in figure 7–23. This part of the graph is drawn over that part of the x-axis represented by $\{x|x < 2\}$.

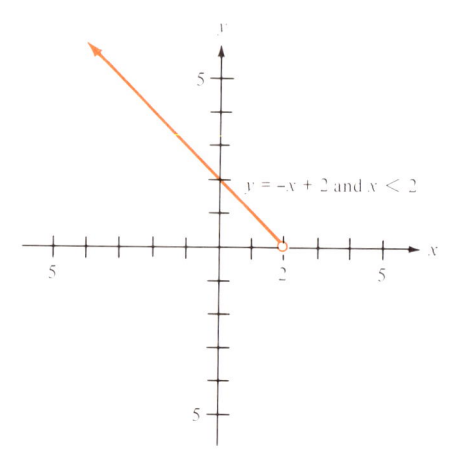

Figure 7–23

If both boundary lines are drawn on one set of coordinate axes and the proper half-planes shaded, the graph of the solution set is represented by the shaded region lying above the two boundary lines. This is shown in figure 7–24.

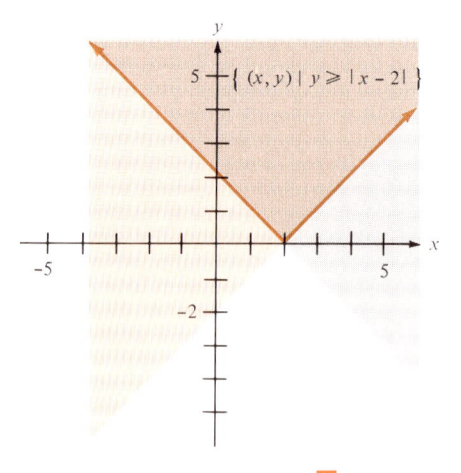

Figure 7–24

7–5 Exercises

In exercises 1–24, draw the graph of the solution set for each inequality. (See examples 1 and 2.)

1. $x + y < 2$
2. $x + 3y > 6$
3. $2x - y \leq 4$
4. $3x - 4y \geq 12$
5. $3x + 2y + 8 > 0$
6. $9x + 6y - 12 \leq 0$
7. $3x \leq y + 6$
8. $y \geq 6x + 6$
9. $\{(x, y) | 5x - 2y \leq 10\}$
10. $\{(x, y) | x + 10 > 5y\}$
11. $\{(x, y) | x - 3y \geq 5\}$
12. $\{(x, y) | 3y - x - 9 \geq 0\}$
13. $\{(x, y) | x \geq 4\}$
14. $\{(x, y) | y \leq -2\}$
15. $\{(x, y) | x \geq 3\} \cap \{(x, y) | y \leq 2\}$
16. $\{(x, y) | |x| < 2\} \cap \{(x, y) | |y| \leq 3\}$
17. $\{(x, y) | x + y > -2\} \cap \{(x, y) | x + y \leq 5\}$
18. $\{(x, y) | 2x - 3y \leq 6\} \cap \{(x, y) | 2x - 3y \geq -12\}$
19. $\{(x, y) | y \geq |x + 3|\}$ *(See example 3.)*
20. $\{(x, y) | y \leq |x - 4|\}$
21. $\{(x, y) | y < |x| + 2\}$
22. $\{(x, y) | y \geq |x| - 1\}$
23. $\{(x, y) | 2x - y \not\geq 8\}$
24. $\{(x, y) | x - 3y \not< 6\}$

In exercises 25–28, find each solution set by graphing. The solution will be shown as any part of the plane shaded either once or twice.

25. $\{(x, y) | x + 2y \geq 8\} \cup \{(x, y) | 2x - 3y \leq 8\}$
26. $\{(x, y) | x > 2y - 4\} \cup \{(x, y) | x + y < 4\}$
27. $\{(x, y) | 5x \geq -15 - 2y\} \cup \{(x, y) | |y| \leq 3\}$
28. $\{(x, y) | x - 4y > -4\} \cup \{(x, y) | |x| > 2\}$

Chapter 7 Summary

Vocabulary

abscissa
component
domain
function
graph of a relation
half-plane
linear function
linear inequality
ordered pair
ordinate
origin
point-slope form
quadrant
range
rectangular coordinate system
relation
slope
slope-intercept form
x-intercept
y-intercept

Definitions

7–1 Function

A function is a relation in which no two ordered pairs have the same first components and different second components.

7–2 Parts of a Function

1. A set of numbers called the *domain of the function*.
2. A set of numbers called the *range of the function*.
3. A rule of correspondence that assigns to each member in the domain of the function one and only one member in the range of the function.

7–3 Slope of a Line

If P_1 and P_2 are two distinct points on a rectangular coordinate system with coordinates (x_1, y_1) and (x_2, y_2), respectively, then the slope of the line (or line segment) joining P_1 and P_2 is

$$m = \frac{y_2 - y_1}{x_2 - x_1}$$

7–4 Linear Inequality

A linear inequality is an inequality of one of the following forms:

$$ax + by < c \quad \text{or} \quad ax + by > c$$
$$ax + by \leq c \quad \text{or} \quad ax + by \geq c$$

where a, b, and c are real numbers.

Rules

7–1 The Distance Formula

If P_1 and P_2 are any two points on a rectangular coordinate system with coordinates (x_1, y_1) and (x_2, y_2), respectively, the distance between P_1 and P_2 is given by

$$d = \sqrt{(x_2 - x_1)^2 + (y_2 - y_1)^2}$$

7–2 Perpendicular and Parallel Lines

Two lines L_1 and L_2 with slopes m_1 and m_2, respectively, are:

1. Parallel if and only if $m_1 = m_2$.
2. Perpendicular if and only if $m_1 = -1/m_2$ ($m_1, m_2 \neq 0$).

7–3 Slope-Intercept Form of an Equation of the Line

The **slope-intercept form** for an equation of the straight line is

$$y = mx + b$$

The slope of the line is given by m and the y-intercept is given by b.

7–4 Point-Slope Form for an Equation of the Straight Line

The **point-slope form** for an equation of the straight line is $y - y_1 = m(x - x_1)$. The slope of the line is given by m and x_1 and y_1 are the coordinates of a point through which the line passes.

Review Problems

Section 7–1
1. Find two ordered pairs that will make $5x + 3y = 4$ a true statement.
2. Write the domain and range for the relation $\{(9, 4), (-7, 5), (9, -6)\}$.

Section 7–2
3. Is it true or false that every relation is a function?
4. Find the function given by $\{(x, y) | 2x - 3y = 7, x \in \{-1, 0, 8\}\}$.
5. Determine if $\{(9, 2), (5, 4), (3, -2)\}$ is a function. Tell why or why not.
6. Give the domain of $f(x) = x\sqrt{3x - 9}$.
7. If $f(x) = 2x - x^2$, find $f(a^2)$.

Section 7–3
8. Find the distance between the points $P_1(5, 7)$ and $P_2(-4, -3)$.
9. Find the perimeter of the triangle whose vertices are located at $P_1(0, 0)$, $P_2(0, 6)$, $P_3(8, 6)$.
10. Find the slope of the line that goes through the points $P_1(-4, -3)$ and $P_2(5, 7)$.
11. Use the distance formula to determine if the three points $P_1(0, 4)$, $P_2(1, 5)$, and $P_3(-1, 2)$ all lie on the same line.
12. A farmer wishes to build a triangular cattle lot by building a new fence to join two other existing east-west and north-south fences, as shown in the figure. The new fence will join the existing east-west fence 80 feet from the point where the two existing fences intersect. It will join the existing north-south fence 60 feet from the point of intersection of the two existing fences. What will be the length of the new fence?

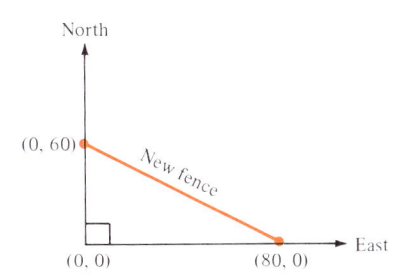

Section 7–4
13. Without drawing the graph, give the slope and y-intercept of the graph of $5x + y = -3$.

14. Write an equation in standard form for a graph with a slope of $\frac{3}{8}$ and y-intercept equal to -3.

15. Use the point-slope form of the equation of a linear function and write an equation in standard form for the line that goes through the points $P_1(8, 2)$ and $P_2(-3, 4)$.

16. Write an equation of the line that goes through $P(-2, 3)$ and is parallel to the line that goes through $P_1(5, 8)$ and $P_2(-10, -7)$.

17. Determine if the graphs of $2x - 3y = 4$ and $3x - 2y = 6$ are perpendicular without drawing any graphs.

18. Draw the graph of $9x - 3y = 12$ by using the intercept method of graphing.

Section 7–5

19. Draw the graph of $\{(x, y) | 5x - (1/2)y \geq 4\}$.

20. Draw the graph of $\{(x, y) | 2x - y \geq 4\} \cup \{(x, y) | x + y \leq 5\}$.

21. Draw the graph of $\{(x, y) | |x| \leq 5\} \cap \{(x, y) | |y| \leq 2\}$.

22. Draw the graph of the ordered pairs in which the abscissa of each ordered pair is at least twice as big as the ordinate of each ordered pair.

Cumulative Test for Chapters 1–7

Chapter 1

1. Supply grouping symbols so that the following is true. $3 \cdot 6 + 2 \cdot 3 + 4 = 60$
2. Evaluate the expression $\dfrac{a^3 - 2ab}{a(b-3)}$ if $a = -2$ and $b = -3$.
3. Simplify: $\dfrac{26 - 3^2 + 3[9 - (-2)]}{(-2)^3 - 2}$.

Chapter 2

4. What is the degree of the second term in the polynomial $5x^3 + 3x^2 - 6x + 3$?
5. Simplify: $\dfrac{9^0 x(-2x^3 y)^3}{-4xy^8}$.
6. Collect like terms: $5x^2 - 4ab + 3c^2 - 5ab - 3x^2 + 7c^2$.
7. Multiply and simplify: $(2x - 3)(x^2 - 6x - 2)$.
8. Factor: $2ac - 2ad + 3c - 3d$.

Chapter 3

9. Use the rules regarding the signs of a fraction and replace the question mark with the correct expression in the following: $-\dfrac{-(x-y)}{-3} = -\dfrac{?}{3}$.
10. Reduce the following: $\dfrac{8x^3 - y^3}{2x^2 + xy - y^2}$.
11. Divide and simplify: $\dfrac{x^3 - y^3}{x^2 - y^2} \div \dfrac{x^2 + xy + y^2}{x^2 + xy}$.
12. Add and simplify: $\dfrac{4x^2}{x^2 + 2x} + \dfrac{8x}{x^2 + 2x}$.
13. Simplify: $\dfrac{\frac{2}{3} + \frac{3}{4}}{\frac{3}{5} + \frac{1}{4}}$.
14. Use synthetic division to divide: $(9x^4 + 3x^2 - x) \div (x + 2)$.

Chapter 4

15. Solve for y: $3(8y - 7) - 4y = -2(y - 3) + 17$.
16. Solve for x: $\dfrac{2}{x+3} + \dfrac{3}{x-3} = \dfrac{23}{x^2 - 9}$.
17. Find the solution for $-12x - 6 < -3(x - 10)$.
18. Solve for x: $-2|3x - 8| = -8$.
19. Solve for v: $|3(2v - 4) - 7| < 5$.
20. Solve $E = \frac{1}{2}mv^2$ for m.

291

21. One machine can do a job in 12 hours and a second machine can do the same job in 8 hours. The first machine is started at 6:00 A.M. At 8:00 A.M., the second machine is started and both machines run until the job is finished. What time is the job finished?

22. Find three consecutive integers such that three times the largest integer is 53 more than the sum of the other two smaller integers.

Chapter 5

23. Simplify and leave all exponents positive: $(m^{-3}n^4)^{-3}$.

24. Simplify so that all exponents are positive: $\dfrac{x^{-6}y^4}{z^{-2}}$.

25. Simplify; leave the answer in scientific notation: $(5 \times 10^{-4})(4.4 \times 10^2)$.

26. Simplify; leave the answer in radical form: $\left(\dfrac{x^{-1/3}y^{1/4}}{x}\right)^{-1/3}$.

27. Subtract: $8\sqrt{24} - 2\sqrt{54}$.

28. Simplify; assume $x > 0$: $\dfrac{2\sqrt{x} + 4\sqrt{x^3}}{6\sqrt{x}}$.

29. Solve for x: $\sqrt{x+2} = 2 + \sqrt{x-6}$.

Chapter 6

30. Solve for x by using the quadratic formula: $3x^2 + 4x + 2 = 0$.

31. Write a quadratic equation whose roots are 3 and -5.

32. Add and simplify: $(6 + 3i) + (4 - 8i)$.

33. Simplify: $7\sqrt{-20} + 3\sqrt{-80}$.

34. Solve for x (there will be five solutions): $x^5 - 13x^3 + 36x = 0$.

35. The longer leg of a right triangle is 2 feet more than twice the shorter leg. The length of the hypotenuse is 1 foot more than the longer leg. Find the length of all three sides of the triangle. (Recall that $c^2 = a^2 + b^2$ for a right triangle, where a, b, and c are the lengths of the two legs and hypotenuse, respectively).

Chapter 7

36. Give the domain of the relation $R = \{(1, 3), (2, 4), (-5, 6)\}$.

37. What are the three ordered pairs in the relation given by $2x + 3y = 12$ if $x \in \{-3, 0, 3\}$?

38. Give the domain of the function given by $y = \dfrac{5x}{x^2 - 2x}$.

39. Find the distance between the points whose coordinates are $(3, -6)$ and $(-2, -3)$.

40. Write an equation of the line going through the points having coordinates $(5, 10)$ and $(-1, -2)$. Leave the answer in the form $Ax + By = C$, where A, B, and C are integers.

41. Write an equation for the straight line with y-intercept -6 and slope $-\frac{2}{3}$. Leave the answer in the form $Ax + By = C$, where A, B, and C are integers.

42. Write an equation of the line perpendicular to the graph of $3x - 4y = 6$ and containing the point with coordinates $(-2, 5)$. Leave the answer in the form $Ax + By = C$, where A, B, and C are integers.

43. Draw the graph of $5x - 4y \leq 20$.

44. Draw the graph of $\{(x, y) | x + 2y \geq 10\} \cap \{(x, y) | x < y\}$.

CHAPTER 8
Relations and Functions: Part II

Objectives of Chapter 8

To be able to:

- ☐ Define a parabola. (8–1)
- ☐ Recognize a parabola from its equation. (8–1)
- ☐ Graph a parabola (8–1)
- ☐ Write the equation of a parabola in standard form. (8–1)
- ☐ Locate the vertex of a parabola. (8–1)
- ☐ Determine which direction a parabola opens by examining its equation. (8–1)
- ☐ Write equations having the form $x = ay^2 + by + c$ as two equations so that each of the two equations represents a function. (8–1)
- ☐ Give the definition of a circle. (8–2)
- ☐ Find the center and radius of a circle. (8–2)
- ☐ Write an equation of a circle knowing its center and a point through which it goes. (8–2)
- ☐ Identify a parabola or circle from its equation. (8–2)
- ☐ Give the standard form and the general form of an equation of a circle. (8–2)
- ☐ Graph equations that represent circles. (8–2)
- ☐ Define an ellipse. (8–3)
- ☐ Find the x- and y-intercepts of an ellipse. (8–3)
- ☐ Graph an ellipse. (8–3)
- ☐ Locate the major axis of an ellipse. (8–3)
- ☐ Give the definition of a hyperbola. (8–3)
- ☐ Graph a hyperbola. (8–3)

- ☐ Identify the conic sections from their equations. (8–3)
- ☐ Give the coordinates of the vertices for an ellipse and a hyperbola. (8–3)
- ☐ Draw the graph of quadratic inequalities. (8–4)
- ☐ Define a ratio. (8–5)
- ☐ Solve word problems using ratios. (8–5)
- ☐ Define a proportion. (8–5)
- ☐ List various properties of a proportion. (8–5)
- ☐ Solve proportions. (8–5)
- ☐ Define direct, inverse, and joint variation. (8–6)
- ☐ Solve problems involving variation. (8–6)
- ☐ Define the inverse of a function. (8–7)
- ☐ Find the inverses of relations and functions. (8–7)
- ☐ Sketch the graph of a relation and its inverse. (8–7)
- ☐ Give the domain and range of a relation and the inverse of the relation. (8–7)

The quadratic relations that we study in this chapter are the parabola, circle, ellipse, and hyperbola. Some knowledge of these relations is essential if we are to better understand our physical world. For example, cross sections of most radar and radiotelescope antennae are parabolas. Large searchlights used at the grand openings of stores have reflectors whose cross sections are parabolas. Circles abound profusely in architecture. The orbital path of a planet around the sun is an ellipse.

The first of these important relations which we will examine is the parabola.

8–1 Quadratic Functions and Relations

Definition 8–1

Parabola

A **parabola** is the set of all points in a plane equidistant from a fixed point, called the **focus,** and a fixed line, called the **directrix.**

Points A, B, C, D, and E in figure 8–1 are points that meet definition 8–1. In definition 8–1, we assume the directrix does not contain the focus. The smooth curve drawn through points A, B, C, D, and E is a parabola.

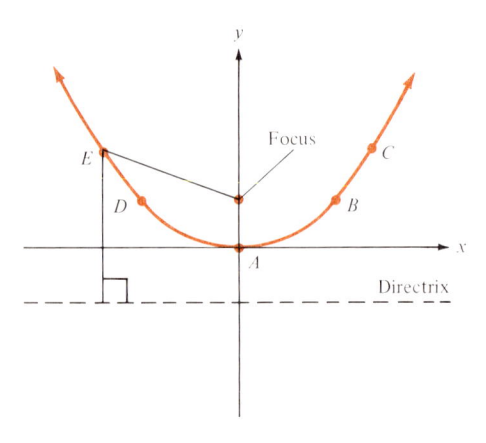

Figure 8–1

Every parabola that we shall study will have equations of the form

$$y = ax^2 + bx + c \qquad (1)$$

or

$$x = ay^2 + by + c \qquad (2)$$

Those parabolas whose equations are of the form of (1) open upward if $a > 0$ and downward if $a < 0$. Those of the form given in (2) open to the right if $a > 0$ and to the left if $a < 0$.

A parabola that opens upward or downward will have a maximum or minimum point called the **vertex.** In figure 8–1, the vertex represents a minimum and is located at point A. If the parabola opens to the right or left, the vertex will be that point furtherest to the left or right on the parabola. Figure 8–2 shows a graphic illustration of parabolas that open in various directions.

Example 1

Determine the direction in which each parabola will open if its equation is given by:

(a) $y = -2x^2 + 3x + 1$
(b) $x = 4y^2 + 2y - 3$
(c) $y = 5x^2 + 2$
(d) $x = -y^2 + 5y$

Solution

(a) Downward, since $a = -2$.
(b) To the right, since $a = 4$.
(c) Upward, since $a = 5$.
(d) To the left, since $a = -1$. ■

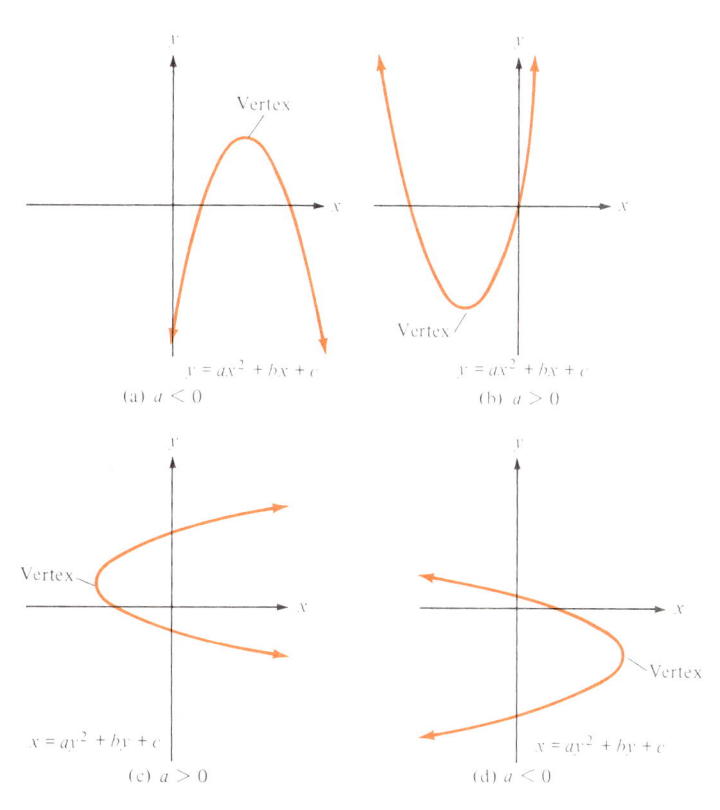

Figure 8–2

Example 2 Draw the graph of $y = (x + 2)^2 - 3$.

Solution Calculate some ordered pairs that make $y = (x + 2)^2 - 3$ true by letting x assume several values. Plot these and draw a smooth curve through the points, as shown in figure 8–3. ■

In figure 8–3, the points with coordinates $(x_1, 0)$ and $(x_2, 0)$ are the x-intercepts for the parabola. They are found (when they exist) by letting y be zero in the given equation and then solving the resulting equation for x. You should verify for the parabola shown that the x-coordinates of these two points are $-2 \pm \sqrt{3}$.

x	-5	-4	-3	-2	-1	0	1
y	6	1	-2	-3	-2	1	6

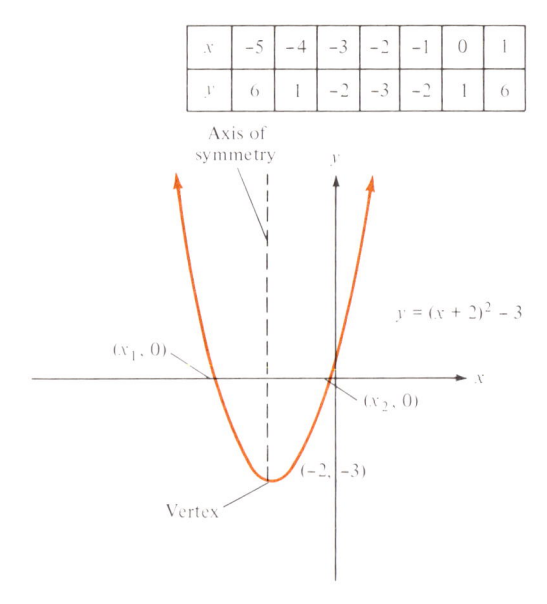

Figure 8–3

The equation for every parabola that opens upward or downward can be written in the form $y = a(x - h)^2 + k$. Those that open to the right or to the left have equations of the form $x = a(y - k)^2 + h$. For equations of parabolas in either of these forms, h and k are the abscissa and ordinate, respectively, of the ordered pair that locates the vertex.

For the graph in figure 8–3, we see $h = -2$ and $k = -3$. Thus the vertex is located at $(-2, -3)$. This fact can be read directly from the equation given in example 2. If the equation is not in the form $y = a(x - h)^2 + k$ or $x = a(y - k)^2 + h$, we must complete the square and factor and then read the values of h and k from the resulting equation.

Example 3 Find the coordinates of the vertex for the parabola whose equation is $y = -3x^2 + 9x + 1$.

Solution

$y = -3x^2 + 9x + 1$

$y = -3(x^2 - 3x) + 1$ Factor -3 from the first two terms on the right side.

$y = -3\left(x^2 - 3x + \dfrac{9}{4}\right) + 1 + \dfrac{27}{4}$ Complete the square. (Note: $\dfrac{9}{4}$ is needed to complete the square. Since $\dfrac{9}{4}$ is multiplied by -3 on the outside of the parentheses, which gives $(-3)\left(\frac{9}{4}\right) = -\frac{27}{4}$, we must add $\frac{27}{4}$ to offset the calculation we made to complete the square.)

$$y = -3\left(x - \frac{3}{2}\right)^2 + \frac{31}{4}$$

The vertex is located at $\left(\frac{3}{2}, \frac{31}{4}\right)$. ■

Every parabola has an **axis of symmetry,** which passes through the vertex. An axis of symmetry is a line that separates a figure into two halves, so that if the figure were folded along the line, the two halves would coincide. There is a more precise definition for an axis of symmetry, but the intuitive idea just expressed will serve our present needs.

Figure 8–3 shows the axis of symmetry for $y = (x + 2)^2 - 3$. The axis of symmetry can be used in drawing the parabola. Any point on the parabola has a "mirror image" on the opposite side of the axis of symmetry. Note that the axis of symmetry is parallel to the y-axis if the parabola opens upward or downward. It is parallel to the x-axis if the parabola opens to the right or left.

Example 4   Draw the graph for $x = 2y^2 - 8y + 3$. Show the axis of symmetry and the coordinates of the vertex; use the axis of symmetry as an aid in drawing the graph.

Solution
$x = 2y^2 - 8y + 3$
$x = 2(y^2 - 4y) + 3$ Factor 2 from the first two terms on the right side.
$x = 2(y^2 - 4y + 4) + 3 - 8$ Complete the square.
$x = 2(y - 2)^2 - 5$ Factor.

The equation now has the form $x = a(y - k)^2 + h$. The vertex is located at $(h, k) = (-5, 2)$. The axis of symmetry will go through $(-5, 2)$ and be parallel to the x-axis, since the parabola opens to the right. If we let $y = 0$, then $x = 3$. This gives $(3, 0)$ as a point on the graph. If the point represented by $(3, 0)$ is reflected across the axis of symmetry we get the point with coordinates $(3, 4)$. The graph may now be sketched with a fair degree of accuracy by using these two points and the point that locates the vertex. See figure 8–4 (a). ■

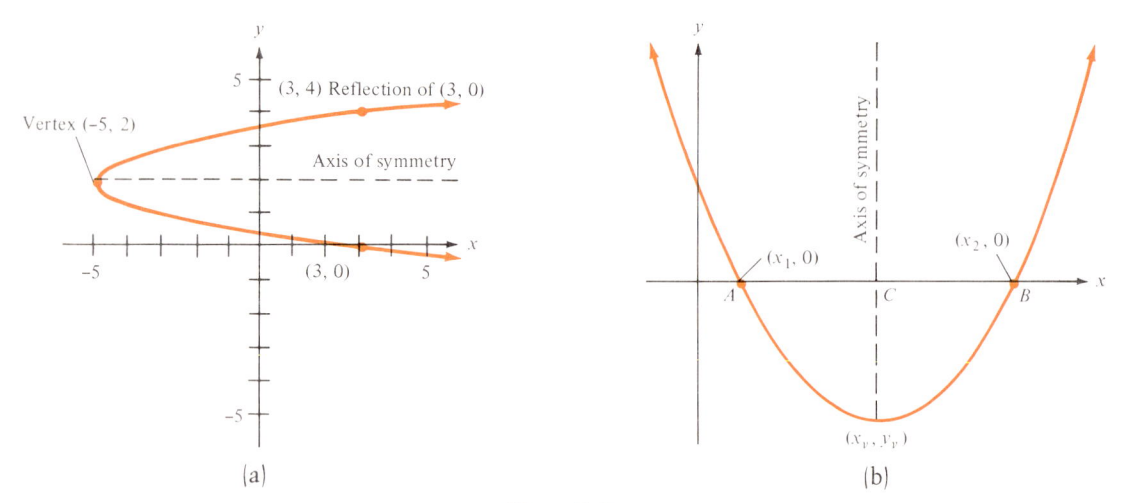

Figure 8-4

The abscissa of the vertex for the graphs of equations of the form $f(x) = ax^2 + bx + c$ can also be found by using the equation $x_v = -b/2a$, where x_v denotes the abscissa of the vertex. The ordinate of the vertex is then given by $f(x_v) = y_v$, where y_v designates the ordinate of the vertex.

To see why $x_v = -b/2a$, examine the graph of $f(x) = ax^2 + bx + c$ in figure 8–4(b). Assume the graph of $f(x)$ intersects the x-axis in two distinct points, A and B. Let the coordinates of A and B be $(x_1, 0)$ and $(x_2, 0)$, respectively. The axis of symmetry intersects the x-axis at a point C, which is midway between points A and B. Because the vertex of the parabola lies on the axis of symmetry, this means x_v is the average of x_1 and x_2, or $x_v = (x_1 + x_2)/2$. Also, x_1 and x_2 are roots of the equation $ax^2 + bx + c = 0$.

$$x_1 = \frac{-b - \sqrt{b^2 - 4ac}}{2a} \quad \text{and} \quad x_2 = \frac{-b + \sqrt{b^2 - 4ac}}{2a}$$

This gives

$$x_v = \frac{x_1 + x_2}{2} = \frac{\dfrac{-b - \sqrt{b^2 - 4ac}}{2a} + \dfrac{-b + \sqrt{b^2 - 4ac}}{2a}}{2}$$

$$= \frac{\dfrac{-2b}{2a}}{2}$$

$$= \frac{-b}{2a}$$

Thus, $x_v = -b/2a$.

If the equation is of the form $g(y) = ay^2 + by + c$, then the

Relations and Functions: Part II

ordinate of the vertex is given by $y_v = -b/2a$. The abscissa is then found by calculating $g(y_v)$.

Example 5   Calculate the coordinates of the vertex for the graph of $f(x) = 2x^2 + 8x + 3$ by each method.
(a) Completing the square.
(b) Using $x_v = -b/2a$.

Solution

(a)
$$f(x) = 2x^2 + 8x + 3$$
$$f(x) = 2(x^2 + 4x) + 3 \quad \text{Factor 2 from the first two terms on the right side.}$$
$$f(x) = 2(x^2 + 4x + 4) + 3 - 8 \quad \text{Complete the square.}$$
$$f(x) = 2(x + 2)^2 - 5 \quad \text{Factor.}$$
$$(h, k) = (-2, -5)$$

(b)
$$f(x) = 2x^2 + 8x + 3$$
$$x_v = \frac{-b}{2a}$$
$$= \frac{-8}{2(2)}$$
$$= -2$$
$$f(x_v) = f(-2) = y_v = 2(-2)^2 + 8(-2) + 3$$
$$= 8 - 16 + 3$$
$$= -5$$

Therefore, $(h, k) = (x_v, y_v) = (-2, -5)$.

Example 6   A village blacksmith has found that the cost of shoeing horses is given by $C(x) = x^2 - 20x + 200$ ($x \geq 0$), where x is the number of horses that the village smithy shoes per day. Find the number of horses per day that the blacksmith should shoe to keep costs at a minimum.

Solution   In examples of this type, the minimum (or maximum) value for $C(x)$ is $C(x_v)$, where x_v is the abscissa of the vertex of the graph of $C(x)$. (This is proven in differential calculus.)

$$x_v = \frac{-b}{2a} = \frac{-(-20)}{2(1)} = 10$$

The village smithy should shoe exactly ten horses per day to keep costs at a minimum. $C(10)$ gives $100. Any $x \neq 10$ will give a cost greater than $100. This can be seen by examining the graph of $C(x)$.

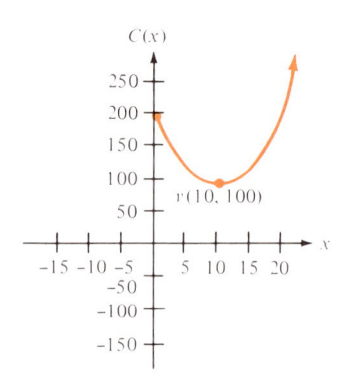

8–1 Exercises

In exercises 1–10, an equation for a parabola is given. Determine the direction which each parabola will open. *(See example 1.)*

1. $y = 3x^2 + 2$
2. $y = 2x^2 - 4x + 2$
3. $f(x) = -2x^2 - 3x + 4$
4. $f(x) = -x^2 - 6x + 3$
5. $f(x) = -2x + 3x^2 + 3$
6. $x = 2y^2 - 6y + 1$
7. $x + 4y^2 = 5$
8. $x + 4y^2 - 7y = -1$
9. $x - 5y = -2y^2 + 3$
10. $x - 3y^2 = 7 - 2y$

In exercises 11–16, complete the square of each equation to find the vertex of each parabola represented by the given equation. *(See example 3.)*

11. $y = x^2 + 6x - 3$
12. $y = x^2 + 2x + 5$
13. $y = 2x^2 - 4x + 5$
14. $y = 5x^2 + 7x + 1$
15. $-2y^2 = -x - 6y + 2$
16. $7y - x = -3y^2 - 4$

In exercises 17–22, find the vertex of the parabola represented by the given equation by using the equation $x_v = -b/2a$ or $y_v = -b/2a$. *(See example 5.)*

17. $y = x^2 + 6x - 3$
18. $y = 2x^2 + 8x + 3$
19. $f(x) = -3x^2 + 12x - 1$
20. $f(x) = -x^2 + 4x - 2$
21. $2y^2 = x + 12y - 5$
22. $x + 3y^2 = 6y + 4$

In exercises 23–28, find the ordered pair for the vertex and two or three other ordered pairs on either side of the vertex. Then draw the graph for each given equation. *(See example 4.)*

23. $y = 2x^2$
24. $x = -y^2$
25. $x = y^2 + 3$
26. $y = x^2 - 3$
27. $y = 2x^2 - 12x + 20$
28. $x^2 = -y + 6x - 2$

Relations and Functions: Part II

In exercises 29–32, write each equation as two equations so that each part is a function. Draw the graph of the part that expresses y as a nonnegative real number.

Example $x = 3y^2 - 4$

Solution We must solve for y^2 and take the square root of both sides of the equation.

$$x = 3y^2 - 4 \qquad \text{Given equation}$$
$$3y^2 = x + 4$$
$$y^2 = \frac{x + 4}{3}$$

$$y = \sqrt{\frac{x + 4}{3}} \quad \text{or} \quad y = -\sqrt{\frac{x + 4}{3}}$$

Since $y = \sqrt{(x + 4)/3}$ gives y as a nonnegative real number, we draw the graph of this part.

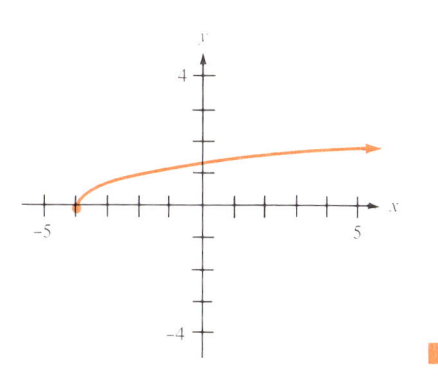

29. $x = 3y^2$ 　　　　30. $x = 2y^2$ 　　　　31. $x = 2y^2 - 3$
32. $x = 5y^2 + 4$

33. Steve and Ross have gone into the lawn-cutting business to earn money to pay for the motor that they broke on their motorboat trip (see problem 28, chapter 6 review). Being good at cutting lawns, but not too good at mathematics, they would like for you to determine how many lawns they should cut per day to maximize their profit if the profit which they make is given by $p(x) = -x^2 + 16x - 5$, where p is the profit and x is the number of lawns cut per day. (See example 6).

34. The total cost, C, (in dollars) for manufacturing x toy trains per day is $C(x) = 1,500 - 6x + 0.03x^2$. How many trains should be manufactured per day to minimize the cost? What would be the cost per train for manufacturing 10 trains per day? How does this compare to the cost per train for the number of trains that gives the minimum cost?

35. A parabola was defined at the beginning of this section. Show that for the fixed point $(p, 0)$ and the fixed line $x = -p$, the equation of the parabola is $y^2 = 4px$. See figure 8–5. (*Hint:* $d_1 = d_2$, and assume $p > 0$.)

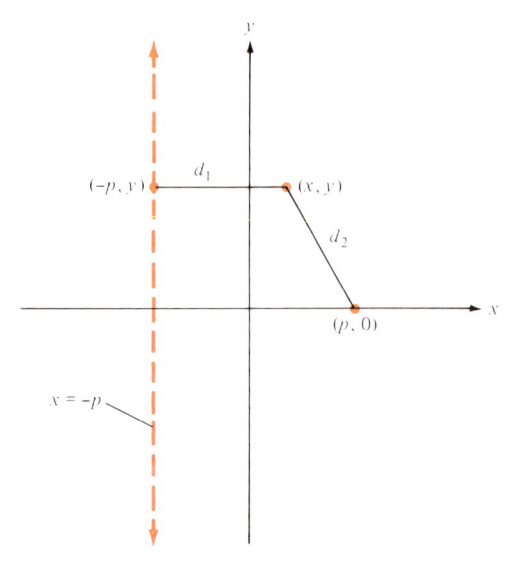

Figure 8–5

36. In a manner similar to exercise 35, show that the parabola given by the set of points the same distance from $(0, p)$ and $y = -p$ has the equation $x^2 = 4py$. Assume $p > 0$.

8–2 The Circle

Another equation of degree two in two variables is the equation whose graph is a circle. The circle is sometimes referred to as one of the conic sections. The parabola of section 8–1, the ellipse, and the hyperbola are also conic sections. **Conic sections** are formed by the intersection of a plane with the surface of a cone, shown in figure 8–6. The manner in which the plane intersects the cone determines which conic section is formed.

Definition 8–2

Circle

A **circle** is the set of all points in a plane located a fixed distance from a fixed point. The measure of the fixed distance is its **radius** and the fixed point is its **center**.

Relations and Functions: Part II

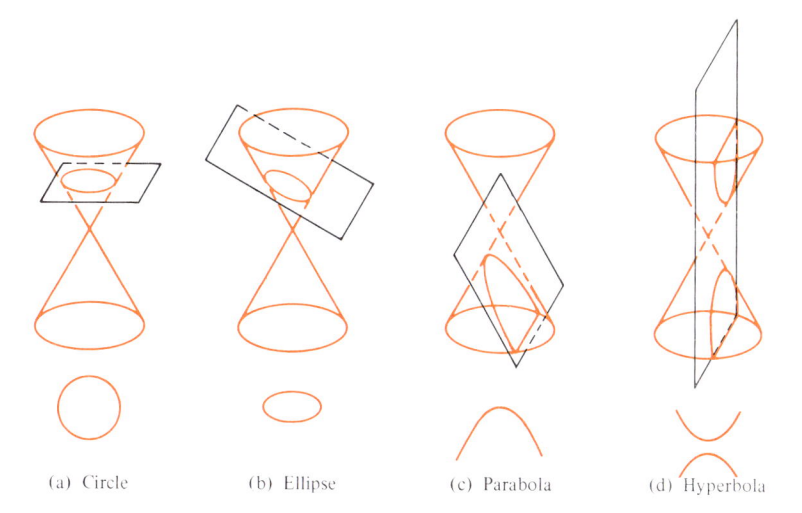

(a) Circle   (b) Ellipse   (c) Parabola   (d) Hyperbola

Figure 8–6

If we let the fixed point be the origin of the coordinate system and if the point $P(x, y)$ is any point on the circle, then we can derive the equation of a circle whose center is at the origin by using the distance formula. See figure 8–7.

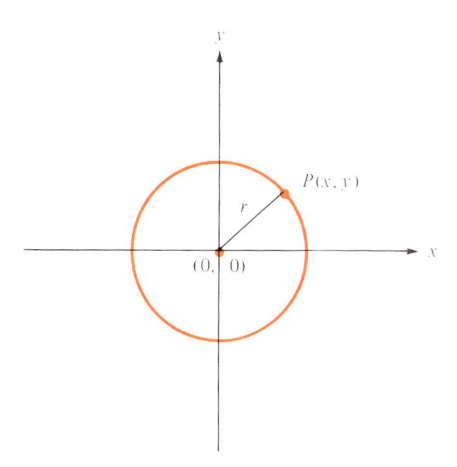

Figure 8–7

Let the distance between the origin and $P(x, y)$ be represented by the letter r (for the radius of a circle).

$$r = \sqrt{(x - 0)^2 + (y - 0)^2} \quad \text{Distance formula.}$$
$$r = \sqrt{x^2 + y^2} \quad \text{Simplify under the radical.}$$
$$r^2 = x^2 + y^2 \quad \text{Square both sides.}$$
$$x^2 + y^2 = r^2 \quad \text{Rearrange the equation.}$$

Any circle whose center is located at the origin and whose radius is r has an equation given by $x^2 + y^2 = r^2$.

If the circle does not have its center at the origin, then we denote the center by the ordered pair $P(h, k)$. Let $Q(x, y)$ be any point on the circle. By applying the distance formula to the points P and Q, we can find an equation of the circle.

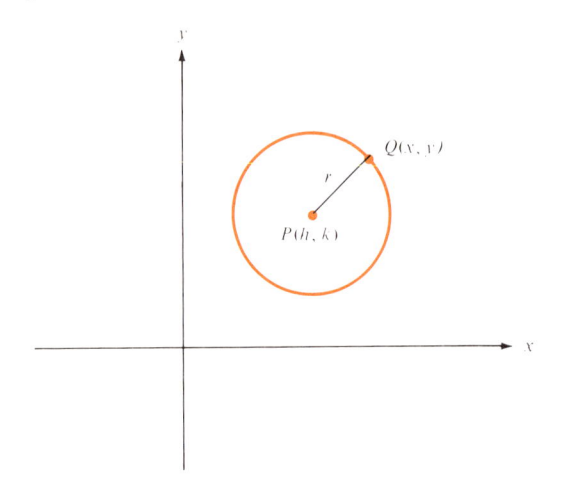

Figure 8-8

The distance, r, between P and Q from figure 8-8 is

$$r = \sqrt{(x - h)^2 + (y - k)^2} \quad \text{distance formula}$$
$$r^2 = (x - h)^2 + (y - k)^2$$
$$(x - h)^2 + (y - k)^2 = r^2$$

Rule 8-1

Standard Form of an Equation of a Circle

Any circle whose radius is $r (r > 0)$ and whose center is located at the origin is given by the equation $x^2 + y^2 = r^2$. If the center is not located at the origin, the equation of the circle has the form $(x - h)^2 + (y - k)^2 = r^2$ $(r > 0)$, where r is the radius and (h, k) is the location of the center. These equations are referred to as the **standard forms** for an equation of a circle.

Example 1   Find the radius and coordinates of the center of the circle given by $x^2 + y^2 = 25$.

Solution   From rule 8-1, we know the circle has its center at $(0, 0)$. We also know $r^2 = 25$; hence $r = 5$. (We do not use $r = -5$ because the radius of a circle is a measure of distance, which is a positive quantity.) ∎

Relations and Functions: Part II

Example 2 Determine the coordinates of the center and radius of the circle given by $(x - 3)^2 + (y + 5)^2 = 9$.

Solution This equation has the form $(x - h)^2 + (y - k)^2 = r^2$. The second part of rule 8–1 states the center is (h, k) and the radius is r. Therefore, the center of this circle is $(3, -5)$ and its radius is $r = \sqrt{9} = 3$. ■

If the standard form of the equation $(x - h)^2 + (y - k)^2 = r^2$ is multiplied out, we have

$$(x - h)^2 + (y - k)^2 = r^2$$
$$x^2 - 2hx + h^2 + y^2 - 2ky + k^2 = r^2 \quad \text{Square the binomials.}$$
$$x^2 + y^2 - 2hx - 2ky + (h^2 + k^2 - r^2) = 0 \quad \text{Rearrange and regroup the terms.}$$

Because h, k, and r are all constants, let $-2h = D$, let $-2k = E$ and let $h^2 + k^2 - r^2 = F$. Then we have

$$x^2 + y^2 + Dx + Ey + F = 0$$

This form is known as the **general form** of an equation of a circle.

Rule 8–2

> **General Form of an Equation of a Circle**
>
> The **general form** of an equation of a circle is
> $$x^2 + y^2 + Dx + Ey + F = 0.$$

The center and radius of a circle cannot be determined by inspection from the general form of an equation of the circle. If these two facts are desired, we must complete the square on the x- and y-terms to put the equation back into the standard form (see example 3).

Example 3 Write $x^2 + y^2 + 6x - 8y - 11 = 0$ in standard form and give the radius and coordinates of the center of the circle.

Solution The given equation is $x^2 + y^2 + 6x - 8y - 11 = 0$. Group the x- and y-terms together and eliminate the constant term on the left side of the equation.

$$(x^2 + 6x + \quad) + (y^2 - 8y + \quad) = 11$$

Complete the square on the x- and y-terms. Be sure to add the numbers used in completing the square to the right side of the equation also.

$$(x^2 + 6x + 9) + (y^2 - 8y + 16) = 11 + 9 + 16$$

Factor and simplify.
$$(x + 3)^2 + (y - 4)^2 = 36$$
From the equation, we see the center of the circle is located at $(-3, 4)$ and the radius is 6. ■

One identifying characteristic of an equation of a circle is the fact that the coefficients of the second degree terms are the same. For example, $x^2 + y^2 + 5x = 3$, $3x^2 + 3y^2 + 6y = -2$, and $-2x^2 - 2y^2 + 6x - 8y = 2$ are all equations of circles, since the coefficients of the two second-degree terms are the same. This is always true and is a big help in determining by inspection if the equation represents a circle or one of the other conic sections. (Recall that the equation of a parabola is of second degree in one variable and of first degree in the other variable.)

Example 4 Tell if the following represent a parabola or a circle.

(a) $x^2 + y = 5$ (b) $x^2 - 5 = -y^2$
(c) $2x^2 + 2y^2 + 6x - 3y = 4$ (d) $x^2 = y - 2$

Solution
(a) Parabola (only one variable is of degree two).
(b) Circle (both second-degree terms have a coefficient of 1 if the equation is put in standard form).
(c) Circle (both second-degree terms have a coefficient of 2).
(d) Parabola (only one variable is of degree two). ■

Example 5 Find an equation of the circle whose center is at $(3, -2)$ and whose radius is 5.

Solution Using $(x - h)^2 + (y - k)^2 = r^2$, we get $(x - 3)^2 + (y + 2)^2 = 25$ as the standard form of the equation. To obtain the general form, we must square and collect terms.
$$(x - 3)^2 + (y + 2)^2 = 25$$
$$x^2 - 6x + 9 + y^2 + 4y + 4 = 25$$
$$x^2 + y^2 - 6x + 4y - 12 = 0$$ ■

Example 6 Find an equation of the circle whose center is at $(-2, 3)$ if it goes through the point with coordinates $(1, 4)$.

Solution

We must first find the radius of the circle, which is the distance between the two given points.

$$r = \sqrt{(-2 - 1)^2 + (3 - 4)^2}$$
$$r = \sqrt{(-3)^2 + (-1)^2}$$
$$r = \sqrt{9 + 1}$$
$$r = \sqrt{10}$$
$$r^2 = 10$$

Using $(x - h)^2 + (y - k)^2 = r^2$, where $(h, k) = (-2, 3)$ and $r^2 = 10$, we find the equation of the circle to be $(x + 2)^2 + (y - 3)^2 = 10$.

8–2 Exercises

In exercises 1–8, identify the graph of the equation as a circle or a parabola. (See example 4.)

1. $-2x^2 - 2y^2 = -10$
2. $x^2 + y = -3$
3. $3x^2 + 3y^2 + 6y = 2$
4. $x^2 + y^2 - 6x + 2y - 5 = 0$
5. $y = x^2 + 2x - 3$
6. $x^2 = 5 - y^2$
7. $x^2 = -y^2 + 6x - 8y + 5$
8. $y^2 + x = 3$

In exercises 9–14, find the center and radius of each circle directly from the equation. (See examples 1 and 2.)

9. $x^2 + y^2 = 16$
10. $x^2 + y^2 = 49$
11. $(x - 3)^2 + (y + 4)^2 = 25$
12. $(x + 5)^2 + (y - 4)^2 = 16$
13. $(x + 7)^2 + (y + 2)^2 = 10$
14. $(x - 4)^2 + (y + 8)^2 = 12$

In exercises 15–20, an equation of a circle is given. Write each equation in standard form and give the coordinates of the center and the length of the radius for each circle. (See example 3.)

15. $x^2 + y^2 - 6x + 8y = 11$
16. $x^2 + y^2 + 4x - 6y - 3 = 0$
17. $x^2 - 6y + y^2 + 10x - 2 = 0$
18. $x^2 + 7x + y^2 + 5y = \frac{7}{4}$
19. $x^2 + 3x + y^2 - 5y = \frac{1}{2}$
20. $-x^2 - y^2 + 6x - 14y = -6$

In exercises 21–26, find an equation of the circle satisfying the given conditions. Give the equation in both standard and general form. (See example 6.)

21. Center at the point with coordinates $(2, 4)$ and passing through the point with coordinates $(-1, -3)$.
22. Center at the point with coordinates $(0, 5)$ and passing through the point with coordinates $(2, -3)$.

23. Center at the point with coordinates $(-3, 0)$ and passing through the point with coordinates $(3, 4)$.

24. Center at the point with coordinates $(-4, -3)$ and passing through the point with coordinates $(8, 3)$.

25. Center at the point with coordinates $(0, -6)$ and passing through the point with coordinates $(0, 3)$.

26. Center at the point with coordinates $(4, -3)$ and passing through the point with coordinates $(0, 0)$.

In exercises 27–30, draw the graph of the given equation.

27. $x^2 + y^2 = 49$

28. $x^2 + y^2 = 25$

29. $x^2 + y^2 + 8x - 6y = 11$

30. $x^2 + y^2 - 2x + 3y = \frac{3}{4}$

In exercises 31–40, the center, C, and radius, r, of a circle are given. Write an equation for each circle. (See example 5.)

31. $C(9, 5), r = 2$

32. $C(-1, 2), r = 5$

33. $C(-4, -3), r = 2\sqrt{5}$

34. $C(-5, 0), r = 3\sqrt{2}$

35. $C(0, 0), r = 8$

36. $C(-10, 0), r = 3\sqrt{7}$

37. $C(0, -6), r = 16$

38. $C(-9, -2), r = \sqrt{10}$

39. $C\left(\frac{3}{4}, 5\right), r = 6$

40. $C\left(-\frac{5}{8}, -\frac{4}{5}\right), r = 7\sqrt{5}$

41. A circle is said to be a *point circle* if its radius is zero. Show that the circle whose equation is $x^2 + y^2 + 4x - 6y + 13 = 0$ is a point circle.

42. A circle is said to be an *imaginary circle* if its radius is an imaginary number. Show that the circle whose equation is $x^2 + y^2 + 8x - 10y + 50 = 0$ is an imaginary circle.

8–3 The Ellipse and Hyperbola

Definition 8–3

Ellipse

An **ellipse** is the set of all points in a plane such that the sum of the measures of the distances from any point in the set to two fixed points is a constant. Each fixed point is called a **focus** (the plural of focus is *foci*).

An ellipse has two axes of symmetry. The longer axis of symmetry is called the **major axis** and the smaller one is called the **minor**

Relations and Functions: Part II

axis. The point where these two axes intersect is called the **center** of the ellipse. Figure 8–9 shows an ellipse with foci at $(c, 0)$ and $(-c, 0)$ x-intercepts $(a, 0)$ and $(-a, 0)$, and y-intercepts $(0, b)$ and $(0, -b)$. The x- and y-intercepts are called the **vertices** of the ellipse.

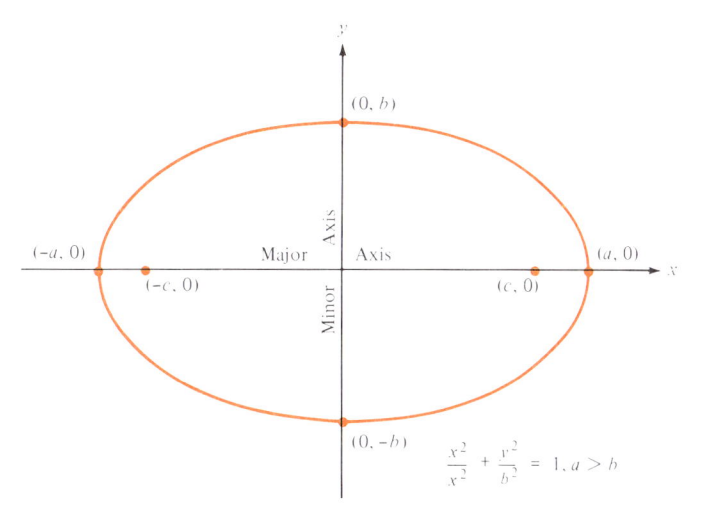

Figure 8–9

Rule 8–3

Standard Form for the Equation of an Ellipse

The **standard form** for the equation of an ellipse whose center is at $(0, 0)$ and whose major axis lies on the x-axis is $\frac{x^2}{a^2} + \frac{y^2}{b^2} = 1$, where $a, b > 0$ and $a > b$.

The general form for the equation of an ellipse may be obtained from the standard form by multiplying $x^2/a^2 + y^2/b^2 = 1$ by a^2b^2 to get $b^2x^2 + a^2y^2 = a^2b^2$.

The major axis of an ellipse may be along the y-axis instead of the x-axis. The equation is $y^2/a^2 + x^2/b^2 = 1$ for this case, where $a > b$.

In this book, we designate half the length of the major axis by a and half the length of the minor axis by b regardless of the axis on which the major axis lies.

Example 1 Draw the graph of $\frac{x^2}{49} + \frac{y^2}{9} = 1$.

Solution Comparing this to the form in rule 8–3, we see

$$a^2 = 49 \qquad b^2 = 9$$
$$a = 7 \quad \text{or} \quad -7 \qquad b = 3 \quad \text{or} \quad -3$$

Therefore, the x-intercepts are (7, 0) and (−7, 0). The y-intercepts are (0, 3) and (0, −3). A fairly good ellipse can be drawn from the intercepts alone, and unless more detail is desired, other ordered pairs are not necessary. See figure 8–10.

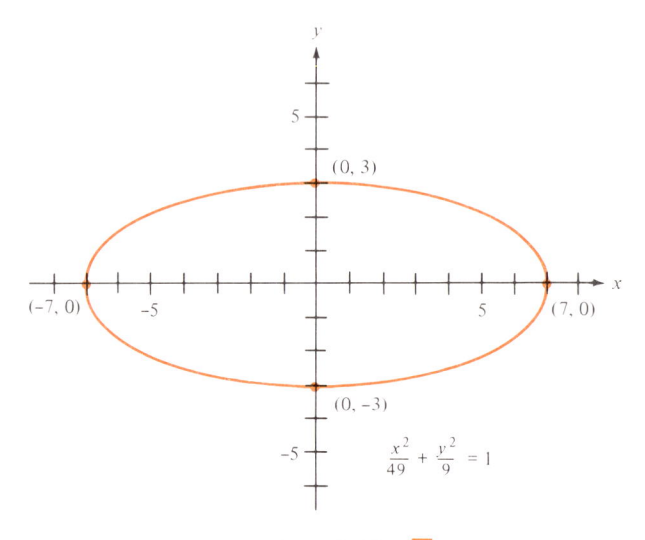

Figure 8–10

Careful examination of example 1 gives us a means of deciding along which axis the major axis of the ellipse will lie. Because 49 is greater than 9 and 49 is the denominator for the term $x^2/49$, the major axis lies on the x-axis. If 49 were the denominator for the y^2-term, then the major axis would be on the y-axis.

Definition 8–4

Hyperbola

A **hyperbola** is the set of all points in a plane such that the difference of the measures of the distances from any point in the set to two fixed points is a constant. The two fixed points are called the **foci**.

Rule 8–4

Standard Form for the Equation of a Hyperbola

The **standard form** for the equation of a hyperbola whose center is at C(0, 0) and whose vertices are (a, 0) and (−a, 0) is
$$\frac{x^2}{a^2} - \frac{y^2}{b^2} = 1.$$

Several important properties are associated with a hyperbola. Figure 8–11 gives a geometric interpretation of these properties, which are listed below.

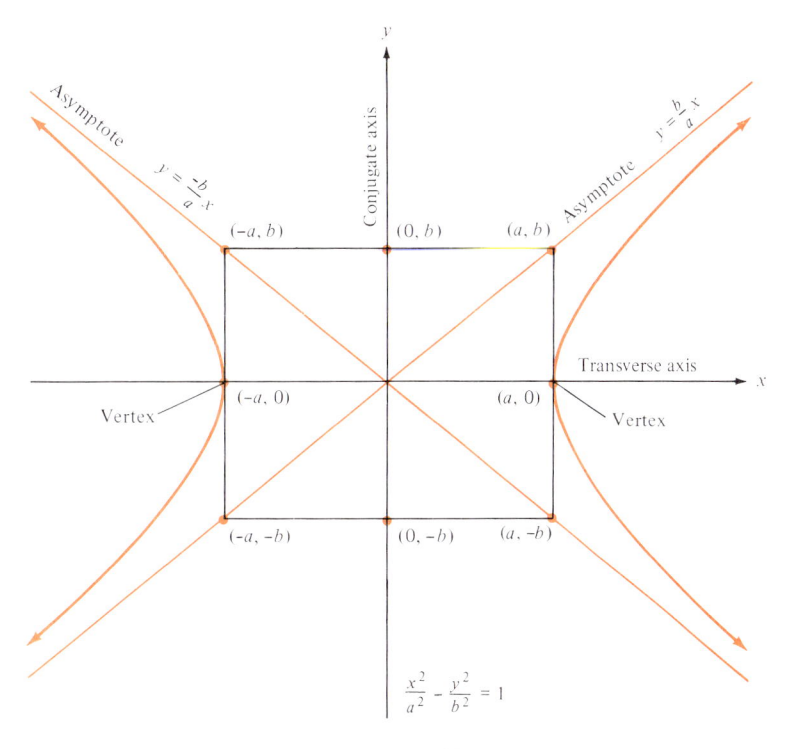

Figure 8–11

1. The hyperbola has two branches. If the x-term is positive when the equation is written in standard form, the branches intersect the x-axis. If the y-term is positive when the equation is written in standard form, the branches intersect the y-axis.
2. The line segment that connects the vertices, $(a, 0)$ and $(-a, 0)$, is called the **transverse axis** and is $2a$ units in length. The perpendicular bisector of the transverse axis is called the **conjugate axis.** The ends of the conjugate axis are located at $(0, b)$ and $(0, -b)$ and it is $2b$ units in length.
3. A reference rectangle, whose dimensions are $2a$ by $2b$ units, can be drawn as an aid in sketching the graph.
4. Associated with the hyperbola is a pair of lines called **asymptotes.** A line is an asymptote if the graph gets closer and closer to the line, but does not intersect it. (There are exceptions to this definition, but we do not encounter them). The asymptotes pass through the center of the rectangle and are extensions of the diagonals of the rectangle. The equations of the asymptotes are found by setting the left side of $x^2/a^2 - y^2/b^2 = 1$ to zero and solving for y.

This will give

$$y = \frac{b}{a}x \quad \text{and} \quad y = \frac{-b}{a}x$$

as the equations of the asymptotes.

5. The vertices of the rectangle are found by plotting all the combinations of ordered pairs that can be formed from positive and negative values of a and b. These are (a, b), $(a, -b)$, $(-a, b)$, and $(-a, -b)$.

The two branches of a hyperbola cross the y-axis if the equation is given by $y^2/a^2 - x^2/b^2 = 1$. The equations of the asymptotes are

$$y = \frac{a}{b}x \quad \text{and} \quad y = \frac{-a}{b}x$$

The vertices of the reference rectangle are (b, a), $(b, -a)$, $(-b, a)$, and $(-b, -a)$. See figure 8–12.

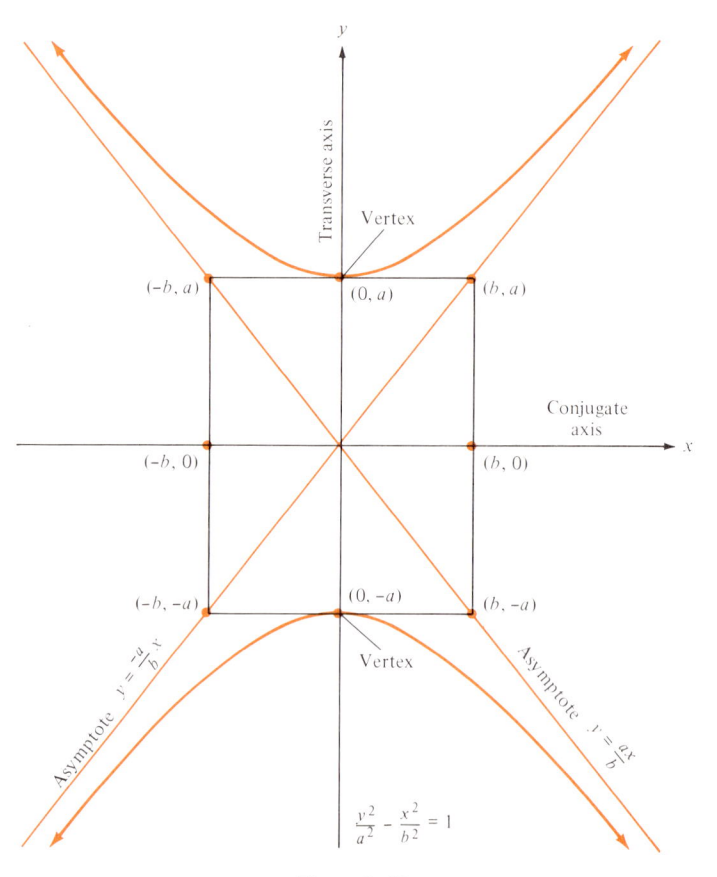

Figure 8–12

Example 2 Draw the graph of $\dfrac{x^2}{25} - \dfrac{y^2}{9} = 1$.

Solution The hyperbola intersects the x-axis, since $x^2/25$ is the positive term. We know $a^2 = 25$, which gives $a = 5$ or -5. Similarly, $b^2 = 9$, which gives $b = 3$ or -3. The rectangle has vertices at (5, 3), (5, -3), (-5, 3) and (-5, -3). The rectangle should be drawn first; the diagonals should then be drawn to give the asymptotes. This is shown in figure 8-13.

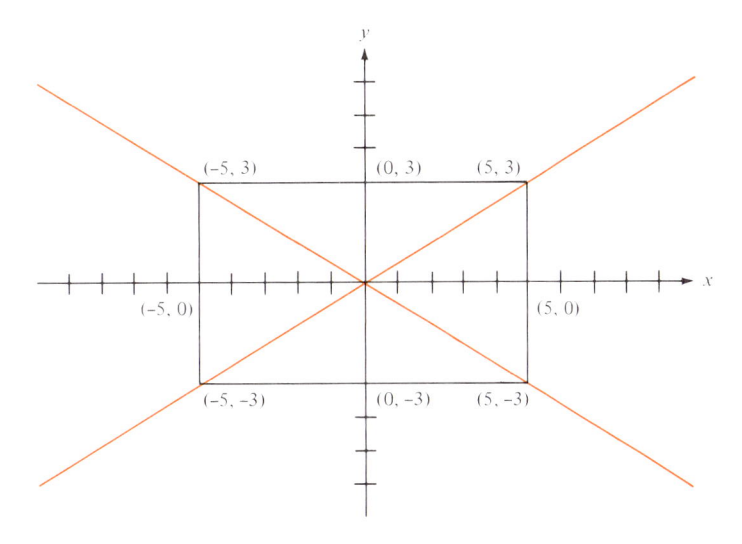

Figure 8-13

We need a few ordered pairs in addition to the asymptotes for an accurate graph. The hyperbola has symmetry with respect to both coordinate axes, so we can find one ordered pair and reflect it as we did in section 8-1 for parabolas. No part of the hyperbola is drawn within the rectangle. The equation is defined only for $x \geq 5$ and $x \leq -5$. If we let $x = 7$, we get

$$\frac{49}{25} - \frac{y^2}{9} = 1$$

$$\frac{-y^2}{9} = 1 - \frac{49}{25}$$

$$\frac{-y^2}{9} = \frac{-24}{25}$$

$$y^2 = \frac{-24}{25}(-9)$$

$$y = \frac{3}{5}\sqrt{24} \approx 2.9 \text{ (The symbol } \approx \text{ is read, "is approximately.")}$$

If we plot (7, 2.9), (7, −2.9), (−7, 2.9) and (−7, −2.9), we can draw the hyperbola. See figure 8–14.

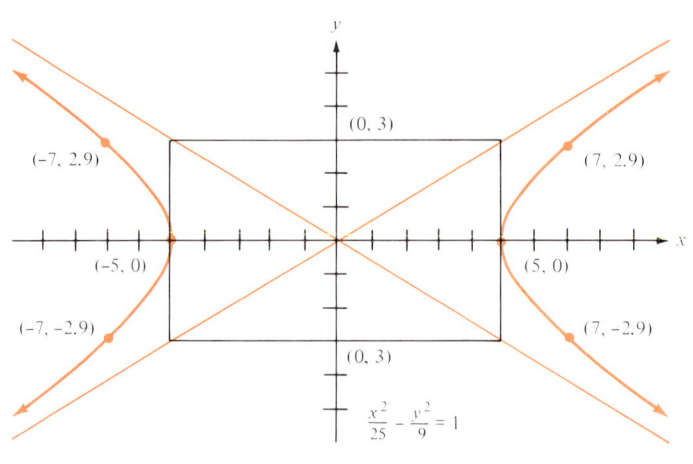

Figure 8–14

8–3 Exercises

In exercises 1–10, identify the graph of the given equation as a parabola, circle, ellipse, hyperbola, or straight line. (Hint: Review example 4 in section 8–2 for those equations that represent a circle or parabola.)

1. $\dfrac{x^2}{4} - \dfrac{y^2}{5} = 1$
2. $y = x^2 - 3x + 2$
3. $3x^2 + 3y^2 + 4x - 3 = 0$
4. $9x^2 + 4y^2 = 36$
5. $3x + 4y = 1$
6. $6x^2 - 12y^2 = 24$
7. $x^2 - y = 1$
8. $x^2 + 9y^2 = 1$
9. $y = 2x + 3$
10. $\dfrac{x^2}{3} - \dfrac{y^2}{4} = 1$

In exercises 11–16, each equation represents an ellipse or hyperbola. Find the coordinates of the vertices. (See examples 1 and 2.)

11. $\dfrac{x^2}{9} - \dfrac{y^2}{49} = 1$
12. $4x^2 + y^2 = 36$
13. $x^2 + 9y^2 = 36$
14. $x^2 - 3y^2 = 27$
15. $3x^2 - 3y^2 = 15$
16. $4x^2 + 5y^2 = 20$

In exercises 17–22, draw the graph of each equation. (See examples 1 and 2.)

17. $\dfrac{x^2}{36} + \dfrac{y^2}{4} = 1$
18. $\dfrac{x^2}{9} + \dfrac{y^2}{25} = 1$
19. $\dfrac{x^2}{9} - \dfrac{y^2}{4} = 1$
20. $\dfrac{x^2}{16} - \dfrac{y^2}{25} = 1$
21. $9y^2 - 16x^2 = 144$
22. $25x^2 + 4y^2 = 100$

23. Show that the equations for the asymptotes of the hyperbola $y^2/a^2 - x^2/b^2 = 1$ are

$$y = \frac{a}{b}x \quad \text{and} \quad y = \frac{-a}{b}x$$

24. Prove the y-intercepts for $x^2/a^2 - y^2/b^2 = 1$ are imaginary.

8–4 Quadratic Inequalities in Two Variables

Linear inequalities are graphed in section 7–5. Recall that to draw the graph of a relation such as $y \leq 2x + 3$, we first find the boundary line given by the equation $y = 2x + 3$. Quadratic inequalities in two variables are graphed in a similar fashion. For example, to draw the graph of $x^2 + 4y^2 < 36$, we first draw the graph of $x^2 + 4y^2 = 36$. Then we pick a point not on the boundary and see if the coordinates of that point make $x^2 + 4y^2 < 36$ true or false. If the coordinates of the test point make the inequality true, we shade that part of the Cartesian plane containing the point. If the coordinates of the test point make the inequality false, we shade that part of the plane on the opposite side of the boundary from the test point. If the boundary is part of the graph, it is drawn as a solid curve. Otherwise, it is shown as a dashed curve.

Example 1 Draw the graph of $x^2 + 4y^2 < 36$.

Solution The boundary is given by $x^2 + 4y^2 = 36$, which is an ellipse. The standard form is $\frac{x^2}{36} + \frac{y^2}{9} = 1$. The graph consists of that part of the plane within the ellipse, as the test point (0, 0) shows. The graph is given in figure 8–15.

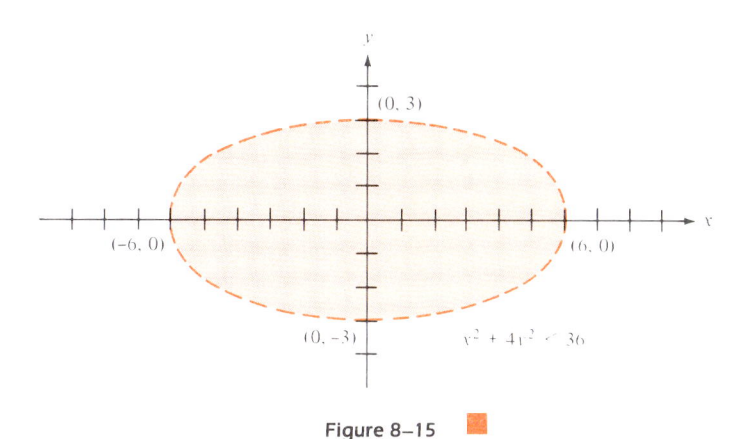

Figure 8–15

Example 2 Draw the graph of $y \geq -2(x + 3)^2 - 2$.

Solution The boundary is a parabola, with its vertex at $(-3, -2)$ and opening downward. The test point $(0, 0)$ makes $y \geq -2(x + 3)^2 - 2$ a true statement. The graph consists of all points on the boundary and outside the boundary in that part of the plane containing $(0, 0)$. See figure 8–16.

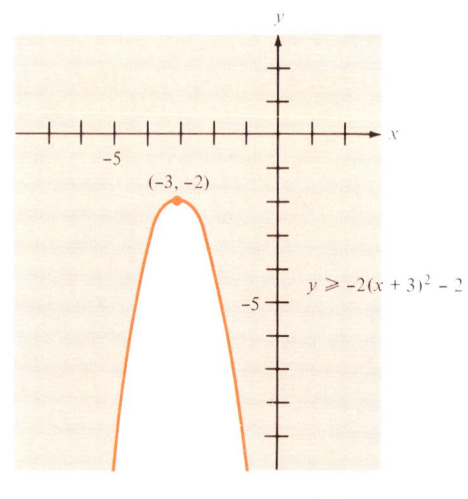

Figure 8–16

Example 3 Draw the graph of $\{(x, y)|x^2 + y^2 \leq 25\} \cap \{(x, y)|x \geq -1\}$.

Solution This graph consists of the intersection of those points within the circle given by $x^2 + y^2 = 25$ and those points to the right (but still within the circle) of the line whose graph is $x = -1$. See figure 8–17.

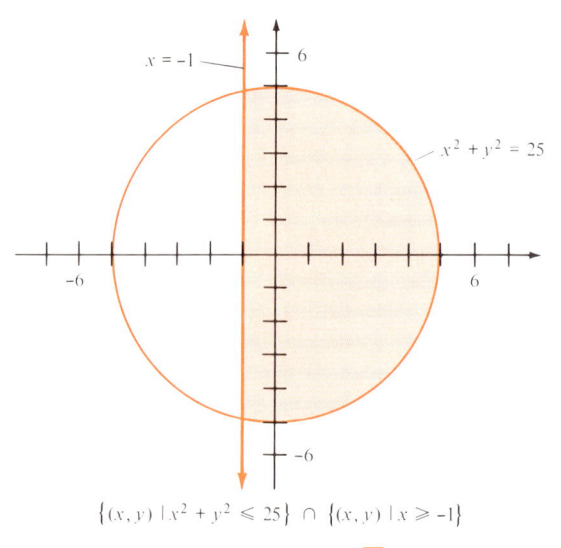

Figure 8–17

8–4 Exercises

In exercises 1–24, draw the graph of each inequality. *(See examples 1 and 2.)*

1. $y \leq x^2 + 1$
2. $x^2 + y^2 > 16$
3. $\dfrac{y^2}{25} - \dfrac{x^2}{4} \leq 1$
4. $4x^2 + 25y^2 \leq 100$
5. $y < x^2 + 6x + 3$
6. $y^2 - 9x^2 < 9$
7. $x^2 \geq -4y^2 + 36$
8. $x^2 > 25y^2 + 25$
9. $y^2 < x^2 + 9$
10. $x^2 - 9 \leq 9y^2$
11. $9x^2 - 36 > -4y^2$
12. $x^2 + y^2 + 6x - 12y \leq 4$
13. $\dfrac{x^2}{\frac{25}{16}} + \dfrac{y^2}{9} < 1$
14. $x^2 + y^2 - 4x + 6y > 12$
15. $\dfrac{x^2}{9} - 1 \geq \dfrac{y^2}{\frac{9}{4}}$
16. $x^2 - 9y^2 \geq 36$
17. $\{(x, y) | x^2 + y^2 \leq 16\} \cap \{(x, y) | x > y\}$ *(See example 3.)*
18. $\{(x, y) | 4x^2 + 9y^2 < 36\} \cap \{(x, y) | x^2 + y^2 < 1\}$ *(See example 3.)*
19. $\{(x, y) | y^2 \geq -x^2 + 4\} \cap \{(x, y) | y \geq x - 6\}$
20. $\left\{(x, y) \left| \dfrac{x^2}{16} + \dfrac{y^2}{36} < 1 \right.\right\} \cap \{(x, y) | x^2 + y^2 > 16\}$
21. $\{(x, y) | x^2 + y^2 \leq 16\} \cup \{(x, y) | x + y < 4\}$
22. $\left\{(x, y) \left| \dfrac{x^2}{49} + \dfrac{y^2}{16} \geq 1 \right.\right\} \cup \{(x, y) | x^2 + y^2 \leq 9\}$
23. $\{(x, y) | y \geq (x - 2)^2 + 3\} \cup \{(x, y) | y - x \geq 3\}$
24. $\{(x, y) | x^2 + y^2 \leq 36\} \cup \{(x, y) | |y| \leq 4\}$

8–5 Ratio and Proportion

The **ratio** of two numbers x and y is the quotient of these two numbers and is represented by x/y. The ratio may also be written as $x:y$; both are read "the ratio of x to y." Ratios can be simplified in the same way we simplify fractions. For example, the ratio of $9:12$ is the same as the ratio of $3:4$.

Example 1

Patricia wants to divide $28,000 between Lori and Rory in a ratio of $3:5$. Rory is to receive the smaller share. How much will each receive?

Solution

The money is to be divided according to the ratio of $3:5$, or into 8 equal parts. Let the amount of each equal part be represented by x.

Rory will receive 3x dollars and Lori will receive 5x dollars. Then

$$3x + 5x = 28{,}000$$
$$8x = 28{,}000$$
$$x = 3{,}500$$

Therefore, $3x = 3(3{,}500) = \$10{,}500$ is Rory's share and $5x = 5(3{,}500) = \$17{,}500$ is Lori's share. ∎

Example 2 The number of linear feet of shelf space in a library allotted for the fictional works of three different authors is in the ratio of $3:5:7$. If there is a total of 75 linear feet of shelf space, how much shelf space is assigned to each author?

Solution The shelf space is to be divided into 15 equal parts. Let x be the amount of linear feet in each of the equal parts.

$$3x + 5x + 7x = 75$$
$$15x = 75$$
$$x = 5$$

Therefore, $3x = 3(5) = 15$ feet of space is used for the first author, $5x = 5(5) = 25$ feet of space is used for the second author, $7x = 7(5) = 35$ feet of space is used for the third author. ∎

A **proportion** is a statement that two ratios are equal. For example, $\frac{3}{7} = \frac{6}{14}$ is a proportion. Also, $a/b = c/d$ is a proportion. Sometimes, $a/b = c/d$ is written as $a:b::c:d$. The numbers represented by a, b, c, and d are called the first, second, third, and fourth terms, respectively, of the proportion. The numbers represented by a and d are called the **extremes** and those represented by b and c are called the **means.** Further, d is called the *fourth proportional* to a, b, and c. If the proportion has equal means such that $a:b::b:c$, then b is called the *mean proportional* between a and c, and c is called the *third proportional* to a and b.

In any proportion, the product of the means equals the product of the extremes, which we now prove.

Let $a/b = c/d$ be the given proportion. If we multiply both sides of the equation by bd, we get

$$(bd)\left(\frac{a}{b}\right) = (bd)\left(\frac{c}{d}\right)$$
$$\frac{abd}{b} = \frac{bcd}{d}$$
$$ad = bc$$

Because ad is the product of the extremes and bc is the product of the means, the proof is finished.

Rule 8–5

Product of the Means and Extremes

In a proportion, the product of the means equals the product of the extremes. That is, if $a/b = c/d$, then $ad = bc$ ($b, d \neq 0$).

Example 3 Solve for x in the proportion

$$\frac{3}{4} = \frac{\frac{1}{12}}{x}$$

Solution From rule 8–5, we know $3x = 4(\frac{1}{12})$. Then

$$3x = \frac{1}{3}$$

$$x = \frac{1}{9}$$

Example 4 Find the mean proportional between 9 and 4.

Solution The mean proportional exists only if the means are the same. Let x represent the mean proportional. Then we have

$$\frac{9}{x} = \frac{x}{4}$$

$$x^2 = 36 \qquad \text{Rule 8–5}$$

$$x = 6 \text{ or } -6$$

Example 5 At 60 miles per hour, an automobile travels 88 feet per second. How many feet per second does the automobile travel at a speed of 36 miles per hour?

Solution Let x represent the number of feet traveled per second at a speed of 36 miles per hour. The ratio

$$\frac{60 \text{(miles per hour)}}{88 \text{(feet per second)}}$$

is the same as the ratio

$$\frac{36 \text{(miles per hour)}}{x \text{(feet per second)}}.$$

This gives

$$\frac{60}{88} = \frac{36}{x}$$
$$60x = 88(36) \quad \text{Rule 8–5}$$
$$x = \frac{88(36)}{60}$$
$$x = 52.8$$

The car travels 52.8 feet per second. ■

Example 6 Amy noticed a recipe called for 3 cups of flour and $\frac{3}{4}$ teaspoon of salt. If she only has 2 cups of flour, how much salt should she use?

Solution Let x be the number of teaspoons of salt needed for 2 cups of flour. Then the amount of salt she needs for 3 cups of flour is proportional to the amount of salt she will need for 2 cups of flour.

$$\frac{3}{\frac{3}{4}} = \frac{2}{x}$$

$$3x = \frac{3}{4}(2) \quad \text{Rule 8–5.}$$

$$3x = \frac{3}{2} \quad \text{Simplify the right side.}$$

$$x = \frac{3}{2} \cdot \frac{1}{3} \quad \text{Multiply each side by 1/3.}$$

$$x = \frac{1}{2}$$

She should use $\frac{1}{2}$ teaspoon of salt for 2 cups of flour. ■

8–5 Exercises

Find the solution for each of the following exercises by using an equation involving a ratio. (See examples 1 and 2.)

1. A piece of lumber 18 feet long is to be divided into two pieces in a ratio of 2:7. How long is each piece?

2. Jared and Jason are to divide an inheritance of $15,500 between them in a ratio of 3:2, with Jason getting the larger share. How much will each receive?

3. Jerry has a collection of two different kinds of pocket knives. If there are 48 knives in the collection and if they are in a ratio of 13:3, how many of each kind of knife does he have?

4. Dave must mix together some concentrated acid and water to get a mixture of acid and water that has a ratio of 5:3. If he started with 130 cubic centimeters of acid, how much water must he add?

5. The ratio of 60-watt light bulbs to 100-watt light bulbs in David's house is 3:4. If his house contains 21 light bulbs, part of which are 60-watt bulbs and the rest 100-watt bulbs, how many of each kind are in his house?

6. The cost of natural gas as compared to electricity for a homeowner for a particular month was in the ratio of 7:13. If the monthly bill for gas and electricity was $140.80, how much was paid for each?

7. Norma and Terri divided $360. If Terri received $160, what is the ratio into which the $360 was divided?

8. The cost per gallon of enamel paint to the cost per gallon for flat latex paint is in the ratio of 7:5. If Don bought $108.00 worth of paint, part of which was enamel paint and the rest of which was flat paint, how much did he spend for each kind of paint?

9. A set of tires and a battery for a certain automobile cost $250. If the tires cost $210, what was the ratio of the cost of the battery to the cost of the tires?

10. Find the ratio of x to y if $2(x + y) = 7(x - y)$.

11. Find the ratio of x to y if $-3(x + y) = 2(x - 2y)$.

12. Find the ratio of y to x if $2(x + y) = 3(2x - y)$.

In exercises 13–18, replace the question mark with the number that will make the proportion true. (See rule 8–5 and example 3.)

13. $\dfrac{2}{5} = \dfrac{8}{?}$

14. $\dfrac{3}{7} = \dfrac{?}{28}$

15. $\dfrac{2a}{5b} = \dfrac{10a}{?}$

16. $\dfrac{5}{1} = \dfrac{1}{?}$

17. $\dfrac{2a}{1} = \dfrac{1}{?}$

18. $\dfrac{0}{6} = \dfrac{?}{7}$

In exercises 19–22, solve for x using the fact that the product of the means equals the product of the extremes. (See examples 3–6.)

19. $\dfrac{3}{5} = \dfrac{x}{15}$

20. $\dfrac{9}{4} = \dfrac{27}{x}$

21. $\dfrac{3}{5 + x} = \dfrac{1}{2}$

22. $\dfrac{7 + x}{27 + x} = \dfrac{1}{3}$

23. Find the mean proportional between 16 and 4. *(See example 4.)*

24. Find the third proportional to 5 and 10.

In exercises 25–30, use a proportion to find the indicated quantity.

25. The cost for renting a care for 3 days is $29.97. Use a proportion to find the cost of renting the car for 7 days, assuming the rental rate for 7 days is proportional to the rental rate for 3 days.

26. Find the number that should be added to 1, 5, 7, and 17 to give four numbers, in that order, that will form a proportion.

27. If 15 loads of clothes can be washed with a $2\frac{1}{2}$-pound box of detergent, then how many loads can be washed with a $6\frac{1}{2}$-pound box if the same amount of detergent is used per load.

28. Norma found that a recipe using $3\frac{3}{4}$ cups of flour called for $1\frac{1}{2}$ cups of sugar. If she wants to increase the number of cups of flour to 5 cups, how much sugar should she use?

29. A baseball team has won 30 games and lost 5. If the team has 5 more games to play, how many of the remaining games must it win to have a final winning average of 0.875? (*Hint:* An average of 0.875 is a ratio of 875/1000).

30. A jet plane traveling 600 miles per hour travels 880 feet per second. What speed in miles per hour is Lori flying her plane if she travels 176 feet per second?

31. If $\dfrac{a}{b} = \dfrac{c}{d}$, show that $\dfrac{d}{c} = \dfrac{b}{a}$.

32. If $\dfrac{a}{b} = \dfrac{c}{d}$, show that $\dfrac{a+b}{b} = \dfrac{c+d}{d}$.

8–6 ■ Variation

Two variables *y* and *x* **vary directly** if their ratio is a constant. This means that if one of the variables becomes twice as large, then the other one becomes twice as large; if one variable is decreased to one-third of its previous value, the other is decreased to one-third of its previous value.

Definition 8–5

Direct Variation

Two variables *y* and *x* **vary directly** if $y/x = k$, where k is a constant and $k \neq 0$.

From definition 8–5, if $y/x = k$, then $y = kx$. This form of the equation is often more convenient. The constant represented by k is called the **constant of variation.** In some problems, k may be known; in others, it may not be known. For example, in $C = \pi d$, π is the known constant of variation, and C varies directly as d.

Example 1

If *y* varies directly as *x* and if $x = 3$ when $y = 12$, find the value of *y* when $x = 5$.

Relations and Functions: Part II

Solution

We must write an equation that represents the facts and find k. The equation we need is $y = kx$. We know $x = 3$ when $y = 12$.

$$y = kx$$
$$12 = k \cdot 3$$
$$k = 4$$

The constant of variation is 4, which gives $y = 4x$. Substitute the second value given for x into $y = 4x$.

$$y = 4x$$
$$y = 4(5)$$
$$y = 20$$

Example 2

The distance an object will fall in a vacuum varies directly as the square of the time. If an object falls 16 feet in 1 second, how far will it fall in 2 seconds?

Solution

We set up an equation to represent the facts and find k. Let s represent the distance and t the time.

$$s = kt^2$$
$$16 = k(1)^2$$
$$k = 16$$

The equation is $s = 16t^2$. Using the last value given for t in the example,

$$s = 16(2)^2$$
$$s = 64$$

The object falls 64 feet in 2 seconds.

A second type of variation is known as *inverse variation*. Two variables x and y **vary inversely** if their product is a constant. This means if one variable is doubled, the other one is halved, or if one is tripled, the other one is only one-third as large.

Definition 8–6

Inverse Variation

Two variables x and y **vary inversely** if $xy = k$, where k is a constant and $k \neq 0$.

From definition 8–6, if $xy = k$ then $y = k/x$. This form of the equation is often more convenient.

Example 3

If y varies inversely as x and if $x = 3$ when $y = 7$, find y when $x = 6$.

Solution   We must find k. Substitute the first two given values of x and y into $y = k/x$.

$$7 = \frac{k}{3}$$
$$k = 21$$

The equation we need is $y = 21/x$. Placing the second value of x into $y = \frac{21}{x}$, we get

$$y = \frac{21}{6}$$
$$y = 3.5 \quad \blacksquare$$

Example 4   The time needed to drive a fixed distance varies inversely as the average rate of speed. If it takes 30 minutes to drive a distance at 40 miles per hour, how long will it take to drive the same distance at 30 miles per hour?

Solution   Let t and r represent the time and rate, respectively. Then the equation is $t = k/r$.

$$30 = \frac{k}{40} \qquad \text{Substitute } t = 30 \text{ and } r = 40.$$
$$k = 1{,}200 \qquad \text{Solve for } k.$$

The equation is $t = 1{,}200/r$. Now substitute $r = 30$ into $t = 1200/r$ to get $t = 40$ minutes. $\blacksquare$

The third type of variation is known as **joint variation.** If y varies as the product of two or more variables, we say y varies jointly as x, w, and z, and we write $y = kxwz$.

Definition 8–7

Joint Variation

The variable y **varies jointly** as the variables x, w, and z if $y = kxwz$, where k is a constant and $k \neq 0$.

By taking direct, inverse, and joint variation in combination, we can write a great assortment of variation problems. Consider the following:

1. $y = kx/z$ states that y varies directly as x and inversely as z.
2. $y = kr^2t/x$ states that y varies jointly as the square of r and t and inversely as x.

Relations and Functions: Part II

Example 5 The volume of a gas varies directly as the absolute temperature and inversely as the pressure. A gas occupies 30 cubic feet at a temperature of 300°K (degrees Kelvin) and a pressure of 40 pounds per square inch. Determine the volume if the temperature is increased to 330°K and the pressure is decreased to 30 pounds per square inch.

Solution Let T, V, and P represent the temperature, volume, and pressure, respectively. The equation we need is $V = KT/P$.

$$V = \frac{kT}{P}$$

$$30 = \frac{k(300)}{40}$$

$$k = 4$$

The equation becomes

$$V = \frac{4T}{P}$$

Substituting $T = 330$ and $P = 30$ into $V = \dfrac{4T}{P}$ gives

$$V = \frac{4(330)}{30}$$

$$= 44$$

The volume is 44 cubic feet. ■

Examples 1–5 illustrate the three steps in solving a variation problem. These steps are given in procedure 8–1.

Procedure 8–1 **Solving a Variation Problem**
1. Set up an equation for direct, inverse, or joint variation from the facts given about the problem.
2. Find k.
3. Find the desired variable.

8–6 Exercises

Solve each of the following exercises. (See examples 1–3.)

1. y varies directly as x. If $y = 10$ when $x = 2$, find y when $x = 7$. (See example 1.)
2. y varies inversely as x. If $y = 3$ when $x = 5$, find y if $x = 30$.
3. y varies inversely as the square of x. If $y = 2$ when $x = 3$, find y when $x = 3\sqrt{2}$.

4. y varies directly as w and inversely as x. If $y = 4$ when $w = 10$ and $x = 12.5$, find y if $w = 2.5$ and $x = 5$.

5. s varies jointly as t and the square of r and inversely as v. If $s = 25$ when $t = 2$, $r = 5$, and $v = 4$, find v when $t = 4$, $r = 3$, and $s = 12$.

6. q varies jointly as m and the square root of n and inversely as p. If $q = 3$ when $p = 10$, $m = 3$, and $n = 4$, find n when $q = 8$, $p = \frac{1}{2}$, and $m = \frac{1}{10}$.

7. x varies directly as the cube of y and inversely as the square root of z. If $x = 32$ when $z = 16$ and $y = 4$, find y when $x = 8$ and $z = 81$.

8. If y varies directly as x and inversely as w, what is the effect on y if x is doubled and w is decreased to one-third of its original value?

9. If p varies directly as the square of q and inversely as r, what is the effect on p if q is doubled and r is quadrupled?

10. If y varies directly as the square root of x and inversely as z, what is the effect on y if x is quadrupled and z is tripled?

11. y varies jointly as the square of x and the cube root of z. What is the effect on y if x is halved and z is made eight times as large?

12. p varies jointly as q and r and inversely as s. What is the effect on p if q is doubled, r is decreased to one-third of its value and s is decreased to one-fourth of its value?

In exercises 13–30, *see examples 4 and 5.*

13. The velocity of an object falling from rest in a vacuum varies directly as the time. If an object has a velocity of 32 feet per second at the end of 1 second, what will be the velocity at the end of 4 seconds?

14. The time that it takes for a pendulum to make one swing across and back again is called the *period* of the pendulum. If the period (T) of a pendulum varies directly as the square root of its length (L), and if the period is 1 second when the length is 1 foot, what will be the period if the length is 4 feet?

15. The current (I), measured in amperes, varies inversely as the resistance (R), measured in ohms. If a current of 3 amperes is measured when the resistance is 25 ohms, what will be the current if the resistance is increased to 150 ohms?

16. The illumination of a light source varies directly as the strength of the light source and inversely as the square of the distance from the source. If the illumination is 100 foot-candles at a distance of 10 feet from a light source whose strength is 2,000 candlepower, find the illumination produced on a surface located 5 feet from a 1000-candlepower source.

17. The length, L, that a spring will stretch beyond its natural length is directly proportional to the force, F, applied to the spring. If a force of 7 pounds stretches it 3.5 inches, how much force is required to stretch it 9 inches?

18. The force, F, caused by liquid pressure at a point is directly proportional to the depth, d, of the point beneath the surface of the liquid. If a force of 125 pounds is exerted at a depth of 2 feet, what will be the force due to the pressure at 12 feet?

19. The volume, V, of a right circular cylinder varies jointly as the square of the radius, r, and the height, h, of the cylinder. If a cylinder with a height of 2 inches and a radius of 10 inches has a volume of 628 cubic inches, what will be the volume if the radius is kept the same and the height is increased by 5 inches?

20. The maximum uniformly distributed load, L, that a beam can support varies jointly as the width, w, and square of the depth, d, and inversely as its length, l. A beam 10 feet long, 1 foot wide, and 2 feet deep can support a uniformly distributed weight of 80 pounds. What is the depth of a beam made of the same material that can support a weight of 400 pounds if the width is 2 feet and it is 9 feet long?

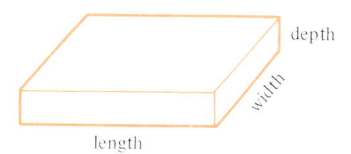

21. The current, I, flowing in an electrical circuit varies directly as the voltage, V, across the circuit and inversely as the resistance, R, of the circuit. If a current of 1 ampere flows when the voltage is 10 volts and the resistance is 10 ohms, find the resistance when the current is 0.1 ampere and the voltage is 200 volts.

22. The inductive reactance, X_L, opposing the flow of alternating current in an electrical circuit varies jointly as the frequency, f, of the current and the inductance, L. If X_L = 628 ohms when f = 100 hertz and L = 1 henry, find X_L if f = 1,000 hertz and L = $\frac{1}{2}$ henry.

23. The amount of interest, I, that a given amount of money will earn varies jointly as the rate of interest, r, and the length of the time, t, in years for which the money is invested. If the interest earned on a certain amount of money at 12% for 3 years is $2,880, how much interest will be earned on the same amount of money in 5 years if the interest rate remains the same?

24. The energy, E, stored in a capacitor varies jointly as the capacitance, C, and the square of the voltage, V, across the capacitor. If the energy is 0.03 joules when C = 0.000006 farads and V = 100 volts, find V when C = 0.000001 farads and E = 0.02 joules.

25. The length of the altitude, h, of a triangle varies directly as the area, A, of the triangle and inversely as the length of the base, b, of the triangle. If h = 5 feet when A = 25 square feet and b = 10 feet, find A when h = 9 feet and b = 8 feet.

26. The distance required to stop an automobile varies directly as the square of the velocity of the automobile. How many times greater is the distance required to stop at 60 miles per hour than at 30 miles per hour?

27. The kinetic energy, K.E., of a moving object varies jointly as the mass, m, of the object and the square of its velocity, v. If the kinetic energy is 32 foot-pounds for an object whose mass is 8 slugs and whose velocity is 16 feet per second, find the kinetic energy for an object whose mass is 128 slugs and whose velocity is 9 feet per second.

28. The force, F, in newtons, between two charged particles varies jointly as the charges, Q_1 and Q_2, measured in coulombs and inversely as the square of the distance, d, measured in meters, between the two particles. If a force of 54 newtons is produced when two charged particles of 2×10^{-5} and 3×10^{-6} coulombs, respectively, are separated by $\frac{1}{10}$ of a meter, find the force when two charged particles of 5×10^{-5} and 6×10^{-5} coulombs, respectively, are separated by $\frac{3}{10}$ of a meter.

29. The weight, w, of an astronaut varies inversely as the square of his or her distance from the center of the earth. If an astronaut weighs 150 pounds on the surface of the earth, how much will he or she weigh at a height of 1,000 miles above the earth's surface? (We are assuming he or she is stationary and not in orbit.) Assume the radius of the earth to be 4,000 miles.

30. If the bases of two triangles are the same, show that the ratio of their respective areas is equal to the ratio of their respective altitudes.

8–7 Inverse of a Function

A relation has already been defined as a set of ordered pairs. The domain of the relation is the set of all first components in the ordered pairs and the range is the set of all second components in the ordered pairs. For example, for the set of ordered pairs

$$R = \{(2, 3), (5, 4), (3, 6)\}$$

the domain is $\{2, 5, 3\}$ and the range is $\{3, 4, 6\}$.

Suppose we interchange the components in the ordered pairs in R. This procedure results in the **inverse** of R, written R^{-1}. The minus one is not an exponent, and R^{-1} is read "R inverse" or "the inverse of R." Making the interchange, we get

$$R^{-1} = \{(3, 2), (4, 5), (6, 3)\}$$

The domain of R^{-1} is $\{3, 4, 6\}$ and the range of R^{-1} is $\{2, 5, 3\}$.

Careful comparison of R and R^{-1} shows the domain of R is the range of R^{-1}, and the range of R is the domain of R^{-1}.

If a relation is a function, the inverse of the relation may or may not be a function. The relations R and R^{-1} just given are both functions, since in each no two ordered pairs have the same first components and different second components.

If we let $S = \{(3, 7), (2, 6), (4, 7)\}$, then $S^{-1} = \{(7, 3), (6, 2), (7, 4)\}$. In this case, S is a function, but S^{-1} is not.

This example shows that if a function has different values of x associated with the same value of y, then the inverse of the function is not a function.

If we want to be assured that every function has an inverse that is also a function, then each y in the range must be associated with a unique value of x in the domain. A function that has the property that each y in the range is associated with a unique value of x in the domain is called a **one-to-one function.** In summary, a function has an inverse that is a function if and only if the function is a one-to-one function.

Example 1 If $R = \{(9, 3), (10, 4) (7, 2)\}$, determine if R^{-1} is a function by inspecting R.

Solution R^{-1} is a function, since R is a one-to-one function. ■

If R is a function, f is customary used rather than R, and the inverse is given by f^{-1}. However, f^{-1} will be a function only if f is a one-to-one function.

Example 2 If $f = \{(3, 9), (2, 4), (8, -1)\}$, find each.

(a) the domain of f^{-1}
(b) the range of f^{-1}

Solution $f^{-1} = \{(9, 3), (4, 2), (-1, 8)\}$

(a) the domain of f^{-1} is $\{9, 4, -1\}$
(b) the range of f^{-1} is $\{3, 2, 8\}$ ■

If a function is given by an equation such as $f = \{(x, y)|x + 2y = 3\}$, then f^{-1} is formed by interchanging x and y in the equation. (This is the same as interchanging the components in the ordered pairs). Thus if

$$f = \{(x, y)|x + 2y = 3\}$$

then

$$f^{-1} = \{(x, y)|y + 2x = 3\}$$

Example 3 If $f = \{(x, y)|3x - 2y = 4\}$, find f^{-1} and solve f^{-1} for y.

Solution $f^{-1} = \{(x, y)|3y - 2x = 4\}$. Solve for y.

$f^{-1} = \left\{(x, y)\Big|y = \dfrac{2x + 4}{3}\right\}$. ■

Example 4 If $f = \{(x, y)|y = x^2\}$, find f^{-1} and solve f^{-1} for y.

Solution $f^{-1} = \{(x, y) | x = y^2\}$. Solve for y.
$f^{-1} = \{(x, y) | y = \pm\sqrt{x}\}$.

In example 4, f is not a one-to-one function; thus f^{-1} is not a function. See figure 8–18.

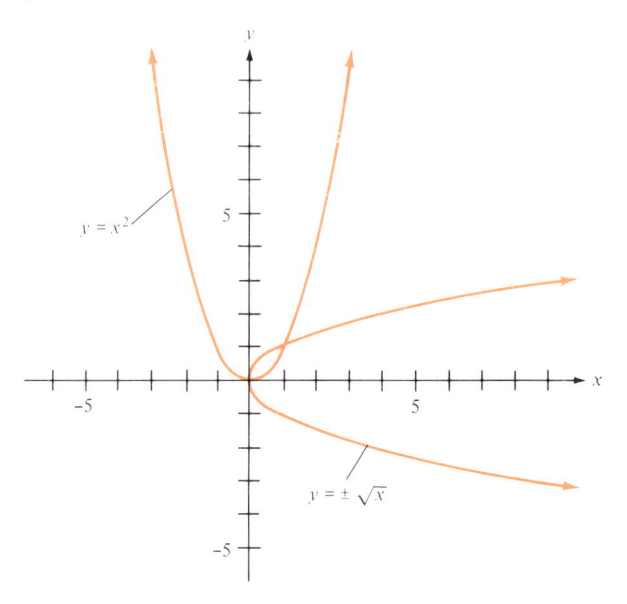

Figure 8–18

The graphs of f and f^{-1} have an unusual relationship. If (x, y) is in f, then (y, x) is in f^{-1}. This means the graphs of (x, y) and (y, x) are symmetrically located on either side of the line given by $y = x$; that is, the line given by $y = x$ is the perpendicular bisector of the line segment connecting the points given by (x, y) and (y, x). See figure 8–19.

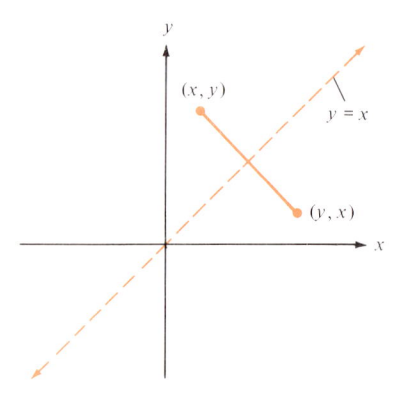

Figure 8–19

Figure 8–20 shows how the graph of a linear function f and its inverse, f^{-1}, are reflected across the line given by $y = x$.

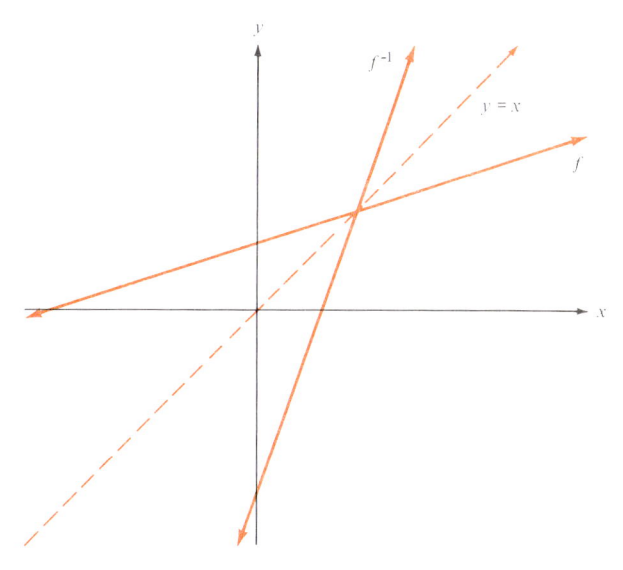

Figure 8–20

If a function f or its inverse, f^{-1}, is to be evaluated for some particular value of x, then we write $f(x)$ for the value of f at x and we write $f^{-1}(x)$ for the value of f^{-1} at x.

Example 5

(a) Evaluate $f(x) = 5x + 2$ for $x = 3$.
(b) Find $f^{-1}(x)$ and evaluate it for $x = 2$.

Solution

(a) $f(3) = 5(3) + 2$
$= 17$

(b) Write $f(x) = 5x + 2$ as $y = 5x + 2$.

$$y = 5x + 2$$
$$x = 5y + 2 \qquad \text{Form the inverse.}$$
$$y = \frac{x - 2}{5} \qquad \text{Solve for } y.$$
$$f^{-1}(x) = \frac{x - 2}{5} \qquad \text{Write } f^{-1}(x) \text{ for } y.$$
$$f^{-1}(2) = \frac{2 - 2}{5} = \frac{0}{5} = 0 \qquad \text{Find } f^{-1}(2).$$

Example 6

If $f(x) = \{(x, y)| y = x^2 + 2, x \geq 0\}$, find $f^{-1}(x)$.

Solution The domain of $f(x)$ is restricted to make it a one-to-one function. Its range is $\{y|y \geq 2\}$.

$$f^{-1}(x) = \{(x, y)|x = y^2 + 2\}$$
$$= \{(x, y)|y = \sqrt{x - 2}\} \quad \text{Solve for } y.$$

The domain of $f^{-1}(x)$ is $\{x|x \geq 2\}$ and the range of $f^{-1}(x)$ is $\{y|y \geq 0\}$. See figure 8–21.

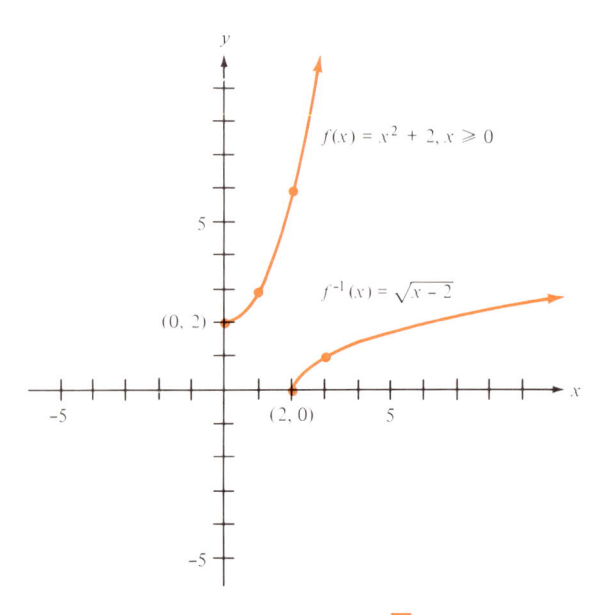

Figure 8–21

Example 7

Let $f(x) = 2x - 3$.

(a) Find $f^{-1}(x)$.
(b) Find $f(2)$.
(c) Find $f^{-1}(f(2))$.

Solution (a) We must first find $f^{-1}(x)$.

$$f(x) = 2x - 3$$
$$y = 2x - 3 \quad \text{Write } y \text{ for } f(x).$$
$$x = 2y - 3 \quad \text{Interchange } x \text{ and } y.$$
$$y = \frac{x + 3}{2} \quad \text{Solve for } y.$$
$$f^{-1}(x) = \frac{x + 3}{2} \quad \text{Write } f^{-1}(x) \text{ for } y.$$

(b) $f(2) = 2 \cdot 2 - 3$
$ = 1$

(c) $f^{-1}(f(2)) = f^{-1}(1)$ See (b), and note $f(2) = 1$.
$$= \frac{1 + 3}{2}$$
$$= 2 \quad \blacksquare$$

8–7 Exercises

In exercises 1–8:
(a) Determine if R is a function.
(b) Give the domain and range of R.
(c) Form R^{-1} and determine if it is a function.
(d) Give the domain and range of R^{-1}.
(See examples 1 and 2.)

1. $R = \{(2, 3), (4, -1), (5, 2)\}$
2. $R = \{(-1, -2), (1, 2), (0, 0)\}$
3. $R = \{(4, 3), (5, 3), (-1, 0)\}$
4. $R = \{(9, 1), (-2, -3), (7, 1)\}$
5. $R = \{(2, 3), (0, 5), (2, -1)\}$
6. $R = \{(3, 1), (4, 6), (3, -3)\}$
7. $R = \{(2, 3), (7, 3), (2, 7)\}$
8. $R = \{(9, -2), (7, -2), (9, 4)\}$

In exercises 9–16, a function f is given.
(a) Find f^{-1} and solve f^{-1} for y.
(b) Determine if f^{-1} is a function.
(c) Sketch the graph of f and f^{-1} on the same set of axes. *(See examples 3 and 4.)*

9. $f = \{(x, y) | 2x - 3y = 6\}$
10. $f = \{(x, y) | 2x + y = -3\}$
11. $f = \{(x, y) | y = x^3\}$
12. $f = \{(x, y) | y = x\}$
13. $f = \{(x, y) | y = x^2\}$
14. $f = \{(x, y) | y = x^2 - 2\}$
15. $f = \{(x, y) | y = x^2 - 1, x \geq 0\}$
16. $f = \{(x, y) | y = x^2 - 4, x \geq 0\}$

In exercises 17–22, a function f(x) is given.
(a) Find $f^{-1}(x)$.
(b) Find $f(2)$.
(c) Find $f^{-1}(f(2))$.
(See example 7.)

17. $f(x) = 3x - 1$
18. $f(x) = 5x + 3$
19. $f(x) = x^3$
20. $f(x) = x^2, x \geq 0$
21. $f(x) = x^2 + 1; \quad x \geq 0$
22. $f(x) = x^4; \quad x \geq 0$

23. If $f(x)$ is a function such that $f(3) = 27$, then what is the value of $f^{-1}(27)$?

24. If a function is not a one-to-one function, then its inverse is not a function. Devise a geometric model or test that explains why this is so.

25. The equation $y = x$ or $f(x) = x$, is called the *identity function*. What is the inverse of the identify function?

26. If f has an inverse function f^{-1}, then what is the meaning of $(f^{-1})^{-1}$?

27. The linear function can be defined as any function of the form $f(x) = ax + b$. Why is it necessary in this function for $a \neq 0$ if $f^{-1}(x)$ is to be a function?

28. The constant function is $y = b$, or $f(x) = b$. What is the graph of $f^{-1}(x)$, and is $f^{-1}(x)$ a function?

Chapter 8 Summary

Vocabulary

asymptote
axis of symmetry
circle
conic sections
conjugate axis
constant of variation
directrix
direct variation
ellipse
extremes
focus
hyperbola
inverse of a relation

inverse variation
joint variation
major axis
means
minor axis
one-to-one function
parabola
proportion
radius
ratio
transverse axis
vertex

Definitions

8-1 Parabola

A **parabola** is the set of all points in a plane equidistant from a fixed point, called the focus, and a fixed line, called the directrix.

8-2 Circle

A **circle** is the set of all points in a plane located a fixed distance from a fixed point. The measure of the fixed distance is its **radius,** and the fixed point is its **center.**

8-3 Ellipse

An **ellipse** is the set of all points in a plane such that the sum of the measures of the distances from any point in the set to two fixed points is a constant. Each fixed point is called a **focus** (the plural of focus is *foci*).

8-4 Hyperbola

A **hyperbola** is the set of all points in a plane such that the difference of the measures of the distances from any point in the set to two fixed points is a constant. The two fixed points are called the **foci.**

8-5 Direct Variation

Two variables y and x **vary directly** if $y/x = k$, where k is a constant and $k \neq 0$.

8–6 Inverse Variation

Two variables x and y **vary inversely** if $xy = k$, where k is a constant and $k \neq 0$.

8–7 Joint Variation

The variable y **varies jointly** as the variables x, w, and z if $y = kxwz$, where k is a constant and $k \neq 0$.

Rules

8–1 Standard Form of an Equation of a Circle

Any circle whose radius is $r (r > 0)$ and whose center is located at the origin is given by the equation $x^2 + y^2 = r^2$. If the center is not located at the origin, the equation of the circle has the form $(x - h)^2 + (y - k)^2 = r^2 \ (r > 0)$, where r is the radius and (h, k) is the location of the center. These equations are referred to as the **standard forms** for an equation of a circle.

8–2 General Form of an Equation of a Circle

The **general form** of an equation of a circle is $x^2 + y^2 + Dx + Ey + F = 0$.

8–3

The **standard form** for the equation of an ellipse whose center is at $(0, 0)$ and whose major axis is along the x-axis is $\dfrac{x^2}{a^2} + \dfrac{y^2}{b^2} = 1$, where $a, b > 0$ and $a > b$.

8–4 Standard Form for the Equation of a Hyperbola

The **standard form** for the equation of a hyperbola whose center is at $C(0, 0)$ and whose vertices are $(a, 0)$ and $(-a, 0)$ is $\dfrac{x^2}{a^2} - \dfrac{y^2}{b^2} = 1.$

8–5 Product of the Means and Extremes

In a proportion, the product of the means equals the product of the extremes. That is, if $\dfrac{a}{b} = \dfrac{c}{d}$, then $ad = bc \quad (b, d \neq 0)$.

Procedures

8–1 Solving a Variation Problem

1. Set up an equation for direct, inverse, or joint variation from the facts given about the problem.
2. Find k.
3. Find the desired variable.

Review Problems

Section 8-1

In exercises 1-3, give the direction in which the parabola represented by each equation will open.

1. $y = -2x^2 + 4$
2. $x = 4y^2 + 6y - 3$
3. $x = -5y^2 + 6$

In exercises 4-6, find the coordinates of the vertex for each parabola represented by the given equation.

4. $y = x^2 + 4x - 2$
5. $y = -2x^2 + 4x - 3$
6. $x = 3y^2 - 6y + 4$

In exercises 7-8, locate the vertex and several other ordered pairs and draw the graph of the given equation.

7. $y = x^2 - 4x + 5$
8. $y = 2x^2 + 6x + 11$

Section 8-2

In exercises 9-12, write the equation in standard form if necessary, and find the coordinates of the center and the length of the radius of each circle represented by the given equation.

9. $x^2 + y^2 = 100$
10. $x^2 + y^2 - 6x + 12y = 4$
11. $x^2 - 8x + y^2 = -10x + 23$
12. $2x^2 + 2y^2 + 8x - 6y = 5$

13. Find an equation of the circle whose center is the point $(-2, -3)$ and which goes through the point $P(3, 7)$.

Section 8-3

In exercises 14-19, identify the conic section represented by the given equation.

14. $\dfrac{x^2}{4} + \dfrac{y^2}{49} = 1$
15. $x^2 + 3y^2 = 7$
16. $y = x^2 + 4$
17. $3x^2 - 4y^2 = 12$
18. $6x^2 + 6y^2 + 7x = 3$
19. $x^2 = -y^2 + 6$

20. What are the equations of the asymptotes for the hyperbola given by $\dfrac{y^2}{9} - \dfrac{x^2}{16} = 1$?

Section 8-4

21. Draw the graph of $x^2 + 6x + y^2 - 8x \le 0$

Section 8-5

22. Towns A and C are 250 miles apart. Town B is located between towns A and C so that it divides the distance between towns A and C into the ratio 2:3. Town B is located closer to town A than it is to town C. Barbara drove nonstop from A to C. If she took 2 hours to drive from A to B, what was her average speed between B and C if it took 6 hours for the entire trip?

Relations and Functions: Part II

Section 8–6

23. p varies directly as the square of q and inversely as r. If $p = 2$ when $r = 25$ and $q = 5$, find p if $q = 10$ and $r = 30$.

24. If x varies directly as y and inversely as z, what is the effect on x if y is decreased to one-third of its previous value and z is decreased to two-fifths of its previous value?

Section 8–7

25. If $R = \{(6, 3), (7, 5), (-2, -4)\}$, find R^{-1} and the domain and range of R^{-1}.

26. If $f = \{(x, y)|8x - 3y = -2\}$, find f^{-1}.

27. If $f(x) = 5x + 6$, find $f^{-1}(f(2))$.

28. If $f(x) = 6x - 4$ and $g(x) = 2x^2 - 1$, find $g(f(3))$.

Cumulative Test for Chapters 1–8

Chapter 1

1. What kind of decimal number represents an irrational number?
2. Is subtraction a cummutative operation?
3. Use the distributive axiom to simplify $98(2) + 98(8)$.
4. Simplify: $\dfrac{-2 - 5 - 10 - 11}{-2^2 + 11}$.

Chapter 2

5. Write the following using a base and an exponent.
$$(2x - 3)(2x - 3)(2x - 3)$$
6. Is it true or false that $5^0 = 0$?
7. Remove the greatest common factor from each of the terms in the expression $3a^3b^4(x - 3)^3 - 15ab^2(x - 3)$.
8. Factor completely: $x^5 - 4x^3 - 8x^2 + 32$.

Chapter 3

9. What values may x not take in the fraction $\dfrac{6x + 2}{x^2 + 2x - 3}$?
10. Reduce: $\dfrac{5x^2 + 17x - 12}{25x^3 - 9x}$.
11. Add: $\dfrac{x^2}{x - y} + \dfrac{4y^2}{y - x} + \dfrac{3xy}{x - y}$.
12. Simplify: $\dfrac{\dfrac{1}{y} - \dfrac{1}{x}}{\dfrac{1}{x^2} - \dfrac{1}{y^2}}$.
13. Divide: $3x + 5 \overline{\smash{\big)}\, 3x^3 + 23x^2 + 21x - 15}$.

Chapter 4

14. Solve for x: $-6(x - 5) + 3x = 3(-x - 6) + 2x$.
15. Solve for t: $\dfrac{5}{3t} - \dfrac{2}{3} = \dfrac{4}{2t}$.
16. Solve for x: $\dfrac{2x}{5} - 4 < \dfrac{6x}{5}$.
17. Solve for x: $\left|\dfrac{6x - 2}{3}\right| = 14$.
18. Solve for x: $|4x - 2| + 6 > 16$.
19. The value of a house and lot is \$110,500. The house has a value $7\frac{1}{2}$ times the value of the lot. What is the lot worth?

Chapter 5

20. Simplify and leave all exponents positive: $\left(\dfrac{9x^4y^{-2}}{3x^{-2}y^4}\right)^{-2}$.
21. Simplify: $(x^{-1} + y^{-1})^{-1}$.

340

22. Simplify $x^{4/5}x^{2/3}$. Leave the answer in simplified radical form.

23. Simplify $\dfrac{\sqrt{8x^3y^4}}{\sqrt{12x^9y^5}}$. Assume $x, y > 0$.

24. Add: $3\sqrt{\tfrac{3}{5}} + 2\sqrt{\tfrac{5}{3}}$.

25. Reduce the following to lowest terms. Assume $x > 0$.
$$\frac{4x - \sqrt{12x^3}}{6x}$$

26. Solve for x: $2x + 1 = \sqrt{4x^2 + 2x + 3}$.

Chapter 6

27. Solve $p^2 + \dfrac{14p}{3} = 8$ by any method.

28. Simplify: $i^2(i^{15})$.

29. Divide and simplify: $\dfrac{3 + 4i}{2 - 3i}$.

30. A rectangular patio measures 10 feet by 12 feet. A uniform walkway is to be built around all four sides of the patio. How wide should the walkway be if the area of the walkway is 104 square feet?

31. Find the solution and draw the graph of the solution set for $x^2 > x + 20$.

32. A boat can travel 20 miles per hour in still water. It takes the boat 2 hours longer to travel 48 miles upriver than it does to travel 24 miles downriver. Find the speed of the current in the river.

Chapter 7

33. Which number is the ordinate in the ordered pair $(-6, 9)$?

34. Give the definition of a relation.

35. Without drawing a graph, give the slope of the line represented by $2x + 5y = -4$.

36. What is the slope of any line perpendicular to the graph of $3x - 4y = 7$?

37. What is an equation of the line that goes through the point with coordinates $(-7, 4)$ and with a slope $-\tfrac{1}{2}$? Leave your answer in the form $Ax + By = C$, where A, B, and C are integers.

38. Are the two lines given by $4x - 3y = 10$ and $3x + 4y = -8$ parallel, perpendicular, or neither?

39. Draw the graph of $5x - 2y > 10$

Chapter 8

40. Give the coordinates of the vertex of the parabola represented by $y = 2x^2 - 8x + 2$.

41. Complete the square on the x- and y-terms of the equation $x^2 + 6x + y^2 - 8y = 0$. Give the coordinates of the center and the radius of the circle represented by the equation.

42. Write an equation of the circle with its center at $(6, 5)$ and which goes through the point $(-2, -5)$.

43. Give the length of the major axis for the ellipse given by the equation $\dfrac{x^2}{16} + \dfrac{y^2}{4} = 1$.

44. Draw the graph of $\dfrac{x^2}{36} - \dfrac{y^2}{4} = 1$.

45. The distance that an object will fall in a given time in a vacuum is directly proportional to the square of the time. If an object falls 64 feet in 2 seconds, how far will it fall in 5 seconds?

46. Write the inverse of the function given by $y = 3x + 6$ and solve the inverse function for y.

CHAPTER 9
Exponential and Logarithmic Functions

Objectives of Chapter 9

To be able to:

☐ Identify an exponential function. (9–1)

☐ Draw the graphs of exponential functions. (9–1)

☐ List the restrictions on the variables in exponential and logarithmic functions. (9–1)

☐ Solve exponential equations. (9–1)

☐ Show that logarithmic and exponential functions are inverses of each other. (9–2)

☐ Write an exponential equation as a logarithmic equation. (9–2)

☐ Write a logarithmic equation as an exponential equation. (9–2)

☐ Use the properties of logarithms to simplify expressions. (9–3)

☐ Fully expand a logarithmic expression by using the properties of logarithms. (9–3)

☐ Find the characteristic of a logarithm. (9–4)

☐ Use a table of logarithms. (9–4)

☐ Interpolate. (9–4)

☐ Make computations using logarithms. (9–4)

☐ Solve exponential and logarithmic equations. (9–5)

☐ Solve word problems by using logarithms. (9–6)

We saw in chapter 5 how exponents and radicals are related. For example, we saw that $b^{1/2} = \sqrt{b}$ for $b > 0$. That is, an exponential statement can be expressed as a radical or a radical statement can be expressed using an exponent.

In this chapter, we explore the properties of exponential and logarithmic functions, as well as the special relationship that exists between them. We see that these two functions are inverses of each other and that an exponential function may be expressed as a logarithmic function, and visa versa.

9–1 The Exponential Function

We have already taken a look at functions such as $f(x) = x^2$, $f(x) = x^3$ and $f(x) = x^{1/2}$. In these functions, x is called the *base* and 2, 3, and $\frac{1}{2}$ are *exponents*. An **exponential function** has the form $g(x) = b^x$, where b is a constant and the exponent is a variable. (We give a formal definition for an exponential function shortly). While $f(x) = x^2$ is a polynomial function, $g(x) = 2^x$ is an exponential function. In a like manner, $f(x) = 3^x$ and $h(x) = (\frac{1}{2})^x$ are examples of exponential functions.

In general, an exponential function can be expressed as $f(x) = b^x$. We must now find what restrictions, if any, to place on b and x.

Let us write $f(x) = b^x$ as $y = b^x$. If $b = 0$ and $x \neq 0$, then $y = b^x$ becomes $y = 0^x$ which simplifies to $y = 0$. While $y = 0$ (or $f(x) = 0$) is a function, it is neither very interesting nor very useful. Furthermore, the inverse of $f(x) = 0$ is not a function, since $f(x) = 0$ is not a one-to-one function. For example, notice (5, 0) and (8, 0) both belong to $f(x) = 0$. This means (0, 5) and (0, 8) both belong to the inverse of $f(x)$, which shows the inverse of $f(x) = 0$ is not a function. Therefore, b cannot be zero if we want the inverse of $y = b^x$ to be a function.

What happens if $b < 0$? For some values of x, $f(x) = b^x$ would make sense if $b < 0$. For example, if $x = \frac{1}{3}$ and $b = -\frac{1}{8}$, then $f(\frac{1}{3}) = (-\frac{1}{8})^{1/3} = -\frac{1}{2}$. This gives a real number for $f(x)$. But if $b = -2$ and $x = \frac{1}{2}$, we have $f(-2) = (-2)^{1/2}$, which is the same as $f(-2) = \sqrt{-2}$. This value of $f(x)$ is imaginary. Because we want all values of $f(x)$ to be real numbers, we insist that b not be negative.

Finally, suppose $b = 1$. Then $f(x) = b^x$ becomes $f(x) = 1^x = 1$. The graph of $y = 1$ (or $f(x) = 1$) is a straight line parallel to the x-axis. This function does not have an inverse that is a function. Therefore, we exclude 1 as a value for b.

To summarize, the function $f(x) = b^x$ is defined for $b > 0$ and $b \neq 1$.

We must now see what values x may take in $f(x) = b^x$. In chapter 5, b^x was defined for x a rational number. We must now see if $f(x) = b^x$ is defined when x is irrational. For example, does $f(\sqrt{5}) = 2^{\sqrt{5}}$ have any meaning for the irrational number $\sqrt{5}$ and $b = 2$? To see, we state without proof that if $b > 1$ and $u < v$, then $b^u < b^v$. For example, if $3 < 4$ and $b = 2$, then $2^3 < 2^4$, or $8 < 16$. Then notice the following:

Whenever	$f(\sqrt{5}) = 2^{\sqrt{5}}$ satisfies
$2 < \sqrt{5} < 3$	$2^2 < 2^{\sqrt{5}} < 2^3$
$2.2 < \sqrt{5} < 2.3$	$2^{2.2} < 2^{\sqrt{5}} < 2^{2.3}$
$2.23 < \sqrt{5} < 2.24$	$2^{2.23} < 2^{\sqrt{5}} < 2^{2.24}$
$2.236 < \sqrt{5} < 2.237$	$2^{2.236} < 2^{\sqrt{5}} < 2^{2.237}$
.	.
.	.
.	.

This "squeeze play" could be carried on indefinitely. The two numbers on either side of $\sqrt{5}$ are getting closer and closer together. Consequently, as $\sqrt{5}$ is squeezed between two numbers that better and better approximate its value, $2^{\sqrt{5}}$ is likewise placed between two numbers that better and better approximate its value. The approximate value of $2^{\sqrt{5}}$ is 4.711113133.

From this we conclude that $f(x) = b^x$, $b > 0$ and $b \neq 1$ is defined for all real numbers x. This is stated as definition 9–1.

Definition 9–1

Exponential Function

If $b > 0$ and $b \neq 1$, then $f(x) = b^x$ is called the base b **exponential function.** Its domain is the set of real numbers.

We also need two other facts, which we will state without proof as rule 9–1. These two facts are given as theorems with proofs in more advanced texts.

Rule 9–1

Properties of the Exponential Function

If $b > 0$ and if $b \neq 1$, then:

1. If $b^x = b^y$, then $x = y$ for any real numbers x and y.
2. If $x = y$, then $b^x = b^y$ for any real numbers x and y.

Let us now take a look at some graphs and examples.

Example 1 Draw the graphs of $f(x) = 2^x$ and $f(x) = (\frac{1}{2})^x$ on the same set of axes.

Solution Find several ordered pairs for each equation and draw the graphs. See figure 9–1.

346   Intermediate Algebra

$f(x) = 2^x$

x	-3	-2	-1	0	1	2	3
y	$\frac{1}{8}$	$\frac{1}{4}$	$\frac{1}{2}$	1	2	4	8

$f(x) = (\frac{1}{2})^x$

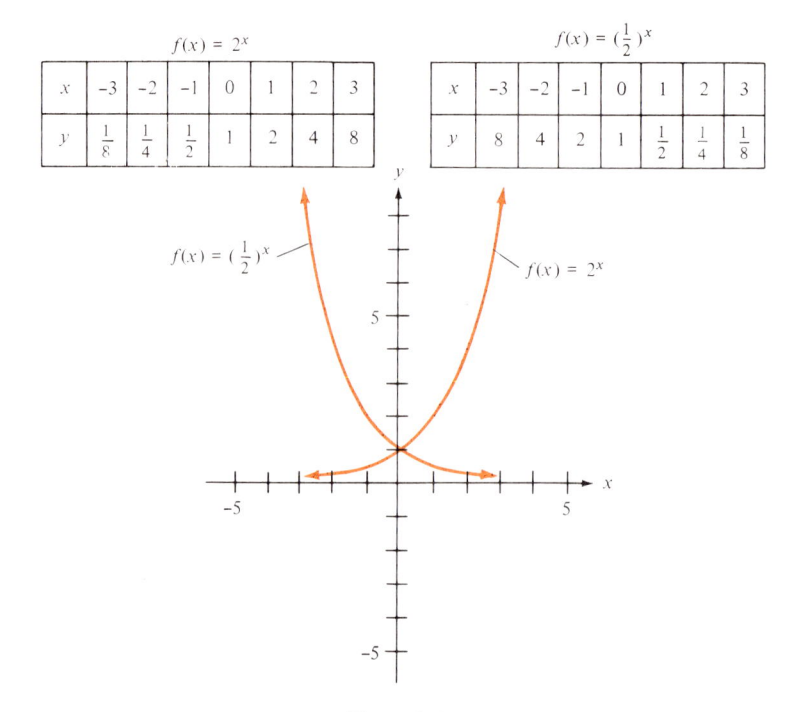

x	-3	-2	-1	0	1	2	3
y	8	4	2	1	$\frac{1}{2}$	$\frac{1}{4}$	$\frac{1}{8}$

Figure 9–1

From figure 9–1, what conclusion do you reach if $0 < b < 1$? What is true if $b > 1$?

Example 2   Draw the graph of $f(x) = 2^{x+1}$. How does this graph differ from the graph of $f(x) = 2^x$?

Solution   Find some ordered pairs and draw the graph. See figure 9–2.

Example 3   Solve $3^x = 27$.

Solution
$3^x = 27$
$3^x = 3^3$ Write 27 as 3^3.
$x = 3$ Apply part 1, rule 9–1.

Example 4   Solve $2^{x+1} = 16$.

Solution
$2^{x+1} = 16$
$2^{x+1} = 2^4$ Write 16 as 2^4.
$x + 1 = 4$ Apply part 1, rule 9–1.
$x = 3$ Solve for x.

Exponential and Logarithmic Functions

x	-3	-2	-1	0	1	2
y	$\frac{1}{4}$	$\frac{1}{2}$	1	2	4	8

$f(x) = 2^{x+1}$

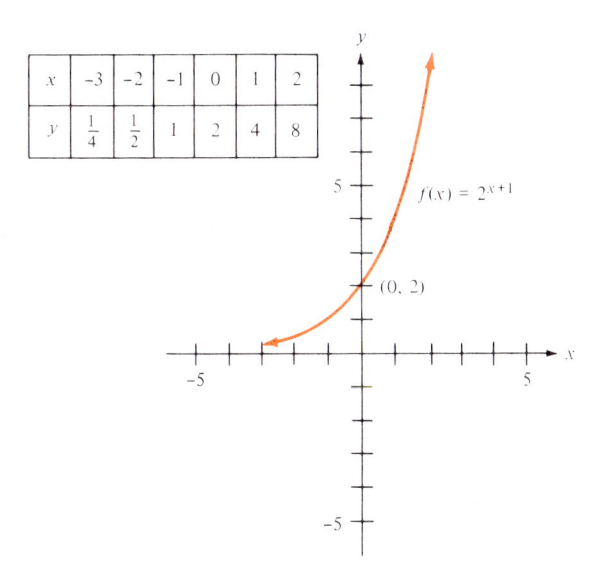

(0, 2)

Figure 9–2

Example 5 Solve $3^{2-x} = \frac{1}{81}$.

Solution Write $\frac{1}{81}$ as $\frac{1}{3^4} = 3^{-4}$.

$$3^{2-x} = \frac{1}{81}$$
$$3^{2-x} = 3^{-4}$$
$$2 - x = -4 \qquad \text{Apply part 1, rule 9–1.}$$
$$x = 6 \qquad \text{Solve for } x.$$

It can be shown in advanced mathematics that the rules for exponents hold for both rational and irrational exponents. We use this fact to simplify problems such as the one shown in example 6.

Example 6 Simplify: $3^{2-\sqrt{5}} \cdot 3^{\sqrt{5}+1}$.

Solution Because the bases are the same, add the exponents.

$$3^{2-\sqrt{5}} \cdot 3^{\sqrt{5}+1} = 3^{2-\sqrt{5}+\sqrt{5}+1}$$
$$= 3^3$$
$$= 27$$

Example 7 Solve: $2^x 3^x = 36$.

Solution

$$2^x 3^x = 36$$
$$2^x 3^x = 6^2$$
$$(2 \cdot 3)^x = 6^2 \quad \text{Rule 2–4.}$$
$$6^x = 6^2$$

Therefore, $x = 2$. ∎

Example 8 Solve $5^{-2x} = 5\sqrt{5}$.

Solution
$$5^{-2x} = 5(5)^{1/2}$$
$$5^{-2x} = 5^{2/2}(5)^{1/2}$$
$$5^{-2x} = 5^{3/2}$$

$$-2x = \frac{3}{2} \quad \text{Rule 9–1.}$$

$$-4x = 3 \quad \text{Multiply each side by 2.}$$

$$x = -\frac{3}{4} \quad \text{Solve for } x.$$ ∎

9–1 Exercises

In exercises 1–10, graph each equation. You may need to use a larger scale on the y-axis than the scale used on the x-axis. (See examples 1 and 2.)

1. $y = 3^x$
2. $y = \left(\dfrac{1}{3}\right)^x$
3. $f(x) = 2^{-x}$
4. $f(x) = 2^{2x}$
5. $f(x) = 2^{x-1}$
6. $f(x) = 2^{x+3}$
7. $f(x) = 2^{|x|}$
8. $f(x) = 2^{|x+1|}$
9. $y = 2^{x^2+1}$
10. $y = 2^{-x+1}$

In exercises 11–34, use part 1 of rule 9–1 to solve for x. (See examples 3–8.)

11. $2^x = 32$
12. $3^{x-1} = 3^{3/2}$
13. $3^{2x-1} = \sqrt{27}$
14. $8^{3x-1} = 4$
15. $10^{3x-4} = 0.001$
16. $5^{-x} = 5\sqrt[5]{5}$
17. $\dfrac{6^x}{2^x} = \dfrac{9^x}{3}$
18. $5^{2x-3} = 1$
19. $8^{2x+1} = \dfrac{1}{32}$
20. $5^{x-2} = \sqrt{\dfrac{1}{5}}$
21. $6^x = 3^x \cdot 2^5$
22. $(10)^x = 5^6 \cdot 2^x$
23. $(81)^x = 9^x \cdot 9^2$
24. $\dfrac{17}{2^x} = 34$
25. $\dfrac{19}{3^x} = 57$
26. $7^{-x} = (28)^{-x}$
27. $11^{-x} = (55)^{-x}$
28. $2^x \cdot 3^4 = 162$
29. $5^x \cdot 3^2 = 45$
30. $\left(\dfrac{5}{8}\right)^x = \dfrac{25}{64}$
31. $\left(\dfrac{2}{3}\right)^x = \dfrac{8}{27}$
32. $\left(\dfrac{1}{3}\right)^{-x} = 27$
33. $\left(\dfrac{1}{5}\right)^{-x} = 125$
34. $(0.01)^x = 100$

In exercises 35–40, simplify by combining exponents. (See example 6.)

35. $3^{1-\sqrt{3}} \cdot 3^{2+\sqrt{3}}$

36. $\dfrac{5^{1-\sqrt{5}} \cdot 3^{\sqrt{2}+1}}{5^{-\sqrt{5}-1} \cdot 3^{\sqrt{2}}}$

37. $\dfrac{7^{1+\sqrt{5}} \cdot 5^{3-\sqrt{2}}}{7^{-1+\sqrt{5}} \cdot 5^{-\sqrt{2}+1}}$

38. $\dfrac{3^{-(1/4)+\sqrt{2}}}{3^{(3/4)+\sqrt{2}}}$

39. $\dfrac{5^{(3/8)+\sqrt{5}}}{5^{(-5/8)+\sqrt{5}}}$

40. $\dfrac{27^{(1/3)+\sqrt{2}}}{27^{(-1/3)+\sqrt{2}}}$

41. A bacteria culture is growing according to the equation $P = 20{,}000(1.1)^{0.04t}$, where t is the number of hours elapsed after the culture was begun. How large will the population be at the end of 25 hours?

42. If P dollars is invested for t years at an interest rate r compounded yearly, the amount of money A at the end of t years is $A = P(1 + r)^t$. How much will $10,000 be worth at the end of 15 years if it is invested at 11%?

9–2 Logarithmic Functions

The graph of $y = b^x$ is a one-to-one function, and it therefore has an inverse that is a function. The inverse of $y = b^x$ is $x = b^y$. The graphs of these two functions (for $b > 1$) are shown in figure 9–3.

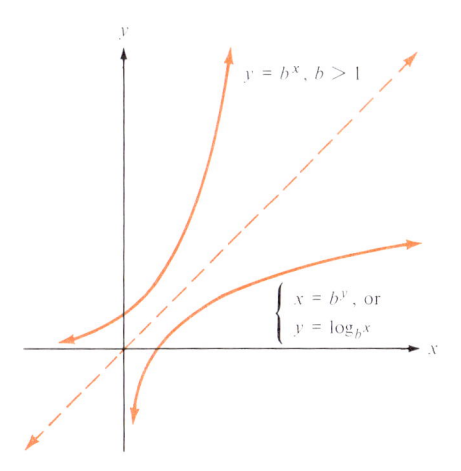

Figure 9–3

It is customary to solve the equation $x = b^y$ for the letter y. Because y is an exponent rather than a factor or term, we cannot solve for it in the usual way. Instead, we "invent" a new notation, $\log_b x$ (where *log* is an abbreviation for *logarithm*), and say that $y = \log_b x$. Whatever $\log_b x$ may be, notice that it is an exponent because it is the same as y in the equation $x = b^y$.

We look at a specific example expressed in exponential form to clarify the idea of a logarithm. We know, for example, that

$$81 = 9^2 \leftarrow \text{exponent, or logarithm}$$

number ↗ ↑ base

This example shows that the number 81 is equal to the base 9 raised to the second power. This same statement can be written in logarithmic form as

$$2 = \log_9 81$$

logarithm ↗ ↑ ↑ number
base

In this form, the equation is solved for the exponent 2, and we say the logarithm of 81 to the base 9 is 2. That is, the power to which 9 must be raised to give 81 is 2. From this, we see a logarithm is an exponent.

Definition 9–2

Logarithm of a Number

If $b > 0$ and $b \neq 1$, then the **logarithm** of a number x is the exponent to which the base b must be raised to equal x. That is, $y = \log_b x$ if and only if $x = b^y$.

Example 1 Write $3^4 = 81$ in logarithmic form.

Solution In logarithmic form, the exponent, 4, will occupy one side of the equation by itself; we write $4 = \log_3 81$. ■

Example 2 Write $2 = \log_5 25$ in exponential form.

Solution We must now solve the equation for the number 25. This is done by writing $25 = 5^2$. ■

The domain of the logarithmic function given by $y = \log_b x$ is $\{x \mid x > 0\}$. The range is the set of all real numbers. This is apparent by examining the graph of $y = \log_b x$ (see figure 9–3).

It is essential to be able to move freely from the exponential form of an equation to the logarithmic form, and visa versa.

Example 3 Write $p = q^r$ in logarithmic form.

Solution $\log_q p = r$ ■

Exponential and Logarithmic Functions

Example 4 Write $\left(\frac{1}{3}\right)^{-3} = 27$ in logarithmic form.
Solution $\log_{1/3} 27 = -3$

Example 5 Write $4 = \log_{1/2} \frac{1}{16}$ in exponential form.
Solution $\left(\frac{1}{2}\right)^4 = \frac{1}{16}$

A special consequence of the logarithmic function is the fact that $b^{\log_b x} = x$ and $\log_b b^y = y$. These can be proved using definition 9–2. To prove $b^{\log_b x} = x$, we know from definition 9–2 that

$$b^y = x \tag{1}$$

and

$$y = \log_b x \tag{2}$$

are equivalent equations. If the value for y in (2) is put into (1), we get

$$b^{\log_b x} = x \tag{3}$$

To prove $\log_b b^y = y$, put the value of x from (1) into (2). We get

$$\log_b b^y = y$$

Rule 9–2

Two Properties Involving Logarithms

1. If $b, x > 0$ and $b \neq 1$, then
$$b^{\log_b x} = x$$
2. If $b > 0$ and $b \neq 1$, then
$$\log_b b^y = y$$

Special cases of the equation in part 2 of rule 9–2 result when $y = 1$ and $y = 0$. For $y = 1$, $\log_b b^1 = 1$, or $\log_b b = 1$. For the case when $y = 0$, we get $\log_b b^0 = 0$, or $\log_b 1 = 0$.

Example 6 Evaluate $\log_{12} 12$.
Solution We know $\log_{12} 12 = 1$ by the discussion following rule 9–2. Let us suppose, however, that we did not know this. Then let $\log_{12} 12 = y$. This implies $12^y = 12^1$. Then by part 1 of rule 9–1, we know $y = 1$. Therefore, $\log_{12} 12 = 1$.

Example 7 Evaluate the following by using rule 9–2.

(a) $\log_5 5^2$
(b) $\log_3 3^9$
(c) $3^{\log_3 5}$
(d) $7^{\log_7 1/2}$

Solution

(a) $\log_5 5^2 = 2$ by part 2, rule 9–2.
(b) $\log_3 3^9 = 9$ by part 2, rule 9–2.
(c) $3^{\log_3 5} = 5$ by part 1, rule 9–2.
(d) $7^{\log_7 1/2} = 1/2$ by part 1, rule 9–2.

Example 8 Simplify $\log_3(\log_2 8)$.

Solution Let $\log_2 8 = y$. Then

$$2^y = 8$$
$$2^y = 2^3$$
$$y = 3 \quad \text{rule 9–1}$$

Therefore,

$$\log_3(\log_2 8) = \log_3 y$$
$$= \log_3 3$$
$$= 1$$

9–2 Exercises

1. Draw the graph of $y = 2^x$ and $x = 2^y$ on the same set of axes. Give the domain and range of each. How are the two equations related?
2. Draw the graph of $y = \log_3 x$ by changing $y = \log_3 x$ into an equivalent exponential equation. Give its domain and range.
3. Draw the graph of $x = \log_2 y$ by changing $x = \log_2 y$ into an equivalent exponential equation. Give its domain and range.
4. Draw the graph of $y = 2^{\log_2 x}$, $x > 0$ (*Hint:* What is the value of $2^{\log_2 x}$?)

In exercises 5–26, write each logarithmic equation in exponential form. (See examples 2 and 5.)

5. $\log_3 9 = 2$
6. $\log_4 2 = \frac{1}{2}$
7. $\log_8 \frac{1}{2} = -\frac{1}{3}$
8. $\log_8 2 = \frac{1}{3}$
9. $\log_4 64 = 3$
10. $\log_{\sqrt{7}} 7 = 2$
11. $\log_{\sqrt[3]{11}} 11 = 3$
12. $\log_{16} 4 = \frac{1}{2}$
13. $\log_8 4 = \frac{2}{3}$

14. $\log_9 3 = \frac{1}{2}$

15. $\log_{10} 0.01 = -2$

16. $\log_9 \frac{1}{3} = -\frac{1}{2}$

17. $\log_{27} 9 = \frac{2}{3}$

18. $\log_{64} \frac{1}{4} = -\frac{1}{3}$

19. $\log_{0.02} 0.0004 = 2$

20. $\log_{1.5} 2.25 = 2$

21. $\log_{1.7} 4.913 = 3$

22. $\log_{5.5} 166.375 = 3$

23. $\log_a b = c$

24. $\log_M P = Q$

25. $\log_a b = \frac{3}{2}$

26. $\log_{1/2} r = s$

In exercises write 27–46, each exponential equation in logarithmic form. (See examples 1, 3, and 4.)

27. $2^5 = 32$

28. $9^2 = 81$

29. $4^{1/2} = 2$

30. $16^{-1/2} = \frac{1}{4}$

31. $27^{-1/3} = \frac{1}{3}$

32. $100^{1/2} = 10$

33. $64^{-1/2} = \frac{1}{8}$

34. $9^0 = 1$

35. $(1.6)^2 = 2.56$

36. $(0.03)^1 = 0.03$

37. $\left(\frac{5}{8}\right)^2 = \frac{25}{64}$

38. $(16)^{-1/2} = \frac{1}{4}$

39. $4^{-3} = \frac{1}{64}$

40. $\left(\frac{1}{2}\right)^0 = 1$

41. $(0.01)^2 = 0.0001$

42. $(5.1)^2 = 26.01$

43. $e^1 = e$

44. $e^2 = b$

45. $a^{1/2} = c$

46. $p^q = r$

In exercises 47–56, use rule 9–2 to determine the value of each. (See example 7.)

47. $\log_8 8^2$

48. $\log_3 3^4$

49. $\log_4 16$ (*Hint:* Write 16 as 4^2.)

50. $\log_5 25$

51. $3^{\log_3 5}$

52. $7^{\log_7 (1/2)}$

53. $6^{\log_6 (1/3)}$

54. $4^{\log_4 \sqrt{2}}$

55. $\log_3 \left(\frac{1}{9}\right)$

56. $\log_{10} 0.01$

In exercises 57–68, solve for x, y, or b. (Hint: Write the given problem in exponential form and apply either the rules of exponents or part 1 of rule 9–1.)

57. $\log_3 9 = y$

58. $\log_b 16 = 2$

59. $\log_2 x = 5$

60. $\log_3 \frac{1}{3} = y$

61. $\log_b 0.01 = -2$

62. $\log_{1/27} 3 = y$

63. $\log_{27} x = \frac{2}{3}$

64. $\log_8 4 = y$

65. $\log_b 1 = 0$

66. $\log_b 5 = 1$

67. $\log_{10} 1{,}000^{3.21} = x$

68. $\log_{10} 100^{1.25} = x$

In exercises 69–74, simplify each expression. (See example 8.)

Example

$\log_4(\log_2 16)$

Solution

Let $\log_2 16 = y$. Then
$$2^y = 16$$
$$2^y = 2^4$$

Therefore, $y = 4$. We then have
$$\log_4(\log_2 16) = \log_4(y)$$
$$= \log_4 4$$
$$= 1$$

69. $\log_3(\log_5 125)$
70. $\log_2(\log_2 16)$
71. $\log_5(\log_{10} 10)$
72. $\log_2(\log_2 256)$
73. $\log_5[\log_2(\log_3 9)]$
74. $\log_{10}[\log_3(\log_3 27)]$
75. Why is the number zero not suitable as a base for logarithms?
76. Find the value of p in the equation $p = \log_{10}(x - 5) + 16$ if $x = 15$.

9–3 Properties of Logarithms

Logarithms have several useful properties that allow us to write problems involving multiplying, dividing, and raising to a power into problems involving additions, subtractions and multiplications. These properties are stated in rule 9–3.

Rule 9–3

Properties of Logarithms

If x, y and b ($b \neq 1$) are positive real numbers and if z is any real number, then:

1. $\log_b xy = \log_b x + \log_b y$
2. $\log_b \dfrac{x}{y} = \log_b x - \log_b y$
3. $\log_b x^z = z \log_b x$
4. In part 3, if $z = \dfrac{1}{r}$, where r is a positive integer, then
$$\log_b \sqrt[r]{x} = \log_b x^{1/r} = \dfrac{1}{r} \log_b x.$$

Exponential and Logarithmic Functions **355**

Proof of Part 1 Let $\log_b x = p$ and $\log_b y = q$. Then $x = b^p$ and $y = b^q$. We find their product:
$$xy = b^p \cdot b^q$$
$$= b^{p+q} \quad \text{Add exponents.}$$

Then $\log_b xy = p + q$. We now substitute the values given above for x and y:
$$\log_b xy = \log_b x + \log_b y$$

The proof of part 2 will be left as an exercise.

Proof of Part 3 Let $\log_b x = p$. Then $x = b^p$. Raise both sides of this equation to the zth power.
$$x^z = (b^p)^z = b^{zp}$$
$$\log_b x^z = zp \quad \text{Definition 9–2.}$$

Substitute $p = \log_b x$ into the last equation.
$$\log_b x^z = z \log_b x$$

Proof of Part 4 Let $\log_b x = p$. Then $x = b^p$. Raise both sides of this equation to the $(1/r)$th power.
$$x^{1/r} = (b^p)^{1/r} = (b^{1/r})^p$$
$$\sqrt[r]{x} = (b^{1/r})^p = b^{(1/r)p} \quad \text{Express } x^{1/r} \text{ in radical form.}$$
$$\log_b \sqrt[r]{x} = \left(\frac{1}{r}\right)p \quad \text{Definition 9–2.}$$

Substitute $p = \log_b x$ into the last equation.
$$\log_b \sqrt[r]{x} = \frac{1}{r} \log_b x$$

Example 1 Express $\log_b 15m$ as a sum of logarithms ($m > 0$).

Solution $\log_b 15m = \log_b 15 + \log_b m$ part 1, rule 9–3.

Example 2 Express $\log_b \dfrac{x}{9}$ as a difference of logarithms ($x > 0$).

Solution $\log_b \dfrac{x}{9} = \log_b x - \log_b 9$ part 2, rule 9–3.

Example 3 Express $\log_b r^{2/3}$ as a product ($r > 0$).

Solution $\log_b r^{2/3} = \frac{2}{3} \log_b r$ ■ part 3, rule 9–3.

Example 4   Fully expand $\log_b \frac{2p^3}{3}$ using as many parts of rule 9–3 as necessary $(p > 0)$.

Solution
$$\begin{aligned}
\log_b \frac{2p^3}{3} &= \log_b 2p^3 - \log_b 3 & \text{part 2, rule 9–3} \\
&= \log_b 2 + \log_b p^3 - \log_b 3 & \text{part 1, rule 9–3} \\
&= \log_b 2 + 3 \log_b p - \log_b 3 & \text{■ part 3, rule 9–3}
\end{aligned}$$

Example 5   Fully expand $\log_b \sqrt{xy} \cdot z^{1/3}$ $(x, y, z > 0)$.

Solution
$$\begin{aligned}
\log_b \sqrt{xy} \cdot z^{1/3} &= \log_b \sqrt{xy} + \log_b z^{1/3} & \text{part 1, rule 9–3} \\
&= \log_b (xy)^{1/2} + \log_b z^{1/3} & \text{part 4, rule 9–3} \\
&= \frac{1}{2} \log_b (xy) + \frac{1}{3} \log_b z & \text{part 4, rule 9–3} \\
&= \frac{1}{2} [\log_b x + \log_b y] + \frac{1}{3} \log_b z & \text{■ part 1, rule 9–3}
\end{aligned}$$

Example 6   Express $\frac{1}{2}[\log_b x + \log_b y] - 2 \log_b z$ as a single logarithm.

Solution   In this example, we must do the opposite of what we do in examples 1–5.

$$\begin{aligned}
\frac{1}{2} [\log_b x + \log_b y] - 2 \log_b z &= \frac{1}{2} (\log_b xy) - 2 \log_b z & \text{part 1, rule 9–3} \\
&= \log_b (xy)^{1/2} - \log_b z^2 & \text{part 3, rule 9–3} \\
&= \log_b \sqrt{xy} - \log_b z^2 & \text{part 4, rule 9–3} \\
&= \log_b \frac{\sqrt{xy}}{z^2} & \text{■ part 2, rule 9–3}
\end{aligned}$$

9–3 Exercises

In exercises 1–14, use rule 9–3 to expand each fully. (See examples 1–5.)

1. $\log_b 2x$
2. $\log_b 5x$
3. $\log_b \frac{3m}{4}$
4. $\log_b \frac{6n}{5}$
5. $\log_b 5t^2$
6. $\log_b 7p^6$
7. $\log_b \sqrt{3x}$
8. $\log_b \sqrt[3]{4x}$
9. $\log_b \frac{\sqrt[5]{4p}}{r^2}$
10. $\log_b \frac{(xy)^{2/3}}{z^4}$
11. $\log_{10} \frac{\sqrt[3]{4r^2}}{s^{1/2}}$
12. $\log_{10} \frac{\sqrt[3]{12t^2}}{\sqrt{7v}}$
13. $\log_3 \sqrt{5\sqrt{2u}}$
14. $\log_5 \sqrt[3]{2x\sqrt{5x}}$

Exponential and Logarithmic Functions

In exercises 15–24, write each as a single logarithm. (See example 6.)

15. $\log_b 5 + \log_b 3$
16. $\log_b 8 + \log_b p$
17. $\log_b m - \log_b 2$
18. $\log_b x - \log_b 5$
19. $2 \log_b x + 3 \log_b y$
20. $\frac{1}{2} \log_b r + \frac{1}{3} \log_b s$
21. $\frac{1}{2} [\log_b t + \log_b u] - 3 \log_b t$
22. $\log_b(x + 1) + \log_b(x - 1)$
23. $\frac{1}{2}(\log_b x + \log_b y) - \frac{1}{3}(\log_b 3 + \log_b x)$
24. $2 \log_b(m - 1) - \log_b(2m + 3)$

In exercises 25–36, find the logarithm of each of the numbers if $\log_b 5 = 1.61$ and $\log_b 2 = 0.69$.

Example Find $\log_b 20$.

Solution We must express 20 as factors of 2's and 5's.
$$\begin{aligned}\log_b 20 &= \log_b(5 \cdot 4) \\ &= \log_b 5 + \log_b 4 \\ &= \log_b 5 + \log_b 2^2 \\ &= \log_b 5 + 2 \log_b 2 \\ &= 1.61 + 2(0.69) \\ &= 1.61 + 1.38 \\ &= 2.99\end{aligned}$$

Thus, the $\log_b 20 = 2.99$.

25. $\log_b 10$
26. $\log_b 4$
27. $\log_b 25$
28. $\log_b 40$
29. $\log_b 80$
30. $\log_b 100$
31. $\log_b 400$
32. $\log_b 250$
33. $\log_b \frac{2}{5}$
34. $\log_b \frac{1}{25}$
35. $\log_b \frac{1}{8}$
36. $\log_b \frac{5}{8}$

37. Prove part 3 of rule 9–3.
38. Is it true or false that $\log_b(x + 3) = \log_b x + \log_b 3$?
39. Is it true or false that $\log_3(\log_7 35 - \log_7 5) = 0$?
40. Prove $\log_b xy = \log_b x + \log_b y$ by starting with $x = b^{\log_b x}$ and $y = b^{\log_b y}$.

9–4 ■ Common Logarithms and Calculations

Logarithms of the form $\log_{10} x$, where 10 is the base, are called **common logarithms.** If the base is not written when writing the logarithm of a number, it is understood the base is 10. Using this custom, we write

$$\log_{10} 100 = 2 \quad \text{as} \quad \log 100 = 2$$

Logarithms in the past were used to make many kinds of calculations. These calculations can now be made with a computer or calculator, and the need for logarithms today as a computational tool is not as great as it once was. However, we show a few of these calculations primarily to illustrate the properties studied in the last section.

Numbers that are integral powers of ten have integer logarithms:

$$\log 0.001 = \log 10^{-3} = -3$$
$$\log 0.01 = \log 10^{-2} = -2$$
$$\log 0.1 = \log 10^{-1} = -1$$
$$\log 1 = \log 10^{0} = 0$$
$$\log 10 = \log 10^{1} = 1$$
$$\log 100 = \log 10^{2} = 2$$

In general, $\log_{10} 10^n = n$ (or $\log 10^n = n$).

Either a table of common logarithms or an electronic calculator with a LOG function key can be used for finding logarithms of numbers that are not in the form $\log 10^n$. For example, to find the log 236 by using a table of common logarithms, first write the number in scientific notation.

$$\log 236 = \log(2.36 \times 10^2)$$
$$= \log 2.36 + \log 10^2 \qquad (1)$$

In line (1), $\log 10^2 = 2$. The value of log 2.36 is found from a table. A portion of a table of common logarithms is reproduced here as table 9–1. A more complete table is given on the back inside cover of the book.

TABLE 9–1

N	0	1	2	3	4	5	6	7	8	9
2.0	.3010	.3032	.3054	.3075	.3096	.3118	.3139	.3160	.3181	.3201
2.1	.3222	.3243	.3263	.3284	.3304	.3324	.3345	.3365	.3385	.3404
2.2	.3424	.3444	.3464	.3483	.3502	.3522	.3541	.3560	.3579	.3598
2.3	.3617	.3636	.3655	.3674	.3692	.3711	.3729	.3747	.3766	.3784

To find log 2.36, find 2.3 under N in the column labeled N. Now find 6 in the row labeled N at the top of the table. The value of log 2.36 is found at the intersection of the row containing 2.3 and the column containing 6. Thus

$$\log 2.36 = 0.3729$$

Exponential and Logarithmic Functions

Returning now to equation (1), we see

$$\begin{aligned} \log 236 &= \log 2.36 + \log 10^2 \\ &= 0.3729 + 2 \\ &= 2.3729 \end{aligned} \quad (2)$$

The integer 2 in (2) is called the **characteristic** of the logarithm. Every logarithm has a characteristic that is an integer. The nonnegative decimal number in (2) is called the **mantissa** of the logarithm. We can think of $\log_{10} x$ as being the sum of its characteristic and mantissa. That is,

$$\log_{10} x = \text{characteristic} + \text{mantissa}$$

To obtain log 236 from some calculators, make the following key entries:

Display

| Clear | 2 | 3 | 6 | Log | 2.372912003 | ≈ 2.3729

Note: Not all calculators operate in the same way. If the instructions just given do not give you the number 2.372912003, either consult the instruction manual that came with your calculator, or ask your instructor to show you how to make the proper sequence of key entries for your particular calculator. For example, the log 236 is found on some calculators as

| Clear | 2 | 3 | 6 | 2nd | Log | 2.372912003 | ≈ | 2.3729 |

Example 1

Find log 2,150 each way.

(a) Using table 9–1.
(b) Using a calculator.

Solution

(a) $\log 2150 = \log 2.150 \times 10^3$ — Write the number in scientific notation.
$= \log 2.150 + \log 10^3$ — Part 1, rule 9–3.
$= 0.3324 + 3$ — Find the log 2.150 from Table 9–1 to be 0.3324.
$= 3.3324$

(b) Make the following key entries: Display

| Clear | 2 | 1 | 5 | 0 | Log | 3.33243846 | ≈ 3.3324

Example 2

Find log 0.00234 each way.

(a) Using table 9–1.
(b) Using a calculator.

Solution

(a) log 0.00234 = log 2.34 × 10^{-3}
= log 2.34 + log 10^{-3}
= 0.3692 + (−3)   (3)

If the two numbers in line (3) are combined, we get

$$0.3692 + (-3) = -2.6308$$

While −2.6308 is the logarithm of 0.00234, the decimal part of −2.6308 is not the mantissa and −2 is not the characteristic. The positive mantissa and characteristic of −3 can be preserved if we return to line (3) and write

$$-3 + 0.3692 = (7 - 10) + 0.3692$$
$$= (7 + 0.3692) - 10$$
$$= 7.3692 - 10$$

We can still read the mantissa (0.3692) and the sum of 7 and −10 still gives the characteristic, −3.

(b) Make the following key entries:

Display

| Clear | 0 | . | 0 | 0 | 2 | 3 | 4 | Log | −2.630784143 | ≈ −2.6308

If we want a positive mantissa and the characteristic, we can add (10 − 10) to −2.6308 to get

$$-2.6308 + (10 - 10) = (10 - 2.6308) - 10$$
$$= 7.3692 - 10$$

which is the same number as we found in (a).

If a number contains more than three significant digits, the logarithm of that number cannot be found in table 9–1, although it can be found using a calculator. Whenever the logarithm of a number containing more than three significant digits must be found from table 9–1, a process known as **linear interpolation** must be used. See example 3.

Example 3   Find log 2,147 by using interpolation and table 9–1.

Solution   Note that log 2,147 lies between log 2,140 and log 2,150, both of which can be found in table 9–1. That is, if log 2,147 = L, then

$$\log 2{,}150 = 3.3324$$
$$\log 2{,}147 = L$$
$$\log 2{,}140 = 3.3304$$

Exponential and Logarithmic Functions

Because 2,147 lies at a point 0.7 of the distance between 2,140 and 2,150, we assume L lies at a point 0.7 of the distance between 3.3304 and 3.3324. We now set up a direct proportion to find L.

$$10 \left\{ 7 \begin{cases} \log 2{,}150 = 3.3324 \\ \log 2{,}147 = L \\ \log 2{,}140 = 3.3304 \end{cases} x \right\} 0.0020$$

The proportion is

$$\frac{7}{10} = \frac{x}{0.0020}$$

which gives $x = 0.0014$. Now we add 0.0014 to 3.3304 to get log 2,147. Thus

$$\log 2{,}147 = L$$
$$= 3.3304 + x$$
$$= 3.3304 + 0.0014$$
$$= 3.3318$$

(b) The value of the log 2,147 is found from a calculator by making the following key entries:

Display

| Clear | 2 | 1 | 4 | 7 | Log | 3.331832044 ≈ 3.3318

Obviously, using a calculator is easier! ■

Example 4   Find log 0.002368 by linear interpolation and table 9–1.

Solution   The nonzero digits of 0.002368 are 2,368, so we look up log 2.368. In table 9–1, 2.368 lies between 2.370 and 2.360. The characteristic of 0.002368 is -3, which is written as $7 - 10$.

$$0.000010 \left\{ 0.000008 \begin{cases} \log 0.002370 = 7.3747 - 10 \\ \log 0.002368 = L \\ \log 0.002360 = 7.3729 - 10 \end{cases} x \right\} 0.0018$$

Because the characteristics cancel each other, we subtract only the decimal parts. The proportion is

$$\frac{0.000008}{0.000010} = \frac{x}{0.0018}$$

which gives $x = 0.0014$ to four decimal places. Thus

$$\log 0.002368 = L$$
$$= (7.3729 - 10) + 0.0014$$
$$= 7.3743 - 10 \quad ■$$

Example 5

If $\log_{10} N = 1.3560$, find N each way.

(a) Using table 9–1.

(b) Using a calculator.

Solution

(a) Find the mantissa 0.3560 in table 9–1. Note that 0.3560 lies opposite the digits 2.2 at the left of the table and under the digit 7 at the top of the table. Then

$$\begin{aligned} \text{antilog } 1.3560 &= \text{antilog}(0.3560 + 1) \\ &= 2.27 \times 10^1 \\ &= 22.7 \end{aligned}$$

(b) Recall $\log_{10} N = 1.3560$ means

$$N = 10^{1.3560}$$

To find N using a calculator, use the 10^x function key.

Display

| Clear | 1 | . | 3 | 5 | 6 | 0 | 2nd | 10^x | 22.69864852 | ≈ 22.7 |

Sometimes you must use linear interpolation (in the absence of a calculator) when finding antilogs.

Example 6

If $\log N = 1.3531$, find N using table 9–1.

Solution

The mantissa, 0.3531 is not in table 9–1. It lies between the mantissas 0.3522 and 0.3541. We see that $\log 22.5 = 1.3522$ and $\log 22.6 = 1.3541$.

$$0.1 \left\{ \begin{array}{l} \log 22.6 = 1.3541 \\ x \left\{ \begin{array}{l} \log N = 1.3531 \\ \log 22.5 = 1.3522 \end{array} \right\} 0.0009 \end{array} \right\} 0.0019$$

The proportion is

$$\frac{x}{0.1} = \frac{0.0009}{0.0019}$$

which gives $x = 0.04737 \approx 0.05$.

$$\begin{aligned} N &= 22.5 + 0.05 \\ &= 22.55 \end{aligned}$$

In this example, x is rounded to two decimal places, so that N has four significant digits.

Exponential and Logarithmic Functions

We now state rule 9–4 without proof.

Rule 9–4

Equality of Logarithms

If $x, y > 0$, $b > 0$, and $b \neq 1$, then $\log_b x = \log_b y$ if and only if $x = y$.

We now show a calculation using logarithms for the purpose of illustrating how the rules of logarithms and the interpolative process are used. We use the table of logarithms given on the back inside cover of the book.

Example 7 Use logarithms to approximate $(3.6)\sqrt[3]{83.4}$.

Solution Let $N = (3.6)\sqrt[3]{83.4}$ and first apply rule 9–4. Then apply rule 9–3 to expand the right side of the equation fully.

$$\begin{aligned}\log N &= \log(3.6)\sqrt[3]{83.4} \\ &= \log 3.6 + \frac{1}{3}\log 83.4 \\ &= 0.5563 + \frac{1}{3}(1.9212) \\ &= 0.5563 + 0.6404 \\ &= 1.1967\end{aligned}$$

Thus $\log N = 1.1967$. Then, to four significant digits,

$$\begin{aligned}N &= \text{antilog}(0.1967 + 1) \\ &\approx 1.573 \times 10^1 \qquad \text{by interpolation} \\ &= 15.73\end{aligned}$$

Example 8 Use logarithms to approximate $\dfrac{(5.03)(0.00436)^{1/3}}{963}$.

Solution Let N equal the given expression.

$$\begin{aligned}\log N &= \log \frac{(5.03)(0.00436)^{1/3}}{963} \\ &= \log 5.03 + \frac{1}{3}\log 0.00436 - \log 963 \\ &= 0.7016 + \frac{1}{3}(7.6395 - 10) - 2.9836\end{aligned}$$

Now write $\frac{1}{3}(7.6395 - 10)$ as $\frac{1}{3}(27.6395 - 30)$

$$\log N = 0.7016 + \frac{1}{3}(27.6395 - 30) - 2.9836$$
$$= 0.7016 + 9.2132 - 10 - 2.9836$$
$$= (9.9148 - 10) - 2.9836$$
$$= (9.9148 - 2.9836) - 10$$
$$= 6.9312 - 10$$
$$N = \text{antilog}(6.9312 - 10)$$
$$= \text{antilog}(0.9312 - 4)$$
$$= 8.535 \times 10^{-4} \qquad \text{by interpolation}$$
$$= 0.0008535 \quad \blacksquare$$

9–4 Exercises

In exercises 1–12, give the characteristic of the logarithm of each number. (*Hint:* Write each number in scientific notation.)

1. 732
2. 1,630
3. 14,800
4. 863
5. 0.005
6. 0.0123
7. 0.000726
8. 0.00194
9. 0.0000087
10. 0.08672
11. 8.35
12. 9.62

In exercises 13–22, find the logarithm for each number. If the characteristic is negative, write the logarithm as shown in example 2. Check with your instructor to see if calculators may be used or use the table on the back inside cover. *(See examples 1 and 2.)*

13. log 642
14. log 73
15. log 87.60
16. log 8.23
17. log 0.216
18. log 0.00916
19. log 0.00075
20. log 0.0016
21. log $(8.73 \cdot 10^6)$
22. log $(2.73 \cdot 10^{-4})$

In exercises 23–34, find each antilogarithm. Check with your instructor to see if calculators may be used or use the table on the back inside cover. *(See examples 5 and 6.)*

23. antilog 3.9731
24. antilog 2.6513
25. antilog 2.9031
26. antilog 6.5263
27. antilog 8.5403 − 10
28. antilog 6.4048 − 10
29. antilog 7.4742 − 10
30. antilog 9.6064 − 10
31. antilog 0.5527
32. antilog 0.9248
33. antilog (−3.2848)   (*Hint:* Write −3.2848 as −3 + (−0.2848); then add and subtract 10.)
34. antilog(−4.5100)

Exponential and Logarithmic Functions

In exercises 35–44, use linear interpolation to find the logarithm of each number. Calculators should not be used. (See examples 3 and 4.)

35. log 56.23   **36.** log 93.42   **37.** log 7.412
38. log 5.123   **39.** log 9,656   **40.** log 74,210
41. log 0.006125   **42.** log 0.0008772   **43.** log 0.0007636
44. log 0.08527

In exercises 45–54, find the antilogarithm by interpolation. (See example 6.)

45. antilog 2.3406   **46.** antilog 3.5265
47. antilog 1.9040   **48.** antilog 1.8410
49. antilog 0.7725   **50.** antilog 0.4892
51. antilog (7.4070 − 10)   **52.** antilog (8.3950 − 10)
53. antilog (6.4836 − 10)   **54.** antilog (5.3860 − 10)

In exercises 55–70, use logarithms to calculate each answer to four significant digits. Check with your instructor to see if calculators may be used. Answers to these exercises were obtained without the use of a calculator. (See examples 7 and 8.)

55. (6.93)(7.62)   **56.** (12.6)(8.17)   **57.** $\dfrac{92.5}{8.16}$

58. $\dfrac{78.6}{5.14}$ **59.** $(17.8)^2$ **60.** $(6.13)^3$

61. $\dfrac{(7.16)(27.3)}{5.14}$ **62.** $\dfrac{(16.8)(9.15)}{22.3}$ **63.** $\sqrt[4]{9.62}$

64. $\sqrt[5]{28.6}$ **65.** $\dfrac{(0.058)^{1/3}}{0.02343}$ **66.** $\dfrac{(16.3)^{1/2}}{(75.4)^{1/3}}$

67. $\dfrac{(6{,}343)\sqrt[3]{0.05614}}{\sqrt{462}}$ **68.** $\dfrac{(543)\sqrt[4]{0.04336}}{(5.2)^2}$ **69.** $(2.56)^{1.27}\sqrt{0.05893}$

70. $(3.14)^{0.15}\sqrt[5]{29.24}$

71. The distance S that an object falls in a vacuum is given by $S = 490\, t^2$, where S is measured in centimeters and t is measured in seconds. Use logarithms to calculate how far an object will fall in 4 seconds.

72. The force F produced by two charged bodies Q_1 and Q_2 separated by a distance d is given by $F = Q_1 Q_2 / d^2$, where Q is measured in electrostatic units of charge (esu), d is measured in centimeters, and f is measured in dynes per esu. Use logarithms to calculate F if Q_1 has a charge of 325 esu, Q_2 a charge of 28 esu, and d is 6.3 centimeters.

9-5 Exponential and Logarithmic Equations

An **exponential equation** is an equation in which a variable appears in an exponent. For example $3^x = 16.5$ is an exponential equation. Rule 9-4 is used to solve exponential equations.

Example 1 Solve: $3^x = 16.5$.

Solution Apply rule 9-4.

$$3^x = 16.5$$
$$\log 3^x = \log 16.5 \qquad \text{Rule 9-4.}$$
$$x \log 3 = \log 16.5 \qquad \text{Part 3, rule 9-3.}$$
$$x = \frac{\log 16.5}{\log 3} \tag{1}$$

$x = (\log 16.5)/(\log 3)$ is the exact answer. An approximate answer is found by looking up the logarithms of 16.5 and 3 to get, to four significant digits,

$$x = \frac{\log 16.5}{\log 3}$$
$$= \frac{1.2175}{0.4771}$$
$$= 2.552 \quad ■$$

Equation (1) in example 1 may be solved using a calculator by making the following key entries:

$$\boxed{\text{Clear}}\ \boxed{1}\ \boxed{6}\ \boxed{.}\ \boxed{5}\ \boxed{\text{Log}}\ \boxed{\div}\ \boxed{3}\ \boxed{\text{Log}}\ \boxed{=}\ \boxed{\text{Display: } 2.551728585} \approx 2.552$$

Example 2 Solve $5^{x-3} = 22$ for x.

Solution
$$5^{x-3} = 22$$
$$\log 5^{x-3} = \log 22 \qquad \text{Rule 9-4.}$$
$$(x - 3)\log 5 = \log 22 \qquad \text{Part 3, rule 9-3.}$$
$$x - 3 = \frac{\log 22}{\log 5} \qquad \text{Divide each side by log 5.}$$
$$x = \frac{\log 22}{\log 5} + 3 \qquad \text{Add 3 to each side.} \tag{2}$$

Exponential and Logarithmic Functions

$$= \frac{1.3424}{0.6990} + 3 \quad \text{Find log 22 and log 5 from the table of logarithms.}$$
$$= 1.920 + 3$$
$$= 4.920 \quad \blacksquare$$

Equation (2) in example 2 is solved using a calculator by making the following key entries:

Display

| Clear | 2 | 2 | Log | ÷ | 5 | Log | + | 3 | = | | 4.920572661 | ≈ 4.921

An equation such as
$$\log(x + 3) - \log(x - 1) = \log 5$$
is called a **logarithmic equation.** Rule 9–4 is also used to solve logarithmic equations.

Example 3 Solve: $\log(x + 3) - \log(x - 1) = \log 5$.

Solution $\log(x + 3) - \log(x - 1) = \log 5$

$$\log \frac{(x + 3)}{x - 1} = \log 5 \qquad \text{Part 2, rule 9–3.}$$

$$\frac{x + 3}{x - 1} = 5 \qquad \text{Rule 9–4 (If log } a = \log b\text{, then } a = b.\text{)}$$

$$x + 3 = 5x - 5$$
$$4x = 8$$
$$x = 2$$

Check: $\log(2 + 3) - \log(2 - 1) = \log 5$
$$\log 5 - \log 1 = \log 5$$
$$\log 5 - 0 = \log 5$$
$$\log 5 = \log 5 \quad \blacksquare$$

Example 4 Solve: $\log 2 + \log(x + 1) = 2$.

Solution $\log 2 + \log(x + 1) = 2$

$$\log 2(x + 1) = 2 \qquad \text{Part 1, rule 9–3.}$$
$$2(x + 1) = 10^2 \qquad \text{Definition 9–2.}$$
$$2x + 2 = 100$$
$$x = 49$$

This example could also be worked as follows:

$$\log 2 + \log(x + 1) = 2$$
$$\log 2 + \log(x + 1) = \log 100 \quad \text{Log 100 = 2.}$$
$$\log 2(x + 1) = \log 100 \quad \text{Part 1, rule 9–3.}$$
$$2(x + 1) = 100 \quad \text{Rule 9–4.}$$
$$2x + 2 = 100$$
$$x = 49$$

We leave the check for you.

Example 5 Solve: $(3.2)^{-x} = 0.143$.

Solution
$$(3.2)^{-x} = 0.143$$
$$-x \log 3.2 = \log 0.143$$
$$-x = \frac{\log 0.143}{\log 3.2}$$
$$-x = \frac{-1 + 0.1553}{0.5051}$$
$$-x = \frac{-0.8447}{0.5051}$$
$$x = 1.672$$

Here, log 0.143 has a characteristic of -1. The value -1 should be combined with the positive mantissa, 0.1553, to obtain -0.8447. In problems such as this, do not write the logarithm as $9.1553 - 10$, but actually combine -1 and 0.1553 to get -0.8447.

9–5 Exercises

In exercises 1–30, solve for the variable, correct to four significant digits. The answers given for these exercises were obtained without using a calculator. Check with your instructor to see if calculators may be used. (See examples 1, 2, and 5.)

1. $(3.5)^m = 29.3$
2. $(2.6)^n = 35.4$
3. $(6.1)^{x+1} = 65$
4. $(5.3)^{2x-1} = 74.3$
5. $(9.2)^p = 0.643$ (See example 5.)
6. $(16.3)^z = 0.743$
7. $(4.3)^{-x} = 16.7$
8. $(9.2)^{-x/2} = 63.5$
9. $\dfrac{(2.03)^t + 1}{5.3} = 9$
10. $\dfrac{(5.02)^y - 3}{9.8} = 10$
11. $x = \log_5 27$ (*Hint:* First write the equation in exponential form.)
12. $x = \log_6 54.7$ (See exercise 11.)
13. $8^{2x-2} = \left(\dfrac{1}{32}\right)^{x-1}$ (*Hint:* Take the base 2 logarithms of both sides of the equation.)

Exponential and Logarithmic Functions

14. $3^{2x-4} = \left(\dfrac{1}{9}\right)^{x-1}$ (*Hint:* Take the base 3 logarithms of both sides of the equation.)

15. $\log_b 92.4 = 4$ **16.** $\log_b 0.0155 = -5$

17. $\log(x + 1) - \log(x - 2) = \log 2$ (See examples 3 and 4.)

18. $\log(2m - 1) - \log(m + 1) = \log 1$

19. $\log 2 + \log(n - 3) = \log 12$ **20.** $\log 5 + \log(2x - 3) = \log 15$

21. $\log_3 3^x = \log_3 27$ **22.** $\log_2 2^p = \log_2 16$

23. $\log(2t + 3) = \log(t - 3) + \log 5$

24. $\log(5x - 3) = \log 2 + \log(x + 3)$

25. $10^{3\log x} = 8$ **26.** $5^{5\log_5 z} = 32$

27. $7^{-\log_7 t} = 49$ **28.** $6^{-3\log_6 r} = 8$

29. $\log(x + 4)^2 = \log(18x + 7) + \log 1$

30. $\log(x + 6)^2 = \log(9x + 14) + \log 2$

In exercises 31–34, use this information: If P dollars are invested at an annual interest rate r, the amount, A, that P will be worth after t years is given by $A = P(1 + r)^t$. Also, use logarithms.

31. If $1,000 yields $1,790 at the end of 10 years, at what rate was it invested?

32. How many years (to the nearest year) will it take for $1,000 to yield $1,796 if it is invested at an annual rate of 5%?

33. How much will $100 be worth at the end of 7 years if it is invested at an annual rate of 8%?

34. How much money must be invested today at a rate of 6% if it is to be worth $4,000 at the end of 10 years?

9–6 ◼ Applications

Exponential and logarithmic functions have a wide application in the study of population growth, compound interest, decay of radioactive elements, and the pH of a solution. We have mentioned just four of the many areas where these functions accurately describe a situation. We take a look at a few of these application in this section.

In chemistry the pH of a solution is defined as

$$\text{pH} = -\log[\text{H}^+]$$

where $[\text{H}^+]$ gives the concentration of hydrogen ions in moles per liter. (See your favorite chemistry teacher for the definition of a mole. *Hint:* It's not an underground subject). A pH of 7 is neutral. A pH less than 7 is acidic and a pH greater than 7 is alkaline.

Example 1 If a solution has a hydrogen ion concentration of 3.2×10^{-5}, calculate the pH of the solution to the nearest tenth.

Solution
$$\begin{aligned}
\text{pH} &= -\log[\text{H}^+] \\
&= -\log(3.2 \times 10^{-5}) \\
&= -[\log 3.2 + \log 10^{-5}] \\
&= -[(0.5051) + (-5)] \\
&= -0.5051 + 5 \\
&= 4.4949 \\
&\approx 4.5
\end{aligned}$$

If P dollars are invested for t years at an interest rate r compounded yearly, the amount of money A at the end of t years is

$$A = P(1 + r)^t$$

If the interest is compounded m times per year, the compound interest formula is

$$A = P\left(1 + \frac{r}{m}\right)^{mt}$$

Example 2 How much will $1000 be worth at the end of 20 years if it is invested at 8% compounded at each interval?

(a) yearly

(b) quarterly (four times per year)

Solution (a) The formula is $A = P(1 + r)^t$, where $P = \$1{,}000$, $r = 0.08$, and $t = 20$.

$$\begin{aligned}
A &= 1{,}000(1 + 0.08)^{20} \\
&= 1{,}000(1.08)^{20} \\
\log A &= \log 1{,}000 + 20 \log 1.08 \qquad \text{Take base 10 logarithms} \\
&\qquad\qquad\qquad\qquad\qquad\qquad \text{of both sides.} \\
&= 3 + 20(0.0334) \\
&= 3 + 0.6680 \\
&= 3.6680 \qquad\qquad\qquad\qquad \text{Find the antilogarithm.} \\
A &= \$4{,}656
\end{aligned}$$

(b) For interest compounded quarterly, the formula is $A = P(1 + r/m)^{mt}$, where $P = \$1{,}000$, $r = 0.08$, $t = 20$, and $m = 4$.

$$\begin{aligned}
A &= 1{,}000\left(1 + \frac{0.08}{4}\right)^{4 \cdot 20} \\
&= 1{,}000(1 + 0.02)^{80} \\
&= 1{,}000(1.02)^{80}
\end{aligned}$$

Exponential and Logarithmic Functions

$$\log A = \log 1{,}000 + 80 \log 1.02 \quad \text{Take base 10 logarithms of both sides.}$$
$$= 3 + 80(0.0086)$$
$$= 3 + 0.6880$$
$$= 3.6880 \quad \text{Find the antilogarithm.}$$
$$A = \$4{,}876 \quad \blacksquare$$

The probability P of an event that happens very infrequently is given by $P = (e^{-m}m^r)/r!$, where $e = 2.718$, m is the mean (or average) number of times the event occurs in a given time, r is the number of times the event actually occurs in the given time, and $r!$ means $r(r-1)(r-2)\cdots 3 \cdot 2 \cdot 1$. If $r = 5$, then $r! = 5! = 5 \cdot 4 \cdot 3 \cdot 2 \cdot 1$.

Example 3 The mean number of accidents occuring yearly at a certain intersection is 5. What is the probability that 7 accidents will occur in a given year at this intersection?

Solution The equation is

$$P = \frac{e^{-m}m^r}{r!}$$
$$= \frac{e^{-5}5^7}{5{,}040} \quad m = 5 \text{ and } r = 7.$$
$$\log P = -5 \log e + 7 \log 5 - \log 5{,}040 \quad \text{Express in logarithmic form.}$$
$$= -5(0.4343) + 7(0.6990) - 3.7024 \quad \text{Log } e, \text{ or log } 2.718, \text{ is } 0.4343.$$
$$= -2.1715 + 4.8930 - 3.7024$$
$$= -0.9809$$

We must now add and subtract 10 to make the mantissa positive.

$$\log P = (-0.9809 + 10) - 10$$
$$= 9.0191 - 10$$
$$P = 0.104 = 10.4\%$$

There is only about a 10% chance that 7 accidents will occur in a given year at this intersection. $\blacksquare$

Example 4 The amount of radioactive material remaining at any time t of a particular radioactive substance is given by $y = y_0 e^{-0.92t}$, where y_0 is the original amount of material and t is measured in days. Find how long it will take for three-fourths of the material to decay (that is, one-fourth of it remains).

Solution

We want to know when $y = \frac{1}{4}y_0$. Substituting this into $y = y_0 e^{-0.92t}$, we get

$$\frac{1}{4} y_0 = y_0 e^{-0.92t}$$

$$\frac{1}{4} = e^{-0.92t} \qquad \text{Divide both sides by } y_0.$$

$$\log 1 - \log 4 = -0.92t \log e \qquad \text{Take base 10 logarithms of both sides.}$$

$$-\log 4 = -0.92t \log e \qquad \text{Recall } \log 1 = 0.$$

Now we solve for t:

$$t = \frac{-\log 4}{-0.92 \log e}$$

$$= \frac{0.6021}{0.92(0.4343)} \qquad \text{Log } e = \log 2.718 = 0.4343.$$

$$= \frac{0.6021}{0.3996}$$

$$= 1.507$$

Three-fourths of the material will decay in 1.507 days.

9–6 Exercises

All answers for this section have been calculated using logarithm tables and interpolation. All pH calculations are given to the nearest tenth and money is calculated to the nearest dollar.

1. Calculate the pH of a solution whose hydrogen ion concentration is 5.9×10^{-5}.

2. Calculate the pH of a solution whose hydrogen ion concentration is 7.4×10^{-9}.

3. Calculate the pH of a solution whose hydrogen ion concentration is 8.7×10^{-10}.

4. What is the hydrogen ion concentration for a solution whose pH is 8.3?

5. What is the hydrogen ion concentration for a solution whose pH is 3.6?

6. How much will $4,000 be worth at the end of 15 years if it is invested at 6% compounded semiannually?

7. How much will $10,000 be worth at the end of 25 years if it is invested at 8% compounded quarterly?

8. How much money must be invested at 6% compounded semiannually to yield $10,000 at the end of 8 years?

9. How much money must be invested at 10% compounded quarterly to yield $15,000 at the end of 10 years?

10. How many years will it take for a fixed amount of money to triple if it is invested at 8% compounded yearly?

11. If money is compounded continuously, the amount, A, after t years is $A = Pe^{rt}$, where r is the rate of interest and $e = 2.718$. Rework exercise 7 and see how compounding continuously compares to compounding quarterly for a 25-year-period.

For exercises 12–13, use the formula $P = e^{-m}m^t/r!$ (See example 3 for a model of this type of problem.)

12. Suppose the mean number of serious accidents in a certain construction industry is 4 accidents per year. Find the probability of 3 accidents per year.

13. Steve and Ross were so industrious at cutting lawns (see exercise 33, section 8.1) that they earned enough money to buy a car. However, they had an accident at the intersection of two busy streets. (It dented their pride and their insurance rate increased, but they were not hurt). The investigating officer told them this accident was the fifth one to occur there this year. If the average number of accidents at that intersection is 3 per year, find the probability of 5 accidents per year at this intersection.

14. The number of bacteria in a culture is given by the equation $y = 800e^{0.5t}$, where t is measured in hours. How many bacteria will be present at the end of 10 hours?

15. The population of a city is increasing according to the equation $y = 40,000e^{0.08t}$. Calculate each of the following.
  (a) The current population (this is done by letting $t = 0$).
  (b) The population at the end of 6 years.

16. The amount of radioactive material of a particular radioactive substance remaining at any time t is given by $y = y_0 e^{-0.10t}$. If y_0 (the initial amount) is 40 grams and t is measured in years, find the number of grams of material remaining after 3 years.

17. The amount of radioactive material left after a certain time t is given by $y = y_0 e^{-0.06t}$, where y_0 is the original amount and t is measured in years. Find the half-life of the substance. (The half-life is the time it takes for one-half of the material to decay).

18. We do not always remember all that we have learned, as any student can verify. Psychologists have been able to develop "forgetting curves," which depict the amount of knowledge retained after t months have elapsed from the time the knowledge was gained. Suppose the percent of the amount of knowledge, y, retained after t months for a certain subject is given by $y = 0.92e^{-0.12t}$. Find the percent of knowledge retained after 3 months.

19. If the rate of inflation in the United States is averaging 8% per year and if this rate continues indefinitely, how long will it take for average prices to double (and buying power to be cut in half)? Work this problem using $A = P(1 + r)^t$, where $A = 2P$ and $r = 0.08$.

20. A certain South American country had an average inflation rate of 178% for 1 year. If this rate of inflation continued indefinitely, how long would it take for average prices to double? *(See exercise 19.)*

Chapter 9 Summary

Vocabulary

antilogarithm
characteristic
common logarithm
exponential equation
exponential function

interpolation
logarithmic equation
logarithmic function
mantissa

Definitions

9–1 Exponential Function

If $b > 0$ and $b \neq 1$, then $f(x) = b^x$ is called the base b **exponential function**. Its domain is the set of real numbers.

9–2 Logarithm of a Number

If $b > 0$ and $b \neq 1$, then the **logarithm** of a number x is the exponent to which the base b must be raised to equal x. That is, $y = \log_b x$ if and only if $x = b^y$.

Rules

9–1 Properties of the Exponential Function

If $b > 0$ and if $b \neq 1$, then:

1. If $b^x = b^y$, then $x = y$ for any real numbers x and y.
2. If $x = y$, then $b^x = b^y$ for any real numbers x and y.

9–2 Two Properties Involving Logarithms

1. If $b, x > 0$, $b \neq 1$, then $b^{\log_b x} = x$.
2. If $b > 0$ and $b \neq 1$, then $\log_b b^x = x$.

9–3 Properties of Logarithms

If x, y, and b ($b \neq 1$) are positive real numbers and if z is any real number, then:

1. $\log_b xy = \log_b x + \log_b y$
2. $\log_b \dfrac{x}{y} = \log_b x - \log_b y$
3. $\log_b x^z = z \log_b x$
4. In part 3, if $z = \dfrac{1}{r}$, where r is a positive integer, then $\log_b \sqrt[r]{x} = \log_b x^{1/r} = \dfrac{1}{r} \log_b x$.

Exponential and Logarithmic Functions

9–4 Equality of Logarithms

If $x, y > 0$, $b > 0$, and $b \neq 1$, then $\log_b x = \log_b y$ if and only if $x = y$.

Review Problems

Section 9–1

Solve for x.

1. $5^{2x} = 125$
2. $3^{2x-1} = 27$
3. $10^{x-3} = 0.001$

Section 9–2

Write each equation in exponential form.

4. $\log_8 4 = \dfrac{2}{3}$
5. $\log_{1/2} 16 = -4$

Write each equation in logarithmic form.

6. $4^{1/2} = 2$
7. $49^{-1/2} = \dfrac{1}{7}$

Solve for the letter x, y, or b.

8. $\log_{36} \dfrac{1}{6} = y$
9. $\log_7 x = -2$
10. $\log_b 0.001 = -3$

Section 9–3

Use the rules of logarithms to expand each expression.

11. $\log_b (2.3)^{1/3} \sqrt{19.6}$
12. $\log_b \dfrac{(16.3)^4 \sqrt{2x}}{y}$
13. $\log_b \dfrac{\sqrt{x}\sqrt{y}}{z}$

Write each as a single logarithm.

14. $2 \log_b x + \log_b 5 - \log_b y$
15. $\dfrac{1}{3}(\log_b 10x - \log_b 2) - (\log_b 5 + \log_b x)$

Section 9–4

Find each logarithm by using tables and interpolation if necessary. (Use a calculator if one is permitted.)

16. log 983
17. log 0.05843

Find each antilogarithm by using the tables and interpolation if necessary. (Use a calculator if one is permitted.)

18. antilog 1.9528
19. antilog(8.4123 − 10)

Section 9–5

20. Use logarithms to calculate the value of $(5.8)^{2/3}(0.0143)^3$. (Check with your instructor to see if calculators may be used.)

21. Solve for m: $6^{2m+1} = 83.4$.

22. Solve for r: $\log(r + 3) + \log 2 = 3$.

Section 9–6

23. How much will $6,000 be worth if it is invested for 7 years at 8% compounded quarterly?

24. The average number of times that a machine on an assembly line is expected to break down per year is three times. Find the probability that it will break down four times, using $P = e^{-m}m^r/r!$.

Cumulative Test for Chapters 1–9

Chapter 1

1. The statement $5 \cdot \frac{1}{5} + 7 = 1 + 7$ illustrates which axiom?

2. Simplify: $\dfrac{-18 - 3^2 - [5(4^2 - 2)] - 3}{[2 - (-3)]^2 - 3(5)}$.

3. Simplify the expression $\dfrac{ab - 2a^3}{-3b}$ if $a = -4$ and $b = 5$.

Chapter 2

4. Simplify: $(-2x^3y^2)(-6x^4y^3)$.

5. Simplify: $\dfrac{(9x - 3)^2(7x - 2)^2}{(9x - 3)^4(7x - 3)^5}$.

6. Factor: $9x^2 + 27x^3$.

7. Factor: $4a^2 - 16$.

8. Factor: $8x^3 - 27$.

Chapter 3

9. Reduce: $\dfrac{2x^2 + 5x - 12}{10x^2 - 21x + 9}$.

10. Multiply: $\dfrac{5x}{3y} \cdot \dfrac{9y^2}{10x^3}$.

11. Divide: $\dfrac{6x^2 + 13x - 5}{8x + 20} \div \dfrac{3x^2 + 2x - 1}{16x^2 + 56x + 40}$.

12. Simplify: $\dfrac{x}{x - 6} - \dfrac{6}{x + 6} - \dfrac{72}{x^2 - 36}$.

13. Simplify: $\dfrac{1 - \dfrac{16}{k^2}}{1 + \dfrac{4}{k}}$.

14. Divide, using synthetic division: $(3x^3 + 5x - 2) \div (x + 2)$.

Chapter 4

15. Solve for x: $3(x - 2) + 5x = -2(x - 10) + 4$.

16. Solve for k: $-4(k - 5) < 2(k - 3) - 4$.

17. Solve for v: $|3v - 4| = 17$.

18. Solve for d: $|-4 + d| < 7$.

19. Solve for n: $\dfrac{7}{n - 4} - \dfrac{8}{n + 5} = \dfrac{3}{n - 4}$.

20. Solve $\dfrac{1}{a} = \dfrac{2}{b} + \dfrac{3}{c}$ for b.

21. The length of a rectangle is 1 centimeter more than three times the width. If the length is decreased by 3 centimeters and the width increased by 6 centimeters, then the length will be the same as the width. Find the length of the rectangle.

22. In a famous automobile race, Zip Adoodi finished the race 48 minutes ahead of Slo Isme. If Zip's average speed was 150 miles per hour and Slo's average speed was 125 miles per hour, what was the time for each, and what was the distance of the race?

Chapter 5

23. Simplify: $2^{-2} + 3^{-1}$.

24. Simplify: Leave all exponents positive. $\left(\dfrac{a^{-5}b^4}{a^2b^{-2}}\right)^3$.

25. Write the number 0.00093 in scientific notation.

26. If $x > 0$, simplify $\dfrac{x^{5/6}}{x^{1/3}}$. Leave your answer in simplified radical form.

27. Simplify $(u^{1/6} - u^{1/2})^2$. Assume $u > 0$. Leave the answer in reduced exponential form.

28. Rationalize the denominator and simplify: $\dfrac{2\sqrt{3} + 2}{\sqrt{6} + \sqrt{2}}$.

29. Solve for k: $\sqrt{5k + 5} - \sqrt{8 - k} = 0$.

Chapter 6

30. Solve for x by completing the square: $2x^2 + 5x - 7 = 0$.

31. Solve for x by using the quadratic formula: $x^2 = 2x - 4$.

32. The sum of two positive numbers is 7. Find the numbers if twice the square of the larger number is 5 more than three times the square of the smaller number.

33. Find the solution and draw the graph of the solution set for $x^2 < -x + 30$.

Chapter 7

34. Is the following set of ordered pairs a function?
$$\{(4, 9), (-2, 3), (6, 6), (5, 3), (4, 2)\}$$

35. Without drawing the graph, give the y-intercept of the line represented by the equation $y + 6x - 4 = 0$.

36. Write an equation of the line going through the two points with coordinates (4, 0) and (0, -6). Leave your answer in the form $y = mx + b$.

37. Draw the graph of $3x - 2y < 6$.

Chapter 8

38. Does the equation $3x^2 - 4y^2 = 5$ represent an ellipse, a hyperbola, or neither?

39. What is the radius of the circle represented by $x^2 + 10x + y^2 - 6y - 6 = 0$?

40. A board 20 feet long is cut into two pieces that have a ratio of 2:3. How long is each piece?

41. If $f(x) = 3x - 2$, what is the value of $f^{-1}(7)$?

Chapter 9

42. Solve for x: $2^{2x-1} = 32$.

43. What is the equivalent exponential form of $\log_9 3 = \frac{1}{2}$?

44. Solve for x: $\log_6 x = -2$.

45. Use the rules of logarithms to expand $\log \dfrac{x^2\sqrt{y}}{2z}$ fully.

46. If $\log_{10} 2 = 0.3010$ and $\log_{10} 7 = 0.8451$, find $\log_{10} 1,400$ without using a calculator or tables.

47. Solve $5^{3n+1} = 8$ for n. Check with your instructor to see if calculators may be used. Keep all numbers, including logarithms, rounded to four decimal places.

48. How much will $5,000 be worth at the end of 3 years if it is invested at 9%, compounded yearly? Use $A = P(1 + r)^t$.

CHAPTER 10 | Systems of Equations

Objectives of Chapter 10

To be able to:

- ☐ Define a system of equations. (10–1)
- ☐ Recognize if a linear system of equations is inconsistent or dependent. (10–1)
- ☐ Solve a linear system of two equations in two variables using the addition method. (10–1)
- ☐ Solve a linear system of two equations in two variables using the substitution method. (10–1)
- ☐ Solve a linear system of three equations in three variables. (10–2)
- ☐ Define a determinant. (10–3)
- ☐ Evaluate a second-order determinant. (10–3)
- ☐ Use Cramer's rule to solve a linear system of two equations. (10–3)
- ☐ Evaluate third- and fourth-order determinants. (10–4)
- ☐ Use minors to evaluate determinants. (10–4)
- ☐ Use determinants to solve a linear system of three or four equations. (10–5)
- ☐ Draw the graphs of quadratic systems of equations. (10–6)
- ☐ Solve a system of quadratic equations. (10–6)
- ☐ Solve word problems using a system of linear equations. (10–7)

We found the solution set for an equation in one variable in chapters 4 and 6. In this chapter we learn how to find the solution set for a system of equations in two or more variables. A **system of equations** can be thought of as a set of two or more equations in two or more variables.

10–1 Linear Systems of Two Equations

A linear system of two equations in the variables x and y has the form

$$\begin{cases} a_1x + b_1y = c_1 \\ a_2x + b_2y = c_2 \end{cases}$$

where a_1 and b_1 are not both equal to 0 and a_2 and b_2 are not both equal to 0. The solution set for such a system consists of those ordered pairs, if any, that make both equations true.

The graph of the system just given is two straight lines. These two lines could be related in three ways: The graphs could intersect in one and only one point, the graphs could be parallel, or the graphs could coincide. These three situations are pictured in figure 10–1.

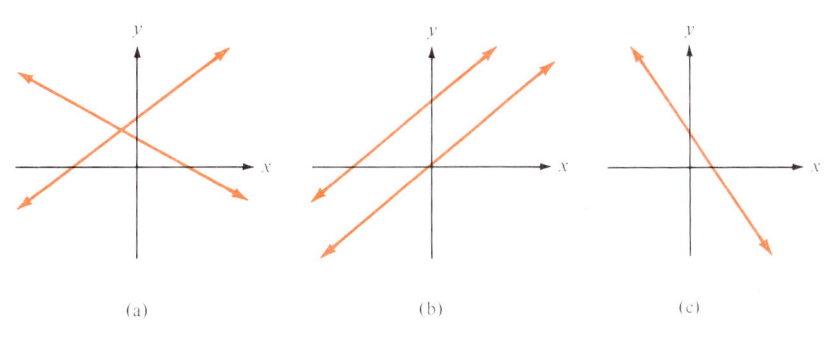

Figure 10–1

The two equations whose graphs are shown in figure 10–1(a) intersect in exactly one point. The coordinates of this point determine the solution for the system.

The two equations whose graphs are shown in figure 10–1(b) are said to be **inconsistent,** and the solution set for the system is empty.

The two equations whose graphs are shown in figure 11–1(c) are said to be **dependent,** and the solution set for either equation is the solution for the system.

One means of solving a system of linear equations uses the **addition method,** or elimination method. In this method, the two equations are added together in a manner that will eliminate one of the variables. One equation containing one variable is left, and we then solve for that variable. Examples 1, 2, and 3 illustrate this procedure.

Example 1

Find the solution set for

$$\begin{cases} 2x + y = 3 & (1) \\ x - 3y = -9 & (2) \end{cases}$$

Solution

We must add the two equations together in a manner that eliminates one of the variables. If we multiply equation (1) by 3 and add it to equation (2), we eliminate the y-variable. This technique is shown in the following table.

Multiplier	Original Equation	Resulting Equation
3	$2x + y = 3$	$6x + 3y = 9$
1	$x - 3y = -9$	$x - 3y = -9$

We now add the resulting equations.

$$6x + 3y = 9$$
$$\underline{x - 3y = -9}$$
$$7x \quad\quad = 0$$
$$x \quad\quad = 0$$

We find the value of y to be 3 by substituting $x = 0$ into equation (1) or equation (2). The solution set is $\{(0, 3)\}$. ■

Example 2

Find the solution set for

$$\begin{cases} 2x - 3y = 5 & (3) \\ 6x - 9y = -2 & (4) \end{cases}$$

Solution

Multiply (3) by -3 and add this result to (4).

Multiplier	Original Equation	Resulting Equation
-3	$2x - 3y = 5$	$-6x + 9y = -15$
1	$6x - 9y = -2$	$6x - 9y = -2$

Add the resulting equations.

$$-6x + 9y = -15$$
$$\underline{6x - 9y = -2}$$
$$0 + 0 = -17$$

Systems of Equations

Both variables cancel and we are left with the false statement $0 = -17$. This means the equations are inconsistent and their graphs are parallel. The solution set is $\emptyset$. ■

Example 3 Find the solution set for
$$\begin{cases} x + 4y = -3 & (5) \\ 3x + 12y = -9 & (6) \end{cases}$$

Solution Multiply equation (5) by -3 and add it to equation (6) to eliminate x.

Multiplier	Original Equation	Resulting Equation
-3	$x + 4y = -3$	$-3x - 12y = 9$
1	$3x + 12y = -9$	$3x + 12y = -9$

Add the resulting equations.
$$-3x - 12y = 9$$
$$\underline{3x + 12y = -9}$$
$$0 + 0 = 0$$

In this example, not only do the variable terms disappear, but the constant terms have a sum of zero also. We are left with the true statement $0 = 0$. This means the equations are dependent and the graphs coincide. The solution set is the set of ordered pairs which makes either equation (5) or equation (6) true. ■

Equations that are inconsistent or dependent can be recognized by observing the following.

$$\frac{a_1}{a_2} = \frac{b_1}{b_2} \neq \frac{c_1}{c_2} \qquad \text{for inconsistent equations}$$

$$\frac{a_1}{a_2} = \frac{b_1}{b_2} = \frac{c_1}{c_2} \qquad \text{for dependent equations}$$

Example 4 Examine the following systems of equations and determine if they are inconsistent, dependent, or neither.

(a) $\begin{cases} 3x - 2y = 4 \\ 12x - 8y = -7 \end{cases}$ (b) $\begin{cases} x - 8y = 4 \\ 5x - 40y = 20 \end{cases}$ (c) $\begin{cases} 3x + 2y = 8 \\ 5x - 3y = 4 \end{cases}$

Solution (a) $\dfrac{3}{12} = \dfrac{-2}{-8} \neq \dfrac{4}{-7}$ (b) $\dfrac{1}{5} = \dfrac{-8}{-40} = \dfrac{4}{20}$ (c) $\dfrac{3}{5} \neq \dfrac{2}{-3} \neq \dfrac{8}{4}$

or	or	or
$\dfrac{1}{4} = \dfrac{1}{4} \neq \dfrac{4}{-7}$	$\dfrac{1}{5} = \dfrac{1}{5} = \dfrac{1}{5}$	$\dfrac{3}{5} \neq \dfrac{2}{-3} \neq \dfrac{2}{1}$
This system is inconsistent.	This system is dependent.	This system is neither inconsistent nor dependent.

A second method for solving a system of equations uses the **substitution method.** It is based on the substitution axiom given in section 1–4. One of the equations is solved for one of its variables, and this result is substituted into the other equation.

Example 5 Find the solution set for

$$\begin{cases} 3x - y = 7 & (7) \\ 4x + y = 14 & (8) \end{cases}$$

using the substitution method.

Solution Solve equation (8) for y and substitute this result into equation (7).

$$4x + y = 14$$
$$y = 14 - 4x$$
$$3x - (14 - 4x) = 7 \qquad \text{Substitute into equation (7).}$$
$$3x - 14 + 4x = 7 \qquad \text{Collect like terms and solve for } x.$$
$$7x = 21$$
$$x = 3$$

Substitute $x = 3$ into either of equations (7) or (8) to find $y = 2$. The solution set is $\{(3, 2)\}$.

Example 6 Find the solution set for

$$\dfrac{1}{x} + \dfrac{1}{y} = 1 \qquad x, y \neq 0 \qquad (9)$$

$$\dfrac{4}{x} - \dfrac{1}{y} = 9 \qquad x, y \neq 0 \qquad (10)$$

Solution

The fractional term $1/y$ can be eliminated from equations (9) and (10) by adding the two equations. Doing this, we get

$$\frac{1}{x} + \frac{1}{y} = 1$$

$$\frac{4}{x} - \frac{1}{y} = 9$$

$$\frac{5}{x} = 10 \qquad \text{Solve for } x.$$

$$x = \frac{1}{2}$$

Substitute $x = \frac{1}{2}$ into equation (9) to find y.

$$\frac{1}{1/2} + \frac{1}{y} = 1$$

$$2 + \frac{1}{y} = 1$$

$$\frac{1}{y} = -1$$

$$y = -1$$

The solution set is $\{(\frac{1}{2}, -1)\}$. ∎

Example 7

The sum of two numbers is sixteen. One number is two more than the other number. Find the two numbers.

Solution

Let x and y represent the two numbers, and set up a system of two equations in x and y. From the first sentence of the example, we see

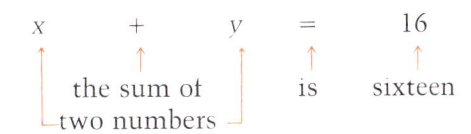

$$\underbrace{x + y}_{\text{the sum of two numbers}} \underbrace{=}_{\text{is}} \underbrace{16}_{\text{sixteen}}$$

The second sentence in the example yields a second equation.

$$\underbrace{x}_{\substack{\text{one} \\ \text{number}}} \underbrace{=}_{\text{is}} \underbrace{2}_{\text{two}} \underbrace{+}_{\text{more}} \underbrace{y}_{\substack{\text{than the} \\ \text{other number}}}$$

The system of equations is

$$\begin{cases} x + y = 16 \\ x - y = 2 \end{cases}$$

The addition method can be used to show that $x = 9$ and $y = 7$. ∎

10–1 Exercises

In exercises 1–6, determine whether each system of equations is dependent, inconsistent, or has exactly one solution. *(See example 4.)*

1. $\begin{cases} x - 3y = 4 \\ 5x - 15y = -3 \end{cases}$

2. $\begin{cases} x + 2y = -6 \\ 5x - y = 3 \end{cases}$

3. $\begin{cases} 3x - y = 4 \\ x + 7y = 2 \end{cases}$

4. $\begin{cases} 2x - 3y = 1 \\ 4x - 6y = 2 \end{cases}$

5. $\begin{cases} 3x - y = 2 \\ 6x - 2y = 4 \end{cases}$

6. $\begin{cases} 2x + y = 4 \\ 2x + y = -8 \end{cases}$

In exercises 7–24, find the solution set for each system of equations by the addition method. *(See examples 1–3.)*

7. $\begin{cases} 2x + y = 4 \\ x - y = 5 \end{cases}$

8. $\begin{cases} x - 4y = 10 \\ -x + y = -4 \end{cases}$

9. $\begin{cases} 3x - 2y = 2 \\ x + 3y = -3 \end{cases}$

10. $\begin{cases} 2x + 5y = 2 \\ 5x - y = 5 \end{cases}$

11. $\begin{cases} 5x - 2y = 7 \\ 3x + 4y = 25 \end{cases}$

12. $\begin{cases} 2x + 3y = 3 \\ 5x + 2y = -9 \end{cases}$

13. $\begin{cases} 5x + 3y = 4 \\ 3x - 2y = 10 \end{cases}$

14. $\begin{cases} 7x - 2y = 7 \\ 3x + 5y = 3 \end{cases}$

15. $\begin{cases} 6x + 4y = 3 \\ 3x + 2y = -5 \end{cases}$

16. $\begin{cases} x - 4y = -8 \\ 5x - 20y = 3 \end{cases}$

17. $\begin{cases} \dfrac{x}{7} - \dfrac{3}{7}y = \dfrac{4}{7} \\ x - \dfrac{1}{2}y = \dfrac{23}{2} \end{cases}$

(*Hint:* Clear both equations of fractions. Then write each equation in the form $Ax + By = C$.)

18. $\begin{cases} \dfrac{2}{3}x - y = \dfrac{5}{3} \\ \dfrac{x}{5} - \dfrac{4}{5}y = -1 \end{cases}$

19. $\begin{cases} x = \dfrac{y}{12} + \dfrac{1}{3} \\ x + \dfrac{5}{2}y = 52 \end{cases}$

20. $\begin{cases} \dfrac{3}{2}x = \dfrac{1}{2}y - \dfrac{1}{4} \\ \dfrac{2}{5}x + \dfrac{3}{5}y = \dfrac{5}{2} \end{cases}$

21. $\begin{cases} 2(2x - 3y) - 3(x - y) = -1 \\ -2(3x + y) + 3(x + y) = 3 \end{cases}$

22. $\begin{cases} 5(x - y) + 2(2x - y) = 3 \\ -3(2x - y) - 2(2x + y) = 17 \end{cases}$

23. $\begin{cases} \dfrac{2x - y}{3} - \dfrac{x + 2y}{2} = \dfrac{x - 3}{2} \\ \dfrac{x - y}{4} + \dfrac{2x + y}{3} = \dfrac{2x - 1}{3} \end{cases}$

24. $\begin{cases} \dfrac{x - y}{2} + \dfrac{x}{2} = \dfrac{3}{4} \\ \dfrac{x + y}{5} - y = \dfrac{x - 9}{2} \end{cases}$

In exercises 25–32, find the solution set for each system of equations by using the substitution method. (See example 5.)

25. $\begin{cases} x - 2y = 3 \\ x + y = -3 \end{cases}$

26. $\begin{cases} x - y = 3 \\ 2x - y = 5 \end{cases}$

27. $\begin{cases} x - 3y = 4 \\ 2x + y = 1 \end{cases}$

28. $\begin{cases} 5x - 3y = 2 \\ 2x - y = 3 \end{cases}$

29. $\begin{cases} y = \dfrac{1}{2}x + 3 \\ 3x + 2y = -2 \end{cases}$

30. $\begin{cases} x = \dfrac{-3}{2}y + 4 \\ 4x + 3y = -8 \end{cases}$

31. $\begin{cases} 2x - 3y = -15 \\ 4x + 5y = 14 \end{cases}$

32. $\begin{cases} 8x - 7y = 28 \\ 5x + 2y = -8 \end{cases}$

In exercises 33–37, solve each system of equations for x and y. (See example 6.)

33. $\begin{cases} \dfrac{1}{x} + \dfrac{3}{y} = \dfrac{1}{2} \\ \dfrac{1}{x} - \dfrac{2}{y} = 0 \end{cases}$

34. $\begin{cases} \dfrac{2}{x} - \dfrac{3}{y} = \dfrac{10}{3} \\ \dfrac{-3}{x} + \dfrac{3}{y} = \dfrac{-7}{2} \end{cases}$

35. $\begin{cases} \dfrac{4}{x} - \dfrac{3}{y} = \dfrac{3}{2} \\ \dfrac{8}{x} + \dfrac{6}{y} = 5 \end{cases}$

36. $\begin{cases} \dfrac{3}{2x} - \dfrac{5}{3y} = \dfrac{7}{3} \\ \dfrac{2}{5x} + \dfrac{1}{2y} = 1 \end{cases}$ (*Hint:* Multiply the first equation by 6 and the second equation by 10. Then eliminate one of the variables.)

37. $\begin{cases} \dfrac{5}{3x} - \dfrac{2}{5y} = 1 \\ \dfrac{1}{2x} + \dfrac{5}{3y} = \dfrac{-4}{9} \end{cases}$ (See exercise 36.)

38. Six years ago Julie was one-fifth as old as her mother. Her mother is now three times as old as Julie. How old was each 6 years ago?

39. A man has part of $8,500 invested at 6% and the remainder invested at 8%. If his annual income from the 8% investment exceeds the annual income from the 6% investment by $190, how much is invested at each rate?

40. If A were to give B $100, B would have the same amount of money as A. If B were to give A $120, A would have 12 times as much money as B. How much does each have?

41. A turtle and a hare had a race. The turtle was given a head start of $9\tfrac{7}{12}$ miles. If the hare traveled at an average rate that was 24 times as great as the average rate for the turtle, which one won the race if the race was to cover a distance of 10 miles and the hare overtook the turtle 50 minutes after the hare began the race? (Do not assume the hare won the race just because it overtook the turtle after 50 minutes. The hare could stop for a carrot and possibly still lose the race.)

42. Two boards of unequal length placed end to end along a straight line will reach 32 feet. Half the length of the longer board is 1 foot more than the length of the shorter board. How long is each board?

43. A certain tank has a capacity of 2,000 liters. Water is pumped into the tank by means of two pumps. The first pump can pump water at a rate of 20 liters per minute and the second can pump at a rate of 40 liters per minute. The first pump operates 10 minutes before the second pump is started. How many liters of water are pumped into the tank by each pump?

44. Bobbie can row 20 miles down the Feather River and back in 18 hours. She can row 5 miles downstream in the same time it takes to row 1 mile upstream. Find the time she spent rowing down the river and the time she spent rowing up the river on the 40-mile round trip.

45. If the numerator of a fraction is tripled and the denominator diminished by 4, its value is $\frac{3}{4}$. If its numerator is doubled and the denominator increased by 6, its value is $\frac{1}{3}$. Find the value of the fraction.

10-2 Linear Systems of Three Equations in Three Variables

A linear equation in three variables is one in which each variable is raised to the first power. A linear equation in three variables has the form

$$ax + by + cz = k$$

where a, b, c, and k are real numbers. The solution set for a linear equation consists of all those ordered triples (x, y, z) whose coordinates will satisfy the equation.

The equation $2x + y + z = 6$ is a true statement for the ordered triples $(0, 1, 5)$, $(-2, 0, 10)$, and $(2, 1, 1)$. The solution set is infinite and we cannot list all its members.

The solution for the system

$$\begin{cases} x + y + z = 5 \\ 2x - y = 6 \\ x - 2y + z = 5 \end{cases}$$

is the ordered triple $(3, 0, 2)$, as direct substitution into the system will show.

Example 1 shows how to solve a system of three equations in three variables.

Example 1

Find the solution set for the system

$$\begin{cases} 2x + 3y - z = 4 & (1) \\ 3x - 2y + z = 6 & (2) \\ 5x - y + 2z = 58 & (3) \end{cases}$$

Solution

We must eliminate one of the variables between two of the equations. Then, we must eliminate that same variable between the remaining equation and one of the pair used in the first elimination. Let us begin by eliminating z from equations (1) and (2) by adding these equations.

$$\begin{aligned} 2x + 3y - z &= 4 \\ 3x - 2y + z &= 6 \\ \hline 5x + y &= 10 \end{aligned} \quad (4)$$

Now eliminate z between equations (1) and (3) by multiplying equation (1) by 2 and adding this result to equation (3).

$$\begin{aligned} 4x + 6y - 2z &= 8 \\ 5x - y + 2z &= 58 \\ \hline 9x + 5y &= 66 \end{aligned} \quad (5)$$

Equations (4) and (5) now form a system of two equations in two variables. The addition method can be used to solve this system. Multiply equation (4) by -5 and add to equation (5) to eliminate the y-variable.

$$\begin{aligned} -25x - 5y &= -50 \\ 9x + 5y &= 66 \\ \hline -16x &= 16 \\ x &= -1 \end{aligned}$$

Substitute $x = -1$ into equation (4) to find $y = 15$. Substitute $x = -1$ and $y = 15$ into any one of equations (1), (2), or (3) to find $z = 39$. The solution set is $\{(-1, 15, 39)\}$.

A system of three equations in three variables may be dependent or inconsistent, just as a system of two equations in two variables may be dependent or inconsistent. Consider the following system.

$$\begin{cases} 3x + 2y + 2z = 3 & (6) \\ 4x - 3y + z = 2 & (7) \\ 10x + y + 5z = 8 & (8) \end{cases}$$

This system appears to be in good order for a unique solution. However, if we multiply equation (7) by -2 and add it to equation (6) to eliminate z, we get

Intermediate Algebra

$$-5x + 8y = -1 \qquad (9)$$

Now multiply equation (7) by -5 and add it to equation (8) to eliminate z. We get

$$-10x + 16y = -2 \qquad (10)$$

Because equation (10) is a multiple of equation (9), we conclude that the system is dependent and has an infinite number of solutions.

Example 2 A company paid $163 for 19 rolls of three different kinds of stamps. One kind cost $2.00 per roll, another kind cost $8.00 per roll, and the third kind cost $15.00 per roll. The number of $15.00 rolls exceeds the number of $2.00 rolls by 1. How many rolls of each kind were bought?

Solution We must analyze the problem for information that will give us three equations in three variables. Let

$$T = \text{the number of \$2.00 rolls bought}$$
$$E = \text{the number of \$8.00 rolls bought}$$
$$F = \text{the number of \$15.00 rolls bought}$$

The sum of the number of rolls is 19. This gives one equation.

$$T + E + F = 19$$

the total number of rolls bought is 19

We can write another equation based on the total cost of the stamps. If the number of rolls of each kind of stamp is multiplied by its respective cost, then the sum of these costs equals $163.

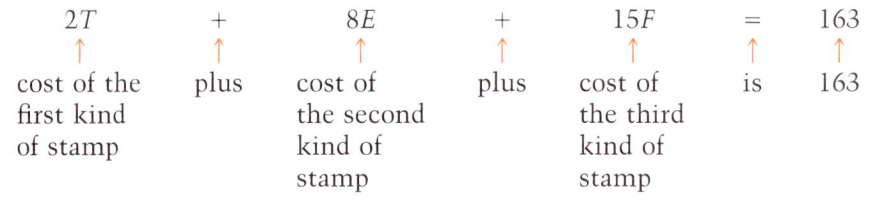

$$2T + 8E + 15F = 163$$

cost of the first kind of stamp plus cost of the second kind of stamp plus cost of the third kind of stamp is 163

The third equation can be obtained from the fact that the number of $15.00 rolls exceeds the number of $2.00 rolls by 1.

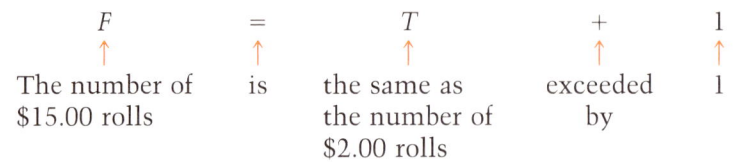

$$F = T + 1$$

The number of $15.00 rolls is the same as the number of $2.00 rolls exceeded by 1

These equations, with the last one rearranged, give the system

Systems of Equations

$$T + E + F = 19 \quad (11)$$
$$2T + 8E + 15F = 163 \quad (12)$$
$$-T + F = 1 \quad (13)$$

We add equations (11) and (13) to eliminate T.

$$E + 2F = 20 \quad (14)$$

Multiply equation (13) by 2 and add it to equation (12) to eliminate T.

$$2T + 8E + 15F = 163$$
$$\underline{-2T + 2F = 2}$$
$$8E + 17F = 165 \quad (15)$$

Equations (14) and (15) now form the system

$$\begin{cases} E + 2F = 20 & (16) \\ 8E + 17F = 165 & (17) \end{cases}$$

Multiply equation (16) by -8 and add it to equation (17) to eliminate E.

$$-8E - 16F = -160$$
$$\underline{8E + 17F = 165}$$
$$F = 5$$

From equation (16), for $F = 5$, then $E = 10$. From equation (13), for $F = 5$, then $T = 4$. The company bought 5 rolls of stamps that cost $15 per roll, 10 rolls that cost $8 per roll, and 4 rolls that cost $2 per roll. ∎

10–2 Exercises

In exercises 1–14, solve each system of equations. *(See example 1.)*

1. $\begin{cases} x + y - z = -6 \\ 2x - 3y + z = 8 \\ x + y + 2z = 9 \end{cases}$

2. $\begin{cases} x + y + z = 6 \\ x - y + z = 2 \\ x + y - z = -4 \end{cases}$

3. $\begin{cases} 2x - 3y + 4z = -1 \\ x + y - 2z = 1 \\ 3x - y + 5z = 19 \end{cases}$

4. $\begin{cases} 3x - 2y + z = -4 \\ 5x - 4y - 2z = -16 \\ -4x + y - 3z = 12 \end{cases}$

5. $\begin{cases} 2x + z = 3 \\ y - 2z = -2 \\ x - y = 6 \end{cases}$

6. $\begin{cases} 3x + 2y = 3 \\ y - z = -2 \\ -5x + z = 10 \end{cases}$

7. $\begin{cases} 2x - 3y + z = -1 \\ -4x + y - 5z = 5 \\ 6x - 9y + 3z = -3 \end{cases}$

8. $\begin{cases} 9x - 5y + 7z = 8 \\ x + 7y - 4z = 10 \\ 2x + 14y - 8z = 3 \end{cases}$

9. $\begin{cases} 5x + 3y = -6 \\ 7y + 2z = -18 \\ 9x - 7z = 14 \end{cases}$
10. $\begin{cases} 6x - 5z = 8 \\ 5x - 3y = -7 \\ 4y - 2z = 4 \end{cases}$

11. $\begin{cases} -2x + 5y + 3z = 5 \\ 3x - 2y + 2z = -2 \\ 5x + 3y - 5z = 3 \end{cases}$
12. $\begin{cases} 6x - y + 5z = 4 \\ x + 3y - 4z = 15 \\ 3x + 4y - z = 19 \end{cases}$

13. $\begin{cases} x + \dfrac{3}{2}y - \dfrac{z}{2} = 10 \\ \dfrac{x}{3} + \dfrac{4}{3}y + \dfrac{z}{3} = 1 \\ x - \dfrac{y}{3} + \dfrac{2}{3}z = -3 \end{cases}$
14. $\begin{cases} \dfrac{5}{3}x - \dfrac{1}{3}y - \dfrac{1}{3}z = 1 \\ -\dfrac{1}{2}x + \dfrac{3}{4}y + \dfrac{3}{4}z = 1 \\ x + \dfrac{4}{7}y - \dfrac{1}{7}z = -\dfrac{5}{7} \end{cases}$

15. A collection of nickels, dimes, and quarters has a value of $1.85. There is a total of 18 coins in the collection, and the number of nickels is one more than the number of dimes. How many of each kind of coin is in the collection? (See example 2.)

16. A half-gallon of milk, a dozen eggs, and a pound of cheese cost $3.34. The cost of the cheese was 24¢ more than the combined cost of the milk and eggs. The milk cost 9¢ less than the eggs. Find the cost of each.

17. A truckload of scrap metal consisting of aluminum, iron, and tin weighed 2,260 pounds, and the entire truckload sold for $182.75. The iron sold for 8¢ per pound, the aluminum for 11¢ per pound, and the tin for 6¢ per pound. The number of pounds of aluminum was 11 pounds less than twice the number of pounds of tin. How many pounds of each kind of metal were on the truck?

18. Three snails ran (crawled?) a relay of 84 centimeters. The first snail traveled a distance that was 4 centimeters less than the distance traveled by the second snail. The second snail traveled a distance that was 4 centimeters less than the distance traveled by the third snail. Find the distance each snail crawled. How long did it take to run the relay if each snail crawled its part of the relay with an average speed of 2.1 centimeters per second?

10–3 Determinants

Systems of equations can be solved using determinants. A **determinant** is a number represented by a square array of numbers with two or more rows and two or more columns. A determinant with two rows and two columns of numbers is called a *second-order determinant*. If a determinant has three rows and three columns of numbers, it is called a *third-order determinant*; in general, if it has n rows and n columns of numbers, it is called an *nth-order determinant*. A second-order determinant is given by the symbol

Systems of Equations

$$\begin{vmatrix} a_1 & b_1 \\ a_2 & b_2 \end{vmatrix}$$

where a_1, a_2, b_1, and b_2 represent real numbers.

The procedure for evaluating a second-order determinant is given in definition 10–1.

Definition 10–1

Second-Order Determinant

$$\begin{vmatrix} a_1 & b_1 \\ a_2 & b_2 \end{vmatrix} = a_1 b_2 - a_2 b_1$$

The multiplication shown in definition 10–1 is often depicted by diagonal lines going through a_1 and b_2 and through a_2 and b_1.

$$\begin{vmatrix} a_1 & b_1 \\ a_2 & b_2 \end{vmatrix}$$

The diagonal marked ① is called the principal diagonal and the one marked ② is called the secondary diagonal. The product of the numbers along diagonal ② is subtracted from the product of the numbers along diagonal ①.

Example 1

Evaluate $\begin{vmatrix} 3 & 4 \\ -2 & 3 \end{vmatrix}$.

Solution

$\begin{vmatrix} 3 & 4 \\ -2 & 3 \end{vmatrix} = 3(3) - (-2)(4) = 9 - (-8) = 9 + 8 = 17$ ■

We now want to show how to use determinants to solve systems of equations that take the form

$$\begin{cases} a_1 x + b_1 y = c_1 & (1) \\ a_2 x + b_2 y = c_2 & (2) \end{cases}$$

We solve the system of equations (1) and (2) for y by the addition method.

Multiplier	Original Equation	Resulting Equation
$-a_2$	$a_1 x + b_1 y = c_1$	$-a_1 a_2 x - a_2 b_1 y = -a_2 c_1$
a_1	$a_2 x + b_2 y = c_2$	$a_1 a_2 x + a_1 b_2 y = a_1 c_2$

Now add the resulting equations.

$$-a_1a_2x - a_2b_1y = -a_2c_1$$
$$a_1a_2x + a_1b_2y = a_1c_2$$

$$a_1b_2y - a_2b_1y = a_1c_2 - a_2c_1$$
$$y(a_1b_2 - a_2b_1) = a_1c_2 - a_2c_1$$
$$y = \frac{a_1c_2 - a_2c_1}{a_1b_2 - a_2b_1}$$

This value of y may be written using determinants as

$$y = \frac{\begin{vmatrix} a_1 & c_1 \\ a_2 & c_2 \end{vmatrix}}{\begin{vmatrix} a_1 & b_1 \\ a_2 & b_2 \end{vmatrix}}$$

By using a manner similar to that used in solving for y, we can show

$$x = \frac{\begin{vmatrix} c_1 & b_1 \\ c_2 & b_2 \end{vmatrix}}{\begin{vmatrix} a_1 & b_1 \\ a_2 & b_2 \end{vmatrix}}$$

This method of solving a system of two linear equations is known as **Cramer's rule**.

Rule 10–1

Cramer's Rule

The two linear equations

$$\begin{cases} a_1x + b_1y = c_1 \\ a_2x + b_2y = c_2 \end{cases}$$

have the solution

$$x = \frac{\begin{vmatrix} c_1 & b_1 \\ c_2 & b_2 \end{vmatrix}}{\begin{vmatrix} a_1 & b_1 \\ a_2 & b_2 \end{vmatrix}} \quad \text{and} \quad y = \frac{\begin{vmatrix} a_1 & c_1 \\ a_2 & c_2 \end{vmatrix}}{\begin{vmatrix} a_1 & b_1 \\ a_2 & b_2 \end{vmatrix}}$$

if $a_1b_2 - a_2b_1 \neq 0$.

Notice in rule 10–1 that the determinants in the denominator for both x and y are the same. This determinant is called the *determinant of the coefficients* and is denoted by D. Its columns are the columns of coefficients for x and y. In the numerators for both x and y, the column of coefficients of the letter for which we are solving is replaced by the column of constants. The numerator for x is denoted by D_x and the numerator for y by D_y.

Example 2

Use determinants to find the solution set for

$$\begin{cases} 2x - 3y = 7 \\ x + 4y = -2 \end{cases}$$

Solution

$$x = \frac{D_x}{D} = \frac{\begin{vmatrix} 7 & -3 \\ -2 & 4 \end{vmatrix}}{\begin{vmatrix} 2 & -3 \\ 1 & 4 \end{vmatrix}} = \frac{7(4) - (-2)(-3)}{2(4) - (1)(-3)} = \frac{22}{11} = 2$$

$$y = \frac{D_y}{D} = \frac{\begin{vmatrix} 2 & 7 \\ 1 & -2 \end{vmatrix}}{\begin{vmatrix} 2 & -3 \\ 1 & 4 \end{vmatrix}} = \frac{2(-2) - 7(1)}{11} = \frac{-11}{11} = -1$$

The solution set is $\{(2, -1)\}$. ■

If $D = 0$ and either $D_x \neq 0$ or $D_y \neq 0$, then the graphs represented by the equations are parallel. The system is inconsistent and has no solution. If $D = 0$ and D_x and D_y are also 0, then the graphs represented by the equations coincide. The system is dependent and has an infinite number of solutions.

10–3 Exercises

In exercises 1–10, find the value of each second-order determinant. *(See example 1.)*

1. $\begin{vmatrix} 3 & 5 \\ 4 & 2 \end{vmatrix}$

2. $\begin{vmatrix} 8 & 3 \\ 2 & 4 \end{vmatrix}$

3. $\begin{vmatrix} 5 & 2 \\ -1 & -3 \end{vmatrix}$

4. $\begin{vmatrix} 7 & -3 \\ -2 & 4 \end{vmatrix}$

5. $\begin{vmatrix} \frac{3}{2} & \frac{2}{3} \\ -3 & 4 \end{vmatrix}$

6. $\begin{vmatrix} \frac{2}{9} & -21 \\ \frac{3}{7} & 18 \end{vmatrix}$

7. $\begin{vmatrix} 7 & 3a \\ 9 & 2a \end{vmatrix}$

8. $\begin{vmatrix} 3 & 4 \\ -x & 2x \end{vmatrix}$

9. $\begin{vmatrix} a_1 & b_1 \\ a_2 & b_2 \end{vmatrix}$

10. $\begin{vmatrix} a_1 & c_1 \\ a_2 & c_2 \end{vmatrix}$

In exercises 11–22, find the solution set for each system of equations by using determinants (Cramer's rule).

11. $\begin{cases} 2x - 3y = -1 \\ 3x - 2y = 1 \end{cases}$

12. $\begin{cases} 3x - 4y = -4 \\ 8x - 3y = -3 \end{cases}$

13. $\begin{cases} 2x + 5y = -11 \\ -x + 6y = -20 \end{cases}$

14. $\begin{cases} 8x - 3y = 1 \\ 7x + 4y = 34 \end{cases}$

15. $\begin{cases} 5x - 4y - 11 = 0 \\ 3x + 5y - 14 = 0 \end{cases}$
16. $\begin{cases} 6x + 5y - 16 = 0 \\ 2x - 7y + 12 = 0 \end{cases}$

17. $\begin{cases} 8x + 5y = 0 \\ 2x = 3y + 34 \end{cases}$
18. $\begin{cases} 6x - y = 0 \\ y = 4x + 2 \end{cases}$

19. $\begin{cases} x + \dfrac{5}{3}y = -\dfrac{1}{3} \\ -\dfrac{5}{2}x + y = \dfrac{-19}{2} \end{cases}$ (*Hint:* First clear the equations of fractions.)

20. $\begin{cases} x + \dfrac{5}{9}y = \dfrac{4}{9} \\ x - \dfrac{2}{3}y = \dfrac{5}{3} \end{cases}$
21. $\begin{cases} x - 2y = 3 \\ \dfrac{4}{5}x - \dfrac{8}{5}y = -1 \end{cases}$

22. $\begin{cases} \dfrac{x}{5} + y = \dfrac{2}{5} \\ \dfrac{3}{7}x + \dfrac{15}{7}y = \dfrac{6}{7} \end{cases}$

In exercises 23 and 24, find the solution set; a and b are nonzero constants.

23. $\begin{cases} 2ax + 3by = 2a^2 + 3a^2b \\ 4ax - by = 4a^2 - a^2b \end{cases}$
24. $\begin{cases} 3ax - 2by = 3a \\ 5ax - 3by = 5a \end{cases}$

25. If the two columns of a second-order determinant are the same, what is the value of the determinant?

26. If the second column of the determinant

$$\begin{vmatrix} 3 & 4 \\ 5 & 10 \end{vmatrix}$$

is multiplied by 3 to give

$$\begin{vmatrix} 3 & 12 \\ 5 & 30 \end{vmatrix}$$

how does the value of the last determinant compare with the first?

27. If the two columns of the determinant

$$\begin{vmatrix} 5 & 4 \\ -1 & -3 \end{vmatrix}$$

are interchanged to form

$$\begin{vmatrix} 4 & 5 \\ -3 & -1 \end{vmatrix}$$

how do the values of the two determinants compare?

28. If the first column of the determinant

$$\begin{vmatrix} 4 & 3 \\ 2 & 7 \end{vmatrix}$$

Systems of Equations

is added to the second column to form the determinant

$$\begin{vmatrix} 4 & 7 \\ 2 & 9 \end{vmatrix}$$

how do the values of the two determinants compare?

10–4 Higher-Order Determinants (Optional)

In this section, we develop ways of evaluating higher-order determinants. We begin with third-order determinants.

A third-order determinant takes the form

$$\begin{vmatrix} a_1 & b_1 & c_1 \\ a_2 & b_2 & c_2 \\ a_3 & b_3 & c_3 \end{vmatrix}$$

The technique for evaluating a third-order determinant is given in definition 10–2.

Definition 10–2

Third-order Determinant

$$\begin{vmatrix} a_1 & b_1 & c_1 \\ a_2 & b_2 & c_2 \\ a_3 & b_3 & c_3 \end{vmatrix} = a_1b_2c_3 + b_1c_2a_3 + c_1a_2b_3 - a_3b_2c_1 - b_3c_2a_1 - c_3a_2b_1$$

The six terms in the expansion of the determinant in definition 10–2 do not need to be memorized. If we recopy the first two columns of the determinant to the immediate right of the determinant, then the six terms in definition 10–2 can be obtained by the multiplications shown by the arrows. The numbers at the tips of the arrows indicate the order in which the multiplications should be done. The multiplications shown by arrows ④, ⑤, and ⑥ are subtracted from the other three multiplications. Also, you are cautioned that only a third-order determinant can be solved by the method of recopying the first two columns.

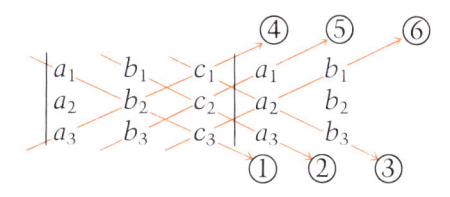

$$\begin{vmatrix} a_1 & b_1 & c_1 \\ a_2 & b_2 & c_2 \\ a_3 & b_3 & c_3 \end{vmatrix} \begin{matrix} a_1 & b_1 \\ a_2 & b_2 \\ a_3 & b_3 \end{matrix}$$

$$= a_1b_2c_3 + b_1c_2a_3 + c_1a_2b_3 - a_3b_2c_1 - b_3c_2a_1 - c_3a_2b_1.$$

Example 1

Evaluate $\begin{vmatrix} 1 & 0 & 3 \\ 2 & 1 & -2 \\ -1 & 2 & 2 \end{vmatrix}$.

Solution

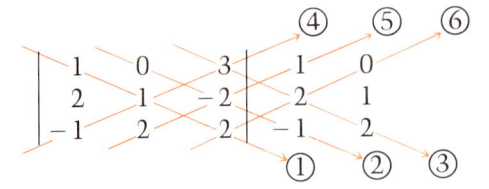

$$= (1)(1)(2) + (0)(-2)(-1) + (3)(2)(2) - (-1)(1)(3) - (2)(-2)(1) - (2)(2)(0)$$
$$= 2 + 0 + 12 + 3 + 4 - 0$$
$$= 21$$

Third-order determinants can also be evaluated by a process called *expansion by minors*. The **minor** for any element of a third-order determinant is the determinant that remains after deleting the row and column containing that element.

Example 2

For the determinant

$$\begin{vmatrix} a_1 & b_1 & c_1 \\ a_2 & b_2 & c_2 \\ a_3 & b_3 & c_3 \end{vmatrix}$$

determine the minor for each.

(a) a_1
(b) b_2

Solution

(a) Delete the row and column containing a_1.

$$\begin{vmatrix} a_1 & b_1 & c_1 \\ a_2 & b_2 & c_2 \\ a_3 & b_3 & c_3 \end{vmatrix}$$

The minor for a_1 is $\begin{vmatrix} b_2 & c_2 \\ b_3 & c_3 \end{vmatrix}$.

(b) Delete the row and column containing b_2.

$$\begin{vmatrix} a_1 & b_1 & c_1 \\ a_2 & b_2 & c_2 \\ a_3 & b_3 & c_3 \end{vmatrix}$$

The minor for b_2 is $\begin{vmatrix} a_1 & c_1 \\ a_3 & c_3 \end{vmatrix}$.

Systems of Equations

If the six terms in the expansion of the determinant in definition 10–2 are selectively grouped, we can write definition 10–2 as

$$\begin{vmatrix} a_1 & b_1 & c_1 \\ a_2 & b_2 & c_2 \\ a_3 & b_3 & c_3 \end{vmatrix} = a_1(b_2c_3 - b_3c_2) - b_1(a_2c_3 - a_3c_2) + c_1(a_2b_3 - a_3b_2) \quad (1)$$

The quantities enclosed in parentheses in (1) may be written as second-order determinants, which gives

$$\begin{vmatrix} a_1 & b_1 & c_1 \\ a_2 & b_2 & c_2 \\ a_3 & b_3 & c_3 \end{vmatrix} = a_1 \begin{vmatrix} b_2 & c_2 \\ b_3 & c_3 \end{vmatrix} - b_1 \begin{vmatrix} a_2 & c_2 \\ a_3 & c_3 \end{vmatrix} + c_1 \begin{vmatrix} a_2 & b_2 \\ a_3 & b_3 \end{vmatrix} \quad (2)$$

The expansion in (2) is the expansion of the determinant by minors about the first row.

By properly grouping the six terms in the expansion of the determinant in definition 10–2, we can also expand the determinant by minors about the first column; this gives

$$\begin{vmatrix} a_1 & b_1 & c_1 \\ a_2 & b_2 & c_2 \\ a_3 & b_3 & c_3 \end{vmatrix} = a_1 \begin{vmatrix} b_2 & c_2 \\ b_3 & c_3 \end{vmatrix} - a_2 \begin{vmatrix} b_1 & c_1 \\ b_3 & c_3 \end{vmatrix} + a_3 \begin{vmatrix} b_1 & c_1 \\ b_2 & c_2 \end{vmatrix} \quad (3)$$

In general, a determinant may be expanded by minors about any row or any column. If we look at the expansions of the determinant in (2) and (3), we notice that some terms are preceeded by a plus sign and some by a minus sign. The sign that preceeds any given term in the expansion is given by the array

$$\begin{vmatrix} + & - & + \\ - & + & - \\ + & - & + \end{vmatrix}$$

which is sometimes called the *array of signs* for a third-order determinant. Notice that the signs alternate between plus and minus, beginning with a plus sign in the upper left-hand position.

Example 3   Expand the determinant

$$\begin{vmatrix} 5 & -2 & 1 \\ -1 & -3 & 2 \\ 2 & 0 & -4 \end{vmatrix}$$

by minors about the second column.

Solution

$$\begin{vmatrix} 5 & -2 & 1 \\ -1 & -3 & 2 \\ 2 & 0 & -4 \end{vmatrix} = -(-2)\begin{vmatrix} -1 & 2 \\ 2 & -4 \end{vmatrix} + (-3)\begin{vmatrix} 5 & 1 \\ 2 & -4 \end{vmatrix} - 0\begin{vmatrix} 5 & 1 \\ -1 & 2 \end{vmatrix}$$
$$= 2(4 - 4) - 3(-20 - 2) - 0(10 + 1)$$
$$= 2(0) - 3(-22) - 0(11)$$
$$= 66 \quad \blacksquare$$

400 Intermediate Algebra

The minor for the element zero does not need to be shown. We have included it only to show the complete expansion and to demonstrate how the array of signs is used.

Fourth-order determinants are evaluated by expansion by minors. The minor for any element will be a third-order determinant, which may be evaluated by either of the two methods just discussed.

Example 4 Expand the determinant

$$\begin{vmatrix} 1 & 0 & 5 & 2 \\ -2 & 1 & 1 & -3 \\ 3 & 2 & 0 & 4 \\ 4 & 0 & -1 & 0 \end{vmatrix}$$

by minors about the first row.

Solution The array of signs for a fourth-order determinant has four rows and four columns of alternating plus and minus signs, beginning with a plus sign in the upper left corner.

$$\begin{vmatrix} 1 & 0 & 5 & 2 \\ -2 & 1 & 1 & -3 \\ 3 & 2 & 0 & 4 \\ 4 & 0 & -1 & 0 \end{vmatrix}$$

$$= 1\begin{vmatrix} 1 & 1 & -3 \\ 2 & 0 & 4 \\ 0 & -1 & 0 \end{vmatrix} - 0\begin{vmatrix} -2 & 1 & -3 \\ 3 & 0 & 4 \\ 4 & -1 & 0 \end{vmatrix} + 5\begin{vmatrix} -2 & 1 & -3 \\ 3 & 2 & 4 \\ 4 & 0 & 0 \end{vmatrix}$$

$$- 2\begin{vmatrix} -2 & 1 & 1 \\ 3 & 2 & 0 \\ 4 & 0 & -1 \end{vmatrix}$$

$$= 1\begin{vmatrix} 1 & 1 & -3 & 1 & 1 \\ 2 & 0 & 4 & 2 & 0 \\ 0 & -1 & 0 & 0 & -1 \end{vmatrix}$$

$$+ 5\begin{vmatrix} -2 & 1 & -3 & -2 & 1 \\ 3 & 2 & 4 & 3 & 2 \\ 4 & 0 & 0 & 4 & 0 \end{vmatrix}$$

$$- 2\begin{vmatrix} -2 & 1 & 1 & -2 & 1 \\ 3 & 2 & 0 & 3 & 2 \\ 4 & 0 & -1 & 4 & 0 \end{vmatrix}$$

$$= 1[0 + 0 + 6 - 0 - (-4) - 0]$$
$$+ 5[0 + 16 + 0 - (-24) - 0 - 0]$$
$$- 2[4 + 0 + 0 - 8 - 0 - (-3)]$$

Systems of Equations

$$= (6 + 4) + 5(16 + 24) - 2(4 - 8 + 3)$$
$$= 10 + 5(40) - 2(-1)$$
$$= 10 + 200 + 2$$
$$= 212 \quad \blacksquare$$

10–4 Exercises

In exercises 1–6, evaluate the determinants by recopying the first two columns. (See example 1.)

1. $\begin{vmatrix} 2 & -1 & 2 \\ 3 & 0 & -2 \\ 1 & 3 & 4 \end{vmatrix}$

2. $\begin{vmatrix} 3 & 0 & 2 \\ 1 & 5 & -3 \\ -2 & 1 & 2 \end{vmatrix}$

3. $\begin{vmatrix} 2 & 0 & 3 \\ -5 & 1 & -1 \\ 4 & 0 & 2 \end{vmatrix}$

4. $\begin{vmatrix} 2 & 0 & 1 \\ 3 & -2 & 2 \\ -1 & 2 & -3 \end{vmatrix}$

5. $\begin{vmatrix} x & y & 2 \\ x & y & -1 \\ 2 & 2 & 2 \end{vmatrix}$

6. $\begin{vmatrix} x & x & 3 \\ y & x & -1 \\ -1 & -2 & 0 \end{vmatrix}$

In exercises 7–12, evaluate the determinants by expanding by minors about the column or row indicated. (See example 2.)

7. $\begin{vmatrix} 2 & 4 & 0 \\ 0 & -1 & 3 \\ 1 & 2 & -1 \end{vmatrix}$, about the first column

8. $\begin{vmatrix} 2 & 0 & 2 \\ -1 & 1 & 1 \\ 5 & 0 & 3 \end{vmatrix}$, about the second column

9. $\begin{vmatrix} 2 & -1 & -2 \\ 3 & 0 & 2 \\ 5 & -1 & 1 \end{vmatrix}$, about the second row

10. $\begin{vmatrix} 3 & -1 & -2 \\ 0 & 1 & 2 \\ -1 & 2 & -3 \end{vmatrix}$, about the third row

11. $\begin{vmatrix} 0 & x & 0 \\ x & 0 & 0 \\ 0 & 0 & x \end{vmatrix}$, about the first column

12. $\begin{vmatrix} a & 0 & 0 \\ 0 & a & b \\ b & b & 0 \end{vmatrix}$, about the first row

13. Solve for x.

$$\begin{vmatrix} x & 2 & 1 \\ 1 & 0 & 2 \\ 3 & x & -1 \end{vmatrix} = 13$$

14. Solve for x.
$$\begin{vmatrix} 0 & 0 & 1 \\ x & 1 & 0 \\ 1 & x & 1 \end{vmatrix} = 0$$

15. Evaluate by expanding on the first column.   (See example 3.)
$$\begin{vmatrix} 0 & 1 & 1 & 1 \\ 2 & -1 & 0 & -1 \\ 3 & 0 & 2 & 0 \\ 0 & 0 & 1 & 2 \end{vmatrix}$$

16. Evaluate by expanding on the third row.   (See example 3.)
$$\begin{vmatrix} 3 & -1 & 0 & -2 \\ 1 & 2 & 4 & 4 \\ 0 & 1 & -1 & 2 \\ 2 & -1 & 2 & -2 \end{vmatrix}$$

17. Evaluate by expanding on the third column.
$$\begin{vmatrix} 2 & 1 & 2 & 2 \\ 1 & 0 & 1 & 1 \\ 0 & 5 & -1 & 0 \\ 3 & -2 & 0 & 3 \end{vmatrix}$$

18. Evaluate by expanding on the first column.
$$\begin{vmatrix} 0 & 1 & 1 & 1 \\ 6 & -1 & 0 & -1 \\ 9 & 0 & 2 & 0 \\ 0 & 0 & 1 & 2 \end{vmatrix}$$

19. Examine the answers for exercises 15 and 18 and make a statement about determinants that are identical except one contains a column that is a multiple of one of the columns in the other determinant.

20. Examine the answer for exercise 17 and make a statement about a determinant containing two identical columns.

10–5 Solution of Systems in Three Variables by Determinants (Optional)

Cramer's rule can be extended to find the solution set for a system of three linear equations in three variables.

Rule 10–2

Cramer's Rule for a System of Three Equations

The three linear equations
$$\begin{cases} a_1x + b_1y + c_1z = d_1 \\ a_2x + b_2y + c_2z = d_2 \\ a_3x + b_3y + c_3z = d_3 \end{cases}$$

Systems of Equations

have the solution

$$x = \frac{D_x}{D}, \quad y = \frac{D_y}{D} \quad \text{and} \quad z = \frac{D_z}{D}$$

where

$$D_x = \begin{vmatrix} d_1 & b_1 & c_1 \\ d_2 & b_2 & c_2 \\ d_3 & b_3 & c_3 \end{vmatrix}, \quad D_y = \begin{vmatrix} a_1 & d_1 & c_1 \\ a_2 & d_2 & c_2 \\ a_3 & d_3 & c_3 \end{vmatrix}$$

$$D_z = \begin{vmatrix} a_1 & b_1 & d_1 \\ a_2 & b_2 & d_2 \\ a_3 & b_3 & d_3 \end{vmatrix}, \quad D = \begin{vmatrix} a_1 & b_1 & c_1 \\ a_2 & b_2 & c_2 \\ a_3 & b_3 & c_3 \end{vmatrix}$$

Example 1 Use Cramer's rule to find the solution set for

$$\begin{cases} 2x + 3y - z = 18 \\ x - 3y = 0 \\ -x + 2y + 3z = -2 \end{cases}$$

Solution We first calculate D, D_x, D_y, and D_z. Then we use rule 10–2 to calculate x, y, and z. Place zeros in the determinant for any terms that are missing from any of the equations.

$$D = \begin{vmatrix} 2 & 3 & -1 \\ 1 & -3 & 0 \\ -1 & 2 & 3 \end{vmatrix} \begin{matrix} 2 & 3 \\ 1 & -3 \\ -1 & 2 \end{matrix}$$
$$= -18 + 0 - 2 + 3 - 0 - 9$$
$$= -26$$

$$D_x = \begin{vmatrix} 18 & 3 & -1 \\ 0 & -3 & 0 \\ -2 & 2 & 3 \end{vmatrix} \begin{matrix} 18 & 3 \\ 0 & -3 \\ -2 & 2 \end{matrix}$$
$$= -162 + 0 + 0 + 6 - 0 - 0$$
$$= -156$$

$$D_y = \begin{vmatrix} 2 & 18 & -1 \\ 1 & 0 & 0 \\ -1 & -2 & 3 \end{vmatrix} \begin{matrix} 2 & 18 \\ 1 & 0 \\ -1 & -2 \end{matrix}$$
$$= 0 + 0 + 2 - 0 - 0 - 54$$
$$= -52$$

$$D_z = \begin{vmatrix} 2 & 3 & 18 \\ 1 & -3 & 0 \\ -1 & 2 & -2 \end{vmatrix} \begin{matrix} 2 & 3 \\ 1 & -3 \\ -1 & 2 \end{matrix}$$
$$= 12 + 0 + 36 - 54 - 0 + 6$$
$$= 54 - 54$$
$$= 0$$

$$x = \frac{D_x}{D} = \frac{-156}{-26} = 6$$

$$y = \frac{D_y}{D} = \frac{-52}{-26} = 2$$

$$z = \frac{D_z}{D} = \frac{0}{-26} = 0$$

The solution set is $\{(6, 2, 0)\}$. ■

If $D = 0$, Cramer's rule cannot be used. If $D = 0$ and if $D_x \neq 0$ or $D_y \neq 0$ or $D_z \neq 0$, the system is inconsistent. If $D = 0$ and if $D_x = 0$, $D_y = 0$, and $D_z = 0$, the system is dependent. A more thorough discussion of this topic is beyond the scope of this book.

10–5 Exercises

In exercises 1–14, use Cramer's rule to solve each system of equations. If Cramer's rule does not apply, make a statement to that effect. (See example 1.)

1. $\begin{cases} x + 4y - 2z = 0 \\ 2x - y + z = -3 \\ 3x + y - z = -7 \end{cases}$

2. $\begin{cases} 2x - y + z = 3 \\ x + 3y + 4z = -2 \\ -x + 2y - z = -1 \end{cases}$

3. $\begin{cases} 2x + 3y = 0 \\ 2y - 3z = -4 \\ -3x + 4z = -9 \end{cases}$

4. $\begin{cases} x - 2z = -1 \\ x - 3y + z = 2 \\ 2x - 4y - 3z = -1 \end{cases}$

5. $\begin{cases} 2x - y = 0 \\ 3x - 2y + z = -1 \\ 5x - y + 2z = 3 \end{cases}$

6. $\begin{cases} 5x - 2y + 3z = 4 \\ 2x + 4y - z = 2 \\ 3x - 2y + 5z = 8 \end{cases}$

7. $\begin{cases} 2x - 3y + z = 1 \\ 5x - 4y - 2z = -3 \\ 3x + y - 3z = -2 \end{cases}$

8. $\begin{cases} 8x - 4y + z = 5 \\ 9x + 2y - 5z = 6 \\ x - 4y + 2z = -1 \end{cases}$

9. $\begin{cases} 2x - 3y - 4z = 5 \\ x + y + z = -4 \\ -4x + 6y + 8z = -10 \end{cases}$

10. $\begin{cases} 5x - 2y + 3z = 10 \\ -2x + 4y + z = 9 \\ 10x - 4y + 6z = 20 \end{cases}$

11. $\begin{cases} 2x + 3y + z = 5 \\ y - z = -1 \\ x + 2z = 3 \end{cases}$

12. $\begin{cases} 2x + y + 3z = -5 \\ 2x + z = -2 \\ y + 2z = -3 \end{cases}$

13. $\begin{cases} 2x + 3y - z + w = -1 \\ -x + y + z - 2w = -7 \\ 2x - y + z - 2w = 0 \\ x + y - 2z + w = 4 \end{cases}$

14. $\begin{cases} 2x + 3y = 2 \\ 3y + 2z = -2 \\ 4x - 2y + z = 3 \\ 2y + z - w = 0 \end{cases}$

A circle whose general form is given by $x^2 + y^2 + Dx + Ey + F = 0$ is completely determined if the constants D, E, and F are known. Find both the general and standard forms of an equation of the circle that passes through the three given points. (Hint: Substitute the coordinates of the three ordered pairs into the general form of the equation for a circle to find three equations in D, E, and F.)

15. (4, 7), (8, 3), (0, 3).   **16.** (0, 0), (2, 2), (4, 0)

10–6 Quadratic Systems in Two Variables

A quadratic system of two equations in two variables includes at least one quadratic equation. For the systems that we consider, either the substitution method or the addition method can be used to solve the system.

If one equation is of degree two and one is linear, then the substitution method can be used. If both equations are of degree two, then the addition method usually works better, although there are times when substitution works equally well. A few examples will help to explain the techniques.

Example 1   Find the solution set for the system

$$\begin{cases} x^2 + y^2 = 40 & (1) \\ \quad\quad y = 3x & (2) \end{cases}$$

Solution   A graph is not necessary, but it is a help in determining how many ordered pairs are in the solution set. The graph of the system is shown in figure 10–2.

The graph of the system shows the circle and line intersect in two points. The coordinates of these two points give us the solution set of the system. Equation (2) is a first-degree equation and we substitute it into equation (1).

$$x^2 + (3x)^2 = 40$$
$$x^2 + 9x^2 = 40$$
$$10x^2 = 40$$
$$x^2 = 4$$
$$x = 2 \text{ or } -2$$

From equation (2), we find

$$y = 3(2) = 6 \quad \text{or} \quad y = 3(-2) = -6$$

The points of intersection are (2, 6) and (−2, −6). The solution set is {(2, 6), (−2, −6)}.

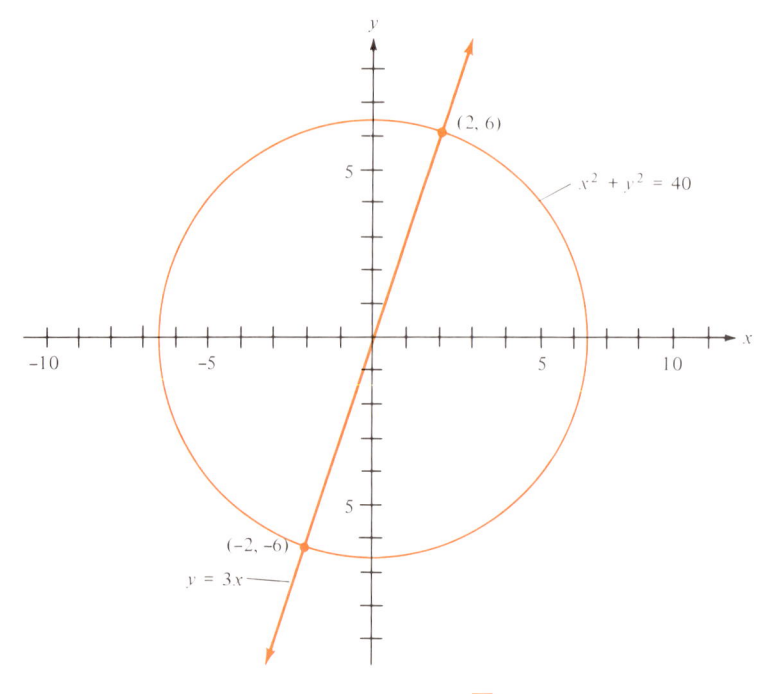

[Figure 10-2: Graph showing circle $x^2 + y^2 = 40$ and line $y = 3x$ intersecting at $(2, 6)$ and $(-2, -6)$.]

Figure 10-2

Example 2 Find the solution set for the system.

$$\begin{cases} 5x^2 + y^2 = 23 & (3) \\ x^2 - y^2 = 1 & (4) \end{cases}$$

Solution The equations can be added to eliminate the y^2-terms. Adding equations (3) and (4), we get

$$6x^2 = 24$$
$$x^2 = 4$$
$$x = 2 \text{ or } -2$$

Substitute these values of x into equation (4),

$$(2)^2 - y^2 = 1$$
$$-y^2 = -3$$
$$y^2 = 3$$
$$y = \sqrt{3} \text{ or } -\sqrt{3}$$

Thus the value $x = 2$ gives two values for y. Two points of intersection for the system are $(2, \sqrt{3})$ and $(2, -\sqrt{3})$. If $x = -2$ is substituted into (4), we also get $y = \sqrt{3}$ or $y = -\sqrt{3}$. The value $x = -2$ gives two more ordered pairs, $(-2, \sqrt{3})$ and $(-2, -\sqrt{3})$. The complete solution set is $\{(2, \sqrt{3}), (2, -\sqrt{3}), (-2, \sqrt{3}), (-2, -\sqrt{3})\}$. See figure 10-3.

Systems of Equations

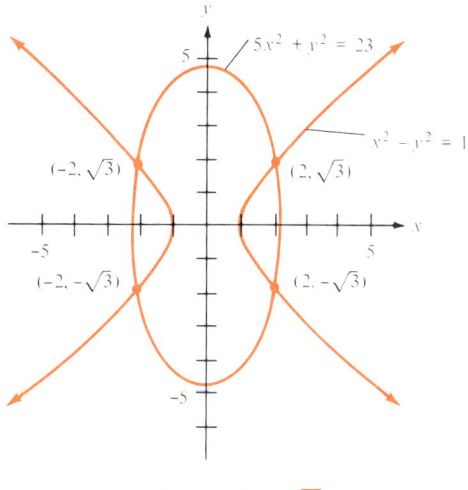

Figure 10-3

Sometimes a combination of both the addition method and substitution must be used to solve a system.

Example 3 Solve the system

$$\begin{cases} 2x^2 + xy - 2y^2 = 34 & (5) \\ -x^2 + y^2 = -15 & (6) \end{cases}$$

Solution The second-degree terms in x and y can be eliminated if equation (6) is multiplied by 2 and added to equation (5). Doing this, we get

$$\begin{aligned} 2x^2 + xy - 2y^2 &= 34 \\ -2x^2 \qquad\quad + 2y^2 &= -30 \\ \hline xy &= 4 \end{aligned} \quad (7)$$

Any solution of the original system is a solution of the new system

$$\begin{cases} -x^2 + y^2 = -15 & (6) \\ xy = 4 & (7) \end{cases}$$

We solve equation (7) for y and substitute it into equation (6).

$$xy = 4$$

$$y = \frac{4}{x}$$

$$-x^2 + \left(\frac{4}{x}\right)^2 = -15 \qquad \text{Substitute } y = \frac{4}{x} \text{ into (6).}$$

$$-x^2 + \frac{16}{x^2} = -15 \qquad \text{Simplify.}$$

$$-x^4 + 16 = -15x^2 \quad \text{Assume } x \neq 0 \text{ and clear the equation of fractions.}$$
$$x^4 - 15x^2 - 16 = 0$$
$$(x^2 - 16)(x^2 + 1) = 0 \quad \text{Factor.}$$

The first factor gives $x = 4$ or -4. The second factor gives $x = i$ or $-i$. To find the values for y, substitute the values found for x into $y = 4/x$. We get, for $x = i$

$$y = \frac{4}{i}$$
$$= \frac{4}{i} \cdot \frac{-i}{-i} \quad \text{Rationalize the denominator.}$$
$$= -4i$$

Continuing in a similar fashion we find the complete solution set to be $\{(i, -4i)(-i, 4i), (4, 1), (-4, -1)\}$.

10–6 Exercises

In exercises 1–15, solve each system by using either the substitution method or the addition method. (See examples 1 and 2.)

1. $\begin{cases} x^2 + y^2 = 29 \\ x + y = 7 \end{cases}$
2. $\begin{cases} x^2 + y^2 = 20 \\ x - y = -2 \end{cases}$
3. $\begin{cases} x^2 + y^2 = 13 \\ xy = 6 \end{cases}$

4. $\begin{cases} 4x^2 + y^2 = 20 \\ x^2 + y^2 = 8 \end{cases}$
5. $\begin{cases} 5x^2 + 2y^2 = 13 \\ x^2 + y^2 = 5 \end{cases}$
6. $\begin{cases} x^2 - y^2 = 5 \\ x^2 + 2y^2 = 17 \end{cases}$

7. $\begin{cases} x^2 + y^2 - 4x - 6y + 4 = 0 \\ x - y = 2 \end{cases}$

8. $\begin{cases} x^2 + y^2 + 4x - 8y + 16 = 0 \\ x - y = -4 \end{cases}$

9. $\dfrac{x^2}{9} + \dfrac{y^2}{4} = 1$

    $\dfrac{2}{5}x - \dfrac{y}{5} = \dfrac{2}{5}$

10. $\dfrac{x^2}{16} + \dfrac{y^2}{2} = 1$

    $\dfrac{x^2}{16} - \dfrac{y^2}{2} = 1$

11. $\begin{cases} -5x^2 + 3xy + 5y^2 = 21 \\ x^2 - y^2 = -3 \end{cases}$ (See example 3.)

12. $\begin{cases} -2x^2 + xy - 2y^2 = -20 \\ x^2 + y^2 = 13 \end{cases}$ (See example 3.)

13. $\begin{cases} 3x^2 + 5xy + 3y^2 = 11 \\ x^2 - 2xy + y^2 = 0 \end{cases}$ (Hint: Factor the second equation and substitute an expression for x or y into the first equation.)

14. $\begin{cases} 5x^2 + 2xy + 3y^2 = 10 \\ x^2 - 2xy + y^2 = 0 \end{cases}$ (See exercise 13.)

15. For the system of equations

$$\begin{cases} \dfrac{x^2}{a^2} + \dfrac{y^2}{b^2} = 1, & a > b \\ \dfrac{x^2}{c^2} - \dfrac{y^2}{d^2} = 1 \end{cases}$$

Draw sketches to illustrate what must be true to have each situation.

$\begin{cases} \text{(a) Exactly four solutions} & (c < a). \\ \text{(b) Exactly two solutions} & (c = a). \\ \text{(c) No solution} & (c > a). \end{cases}$

16. For the system of equations

$$\begin{cases} x^2 + y^2 = r^2 \\ \dfrac{x^2}{a^2} + \dfrac{y^2}{b^2} = 1, & a > b \end{cases}$$

Draw sketches to illustrate what must be true to have each situation.

(a) No solution $(r > a)$.
(b) Two solutions $(r = a)$.
(c) Four solutions $(r < a, r > b)$.

17. The area of a rectangle is 15 square feet. If the perimeter is 16 feet, find the dimensions of the rectangle by using a system of equations.

18. The area of a triangle is 80 square feet. The length of the base is 4 feet more than twice the length of the altitude. Find the length of the base of the triangle by using a system of equations.

10–7 Applications

Most of the word problems with which we have dealt up to this point have reduced to one equation in one variable. However, we have presented a few word problems in various sections of this chapter that have been represented by a system of equations in two or three variables. A review of these problems in sections 10–1 and 10–2 shows that it is frequently convenient (or even necessary) to represent the conditions in a word problem with a system of two or three equations. A unique solution can then be obtained if the system is not dependent or inconsistent.

In this section, we take a more extensive look at word problems that can be represented by a system of equations.

Example 1

The sum of two numbers is 50. Twice the larger number is three times the smaller. Find the two numbers.

Solution We must carefully examine the stated problem for sentences that can be expressed as equations. Also, we must represent the two numbers we are trying to find with two variables before attempting to write the two equations. Let

x represent the smaller number

y represent the larger number

One equation can be written from the first sentence in the example.

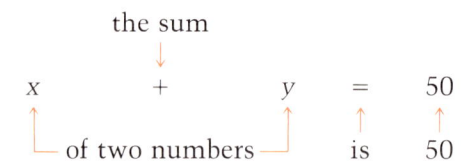

$$\underbrace{x}_{\text{of two numbers}} + y = 50$$

the sum ↓ is 50

The second equation can be obtained from the second sentence in the example.

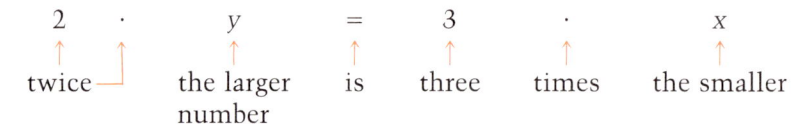

$$\underbrace{2}_{\text{twice}} \cdot \underbrace{y}_{\substack{\text{the larger} \\ \text{number}}} = \underbrace{3}_{\text{three}} \cdot \underbrace{x}_{\text{the smaller}}$$

is   times

These two equations give the system

$$\begin{cases} x + y = 50 & (1) \\ 2y = 3x & (2) \end{cases}$$

If we solve equation (2) for y, we get

$$y = \frac{3x}{2}$$

This equation can now be substituted into equation (1) to get

$$x + \frac{3x}{2} = 50$$

$$2x + 3x = 100$$

$$5x = 100$$

$$x = 20$$

If $x = 20$ is substituted into (1), we get $y = 30$. The two numbers are 20 and 30. ■

Example 2 The sum of two numbers is $\frac{57}{20}$. Five times the smaller number added to four times the larger number is 12. Find the two numbers.

Solution We look for sentences from which we can write two equations in two variables, but first we must represent the two numbers we are trying to find with some convenient variables.

Systems of Equations

Let S represent the smaller number.
Let L represent the larger number.

Then one equation is:

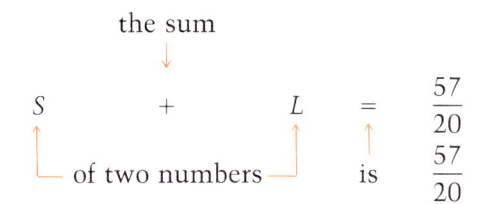

$$S + L = \frac{57}{20}$$

the sum of two numbers is $\frac{57}{20}$

The second equation is:

$$5 \cdot S + 4 \cdot L = 12$$

five times the smaller added to four times the larger is 12

We now have the system

$$\begin{cases} S + L = \dfrac{57}{20} & (3) \\ 5S + 4L = 12 & (4) \end{cases}$$

Multiplying equation (3) by 20 to eliminate fractions gives

$$\begin{cases} 20S + 20L = 57 & (5) \\ 5S + 4L = 12 & (6) \end{cases}$$

To eliminate S, multiply (6) by -4 and add this result to (5).

$$\begin{aligned} 20S + 20L &= 57 \\ -20S - 16L &= -48 \\ \hline 4L &= 9 \\ L &= \frac{9}{4} \end{aligned}$$

Substitute $L = \frac{9}{4}$ into equation (6) to get

$$5S + 4\left(\frac{9}{4}\right) = 12$$
$$5S + 9 = 12$$
$$S = \frac{3}{5}$$

The numbers are $\frac{3}{5}$ and $\frac{9}{4}$. ■

Example 3   The area of a rectangle is 66 square meters and the perimeter is 34 meters. Find the dimensions of the rectangle.

Solution

The area for a rectangle is given by $A = LW$, and the perimeter is given by $P = 2W + 2L$. Careful reading of the problem indicates we can form the following two equations:

$$\underbrace{L \cdot W}_{\text{the area of a rectangle}} \underbrace{=}_{\text{is}} \underbrace{66}_{66}$$

$$\underbrace{2W + 2L}_{\text{the perimeter of the rectangle}} \underbrace{=}_{\text{is}} \underbrace{34}_{34}$$

The system of equations that we must solve is:

$$\begin{cases} LW = 66 & (7) \\ 2W + 2L = 34 & (8) \end{cases}$$

Equation (7) is equivalent to

$$L = \frac{66}{W}$$

If we divide equation (8) by 2 and substitute $L = 66/W$ into it, we get

$$W + \frac{66}{W} = 17$$

$W^2 + 66 = 17W$ Multiply each side by W.
$W^2 - 17W + 66 = 0$ Write the equation in standard form.
$(W - 11)(W - 6) = 0$ Factor.
$W = 11$ or $W = 6$

If $W = 11$, then $L = 66/11 = 6$.
If $W = 6$, then $L = 66/6 = 11$.
Thus, the rectangle is either 11 meters by 6 meters, or it is 6 meters by 11 meters. ■

Example 4   A woman bought a dress, coat, and shoes for a total cost of $200. The cost of the coat was $10 more than the total cost of the dress and shoes. The cost of the dress was $15 more than the cost of the shoes. How much did she pay for each?

Solution   Let d = the cost of the dress.
Let c = the cost of the coat.
Let s = the cost of the shoes.

The first sentence of the problem gives the equation

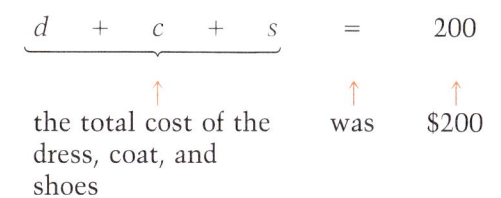

$$\underbrace{d + c + s}_{\text{the total cost of the dress, coat, and shoes}} = \underset{\text{was}}{200} \;\; \underset{\$200}{}$$

The second sentence in the problem gives the equation

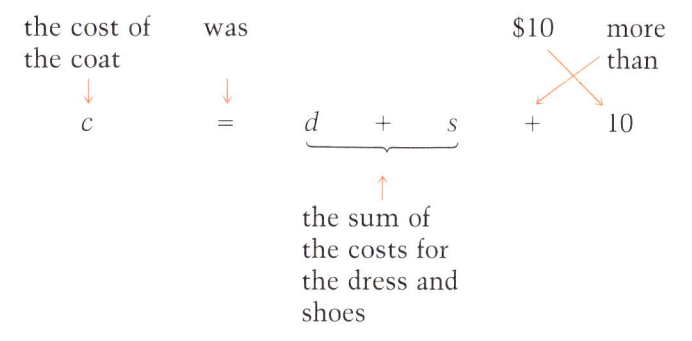

the cost of the coat was $10 more than

$$c = \underbrace{d + s}_{\text{the sum of the costs for the dress and shoes}} + 10$$

The last sentence in the example is represented by

$$\underset{\substack{\uparrow \\ \text{the cost of the dress}}}{d} = \underset{\substack{\uparrow \\ \text{was}}}{} \underset{\substack{\uparrow \\ \$15}}{15} + \underset{\substack{\uparrow \\ \text{more than}}}{} \underset{\substack{\uparrow \\ \text{the cost of the shoes}}}{s}$$

The three equations just written form the system

$$\begin{cases} d + c + s = 200 \\ c = d + s + 10 \\ d = 15 + s \end{cases}$$

If we write the system in standard form we have,

$$\begin{cases} d + c + s = 200 & (9) \\ -d + c - s = 10 & (10) \\ d - s = 15 & (11) \end{cases}$$

If we add equations (9) and (10), we can eliminate both d and s.

$$2c = 210$$
$$c = 105$$

If equations (11) and (9) are added, we can eliminate s.

$$2d + c = 215$$

Substituting $c = 105$ into this last equation, we find

$$d = 55$$

If we substitute $d = 55$ into (11), then $s = 40$. By these calculations we find the dress cost $55, the coat cost $105, and the shoes cost $40. ■

10–7 Exercises

Solve using a system of equations. Any method that we have studied which applies to the system may be used to solve it. (See examples 1–4.)

1. The sum of two numbers is 75 and their difference is 5. Find the two numbers.

2. The sum of two numbers is -30 and their difference is 10. Find the two numbers.

3. The difference of two numbers is 15. The larger one is 20 less than twice the smaller. Find the two numbers.

4. Twice a larger number added to a smaller number is 80. Three times the smaller number added to twice the larger is 120. Find the two numbers.

5. If a smaller number is multiplied by $\frac{5}{2}$ and added to a larger one multiplied by $\frac{3}{4}$, the sum is 29. However, if the smaller number is multiplied by $\frac{3}{4}$ and added to the larger multiplied by $\frac{5}{2}$, the sum is 36. Find the two numbers.

6. Steve and Ross must paint Steve's entire house. (They are now weekend contractors). Steve bought 3 gallons of paint and 2 brushes at one store for $48. Ross bought 5 gallons of paint and 1 brush (he intended for Steve to do all the work) at another store for $66. Assuming the price per gallon and cost per brush were the same at both stores, find what they paid per gallon for paint and the cost per brush.

7. If a first number is divided into 12 and added to a second number divided into 16, the sum is 8. On the other hand, if the first number is divided into 9 and added to the second number divided into 24, the sum is 9. Find the two numbers.

8. The sum of a first number divided into 10 and a second number divided into 12 is 122. However, the sum of the first number divided into 5 and the second number divided into 2 is 37. Find the two numbers.

9. Rory bought 40 gallons of fuel and 5 quarts of oil for his truck for a cost of $52.50. If the price per quart of oil was 30¢ less than the price per gallon for fuel, what was the price per gallon he paid for fuel and the price per quart for oil?

10. The sum of three numbers is 92. The sum of the two smaller numbers is 8 more than the largest one, while twice the smallest number is 10 more than the middle number. Find the three numbers.

11. Pete bought 3 shirts, 2 pairs of pants, and 1 pair of socks for $100. John bought 1 shirt, 1 pair of pants and 1 pair of socks for $42. Kevin bought 2 shirts, 1 pair of pants and 3 pairs of socks for $64. If each paid the same price for a shirt, a pair of pants, and a pair of socks, find the price of each item of clothing.

12. The sum of three integers is 39. The middle integer is 8 less than the largest integer and the sum of the two smaller integers is 1 less than the largest one. Find the three integers.

13. In one week's time, an electrician, an assistant, and a helper together earned $1,050. The sum of the weekly salaries for the assistant and helper was $200 more than the electrician's salary, while twice the helper's salary was $125 more than the electrician's salary. Find the weekly salary for all three.

14. The sum of three integers is 11. The middle integer is 9 more than the first, and the largest integer is twice the middle one. Find the integers.

15. Michelle, Debbie, and Greg worked collectively a total of $9\frac{1}{2}$ hours clearing snow from the sidewalks and driveway of their house. The combined hours that Greg and Debbie worked were $2\frac{4}{5}$ times the number of hours that Michelle worked. The hours that Michelle worked were $2\frac{1}{2}$ times the number of hours that Greg worked. How long did each work?

16. Three tractors were able to plow a field in $12\frac{1}{2}$ hours. The combined number of hours that the first and second tractors worked was $2\frac{1}{8}$ times the number of hours the third tractor worked. The combined number of hours that the first and third tractor worked was $2\frac{1}{2}$ hours more than the number of hours the second tractor worked. How long did each tractor work in the field?

17. A company rents three sizes of automobiles: subcompacts, compacts, and full size. The subcompacts rent for $20 a day, the compacts for $25 a day, and the full size for $40 a day. A total of nine cars are rented and the total daily rental fee for all nine cars is $275. The number of full-size automobiles rented is one more than the number of compacts rented. How many of each kind of automobile are rented?

18. A total of $15,000 was divided into three parts and the three parts were invested at the rates of 9%, 10%, and 11%, respectively. The total interest earned at the end of the first year for all three investments was $1,490. The interest income of $1,490 was not reinvested. In the second year, only the first two amounts were kept invested, but this time the rates of interest were 12% and 14%, respectively. The total interest at the end of the second year on the first two investments was $1,280. How much was invested at each of the three original rates?

19. Maria can row 18 miles down a river and back in 8 hours. Also, she can row 9 miles down river in the same amount of time she can row 3 miles up the river. Find the rate of the boat in still water and the rate of the current in the river.

20. A boat travels downriver at a rate of 10 miles per hour. It takes $1\frac{2}{3}$ times as long to go 1 mile upriver as it does to go 1 mile downriver. Find the speed of the boat and the speed of the current in the river.

Chapter 10 Summary

Vocabulary

addition method
Cramer's rule
dependent equations
determinant
inconsistent equations

minor
principal diagonal
secondary diagonal
substitution method
system of equations

Definitions

10–1 Second-Order Determinant

$$\begin{vmatrix} a_1 & b_1 \\ a_2 & b_2 \end{vmatrix} = a_1 b_2 - a_2 b_1$$

10–2 Third-Order Determinant

$$\begin{vmatrix} a_1 & b_1 & c_1 \\ a_2 & b_2 & c_2 \\ a_3 & b_3 & c_3 \end{vmatrix} = a_1 b_2 c_3 + b_1 c_2 a_3 + c_1 a_2 b_3 - a_3 b_2 c_1 - b_3 c_2 a_1 - c_3 a_2 b_1$$

Rules

10–1 Cramer's Rule

The two linear equations

$$\begin{cases} a_1 x + b_1 y = c_1 \\ a_2 x + b_2 y = c_2 \end{cases}$$

have the solution

$$x = \frac{\begin{vmatrix} c_1 & b_1 \\ c_2 & b_2 \end{vmatrix}}{\begin{vmatrix} a_1 & b_1 \\ a_2 & b_2 \end{vmatrix}} \quad \text{and} \quad y = \frac{\begin{vmatrix} a_1 & c_1 \\ a_2 & c_2 \end{vmatrix}}{\begin{vmatrix} a_1 & b_1 \\ a_2 & b_2 \end{vmatrix}}$$

if $a_1 b_2 - a_2 b_1 \neq 0$.

10–2 Cramer's Rule for a System of Three Equations

The three linear equations

$$\begin{cases} a_1 x + b_1 y + c_1 z = d_1 \\ a_2 x + b_2 y + c_2 z = d_2 \\ a_3 x + b_3 y + c_3 z = d_3 \end{cases}$$

have the solution

$$x = \frac{D_x}{D}, \quad y = \frac{D_y}{D}, \quad \text{and} \quad z = \frac{D_z}{D}$$

where

$$D_x = \begin{vmatrix} d_1 & b_1 & c_1 \\ d_2 & b_2 & c_2 \\ d_3 & b_3 & c_3 \end{vmatrix}, \quad D_y = \begin{vmatrix} a_1 & d_1 & c_1 \\ a_2 & d_2 & c_2 \\ a_3 & d_3 & c_3 \end{vmatrix}$$

$$D_z = \begin{vmatrix} a_1 & b_1 & d_1 \\ a_2 & b_2 & d_3 \\ a_3 & b_3 & d_3 \end{vmatrix}, \quad D = \begin{vmatrix} a_1 & b_1 & c_1 \\ a_2 & b_2 & c_2 \\ a_3 & b_3 & c_3 \end{vmatrix}$$

Miscellaneous Information

1. A linear system of two equations has the form
$$\begin{cases} a_1 x + b_1 y = c_1 \\ a_2 x + b_2 y = c_2 \end{cases}$$

2. For a linear system of two equations, if
$$\frac{a_1}{a_2} = \frac{b_1}{b_2} \neq \frac{c_1}{c_2}$$

    then the system is inconsistent. If
$$\frac{a_1}{a_2} = \frac{b_1}{b_2} = \frac{c_1}{c_2}$$

    then the system is dependent.

3. The minor for any element of a third- (or higher-) order determinant is the determinant that remains after deleting the row and column containing that element. For example, the minor for c_2 in the determinant
$$\begin{vmatrix} a_1 & b_1 & c_1 \\ a_2 & b_2 & c_2 \\ a_3 & b_3 & c_3 \end{vmatrix}$$
    is
$$\begin{vmatrix} a_1 & b_1 \\ a_3 & b_3 \end{vmatrix}$$

4. The array of signs for expanding a determinant of order three or greater by minors about any row or column is
$$\begin{vmatrix} + & - & + & - & + & \cdots \\ - & + & - & + & - & \cdots \\ + & - & + & - & + & \cdots \\ - & + & - & + & - & \cdots \\ \cdot & \cdot & \cdot & \cdot & \cdot & \\ \cdot & \cdot & \cdot & \cdot & \cdot & \\ \cdot & \cdot & \cdot & \cdot & \cdot & \end{vmatrix}$$

Review Problems

Section 10-1

1. Find the solution set for
$$\begin{cases} 5x - 3y = -19 \\ 7x + 4y = -2 \end{cases}$$
by using the addition method.

2. Find the solution set for
$$\begin{cases} 2x - 3y = -14 \\ 5x + 2y = 3 \end{cases}$$
by using the substitution method.

3. Find the solution set for
$$\frac{-2}{x} + \frac{5}{y} = 0$$
$$\frac{3}{x} - \frac{4}{y} = \frac{7}{300}$$
by using the addition method.

Section 10-2

4. Find the solution set for
$$\begin{cases} 2x + 5y - z = -10 \\ -x + y + z = 6 \\ -3x + 2y - z = 10 \end{cases}$$
by using the addition method.

Section 10-3

5. Evaluate the determinant
$$\begin{vmatrix} 2 & 3 \\ -8 & 10 \end{vmatrix}$$

6. Find the solution set for
$$\begin{cases} 2x + y = 12 \\ -4x + 5y = -52 \end{cases}$$
by using Cramer's rule.

Section 10-4 (Optional)

7. Evaluate the determinant
$$\begin{vmatrix} 2 & 4 & 2 \\ 1 & 1 & 0 \\ -3 & 0 & -1 \end{vmatrix}$$
by recopying the first two columns.

8. Evaluate the determinant

$$\begin{vmatrix} 2 & 3 & -3 \\ 1 & -4 & 0 \\ -2 & 1 & 0 \end{vmatrix}$$

by expanding the determinant by minors about the third column.

Section 10–5 (Optional)

9. Find the solution set for

$$\begin{cases} 2x + y & = -3 \\ x & - 4z = -22 \\ x + 3y - 2z = -9 \end{cases}$$

by using determinants.

Section 10–6

10. Find the solution set for

$$\begin{cases} 9x^2 + 16y^2 = 144 \\ y^2 = \dfrac{7x^2}{16} \end{cases}$$

11. Find the solution set for

$$\begin{cases} y = -(x - 4)^2 + 5 \\ x + y = -3 \end{cases}$$

Section 10–7

In problems 12–13, use a system of two equations in two variables to solve.

12. The sum of two numbers is 4 and their difference is $-\frac{14}{5}$. Find the two numbers.

13. On her trip to Bakersfield, Kathy spent $27.10 for 20 gallons of gasoline and 1 quart of oil. On the return trip she spent $34.70 for 25 gallons of gasoline and 2 quarts of oil. Assuming she paid the same price per gallon for gasoline and the same price per quart for oil on both trips, find the cost per gallon for gasoline and the cost per quart for oil.

Use a system of three equations in three variables to solve the following problem.

14. The sum of three integers is 12. If the first is decreased by 13, it equals the third integer. Three times the sum of the second and third one equals the first one. Find the three integers.

Cumulative Test for Chapters 1–10

Chapter 1
1. If $A = \{2, 5, 9\}$ and $B = \{3, 4, 5, 6\}$, find $A \cap B$.
2. Which two axioms does $(3 + 0) + 5 \cdot 1 = 3 + 5$ illustrate?
3. Simplify: $8 - \{4 - 3[-3^2(6 - 8) - 5] + 2\}$.

Chapter 2
4. Simplify: $\dfrac{-9a^5b^4}{-3^2a^8b}$.
5. Give the numerical coefficient of $5x^3y$.
6. Combine like terms: $(9x^3 + 5x - 3) + (-x^3 + 3x^2 - 8)$.
7. Square and simplify: $(8x - 3)^2$.
8. Factor completely: $9x^3 - 27x^2 - x + 3$.

Chapter 3
9. Reduce: $\dfrac{3ax^2 + 3ax - 90a}{12a^2x - 60a^2}$.
10. Divide: $\dfrac{ac + ad + bc + bd}{x^3 - x^2 - x + 1} \div \dfrac{cx + dx}{2x^2 - 2}$.
11. Subtract: $\dfrac{x}{3x - 3y} - \dfrac{y}{3x - 3y}$.

Chapter 4
12. Solve for n: $3(-n - 5) + 2n = -4(n - 5) + 4$.
13. Find the solution for $-5(x - 3) < 3(x - 2) - 1$.
14. Find the solution for $4|2m - 3| < 20$.
15. One machine can do a job in 6 hours and a second machine can do the same job in 4 hours. The slower machine is started at 9 A.M. At 9:30 A.M., the faster machine is started, and both machines run until the job is finished. What time is the job finished?

Chapter 5
16. Simplify: $\dfrac{2^{-3}}{3^2}$.
17. Simplify, and leave all exponents positive: $(-2x^0y^{-3}z^2)^{-2}$.
18. Which is larger: 9.63×10^{-3} or 2×10^3?
19. Add: $3x\sqrt{18x} + 2\sqrt{50x^3}$ $(x > 0)$.
20. Simplify: $\dfrac{8 + 2\sqrt{12}}{6}$.
21. Solve for x: $\sqrt{5x - 19} + \sqrt{x - 3} = 0$. Check your answer.

Chapter 6
22. Solve for s: $(s + 6)(2s^2 + s - 28) = 0$.
23. Solve for x by completing the square: $x^2 = \dfrac{6x + 4}{3}$.

24. Solve for x by using the quadratic formula: $3x^2 = 6x - 2$.
25. Simplify: $5\sqrt{-24} - 3\sqrt{-6}$.
26. Find the values of k so that $2x^2 + kx + 2 = 0$ will have imaginary solutions.
27. Solve for x: $\sqrt{2x - 3} + x = 9$.
28. Find all numbers such that the square of the number is the same as the given number increased by 2.

Chapter 7

29. Is it true or false that every function is a relation?
30. Find the function given by $y = 5x - 3$ if $x \in \{-2, 0, 3\}$.
31. What will be the slope of any line parallel to the graph of $5x + 6y = -3$?
32. What is an equation of the line that goes through the point with coordinates $(-5, 8)$ and is perpendicular to the graph of $3x - y = 4$? Leave your answer in the form $y = mx + b$.
33. Draw the graph of

$$\{(x, y) | 2x - y < 8\} \cap \{(x, y) | x > 3\}$$

Chapter 8

34. In which direction does the parabola given by $x = -6y^2 + 2y - 3$ open? Do not draw the graph to answer the question.
35. What are the coordinates of the points of intersection of the graph of $12x^2 - 4y^2 = 36$ with the x-axis?
36. Does the major axis of the graph of $x^2 + 4y^2 = 36$ lie on the x-axis or the y-axis?
37. Write an equation of the circle with center located at $(-8, 5)$ and with radius $\sqrt{5}$ feet.
38. Suppose x varies directly as the square of y and inversely as z. What is the effect on x if y is tripled and z is halved?
39. If f is a one-to-one function, then what true statement can be made about the inverse of f?

Chapter 9

40. Solve for x: $5^{3x-1} = \dfrac{1}{25}$.
41. Solve for b: $\log_b 25 = 2$.
42. Write $3 \log x - \tfrac{1}{2} \log y + \log z$ as a single logarithm.
43. Solve for x: $\log(2x + 104) - \log(2x + 5) = 2$.

Chapter 10

44. Solve the following system of equations for x and y by using the substitution method.

$$\begin{cases} 2x - 3y = 13 \\ x + 5y = -13 \end{cases}$$

45. Find the solution for the following system of equations by the addition method.
$$\begin{cases} 4x - 2y = -3 \\ 3x + 5y = 4 \end{cases}$$

46. Evaluate the determinant $\begin{vmatrix} 5 & 3 \\ -4 & 2 \end{vmatrix}$.

47. Solve the following system of equations by using Cramer's rule.
$$\begin{cases} 2x + 5y = -2 \\ x - 2y = -1 \end{cases}$$

48. Solve the quadratic system
$$\begin{cases} 2x^2 + y^2 = 9 \\ x - y = 1 \end{cases}$$

49. Use a system of equations in two variables to solve the following: The sum of two numbers is 70 and their difference is -28. Find the two numbers.

50. The Electronics Store sells two kinds of small radios—model C and model D. The store manager plans on selling a total of 500 small radios next month. If the manager must make a profit of $1,145 on the sale of small radios, how many of each model should be sold if the profits on model C and model D are $1.15 and $3.05, respectively? Use a system of two equations in two variables.

CHAPTER 11 | Sequences and Series

Objectives of Chapter 11

To be able to:
- ☐ Give the definition of a sequence. (11–1)
- ☐ Write a general term for a sequence. (11–1)
- ☐ Find the first n terms of a sequence. (11–1)
- ☐ Work with sigma notation. (11–2)
- ☐ Recognize an arithmetic progression. (11–3)
- ☐ Find a specified term in an arithmetic progression. (11–3)
- ☐ Find the arithmetic means between two terms. (11–3)
- ☐ Find the sum of the first n terms of an arithmetic series. (11–3)
- ☐ Recognize a geometric progression. (11–4)
- ☐ Find a specified term in a geometric progression. (11–4)
- ☐ Find the geometric means between two terms. (11–4)
- ☐ Find the sum of the first n terms of a geometric series. (11–4)
- ☐ Recognize which infinite geometric series have a sum. (11–5)
- ☐ Find the sum of an infinite geometric series. (11–5)
- ☐ Write a repeating decimal as a fraction. (11–5)
- ☐ Simplify expressions containing factorial notation. (11–6)
- ☐ Expand a binomial using the binomial theorem. (11–6)
- ☐ Find a specified term in the expansion of a binomial. (11–6)

424 Intermediate Algebra

Number patterns play a very important role in our lives. For example, the number pattern 1, 2, 3, 4, ··· is the set of natural numbers, and your experience with it began the day you learned to count.

Sometimes the sum of the first few numbers in a number pattern is desired. For example, one method of depreciating a house or piece of equipment requires that the sum of the first n natural numbers be found.

In this chapter we learn how to form two special number patterns, and we also learn how to find the sum of the first n numbers in these patterns.

In the last section of the chapter we study a quick and easy way to simplify expressions such as $(x + y)^5$ or $(3x - 4y)^7$.

11–1 Sequences

Suppose this puzzle were proposed to you. Start from where you are and walk halfway to the nearest exit and stop. Now walk one-half of the remaining distance to the exit and stop again. Continue to walk one-half of the remaining distance and stop. What part of the total distance to the exit remains to be walked each time you stop? The first time you stop, one-half of the distance remains. The second time you stop, one-half of one-half, or one-fourth of the distance remains. On the third stop, one-half of one-fourth, or one-eighth of the distance remains. The table shows the pattern formed by the numbers representing the remaining distance to the exit.

Stop	1	2	3	4	5	···
Distance Remaining	$\frac{1}{2}$	$\frac{1}{4}$	$\frac{1}{8}$	$\frac{1}{16}$	$\frac{1}{32}$	···

The numbers $\frac{1}{2}, \frac{1}{4}, \frac{1}{8}, \frac{1}{16}, \frac{1}{32}, \ldots$ form a *sequence*.

Definition 11–1

Sequence Function

A function whose domain is the set of positive integers is called a **sequence function.**

A sequence is formed when the numbers (called terms) in the range of a sequence function are arranged in order. The sequence that is formed may either be a finite sequence of real numbers or an infi-

Sequences and Series

nite sequence of real numbers. The definition for these two kinds of sequences of real numbers is given in definition 11–2.

Definition 11–2

Finite and Infinite Sequences

A **finite sequence** of real numbers is a function whose domain is the set $\{1, 2, 3, \cdots, k\}$ of the first k positive integers.
An **infinite sequence** of real numbers is a function whose domain is the set $\{1, 2, 3, \cdots\}$ of all positive integers.

One notation for a sequence function is $a(n)$. For example, if
$$a(n) = 2n - 3 \tag{1}$$
then by letting $n \in \{1, 2, 3, 4\}$, we find
$$a(1) = 2(1) - 3 = 2 - 3 = -1$$
$$a(2) = 2(2) - 3 = 4 - 3 = 1$$
$$a(3) = 2(3) - 3 = 6 - 3 = 3$$
$$a(4) = 2(4) - 3 = 8 - 3 = 5$$

Thus the sequence function given by equation (1) forms the finite sequence $-1, 1, 3, 5$.

The nth term of a sequence is called the **general term** of the sequence. In the infinite sequence
$$1, 4, 9, \ldots, n^2, \ldots$$
n^2 is a general term. A sequence is completely determined when its general term and domain are known. For example, the sequence just given could be written as $a(n) = n^2$, $n \in \{1, 2, 3, \ldots\}$.

If we return to the sequence formed by the puzzle at the beginning of this section, we see
$$a(n) = \frac{1}{2}, \frac{1}{4}, \frac{1}{8}, \frac{1}{16}, \ldots, \left(\frac{1}{2}\right)^n, \ldots \tag{2}$$

The sequence in (2) can be shortened to read
$$a(n) = \left(\frac{1}{2}\right)^n = \frac{1}{2^n}, \quad n \in N \tag{3}$$

The general term of a sequence is usually given by the symbol a_n rather than $a(n)$. Written in this manner, equation (3) becomes $a_n = 1/2^n$. The first term is then written as a_1, the second as a_2, and so forth.

If the first few terms of a sequence are given, we sometimes can guess a general term. Whatever our guess is, however, it is not unique. For example, suppose we are given the sequence
$$3, 6, 9, \ldots \tag{4}$$

What is the next term? Notice that

$$a_n = 3n \tag{5}$$

or

$$a_n = 3n + (n-1)(n-2)(n-3) \tag{6}$$

or

$$a_n = 3n + (n-1)(n-2)(n-3)(2n-4) \tag{7}$$

would all serve as general terms to give the first three terms in (4). However, equation (5) gives the sequence 3, 6, 9, 12, . . . , a_n, . . . , and equation (6) gives 3, 6, 9, 18, . . . a_n, . . . , and equation (7) gives 3, 6, 9, 36, . . . a_n, ...

Example 1 If $a_n = n + n^2$, find the first four terms of the sequence.

Solution
$a_1 = 1 + (1)^2 = 2$
$a_2 = 2 + (2)^2 = 6$
$a_3 = 3 + (3)^2 = 12$
$a_4 = 4 + (4)^2 = 20$

The first four terms are 2, 6, 12, and 20.

Example 2 If $a_n = \dfrac{(-1)^n(n+1)}{n^2}$, find the first four terms of the sequence.

Solution
$a_1 = \dfrac{(-1)^1(1+1)}{1^2} = \dfrac{-1(2)}{1} = -2$

$a_2 = \dfrac{(-1)^2(2+1)}{2^2} = \dfrac{1(3)}{4} = \dfrac{3}{4}$

$a_3 = \dfrac{(-1)^3(3+1)}{3^2} = \dfrac{-1(4)}{9} = -\dfrac{4}{9}$

$a_4 = \dfrac{(-1)^4(4+1)}{4^2} = \dfrac{1(5)}{16} = \dfrac{5}{16}$

The first four terms are -2, $\tfrac{3}{4}$, $-\tfrac{4}{9}$, and $\tfrac{5}{16}$.

Example 3 Find a general term for the sequence 2, 5, 8, 11, 14, . . . by trial and error.

Solution At this stage of our development, we have no organized method of attack for finding a general term. However, notice the terms are alternately even and odd. This suggests we need to either add or subtract 1 to or from some number. With this suggestion and a little luck (and, no doubt, a few false starts), we find

$$a_n = 3n - 1$$

11–1 Exercises

In exercises 1–23, find the first four terms of the sequence whose general term is given. (See examples 1 and 2.)

1. $a_n = 2n$
2. $a_n = 5n$
3. $a_n = 2n - 1$
4. $a_n = 2n^2 + 1$
5. $a_n = (-1)^n(3n^2 - 1)$
6. $a_n = \dfrac{(-1)^n(2n - 1)}{n}$
7. $a_n = (-1)^n 2^{n+1}$
8. $a_n = (-1)^{n+1} 3^{-n}$
9. $a_n = (-1)^{n-1} 2^{2n-1}$
10. $a_n = \dfrac{(n + 1)^2}{n}$
11. $a_n = n^2 - 2n$
12. $a_n = 5 + \dfrac{1}{n}$
13. $a_n = n + \dfrac{1}{n}$
14. $a_n = (-1)^n$
15. $a_n = 2^{-n}$
16. $a_n = (-1)^n 2^{-n}$
17. $a_n = \dfrac{n^{-1}}{n + 1}$
18. $a_n = 3^{-n}$
19. $a_n = (-2)^n \cdot n$
20. $a_n = (2n)^n$
21. $a_n = n^{-1}$
22. $a_n = (-1)^{n+1} \cdot n$
23. $a_n = n^3 + n^{-2}$

In exercises 24–39, find a general term for each sequence. (See example 3.)

24. $2, 4, 6, 8, \ldots$
25. $4, 7, 10, 13, \ldots$
26. $-1, 1, -1, 1, \ldots$
27. $\dfrac{1}{2}, \dfrac{1}{4}, \dfrac{1}{6}, \dfrac{1}{8}, \ldots$
28. $1, \dfrac{1}{4}, \dfrac{1}{9}, \dfrac{1}{16}, \ldots$
29. $2, \dfrac{3}{2}, \dfrac{4}{3}, \dfrac{5}{4}, \ldots$
30. $-1, \dfrac{4}{3}, -\dfrac{3}{2}, \dfrac{8}{5}, \ldots$
31. $3, \dfrac{9}{2}, 9, \dfrac{81}{4}, \ldots$
32. $-2, -\dfrac{1}{4}, 0, \dfrac{1}{16}, \dfrac{2}{25}, \ldots$
33. $\dfrac{1}{2}, \dfrac{\sqrt{2}}{3}, \dfrac{\sqrt{3}}{4}, \dfrac{2}{5}, \ldots$
34. $1, \dfrac{8}{\sqrt{2}}, \dfrac{27}{\sqrt{3}}, 32, \ldots$
35. $-\dfrac{1}{2}, \dfrac{4}{3}, -\dfrac{9}{4}, \dfrac{16}{5}, \ldots$
36. $x^4, x^5, x^6, x^7, \ldots$
37. $x, \dfrac{x^2}{4}, \dfrac{x^3}{9}, \dfrac{x^4}{16}, \ldots$
38. $\dfrac{-x^2}{2}, \dfrac{x^3}{4}, \dfrac{-x^4}{8}, \dfrac{x^5}{16}, \ldots$
39. $\dfrac{\sqrt{x}}{3}, \dfrac{\sqrt[3]{x}}{9}, \dfrac{\sqrt[4]{x}}{27}, \dfrac{\sqrt[5]{x}}{81}, \ldots$

40. Suppose at the end of any given year a piece of office equipment is worth nine-tenths of the value it had at the beginning of that year. If the piece of equipment cost $800 when new, what is its value at the end of the third year? What is a general term for the sequence?

41. A certain radioactive substance has a half-life of 1,600 years. If 10 kilograms of the substance are in existence today, how much will remain after 3,200 years? How much would remain after 4,800 years? (The half-life of a radioactive substance is the length of time needed for half of the substance to decay.)

42. A weight is attached to a string, and the string is attached to an overhead support that allows the weight to swing freely. If the weight is pulled off center to the left and released, it travels 10 feet on the first swing to the right. On the return swing to the left, it travels 60% of the first distance. The weight continues to swing in a manner that causes it to travel 60% of the distance traveled on the previous swing. How far will it travel on the fourth swing? What is the total distance traveled at the end of the fourth swing?

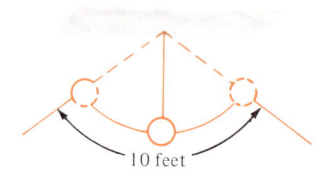

11–2 Sigma Notation and Series

Associated with any sequence is a series. A **series** is the indicated sum of the terms in a sequence. If the sequence is finite, then the number of terms in the series is finite. If the sequence is infinite, then the number of terms in the series is infinite. For example, the infinite sequence

$$4, 9, 14, \ldots, 5n - 1, \ldots$$

has associated with it the infinite series

$$4 + 9 + 14 + \ldots + 5n - 1 + \ldots$$

If a series contains many terms or is infinite, it is not always necessary—or possible—to write all the terms in the series. Instead, an expression that will produce the series is frequently sufficient. Such an expression is called **sigma** (or summation) **notation** for the series. The Greek letter $\sum$ (sigma) is used to designate a sum. For example,

$$\sum_{i=1}^{3} (2i + 3)$$

means

$$\sum_{i=1}^{3} (2i + 3) = [2(1) + 3] + [2(2) + 3] + [2(3) + 3]$$
$$= (2 + 3) + (4 + 3) + (6 + 3)$$
$$= 5 + 7 + 9$$
$$= 21$$

The letter i is called the **index of summation**. The replacement set for i is called the **range of summation**. From the example just given, we see i is to assume natural number values starting with 1 and ending with 3. The series $\sum_{i=1}^{3} (i^2 - 1)$ is the finite series $0 + 3 + 8$.

The designation for an infinite series using sigma notation is

$$\sum_{i=c}^{\infty}$$

where c is some natural number. The symbol ∞ is the symbol for infinity. For example,

$$\sum_{i=3}^{\infty} (2i)^2$$

means

$$\sum_{i=3}^{\infty} (2i)^2 = [2(3)]^2 + [2(4)]^2 + [2(5)]^2 + \cdots$$
$$= 36 + 64 + 100 + \cdots$$

The letter used for the index of summation is chosen arbitrarily, so

$$\sum_{i=1}^{3} 2i = \sum_{j=1}^{3} 2j = \sum_{k=1}^{3} 2k$$

Example 1

(a) Write $\sum_{i=1}^{3} (i + 4)$ in expanded form.

(b) Find the sum of the terms.

Solution

(a) $\sum_{i=1}^{3} (i + 4) = (1 + 4) + (2 + 4) + (3 + 4)$
$= 5 + 6 + 7$

(b) The sum of the terms is $5 + 6 + 7 = 18$

Example 2

Expand: $\sum_{j=1}^{4} \frac{2j + 1}{j^2}$.

Solution

$$\sum_{j=1}^{4} \frac{2j+1}{j^2} = \frac{2(1)+1}{1^2} + \frac{2(2)+1}{2^2} + \frac{2(3)+1}{3^2} + \frac{2(4)+1}{4^2}$$
$$= 3 + \frac{5}{4} + \frac{7}{9} + \frac{9}{16} \quad \blacksquare$$

Example 3 Express the series $1 + \frac{1}{4} + \frac{1}{9} + \frac{1}{16} + \frac{1}{25}$ in sigma notation.

Solution $1 + \frac{1}{4} + \frac{1}{9} + \frac{1}{16} + \frac{1}{25} = \sum_{i=1}^{5} \frac{1}{i^2} \quad \blacksquare$

Example 4 Express the series $1 - 2 + 3 - 4 + 5 - 6 + \cdots$ in sigma notation. (The solution is not unique.)

Solution The series is an infinite series. Consequently, the range of summation will extend to infinity. Also, the terms are alternately positive and negative. This suggests we will need a factor such as $(-1)^i$ or $(-1)^{i+1}$ in the sigma notation. With a little trial and error, we find

$$1 - 2 + 3 - 4 + 5 - 6 + \cdots = \sum_{i=1}^{\infty} (-1)^{i+1} i$$

Also, $\sum_{i=2}^{\infty} (-1)^i (i-1)$ is a solution. $\quad \blacksquare$

11–2 Exercises

In exercises 1–16, write each in expanded form and simplify. (See examples 1 and 2.)

1. $\sum_{i=1}^{4} (2i + 3)$
2. $\sum_{i=1}^{3} (i^2 + 6)$
3. $\sum_{i=1}^{3} (i^2 + i)$
4. $\sum_{i=1}^{4} (3i - i^2)$
5. $3 \sum_{j=1}^{3} (2j - j^3)$
6. $2 \sum_{k=2}^{4} (k^3 - 1)$
7. $\sum_{i=3}^{5} (2i^2 - i)$
8. $\sum_{i=1}^{3} [(i)^i + 4]$
9. $\sum_{i=1}^{3} \frac{i+1}{i^2}$
10. $\sum_{i=1}^{4} \frac{2i+3}{i+1}$
11. $\sum_{k=2}^{5} (-1)^k (4k - 1)$
12. $\sum_{i=1}^{3} (-1)^{i+1} (i^2 + 1)$
13. $\sum_{i=2}^{3} (-1)^i \frac{i^2}{2i+1}$
14. $\sum_{k=8}^{10} k(k+3)$

Sequences and Series

15. $\sum_{i=7}^{9} 2i(i-1)$ **16.** $\sum_{i=2}^{3} i^{(2i)}$

In exercises 17–20, show the expanded form of the given infinite series.

Example

$$\sum_{i=2}^{\infty} (i^2 + 3) = (2^2 + 3) + (3^2 + 3) + (4^2 + 3) + \cdots$$
$$= 7 + 12 + 19 + \cdots$$

17. $\sum_{i=1}^{\infty} (2i + 1)$ **18.** $\sum_{i=1}^{\infty} i^2$

19. $\sum_{i=4}^{\infty} \frac{3i}{i+5}$ **20.** $\sum_{i=3}^{\infty} \frac{i(i+1)}{3}$

In exercises 21–28, write the given series using sigma notation. (See examples 3 and 4.)

21. $2 + 4 + 6 + 8 + 10$ **22.** $1 + 4 + 9 + 16$

23. $\frac{1}{2} + \frac{2}{3} + \frac{3}{4} + \frac{4}{5}$ **24.** $\frac{1}{3} + \frac{4}{3} + \frac{9}{3} + \frac{16}{3}$

25. $-\frac{1}{3} + \frac{2}{4} - \frac{3}{5} + \frac{4}{6}$ **26.** $\frac{2}{5} + \frac{4}{6} + \frac{6}{7} + \frac{8}{8} + \cdots$

27. $16 - 25 + 36 - 49 + \cdots$ **28.** $\frac{2}{3} - \frac{4}{9} + \frac{8}{27} - \frac{16}{81} + \cdots$

29. Show $\sum_{i=1}^{3} (i^2 + 2i) = \sum_{i=1}^{3} i^2 + \sum_{i=1}^{3} 2i$.

30. Show $\sum_{i=1}^{3} i(i + 1) \neq \left[\sum_{i=1}^{3} i\right]\left[\sum_{i=1}^{3} (i + 1)\right]$.

31. Show $\sum_{i=1}^{4} 3 = 12$.

32. Show $\sum_{i=1}^{4} 2i = 2 \sum_{i=1}^{4} i$.

33. Show by example that $\sum_{i=c}^{k} i$ has $(k - c) + 1$ terms in its expansion.

34. Use the ideas expressed in exercises 31 and 33 to prove $\sum_{i=1}^{n} c = nc$, where c is a constant.

35. Use the result of exercise 33 to show that $\sum_{i=0}^{n} i$ has $n + 1$ terms in its expansion.

36. Prove $\sum_{i=1}^{n} (3i + 1) = 3 \sum_{i=1}^{n} i + n$.

11-3 Arithmetic Progressions

If a sequence has the property that any term in the sequence after the first is found by adding the same number to the preceding term, it is called an **arithmetic progression.** The number added to each term is called the **common difference** and is represented by the letter d.

The sequence

$$4, 9, 14, 19 \ldots$$

is an arithmetic progression. This can be verified by observing the common difference between any two adjacent terms is 5. The value of d is found by subtracting any term from the next term in the series.

$$d = 9 - 4 = 5$$
$$d = 14 - 9 = 5$$
$$d = 19 - 14 = 5$$

Example 1

Find the common difference for each arithmetic progression.

(a) 11, 14, 17, 20, . . .
(b) $-4, -7, -10, -13, \ldots$
(c) $-5\frac{1}{2}, -2\frac{1}{2}, \frac{1}{2}, 3\frac{1}{2},$
(d) $2\frac{2}{3}, 3\frac{2}{9}, 3\frac{7}{9}, 4\frac{1}{3}$

Solution

(a) $d = 14 - 11 = 3$
(b) $d = -7 - (-4) = -7 + 4 = -3$
(c) $d = -2\frac{1}{2} - \left(-5\frac{1}{2}\right) = -2\frac{1}{2} + 5\frac{1}{2} = 3$
(d) $d = 3\frac{2}{9} - 2\frac{2}{3} = \frac{29}{9} - \frac{8}{3} = \frac{29}{9} - \frac{24}{9} = \frac{5}{9}$ ■

In general, an arithmetic progression may be written as

$$a_1, a_1 + d, a_1 + 2d, a_1 + 3d, \ldots$$

where a_1 is the first term and d is the common difference. Notice that the number of differences added to a_1 to give any particular term is one less than the number of that term in the sequence. That is, if we want the fourth term, we must add three differences to a_1. This observation leads to the general formula for finding the nth term, which we denote by a_n.

Rule 11-1

_n_th Term of an Arithmetic Progression

The nth term of an arithmetic progression is given by $a_n = a_1 + (n - 1)d$, where a_1 is the first term and d is the common difference.

Example 2 Find the 21st term of the progression

$$-4, 2, 8, 14, \ldots$$

Solution For this example, $a_1 = -4$, $d = 6$, and $n = 21$.

$$\begin{aligned} a_n &= a_1 + (n - 1)d \\ &= -4 + (21 - 1)(6) \\ &= -4 + (20)(6) \\ &= 116 \end{aligned}$$ ■

The terms between two given terms of an arithmetic progression are called the **arithmetic means.** For example, in the progression 5, 8, 11, 14, 17, . . . , the arithmetic means between 5 and 17 are 8, 11, and 14.

Example 3 Find three arithmetic means between -3 and 7.

Solution Let us draw some blanks to hold the place for the missing terms.

$$-3, \underline{\hspace{1em}}, \underline{\hspace{1em}}, \underline{\hspace{1em}}, 7$$

We know $a_1 = -3$, $a_n = a_5 = 7$, and $n = 5$. By using rule 11-1, we can find d.

$$\begin{aligned} a_n &= a_1 + (n - 1)d \\ 7 &= -3 + (5 - 1)d \\ 7 &= -3 + 4d \\ 4d &= 10 \\ d &= 2\tfrac{1}{2} \end{aligned}$$

If we add $2\tfrac{1}{2}$ successively, the progression is

$$-3, -\tfrac{1}{2}, 2, 4\tfrac{1}{2}, 7$$

and the three arithmetic means are $-\tfrac{1}{2}$, 2, and $4\tfrac{1}{2}$. ■

Associated with any arithmetic progression is an **arithmetic series.** For example, associated with

$$3, 7, 11, 15, \ldots$$

is the series
$$3 + 7 + 11 + 15 + \cdots$$

The sum of the first n terms of an arithmetic series is given by the symbol S_n. The method for finding this sum is given in rule 11–2.

Rule 11–2

Sum of an Arithmetic Series

The sum of the first n terms of an arithmetic series is given by

$$S_n = \frac{n}{2}(a_1 + a_n)$$

where a_1 is the first term of the series and a_n is the nth term of the series.

Proof of Rule 11–2

The sum, S_n, of the first n terms can be written as

$$S_n = a_1 + (a_1 + d) + (a_1 + 2d) + \cdots + (a_n - 2d) \quad (1)$$
$$+ (a_n - d) + a_n$$

Reverse the terms in equation (1) to get

$$S_n = a_n + (a_n - d) + (a_n - 2d) + \cdots + \quad (2)$$
$$(a_1 + 2d) + (a_1 + d) + a_1$$

Now add equations (1) and (2).

$$2S_n = (a_1 + a_n) + (a_1 + a_n) + (a_1 + a_n) + \cdots \quad (3)$$
$$+ (a_1 + a_n) + (a_1 + a_n) + (a_1 + a_n)$$

There are n terms in equation (3), all of which are the same. We can write equation (3) as

$$2S_n = n(a_1 + a_n)$$

Then

$$S_n = \frac{n}{2}(a_1 + a_n)$$

Because $a_n = a_1 + (n - 1)d$, an alternate form of rule 11–2 is

$$S_n = \frac{n}{2}[a_1 + a_1 + (n - 1)d]$$
$$= \frac{n}{2}[2a_1 + (n - 1)d] \quad (4)$$

Example 4

Find the sum of the first 12 terms of the series
$$-6 + (-1) + 4 + 9 + \cdots$$

Sequences and Series   435

Solution We find the sum by using equation (4) and noting that $n = 12$, $a_1 = -6$, and $d = 5$.

$$S_{12} = \frac{12}{2}[2(-6) + 11(5)]$$
$$= 6(-12 + 55)$$
$$= 258$$

Example 5 Frank deposited $100 to Amy's bank account on her first birthday, $200 on her second birthday, $300 on her third birthday, and so on, continuing to deposit money in this pattern to her account through her twelfth birthday. What total amount of money was deposited to her account?

Solution If we write the amounts deposited, we get the series

$$100 + 200 + 300 + \cdots + 100n$$

where $100n = 100(12) = 1,200$. We now know the following information:

$$n = 12$$
$$a_1 = 100$$
$$a_{12} = 1,200$$

If we use rule 11–2 to find the sum, we get

$$S_{12} = \frac{12}{2}(100 + 1,200)$$
$$= 6(1,300)$$
$$= 7,800$$

Amy had a total of $7,800 deposited to her account by the time she was 12 years old.

11–3 Exercises

In exercises 1–5, determine if the sequences are arithmetic progressions. Give the value of d for those that are. (See example 1.)

1. 5, 7, 9, 11, . . .

2. 9, 13, 17, 21, . . .

3. $\sqrt{3}, \frac{4\sqrt{3}}{3}, \frac{5\sqrt{3}}{3}, 2\sqrt{3} \ldots$

4. $\sqrt{3}, \sqrt{6}, \sqrt{9}, \sqrt{12}, \ldots$

5. $\frac{1}{2}, \frac{1}{4}, \frac{1}{8}, \frac{1}{16}, \ldots$

6. In the arithmetic progression ———, $\frac{\sqrt{5}}{4}$, $\sqrt{5}$, what is the missing term?

In exercises 7–10, find the indicated term of the arithmetic progression. (See example 2.)

7. 12, 16, 20, 24, . . . ; the tenth term

8. 10, 7, 4, 1, . . . ; the eighth term

9. $\frac{3}{4}, \frac{1}{4}, -\frac{1}{4}, -\frac{3}{4}, \ldots$; the eleventh term.

10. $\frac{5}{8}, \frac{7}{8}, \frac{9}{8}, \frac{11}{8}, \ldots$; the twenty-first term.

11. Find three arithmetic means between -6 and 26. (See example 3.)

12. Find five arithmetic means between $\frac{1}{2}$ and $\frac{5}{2}$.

13. Find four arithmetic means between $\frac{1}{5}$ and $\frac{29}{20}$.

14. In an arithmetic progression, $a_1 = 2$, $a_n = 12$, and $n = 6$. What is the common difference? Write the first four terms of the progression. (*Hint:* Fill in the known values in the equation $a_n = a_1 + (n - 1)d$.)

15. In an arithmetic progression, $a_5 = 3$, $n = 5$, and $d = \frac{2}{3}$. Find a_1 and write the first four terms of the progression. (See exercise 14 for a hint.)

16. How many terms are in an arithmetic progression if the first term is 3, the last term is 87, and the common difference is 4?

17. How many terms are in an arithmetic progression if $a_1 = \frac{3}{2}$, $d = 4$, and $a_n = 129\frac{1}{2}$?

18. Find the sum of the first 12 terms of the arithmetic series $19 + 27 + 35 + 43 + \cdots$. (See example 4.)

19. Find the sum of the first 75 natural numbers. (*Hint:* Write the first five terms of the series and then see example 4.)

20. Find the sum of all the natural numbers between 1 and 721 that are divisible by 5.

21. Find the sum of all the natural numbers between 2 and 104 that are divisible by 3.

22. If $S_n = 10$, $a_1 = \frac{1}{2}$ and $a_n = \frac{7}{2}$, find n.

23. If $S_n = 4.2$, $a_1 = 0.4$, and $n = 6$, find d.

24. Steve gave Lori one yellow rose for their first anniversary and two yellow roses for their second anniversary; he continued giving her roses in this manner through their eighteenth anniversary. How many roses has Lori received altogether from Steve? (See example 5.)

25. Julie's mother deposited $200 in a bank account on her first birthday, $400 on her second birthday, and $600 on her third birthday; this pattern was continued through Julie's twentieth birthday. How much money was deposited into Julie's account? (See example 5.)

26. Suppose someone offered you a job for 1 year at a salary of $50,000 or offered instead to pay you $1 for the first day's work, $2 for the second day's work, and $3 for the third day's work; they continued paying you in this manner for 365 days. Which offer would pay you the greater amount of money?

Sequences and Series

27. Earl bought a piece of equipment with no down payment for $10,000. He agreed to pay $2,000 annually plus 3% interest on the amount unpaid at the end of that year. How long did it take him to pay for the equipment? How much interest did he pay on the $10,000 debt?

28. If $x - 6$, $x + 3$, and $2x$ form an arithmetic progression, find x and give the first four terms of the progression.

29. Three numbers whose sum is 15 and whose product is 80 form an arithmetic progression. Find the three numbers.

30. Nine is the arithmetic mean between two numbers. The product of these two numbers is 77. Find the two numbers and write the first three terms of the progression.

31. At the end of 1 year after being purchased new, the value of an automobile was $800 less than its original price. For each successive year thereafter, the value of the automobile decreased by $500 per year. If the purchase price was $6,000, in how many years will the car be worth $1,200?

32. Four numbers form an arithmetic progression. The sum of the first two terms is 20 and the sum of the second and fourth terms is 38. Find the four numbers.

33. Show that the sum of the first n terms of $3 + 5 + 7 + 9 + \cdots + 2n + 1 + \cdots$ is $2n + n^2$.

34. Show that the sum of the first n terms of $10 + 14 + 18 + 22 + \cdots + 4n + 6 + \cdots$ is $8n + 2n^2$.

11-4 Geometric Progressions

A **geometric progression** is a sequence in which each term after the first one is found by multiplying the preceding term by a fixed number. The fixed number is called the **common ratio.** For example,

$$5, 15, 45, 135, \ldots$$

is a geometric progression. The common ratio is 3 and is found by dividing any term (except the first) by the preceding term. In the example just given, $15 \div 5$ gives 3, $45 \div 15$ gives 3, and $135 \div 45$ gives 3. The common ratio is represented by the letter r.

Example 1   Find the common ratio for each geometric progression.

(a) $2, 4, 8, 16, 32, \ldots$

(b) $\dfrac{1}{2}, -\dfrac{1}{10}, \dfrac{1}{50}, -\dfrac{1}{250}, \ldots$

(c) $\dfrac{1}{\sqrt{3}}, 1, \sqrt{3}, 3, \ldots$

Solution

(a) $r = \dfrac{4}{2} = 2$

(b) $r = \dfrac{-\dfrac{1}{10}}{\dfrac{1}{2}} = \left(-\dfrac{1}{10}\right)\left(\dfrac{2}{1}\right) = -\dfrac{1}{5}$

(c) $r = \dfrac{1}{1/\sqrt{3}} = \sqrt{3}$

In general, a geometric progression may be represented as $a_1, a_1r, a_1r^2, a_1r^3, \ldots$, where a_1 is the first term of the progression and r is the common ratio. Notice that each term is the product of a_1 and a power of r. The exponent of r for any term is always one less than that term. That is, if we want the fourth term, we must multiply the first term by r^3. From this, we see the general term of a geometric progression is

$$a_n = a_1 r^{n-1}$$

Rule 11–3

*n*th **Term of a Geometric Progression**

The nth term of a geometric progression is given by $a_n = a_1 r^{n-1}$, where a_1 is the first term and r is the common ratio.

Example 2

Find the fourth term of the geometric progression whose first term is 3 and whose common ratio is $\tfrac{1}{3}$. Write the first four terms of the progression.

Solution

We know $a_1 = 3$ and $r = \tfrac{1}{3}$. Using rule 11–3, we get

$$a_n = a_1 r^{n-1}$$

$$a_4 = 3\left(\dfrac{1}{3}\right)^3$$

$$= 3\left(\dfrac{1}{27}\right)$$

$$= \dfrac{1}{9}$$

The progression is $3, 1, \tfrac{1}{3}, \tfrac{1}{9} \ldots$

Sequences and Series

The terms between two given terms of a geometric progression are called the **geometric means.** For example, in the progression -4, 12, -36, 108, . . . , the geometric means between -4 and 108 are 12 and -36.

Example 3 Find two geometric means between -8 and 216.

Solution Write the progression with some blanks to hold the place for the missing terms.

$$-8, \text{_____}, \text{_____}, 216$$

We know $a_1 = -8$, $a_n = a_4 = 216$, and $n = 4$. Using rule 11–3, we write

$$a_n = a_1 r^{n-1}$$
$$216 = -8r^3$$
$$r^3 = -27$$
$$r = \sqrt[3]{-27} = -3$$

If we multiply successively by -3, the progression is

$$-8, 24, -72, 216$$

and the geometric means are 24 and -72. ■

Associated with any geometric sequence is a **geometric series.** For example, associated with the geometric progression

$$\frac{1}{2}, 1, 2, 4, \ldots$$

is the geometric series

$$\frac{1}{2} + 1 + 2 + 4 + \ldots$$

The sum of the first n terms of a geometric series is represented by S_n. The method for finding this sum is given in rule 11–4.

Rule 11–4

Sum of a Geometric Series

The sum of the first n terms of a geometric series is

$$S_n = \frac{a_1(r^n - 1)}{r - 1}$$

where a_1 is the first term of the series and r is the common ratio $(r \neq 1)$.

Proof of Rule 11–4

The sum, S_n, of the first n terms of a geometric series can be written as

$$S_n = a_1 + a_1 r + a_1 r^2 + \cdots + a_1 r^{n-1} \qquad (1)$$

We multiply equation (1) by r.

$$r S_n = a_1 r + a_1 r^2 + \cdots + a_1 r^{n-1} + a_1 r^n \qquad (2)$$

Next, subtract equation (1) from equation (2)

$$r S_n - S_n = -a_1 + a_1 r^n$$

Factor both sides of this equation and solve for S_n.

$$S_n(r - 1) = a_1(r^n - 1)$$

$$S_n = \frac{a_1(r^n - 1)}{r - 1} \qquad (r \neq 1)$$

An alternate form of rule 11–4 is

$$S_n = \frac{a_n r - a_1}{r - 1} \qquad (r \neq 1) \qquad (3)$$

The proof of this is left as an exercise.

Example 4

Find the sum of the first five terms of the geometric series $\frac{1}{2} + 1 + 2 + 4 + \cdots$.

Solution

We know $a_1 = \frac{1}{2}$ and $r = 2$.

$$S_n = \frac{a_1(r^n - 1)}{r - 1}$$

$$S_5 = \frac{\frac{1}{2}[(2)^5 - 1]}{2 - 1}$$

$$= \frac{\frac{1}{2}(32 - 1)}{1}$$

$$= \frac{1}{2}(31)$$

$$= 15\frac{1}{2}$$

11–4 Exercises

In exercises 1–6, determine if the sequences are geometric progressions. For those that are, give the value of r. *(See example 1.)*

1. $1, 3, 9, 27, \ldots$

2. $5, 2\frac{1}{2}, 1\frac{1}{4}, \frac{5}{8}, \ldots$

3. $\sqrt{5}, 10, 20\sqrt{5}, 200 \ldots$

4. $x, x^3, x^5, x^6, \ldots$

5. $x^2, x^5, x^8, x^{10}, \ldots$

6. $x^2y, x^3y^2, x^4y^3, x^5y^4, \ldots$

7. In the geometric progression ———, $\frac{1}{24}, \frac{1}{72}$, what is the missing term?

In exercises 8–12, find the indicated term of the geometric progression. (See example 2.)

8. $6, 12, 24, 48, \ldots$; the eighth term

9. $-12, 6, -3, 1\frac{1}{2}, \ldots$; the seventh term

10. $\sqrt{3}, 1, \frac{\sqrt{3}}{3}, \frac{1}{3}, \ldots$; the seventh term

11. $\frac{3}{2}, \frac{1}{2}, \frac{1}{6}, \frac{1}{18}, \ldots$; the sixth term

12. $\frac{8}{5}, -\frac{4}{5}, \frac{2}{5}, -\frac{1}{5}, \ldots$; the ninth term

13. Find three geometric means between 10 and $\frac{5}{8}$ (two possible sets of means). (See example 3.)

14. Find two geometric means between 15 and $\frac{3}{25}$.

15. The eighth term of a geometric progression is $\frac{1}{16}$ and the third term is -2. Find r and the sixth term.

16. The sixth term of a geometric progression is $\frac{1}{27}$ and the second term is 3. Find r and the seventh term (two answers possible).

17. The sum of the second term and sixth term of a geometric progression is 68. The sum of the third term and seventh term is -136. Find the common ratio and write the first four terms of the progression.

In exercises 18–22, find the sum of the indicated number of terms of the given geometric series. (See example 4.)

18. $1 + 2 + 4 + 8 + \cdots$; the first eight terms

19. $27 + 9 + 3 + 1 + \cdots$; the first five terms.

20. $16 + 8 + 4 + 2 + \cdots$; the sum of the third through seventh terms, inclusively.

21. $\frac{1}{9} + \frac{1}{3} + 1 + 3 + \cdots$; the sum of the third through sixth terms, inclusively.

22. $14 + 7 + \frac{7}{2} + \frac{7}{4} + \cdots$; the sum of the second through fifth terms, inclusively.

23. In a geometric progression, $a_n = \frac{3}{8}, r = \frac{1}{2}$, and $n = 6$. Find a_1 and S_6.

24. In a geometric progression, $a_1 = 4, a_n = 256$, and $n = 7$. Find r and S_7 (two sets of answers possible).

25. If a rubber ball is dropped, it will rebound one-third of the height from which it was dropped. If the ball is dropped from a height of 9 feet, how far will the ball have traveled when it strikes the ground for the fourth time?

442 Intermediate Algebra

26. A bag containing 144 marbles was passed among 5 children. Each child, in turn, took one-half of the marbles that remained in the bag when the bag was passed to him or her. How many marbles remained in the bag when it reached the fifth child? Could the bag be passed in this manner to a sixth child?

27. Prove equation (3).

28. Show that the sum of the first n terms for a geometric series is $a_1 n$ if $r = 1$.

11–5 Infinite Geometric Series

Let us return to the puzzle proposed at the beginning of section 11–1. Recall the puzzle required you walk one-half of the distance to an exit and stop. Then you had to walk one-half of the remaining distance to the exit and stop again. Suppose you had walked a distance of 10 feet when you stopped the first time. Since this 10-foot walk was half the distance to the exit, then the exit was originally 20 feet away. This pattern of walking would create the geometric series,

$$10 + 5 + \frac{5}{2} + \frac{5}{4} + \frac{5}{8} + \cdots$$

We know the exit lies 20 feet away from where you started walking, and it seems reasonable to say that the sum to infinity for the series $10 + 5 + \frac{5}{2} + \frac{5}{4} + \frac{5}{8} + \cdots$ is 20 feet. We prove this is the case shortly.

Consider the equation

$$S_n = \frac{a_1(r^n - 1)}{r - 1} \tag{1}$$

developed in the last section. If we multiply the numerator and denominator by -1, we can write equation (1) as

$$S_n = \frac{a_1(1 - r^n)}{1 - r} \tag{2}$$

Furthermore, equation (2) may now be written as

$$S_n = \frac{a_1}{1 - r}(1 - r^n) \tag{3}$$

By using advanced methods, it is possible to show that geometric series for which $|r| < 1$ have sums, while those for which $|r| \geq 1$ do not have sums.

We can see this intuitively by observing that $|r| < 1$ means $-1 < r < 1$. That is, r is a proper fraction. If any proper fraction is raised to

Sequences and Series

larger and larger positive integral powers, the result becomes smaller and smaller. To see this let $r = \frac{2}{3}$. Then

$$\left(\frac{2}{3}\right)^1 = \frac{2}{3}$$

$$\left(\frac{2}{3}\right)^2 = \frac{4}{9}$$

$$\left(\frac{2}{3}\right)^3 = \frac{8}{27}$$

$$\left(\frac{2}{3}\right)^4 = \frac{16}{81}$$

$$\vdots$$

$$\left(\frac{2}{3}\right)^{10} = \frac{1{,}024}{59{,}049}$$

By taking the exponent as large as we please, we can force $(\frac{2}{3})^n$ (n a natural number) to become as small as we wish.

If we apply this idea to equation (3), the factor $1 - r^n$ can be made to become as close to 1 as we please for $|r| < 1$ and n a positive integer. If n is extremely large, $1 - r^n$ has a negligible effect upon S_n in equation (3). If we ignore this negligible effect of $1 - r^n$, equation (3) can be written as

$$S_n = \frac{a_1}{1 - r} \qquad (4)$$

It is customary to write equation (4) as

$$S_\infty = \frac{a_1}{1 - r}, \qquad |r| < 1 \qquad (5)$$

Rule 11–5

Sum of an Infinite Geometric Series

If $|r| < 1$ and $S_\infty = a_1 + a_1 r + a_1 r^2 + a_1 r^3 + \cdots$, then

$$S_\infty = \frac{a_1}{1 - r}$$

The symbol S_∞ means we are to find the "sum" (more properly, the limit of the sum) of all the terms in an infinite geometric series.

Example 1

Prove that the distance which you walked to get to the exit is 20 feet. That is, prove that the series

$$10 + 5 + \frac{5}{2} + \frac{5}{4} + \cdots$$

has a sum of 20.

Solution

We know $a_1 = 10$ and $r = \frac{1}{2}$.

$$S_\infty = \frac{a_1}{1-r} = \frac{10}{1-\frac{1}{2}} = \frac{10}{\frac{1}{2}} = 20.$$

Example 2

Determine if the geometric series

$$10 + 11 + \frac{121}{10} + \cdots$$

has a sum.

Solution

We must determine if $-1 < r < 1$. If we divide the second term by the first one, we get

$$r = \frac{11}{10}$$

This value for r is greater than or equal to 1. Therefore, the series does not have a sum.

Example 3

Find the sum for the series

$$2 + 1 + \frac{1}{2} + \frac{1}{4} + \cdots$$

Solution

$r = \frac{1}{2}$ and $a_1 = 2$. Therefore

$$S_\infty = \frac{a_1}{1-r} = \frac{2}{1-\frac{1}{2}} = \frac{2}{\frac{1}{2}} = 4$$

Example 4

A decimal number is called a *repeating decimal* if it has a group of digits that repeats. Prove that the repeating decimal

$$0.171717\ldots$$

can be written as the fraction $\frac{17}{99}$.

Solution

The repeating decimal can be written as

$$0.17 + 0.0017 + 0.000017 + \cdots$$

We find

$$r = \frac{0.0017}{0.17} = 0.01 = \frac{1}{100}$$

Sequences and Series

If we use $a_1 = 0.17$ and $r = \frac{1}{100}$, we get

$$S_\infty = \frac{0.17}{1 - \frac{1}{100}} = \frac{\frac{17}{100}}{1 - \frac{1}{100}} = \frac{\frac{17}{100}}{\frac{99}{100}}$$

$$= \frac{17}{100} \cdot \frac{100}{99} = \frac{17}{99}$$

You can check the solution by dividing 17 by 99.

In example 4, the decimal number 0.171717 . . . is frequently written as $0.\overline{17}$. The bar gives a concise way to indicate the repeating digits.

Example 5

If the sum of an infinite geometric series is $\frac{9}{2}$ and $r = \frac{1}{3}$, find a_1 and list the first four terms of the series.

Solution

$$S_\infty = \frac{a_1}{1 - r}$$

$$\frac{9}{2} = \frac{a_1}{1 - \frac{1}{3}}$$

$$\frac{9}{2} = \frac{a_1}{\frac{2}{3}}$$

$$\left(\frac{9}{2}\right)\left(\frac{2}{3}\right) = a_1$$

$$a_1 = 3$$

The series is $3 + 1 + \frac{1}{3} + \frac{1}{9} + \cdots$.

11–5 Exercises

In exercises 1–14, find the sum of the infinite geometric series. If the series has no sum, explain why. (See examples 1–3.)

1. $\frac{1}{8} + \frac{1}{24} + \frac{1}{72} + \cdots$

2. $\frac{9}{16} + \frac{9}{32} + \frac{9}{64} + \cdots$

3. $-20 + \frac{25}{2} - \frac{125}{16} + \cdots$

4. $1 + \frac{3}{2} + \frac{9}{4} + \cdots$

5. $\frac{3}{8} + \frac{1}{2} + \frac{2}{3} + \cdots$

6. $\frac{1}{3} + \frac{1}{3\sqrt{3}} + \frac{1}{9} + \cdots$

7. $\frac{1}{5} + \frac{1}{5\sqrt{5}} + \frac{1}{25} + \cdots$

8. $0.02 + 0.0002 + 0.000002 + \cdots$

9. $0.05 + 0.001 + 0.00002 + \cdots$

10. $\dfrac{5}{8} - \dfrac{1}{8} + \dfrac{1}{40} - \dfrac{1}{200} + \cdots$

11. $\dfrac{3}{7} - \dfrac{1}{7} + \dfrac{1}{21} - \dfrac{1}{63} + \cdots$

12. $-\dfrac{14}{3} + \dfrac{7}{3} - \dfrac{7}{6} + \cdots$

13. $-\dfrac{18}{5} + \dfrac{6}{5} - \dfrac{2}{5} + \cdots$

14. $\sqrt{2} - 1 + \dfrac{1}{\sqrt{2}} - \dfrac{1}{2} + \cdots$

In exercises 15–18, find a fraction equivalent to each repeating decimal. (See example 4.)

15. $0.\overline{18}$
16. $0.\overline{132}$
17. $2.31\overline{616}$
18. $5.2\overline{231}$

19. If the sum of an infinite geometric series is $\tfrac{5}{4}$ and $a_1 = \tfrac{5}{8}$, find r and list the first four terms of the series. (See example 5.)

20. If the sum of an infinite geometric series is $-\tfrac{4}{9}$ and $r = -\tfrac{1}{2}$, find a_1 and list the first four terms of the series. (See example 5.)

21. If the sum of an infinite geometric series is $\tfrac{9}{2}$, and if the second term has a value of 1, find r and a_1 and list the first four terms of the series (two answers possible).

22. An infinite geometric series, which has a sum, has terms that alternate positive and negative with the first term being positive. If the third term has a value of $\tfrac{1}{21}$ and if the fifth term has a value of $\tfrac{1}{189}$, find r, list the first four terms of the series, and find the sum.

23. In an infinite geometric series that has a sum, the first term has a value of $\tfrac{5}{9}$ and the fourth term has a value of $-\tfrac{40}{243}$. Find r, list the first four terms of the series, and find the sum of the series.

24. Water is removed from a container in such a manner that each minute, 0.2 of the remaining water in the container is removed. What percent of the water is left in the container at the end of 4 minutes?

25. A pendulum travels 10 feet on the first swing. Each swing thereafter is 0.7 of the distance of the previous swing. How many feet will the pendulum have traveled when it comes to rest?

26. A woman gave $40,000 to charity in one year. The next year she gave $20,000. The year after that she gave $10,000. If she (or her descendents) could continue this pattern of giving indefinitely, what would be the total amount that would be given to charity?

11–6 The Binomial Theorem

In this section, we learn how to expand a binomial raised to any positive integral power without having to multiply the problem factor by factor, which is often very tedious.

Before looking at the expansion of a binomial, we need to look at **factorial notation.** We frequently need to write quantities such as $5 \cdot 4 \cdot 3 \cdot 2 \cdot 1$ or $7 \cdot 6 \cdot 5 \cdot 4 \cdot 3 \cdot 2 \cdot 1$. These products always start with a positive integer and each succeeding factor is one less than the preceding one. This pattern continues until a factor of 1 is reached. We will agree to write $5 \cdot 4 \cdot 3 \cdot 2 \cdot 1$ as $5!$ and $7 \cdot 6 \cdot 5 \cdot 4 \cdot 3 \cdot 2 \cdot 1$ as $7!$. In general $n(n-1)(n-2) \cdots (2)(1) = n!$; $n!$ is read "n factorial."

Example 1 Evaluate $4!$

Solution $4! = 4 \cdot 3 \cdot 2 \cdot 1 = 24$

Example 2 Evaluate $\dfrac{7!}{3!4!}$

Solution $\dfrac{7!}{3!4!} = \dfrac{7 \cdot 6 \cdot 5 \cdot 4!}{(3 \cdot 2 \cdot 1)(4!)} = \dfrac{7 \cdot 6 \cdot 5}{3 \cdot 2 \cdot 1} = 35$

Notice in example 2 the last four factors of $7!$ can be written as $4!$ and divided out (or canceled) with the $4!$ in the denominator.

We must make the two special definitions:

$$1! = 1 \quad \text{and} \quad 0! = 1$$

To see why $0! = 1$, write $n!$ as $n(n-1)!$ Then for $n = 1$, $n! = n(n-1)!$ becomes

$$1! = 1(1-1)!$$
$$1 = 1(0!)$$

If $1(0!)$ is to be equal to 1, we must define $0!$ to be 1.

Let us now return to raising a binomial to a power. If we take the time to multiply $(a + b)$ by itself repeatedly, we obtain the following expansions.

$$(a + b)^0 = 1$$
$$(a + b)^1 = a + b$$
$$(a + b)^2 = a^2 + 2ab + b^2$$
$$(a + b)^3 = a^3 + 3a^2b + 3ab^2 + b^3$$
$$(a + b)^4 = a^4 + 4a^3b + 6a^2b^2 + 4ab^3 + b^4$$
$$(a + b)^5 = a^5 + 5a^4b + 10a^3b^2 + 10a^2b^3 + 5ab^4 + b^5$$

Careful study of these expansions reveals several patterns and characteristics, which we list. We let any binomial raised to its respective power in the list be represented by $(a + b)^n$:

1. Each expansion contains $n + 1$ terms.
2. The first and last terms are a^n and b^n, respectively.

3. The powers of a begin with n in the first term and decrease by 1 in each succeeding term until the power is 0 in the last term. (Because $a^0 = 1$ for $a \neq 0$, the factor of a^0 is not usually shown in the last term.)
4. The powers of b begin with 0 in the first term (although it is not shown, since $b^0 = 1$ for $b \neq 0$) and increase by 1 for each succeeding term; the power is n in the last term.
5. The sum of the powers on a and b in any given term equals n.
6. The numerical coefficients of the first and last terms are 1.
7. The numerical coefficient of the second term and the next to the last term is n. (*Numerical coefficient* is implied whenever the word *coefficient* is used.).
8. The coefficient of any term after the first can be found in the following way: Take the coefficient of the preceding term, multiply it by the exponent on a in that term and divide this product by one more than the exponent on b in that term. For example, in $(a + b)^5$, we know the second term is $5a^4b$. Therefore, the coefficient of the third term is $5(4)/2 = 10$. This pattern can be followed to write the coefficient of any term from the one preceding it.

Example 3

The first three terms of the expansion of $(a + b)^8$ are $a^8 + 8a^7b + 28a^6b^2$. What is the fourth term?

Solution

From the eighth statement just given, the coefficient of the next term is $28(6)/3 = 56$. From the third and fourth statements, the exponents on a and b are 5 and 3, respectively. Therefore, the fourth term is $56a^5b^3$.

Example 4

Expand and simplify: $(x + 2)^4$.

Solution

$$(x + 2)^4 = x^4 + 4x^3(2) + \frac{4(3)}{2} x^2(2)^2 + \frac{4(3)(2)}{1 \cdot 2 \cdot 3} x(2)^3 + \frac{4(3)(2)(1)}{1 \cdot 2 \cdot 3 \cdot 4} (2)^4$$
$$= x^4 + 8x^3 + 24x^2 + 32x + 16$$

Careful examination of the first line in the solution of example 4 shows that, in general, the expansion of a binomial follows the pattern shown by the **binomial theorem** in rule 11–6.

Rule 11–6

The Binomial Theorem

$$(a + b)^n = a^n + \frac{n}{1} a^{n-1}b + \frac{n(n-1)}{1 \cdot 2} a^{n-2}b^2$$
$$+ \frac{n(n-1)(n-2)}{1 \cdot 2 \cdot 3} a^{n-3}b^3 + \cdots + b^n$$

Sequences and Series

The expansion in rule 11–6 can be written using factorials as

$$(a + b)^n = \frac{a^n}{0!} + \frac{na^{n-1}b}{1!} + \frac{n(n-1)}{2!}a^{n-2}b^2 + \frac{n(n-1)(n-2)}{3!}a^{n-3}b^3 + \cdots + \frac{(n)(n-1)(n-2)\cdots(1)}{n!}b^n.$$

Example 5 Expand and simplify: $(x - 2y)^4$.

Solution Write $(x - 2y)^4$ as $[x + (-2y)]^4$

$$[x + (-2y)]^4 = x^4 + 4x^3(-2y)^1 + \frac{4 \cdot 3}{2!}x^2(-2y)^2$$
$$+ \frac{4 \cdot 3 \cdot 2}{3!}x(-2y)^3 + \frac{4 \cdot 3 \cdot 2 \cdot 1}{4!}(-2y)^4$$
$$= x^4 - 8x^3y + 24x^2y^2 - 32xy^3 + 16y^4$$

Notice if the sign between the terms in the binomial is a minus sign, then the terms in the expansion of the binomial have alternating signs, beginning with a positive first term.

Any particular term in the expansion of a binomial is given by

$$r^{\text{th}} \text{ term} = \frac{n(n-1)\cdots(n-r+2)}{(r-1)!}a^{n-r+1}b^{r-1}$$

The expression $r - 1$ occurs twice in this formula; it is the key to writing the entire term. The exponent of the factor b is always one less than the number of the term in which it appears. An example will help to see the procedure.

Example 6 Find the seventh term of $(x + y)^{10}$.

Solution In the seventh term, y is raised to the sixth power. Therefore, x must be raised to the fourth power. We now have

$$\text{seventh term} = \frac{?}{?}x^4y^6$$

The denominator of the numerical coefficient contains six factors, beginning with 1 and increasing by 1 for each factor. We now have

$$\text{seventh term} = \frac{}{1 \cdot 2 \cdot 3 \cdot 4 \cdot 5 \cdot 6}x^4y^6.$$

The numerator of the coefficient also contains six factors, beginning with 10 and decreasing by 1 for each factor. This gives

$$\text{seventh term} = \frac{10 \cdot 9 \cdot 8 \cdot 7 \cdot 6 \cdot 5}{1 \cdot 2 \cdot 3 \cdot 4 \cdot 5 \cdot 6}x^4y^6$$

Simplifying we get,

$$\text{seventh term} = 210x^4y^6$$

11–6 Exercises

In exercises 1–6, simplify each factorial. (See examples 1 and 2.)

1. $\dfrac{5!}{3!}$
2. $\dfrac{8!}{5!}$
3. $\dfrac{7!}{3!4!}$
4. $\dfrac{8!}{5!3!}$
5. $\dfrac{6! - 5!}{4!}$
6. $\dfrac{5! - 4!}{3 - 0!}$

In exercises 7–22, expand and simplify the binomial. (See examples 3–5.)

7. $(a + b)^4$
8. $(x + y)^5$
9. $(x - y)^4$
10. $(a - 2)^5$
11. $(a - 2b)^5$
12. $\left(x + \dfrac{1}{2}\right)^4$
13. $\left(x - \dfrac{1}{3}\right)^4$
14. $\left(2x - \dfrac{y}{2}\right)^5$
15. $\left(2x + \dfrac{y}{3}\right)^4$
16. $(x^2 - 3)^5$
17. $(x^3 + 2)^6$
18. $(x^{1/2} + y)^4$; $x > 0$
19. $\left(x^{1/2} + \dfrac{y}{2}\right)^4$; $x > 0$
20. $(\sqrt{2} + y)^4$ (*Hint:* Write $\sqrt{2}$ as $2^{1/2}$.)
21. $\left(\sqrt{3} - \dfrac{y}{2}\right)^4$ (See the hint in exercise 20.)
22. $(2a^{1/2} + y^{1/3})^4$; $a > 0$ (Leave the answer in reduced radical form.)

In exercises 23–26, write the first three terms and simplify.

23. $(2x + 3y)^{10}$
24. $(x^2 + 2y)^8$
25. $(x^{3/2} - y^{1/4})^8$; $x, y > 0$ (Leave the answer in reduced radical form.)
26. $(x^{-1/3} + y^{1/2})^{10}$; $y > 0$ (Leave the answer in simplified radical form.)

In exercises 27–32, write the specified term and simplify it. (See example 6.)

27. $(2x + y)^6$; the fifth term
28. $(x - 3y)^8$; the fourth term
29. $(x^2 + 2)^6$; the middle term
30. $(x^{1/2} - y^{1/2})^7$; $x, y > 0$; the fifth term
31. $(xy^{1/2} - y)^5$; $y > 0$; the third term
32. $(xy^3 + x^2)^8$; the middle term

A binomial expansion can be used to approximate the value of certain numbers raised to a power.

Example

Approximate $(1.03)^6$ to the nearest hundredth by writing and simplifying the first four terms of the expansion of $(1 + 0.03)^6$.

Solution

$$(1 + 0.03)^6 = 1^6 + 6(1)^5(0.03) + \frac{6 \cdot 5}{1 \cdot 2}(1)^4(0.03)^2 + \frac{6 \cdot 5 \cdot 4}{1 \cdot 2 \cdot 3}(1)^3(.03)^3$$
$$= 1 + 6(0.03) + 15(0.03)^2 + 20(0.03)^3$$
$$= 1 + (0.18) + (0.0135) + (0.00054)$$
$$= 1.19404$$
$$= 1.19$$

In exercises 33–36, approximate each number to the nearest hundredth by writing and simplifying the first four terms of the expansion for each number expressed in the form $(a + b)^n$. (See the example just given.)

33. $(1.01)^6$
34. $(1.04)^5$
35. $(0.99)^7$ (*Hint:* $0.99 = 1 - 0.01$)
36. $(2.02)^6$

The binomial theorem is valid (under certain conditions) if the exponent on the binomial is fractional or negative. The expansion under these conditions contains an infinite number of terms.

Example

Assume the binomial theorem is valid and write the first four terms in the expansion of $(x + y)^{1/2}$.

Solution

$$(x + y)^{1/2} = x^{1/2} + \frac{1}{2}(x^{-1/2})y + \frac{\frac{1}{2}\left(\frac{1}{2} - 1\right)}{1 \cdot 2}(x^{-3/2})y^2$$
$$+ \frac{\frac{1}{2}\left(\frac{1}{2} - 1\right)\left(\frac{1}{2} - 2\right)}{1 \cdot 2 \cdot 3}x^{-5/2}y^3$$
$$= x^{1/2} + \frac{y}{2x^{1/2}} - \frac{y^2}{8x^{3/2}} + \frac{y^3}{16x^{5/2}} - \cdots$$

In exercises 37–40, assume the binomial theorem is valid and write the first four terms in the expansions of each binomial. Simplify and leave all exponents as positive rational numbers.

37. $(x + y)^{-2}$
38. $(2x + y)^{-1}$
39. $(x + y)^{1/3}$
40. $(x + 2y)^{1/4}$

41. The fourth term in the expansion of $(p + q)^7$ gives the probability that if seven coins are tossed that four of them will be heads and three will be tails. Find the probability of seven coins showing four heads and three tails if both p and q have a value of $\frac{1}{2}$. Find the fourth term of the expansion first and then substitute in the values of p and q.

42. The third term in the expansion of $(p + q)^5$ gives the probability that if five dice are rolled, three of them will show a four and the other two will not show a four. Find the probability that five dice will show three fours and two will show anything except a four if $p = \frac{1}{6}$ and $q = \frac{5}{6}$. Find the third term of the expansion first and then substitute in the values for p and q.

Chapter 11 Summary

Vocabulary

arithmetic means
arithmetic progression
arithmetic series
binomial theorem
common difference
common ratio
factorial notation
finite sequence
general term
geometric means
geometric progression
geometric series
index of summation
infinite sequence
range of summation
sequence
sequence function
series
sigma notation

Definitions

11-1 Sequence Function

A function whose domain is the set of positive integers is called a sequence function.

11-2 Finite and Infinite Sequences

A finite sequence of real numbers is a function whose domain is the set $\{1, 2, 3, \cdots, k\}$ of the first k positive integers.

An infinite sequence of real numbers is a function whose domain is the set $\{1, 2, 3, \cdots\}$ of all positive integers.

Rules

11-1 nth Term of an Arithmetic Progression

The nth term of an arithmetic progression is given by $a_n = a_1 + (n - 1)d$, where a_1 is the first term and d is the common difference.

11-2 Sum of an Arithmetic Series

The sum of the first n terms of an arithmetic series is given by

$$S_n = \frac{n}{2}(a_1 + a_n)$$

where a_1 is the first term of the series and a_n is the nth term of the series.

11–3 nth Term of a Geometric Progression

The nth term of a geometric progression is given by $a_n = a_1 r^{n-1}$, where a_1 is the first term and r is the common ratio.

11–4 Sum of a Geometric Series

The sum of the first n terms of a geometric series is

$$S_n = \frac{a_1(r^n - 1)}{r - 1}$$

where a_1 is the first term of the series and r is the common ratio ($r \neq 1$).

11–5 Sum of an Infinite Geometric Series

If $|r| < 1$ and $S_\infty = a_1 + a_1 r + a_1 r^2 + \cdots$, then

$$S_\infty = \frac{a_1}{1 - r}$$

11–6 The Binomial Theorem

$$(a + b)^n = a^n + \frac{n}{1}a^{n-1}b + \frac{n(n-1)}{1 \cdot 2}a^{n-2}b^2 + \frac{n(n-1)(n-2)}{1 \cdot 2 \cdot 3}a^{n-3}b^3 + \cdots + b^n$$

Review Problems

Section 11–1

1. Find the first four terms of the sequence whose general term is $\dfrac{3n - 1}{n}$.

2. Find a general term for the sequence $5, \dfrac{10}{3}, \dfrac{15}{5}, \dfrac{20}{7}, \ldots$

Section 11–2

3. Write in expanded form and simplify the terms of the expansion: $\displaystyle\sum_{i=2}^{5} \dfrac{i^2 - 1}{i + 1}$.

4. Write the following series using sigma notation: $2 + \dfrac{3}{2} + \dfrac{4}{3} + \dfrac{5}{4}$.

Section 11–3

5. What should be the value of x so that $\dfrac{6}{7}, x, \dfrac{13}{7}$ will form an arithmetic progression?

6. The second term of an arithmetic progression is $\dfrac{14}{15}$ and the fourth term is $\dfrac{8}{5}$. Find the sixth term.

7. What are the first four terms of an arithmetic progression if $a_n = 22$, $d = 5$, and $n = 7$?

Section 11-4

8. Find a value of x so $\frac{5}{8}$, x, and $\frac{45}{392}$ form a geometric progression.

9. Find the seventh term of a geometric progression whose first term is $\frac{2}{9}$ and whose fourth term is $\frac{1}{36}$.

10. What is the sum of the first five terms of the geometric series $-10 + 6 - \frac{18}{5} + \ldots$?

11. Suppose the number of bacteria growing in a culture doubles every hour. If the culture started with 1,000 bacteria, how many bacteria will there be at the end of 4 hours?

Section 11-5

12. Find the sum of the infinite geometric series $-10 + 6 - \frac{18}{5} + \cdots$.

13. Find a fraction equivalent to the decimal $2.\overline{16}$.

Section 11-6

14. Evaluate $\dfrac{8!}{6!2!}$.

15. Expand and simplify $(x - 3y)^4$.

16. Find the middle term of $(x + y)^{12}$.

17. Approximate $(0.98)^6$ to the nearest hundredth by finding the value of the first four terms of the expansion $(1 - 0.02)^6$.

18. Write and simplify the first three terms of $(x + y^2)^{1/2}$. Leave all exponents as positive rational numbers.

19. Approximate $(1.03)^6$ by finding the value of the first four terms of $(1 + 0.03)^6$. Round the final answer to the nearest thousandth.

20. The fourth term in the expansion of $(p + q)^6$ gives the probability that if six coins are tossed, they will land showing three heads and three tails. Find this probability if both p and q have a value of $\frac{1}{2}$.

Cumulative Test for Chapters 1–11

Chapter 1

1. Is it true or false that every real number may be classified either as a rational number or an irrational number?
2. Name two operations that are commutative.
3. Evaluate the expression $-a^2 - 4ab^3$ if $a = 3$ and $b = -4$.

Chapter 2

4. Simplify: $\left(\dfrac{-16x^5y^3}{4x^2y^4}\right)^3$.
5. Multiply and combine like terms: $(5x^2 - x)(-2x^2 - 5x + 2)$.
6. Factor completely: $5x^2(x^2 - 9) - 2x(x^2 - 9)$.

Chapter 3

7. Reduce: $\dfrac{x^3 - 6x^2 + 9x}{5x^4 - 16x^3 + 3x^2}$.
8. Simplify: $\dfrac{-32}{x^2 - 16} + \dfrac{x}{x + 4} - \dfrac{x}{x - 4}$.
9. Simplify: $\dfrac{\dfrac{49}{k^2} - 9}{\dfrac{7}{k} + 3}$.
10. Make the following division by using synthetic division: $(9x^5 - 4x^3 + 2x) \div (x - 2)$.

Chapter 4

11. Solve for n: $-7(n - 9) - 2(n - 3) = -n - (2n + 5)$
12. Solve for x: $|2x + 6| = |x - 4|$.
13. Solve for c: $\dfrac{2}{a} - \dfrac{3}{2b} = \dfrac{1}{c}$.
14. How much water must be evaporated from 50 quarts of a 20% salt solution to obtain a 32% salt solution?

Chapter 5

15. Simplify, and leave all exponents positive: $\left(\dfrac{2a^{-3}bc^2}{5ab^{-2}c^0}\right)^{-1}$.
16. Add and simplify: $9x\sqrt{54} + 3\sqrt{24x^2}$, $x > 0$.
17. Add. Leave the answer in simplified radical form.
$$2\dfrac{\sqrt{8}}{\sqrt{3}} + 6\sqrt{\dfrac{3}{8}}$$
18. Rationalize the denominator of $\dfrac{3 + \sqrt{2}}{2\sqrt{6} - 4\sqrt{2}}$.
19. Solve for x: $\sqrt{7x + 2} - \sqrt{5x + 4} = 0$.

Chapter 6

20. Solve for x by factoring: $2x = \dfrac{27}{x} - 15$.
21. Solve for x by completing the square: $3x^2 = 2 - 5x$.

455

22. Multiply and simplify: $(5 - 3i)(4 + 7i)$.

23. Solve for p: $\sqrt{2p - 1} = \sqrt{p - 4} + 2$.

24. The profit, p, in dollars, for selling x chairs per month is $p = 100 + 160x - x^2$. How many chairs must be sold per month to realize a profit of $1,600?

Chapter 7

25. What is the range of the relation given by $R = \{(-6, 2), (0, 4), (5, 7)\}$?

26. For what values of x is the equation given by
$$y = \frac{3x - 2}{6x^2 - 3x}$$
defined?

27. What is the length of the line segment connecting the two points with coordinates $(-7, -6)$ and $(1, 0)$?

28. What is an equation of the line that has x- and y-intercepts given by $(5, 0)$ and $(0, -4)$, respectively? Leave your answer in the form $y = mx + b$.

29. Draw the graph of $y > \frac{3}{5}x + 2$.

Chapter 8

30. Draw the graph of $y = 2x^2 - 8x + 1$.

31. Write an equation of the circle with center located at $(4, -3)$ and radius equal to $\sqrt{17}$. Leave the answer in standard form.

32. Which conic section does the equation $y = 4x^2 - 3$ represent?

33. The distance required to stop a car once the brakes are applied is directly proportional to the square of the speed of the car. If 50 feet are required to stop a car traveling 30 miles per hour, how many feet of stopping distance are needed at 50 miles per hour?

34. If $f(x) = \dfrac{6x - 3}{5x - 2}$, find $f(\frac{1}{4})$.

Chapter 9

35. Solve for x: $8^{5x-3} = \frac{1}{4}$.

36. Find the value of $\log_{36} 6$.

37. Find $\log_3[\log_{10} 1,000]$.

38. Give the expanded form of $\log \dfrac{\sqrt[3]{xy}}{z^2}$.

39. Solve $e^{5t} = 16$ for t. Use 0.4343 for $\log_{10} e$. Ask your instructor whether you should use log tables or a calculator.

Chapter 10

40. Solve the following system of equations by the addition method.
$$\begin{cases} 5x - 4y = -7 \\ x + 3y = 10 \end{cases}$$

41. Solve for y: $10 = \begin{vmatrix} 3 & -2 \\ -1 & y \end{vmatrix}$.

42. Solve the following system of equations for x by using Cramer's rule.
$$\begin{cases} 5x - 2y = 3 \\ x + 3y = 0 \end{cases}.$$

43. Carolyn flew her plane from Bakersfield to Napa in 2 hours. The return trip over the same route took $2\frac{1}{2}$ hours. Find her average speed from Bakersfield to Napa if her average speed for that part of the trip was 30 miles per hour faster than the average speed for the return trip. Use two equations in two variables.

Chapter 11

44. Give the first three terms of the sequence whose general term is $a_n = (-1)^n(2n^2 - 1)$.

45. Expand and simplify: $\sum_{i=1}^{3} [2i^2 - 5]$

46. Find the 25th term of the sequence 7, 10, 13, 16, ...

47. Find the sum of the first ten terms of the series $-4 + 1 + 6 + 11 + \cdots$.

48. Find the first four terms of an arithmetic sequence if $a_1 = \frac{1}{2}$, $a_n = \frac{19}{8}$, and $n = 6$.

49. Find the eighth term of the sequence 5, 10, 20, 40, ...

50. Find r if $a_1 = \frac{5}{3}$, $a_n = -\frac{5}{96}$, and $n = 6$.

51. Find the sum of the infinite series $5 + \frac{5}{3} + \frac{5}{9} + \frac{5}{27} + \cdots$.

52. Evaluate: $\dfrac{7!}{3!4!}$.

53. Expand and simplify $(2a + b)^5$ by using the binomial theorem.

54. Write the sixth term for $(x + y)^{12}$.

Answers to Odd-Numbered Exercises, Review Problems, and Cumulative Tests

Chapter 1

Section 1–1

1. {2, 3, 4, 5}   **3.** {2, 3, 4, 5, 6, 7, 8}   **5.** {3, 4, 5}   **7.** {3, 4, 5}
9. $(A \cap B) \cap C = \{2, 3, 4, 5\} \cap \{3, 4, 5, 6, 7, 8\} = \{3, 4, 5\}$
$A \cap (B \cap C) = \{0, 1, 2, 3, 4, 5\} \cap \{3, 4, 5, 6, 7\} = \{3, 4, 5\}$
The results are the same, so $(A \cap B) \cap C = A \cap (B \cap C)$.
11. $(A \cup B) \cup C = \{0, 1, 2, 3, 4, 5, 6, 7\} \cup \{3, 4, 5, 6, 7, 8\}$
$= \{0, 1, 2, 3, 4, 5, 6, 7, 8\}$
$A \cup (B \cup C) = \{0, 1, 2, 3, 4, 5\} \cup \{2, 3, 4, 5, 6, 7, 8\}$
$= \{0, 1, 2, 3, 4, 5, 6, 7, 8\}$
Therefore, $(A \cup B) \cup C = A \cup (B \cup C)$.
13. $A \cap (A \cup B) = \{0, 1, 2, 3, 4, 5\} \cap \{0, 1, 2, 3, 4, 5, 6, 7\} = \{0, 1, 2, 3, 4, 5\} = A$.   **15.** no
17. Yes; the empty set is a subset of every set.   **19.** no   **21.** no   **23.** yes   **25.** no
27. no   **29.** yes   **31.** yes   **33.** no   **35.** yes   **37.** {5, 7, 11}   **39.** {6}
41. {5, 6}   **43.** {3, 4, 6, 7}   **45.** {3, 4, 5, 6, 7, 8, 9}

Section 1–2

1. yes   **3.** no   **5.** no   **7.** no   **9.** $2.5 = 2\frac{5}{10} = \frac{25}{10}$, which is a ratio of two integers. Therefore, 2.5 is a rational number.   **11.** 0.2727 . . .   **13.** No, division by zero is meaningless.   **15.** $2.4/3 = 2\frac{4}{10}/3 = \frac{24}{10}/3 = \frac{24}{30}$, which is a ratio of two integers. Therefore, 2.4/3 is a rational number.   **17.** Answers may vary; for example, $\frac{5}{8}, \frac{3}{10}$.   **19.** $A \cap B = \{0.75\}$, and $0.75 = \frac{75}{100}$, which is a ratio of two integers. Therefore, 0.75 is a rational number.   **21.** N is a proper subset of W if any $X \in N$ implies $X \in W$ and if W contains at least one element not contained in N. We see for any $X \in N$, $X \in W$. Also, $0 \in W$, but $0 \notin N$. Hence, $N \subset W$.
23. {5}   **25.** {−6.8, −7}   **27.** {−7, 0, 5}   **29.** {1.1313 . . ., −6.8}   **31.** {1.1313 . . .}
33. false   **35.** true   **37.** false   **39.** true   **41.** true   **43.** true   **45.** 0.625; terminating   **47.** 0.666 . . .; repeating   **49.** 0.285714285714 . . .; repeating   **51.** 0.1 ; terminating

Section 1–3

1. 7   **3.** 8   **5.** 3   **7.** 43   **9.** 121   **11.** 9   **13.** 4   **15.** 5504   **17.** 1
19. 53   **21.** 25   **23.** 36   **25.** 4   **27.** 0   **29.** 324   **31.** 1047
33. $5(18 \div 3 \cdot 2) - 3$ **35.** $4[9 + 2(5 \cdot 3) - 2]$ **37.** Your answer will be the number with which you started; for 7:

$7 + 3 = 10$
$10 \cdot 6 = 60$
$60 \div 2 = 30$
$30 - 9 = 21$
$21 \div 3 = 7$

39. The answer will be your age; for 45:
$$45 - 5 = 40$$
$$40(16) = 640$$
$$640 \div 2^3 = 640 \div 8 = 80$$
$$80 + 10 = 90$$
$$90 \div 2 = 45$$

41. 20.83   **43.** 119.6076

Section 1–4

1. symmetric axiom of equality   **3.** substitution axiom   **5.** reflexive axiom of equality   **7.** symmetric axiom of equality   **9.** substitution axiom (applied twice)   **11.** symmetric axiom of equality   **13.** reflexive axiom of equality   **15.** transitive axiom of equality or the substitution axiom   **17.** substitution axiom   **19.** $22 = 22$
21. $\quad 2a + 3 = c \qquad c = 2a + 3$
$\quad\quad 2 \cdot 5 + 3 = 13 \quad 13 = 2 \cdot 5 + 3$
$\quad\quad\quad\; 13 = 13 \quad\quad\; 13 = 13$
symmetric axiom of equality
23. $2x + y$ **25.** 4
27. b **29.** $8 - 3 = 5$ **31.** $9a^2 = c$

Section 1–5

1. trichotomy axiom   **3.** definition of inequality   **5.** definition of inequality (18 lies to the left of 42 on the number line)   **7.** definition of inequality   **9.** transitive axiom of inequality   **11.** trichotomy axiom   **13.** 5   **15.** <   **17.** $3b$ **19.** <   **21.** $x > 5$ or $x < 5$ **23.** $7 < t$ **25.** $9 < (t - 1)$ **27.** symmetric axiom of equality   **29.** definition of inequality   **31.** reflexive axiom of equality   **33.** substitution axiom   **35.** trichotomy axiom   **37.** definition of inequality   **39.** substitution axiom   **41.** substitution axiom   **43.** $s > t$ or $t < s$ **45.** (a) $d = 5(t + 3)^2 + 3(t + 3)$; (b) 140

Section 1–6

1. commutative axiom for multiplication   **3.** additive identity axiom   **5.** distributive axiom   **7.** associative axiom for multiplication; commutative axiom for multiplication   **9.** distributive axiom   **11.** equality multiplication theorem   **13.** commutative axiom for addition   **15.** distributive axiom   **17.** equality multiplication theorem   **19.** equality multiplication theorem   **21.** distributive axiom   **23.** commutative axiom for addition; commutative axiom for multiplication   **25.** commutative axiom for addition; associative axiom for addition; commutative axiom for addition   **27.** multiplicative identity axiom; commutative axiom for multiplication; distributive axiom   **29.** $72c$ **31.** $6t$ **33.** $21 + t$ **35.** $27 + a$ **37.** $8t$ **39.** $\frac{7}{9} + x$ **41.** $45 + 5b$ **43.** $13x$ **45.** $15 + 3x$ **47.** $\frac{3}{5}x$ **49.** $10c$ **51.** $60 + 5c$
53. $7(5 + \frac{1}{7}b + 6c) = 7 \cdot 5 + 7(\frac{1}{7}b) + 7(6c)\quad$ distributive axiom
$\quad\quad\quad\quad\quad\quad\quad\;\; = 35 + 7(\frac{1}{7}b) + 7(6c)\quad$ multiplication fact
$\quad\quad\quad\quad\quad\quad\quad\;\; = 35 + (7 \cdot \frac{1}{7})b + (7 \cdot 6)c\quad$ associative axiom for multiplication
$\quad\quad\quad\quad\quad\quad\quad\;\; = 35 + 1b + (7 \cdot 6)c\quad$ multiplicative inverse axiom
$\quad\quad\quad\quad\quad\quad\quad\;\; = 35 + b + (7 \cdot 6)c\quad$ multiplicative identity axiom
$\quad\quad\quad\quad\quad\quad\quad\;\; = 35 + b + 42c\quad$ multiplication fact

55. $9x + 5x + 5y + 6y = (9x + 5x) + (5y + 6y)$ associative axiom for addition
$= (9 + 5)x + (5 + 6)y$ distributive axiom
$= 13x + 11y$ addition facts
57. $6c + 7c + 8d + 9d + 5 + 3$
$= (6c + 7c) + (8d + 9d) + (5 + 3)$ associative axiom for addition
$= (6 + 7)c + (8 + 9)d + (5 + 3)$ distributive axiom
$= 13c + 17d + 8$ addition facts
59. 5   **61.** 6   **63.** a **65.** c

Section 1–7

1. 8   **3.** -39 **5.** 46   **7.** 10   **9.** 27   **11.** -31 **13.** 32   **15.** 55   **17.** 260
19. 8,200   **21.** -2 **23.** -79 **25.** -18 **27.** 8   **29.** 12   **31.** 65   **33.** 0
35. -25 **37.** 65   **39.** 2   **41.** 25   **43.** 2   **45.** $15 - 5a$ **47.** $18 - 15c$
49. $14a - 7b + 7c$ **51.** $72a + 48b + 12$ **53.** $5x + 2xy$ **55.** $18b - 45$ **57.** 0
59. $-2c + 5$ **61.** $9c + 7d$ **63.** $-11x + 9y$ **65.** $-3x - 2y$ **67.** -5.73
69. 7,131 feet   **71.** $-\$3,124$ (or $\$3,124$ in debt)

Section 1–8

1. 45   **3.** 200   **5.** 0   **7.** 36   **9.** 28   **11.** -160 **13.** -48 **15.** -24
17. -4 **19.** 7   **21.** 8   **23.** -38 **25.** 4   **27.** 278   **29.** 0   **31.** 2   **33.** 144
35. 52   **37.** 32   **39.** 72   **41.** -64 **43.** 1   **45.** $-\frac{3}{2}$ **47.** 84.157

Chapter 1 Review Problems

1. false   **3.** true   **5.** false   **7.** $\{1, 2, 3, 4, 5, 8\}$ **9.** $\{3, 5\}$ **11.** true   **13.** No; the two sets of numbers are disjoint   **15.** multiplication   **17.** $\frac{8}{0}$ **19.** 34   **21.** 50
23. 33   **25.** 2   **27.** 1   **29.** reflexive axiom of equality   **31.** subsitution axiom
33. associative axiom of addition   **35.** identity axiom of multiplication   **37.** -14
39. 10   **41.** 32   **43.** -9 **45.** -12 **47.** 8   **49.** 99   **51.** -21

Chapter 2

Section 2–1

1. $72z^3$ **3.** $(6x - y)^3$ **5.** $-27a^3b^6$ **7.** $-80a^{15}$ **9.** x^5 **11.** $\frac{x^2y^5}{4}$ **13.** $\frac{4}{x^4y^4}$
15. $\frac{1}{r-4}$ **17.** $3b^{2m-3}$ **19.** 5   **21.** $(2x - 3)^4(4x - 3)$ **23.** $4m^6$ **25.** $3a^3b$
27. $28b^2c$ **29.** 1   **31.** $\frac{25}{2}$ **33.** $\frac{27}{512}$ **35.** 1   **37.** 289   **39.** $-(2a - 3)^5$
41. $(2a - 3)^{m+2}$ **43.** $81a^{28}$ **45.** 1   **47.** $x^{10.29}$ **49.** 4   **51.** 475   **53.** 924
55. 500   **57.** 18×10^{20}

Section 2–2

1. binomial   **3.** binomial   **5.** polynomial of four terms   **7.** binomial   **9.** trinomial
11. binomial   **13.** binomial   **15.** $5x^2 - 5x$; binomial   **17.** $3x^2 + 12xy$; binomial
19. $5x^3 + 6x^2 - 9x$; trinomial   **21.** $4x^5 + 3x^4 - 18x$; trinomial   **23.** $24x - 15$; binomial
25. $4x + 16$; binomial   **27.** $-8x^2 - 6x$; binomial   **29.** degrees 3, 1, and 0, respectively

31. 3 **33.** 3 **35.** 3 **37.** 5 **39.** 1 **41.** 0 **43.** $8y^4$ **45.** $8x^3y^4$ **47.** x^3z
49. 8; numerical coefficient **51.** x^3y^4z **53.** $7x$ **55.** $9a + b$ **57.** $10x^2 + y$
59. $7x + 2y - 4$ **61.** $-a^2x^2 + 9ax^2$ **63.** $-4xy + a^2 - 3$ **65.** $10a^2bc - 4ab^2c - 4abc$
67. $-8y$ **69.** $11ab^2 + ab - 6cd$ **71.** $9x^2y - 15x + 1$ **73.** $10x^2y - 2xy - 2y + 9$
75. $6x^2y + 7xy^2 - 13x + 5$ **77.** $13x^2y + 5xy^2 - 5xy - 3x + 9$
79. $7abc^2 - 6ac^2 - 5ac - 5a + 4$ **81.** 2 **83.** undefined (zero denominator) **85.** 7
87. 95 **89.** 10 **91.** $8c^2 + 6c + 5$ **93.** $0.86x^2 + 24.05x - 8.99$

Section 2–3

1. $42a^4 - 28a^3 - 14a^2b$ **3.** $12ax^3 - 24x^2y$ **5.** $40a^4b - 16a^3c$ **7.** $-27a^4b^2 - 18a^3b^3$
9. $20a^4c - 8a^2c^3 - 12a^4c^2$ **11.** $a^2 + ab + ac - a^3c$ **13.** $15x^2 - 29x + 12$
15. $-3x^2 + 22x - 35$ **17.** $6x^2 - 7x - 20$ **19.** $9x^2 + 39x - 30$ **21.** $3x^2 - 2xy - 8y^2$
23. $2u^2 - uv - 3v^2$ **25.** $15 + 2x - 8x^2$ **27.** $5y^2 + 40y - 45$
29. $16a^2b^2 + 36abc - 10c^2$ **31.** $2x^3 + 7x^2 - 21x + 9$ **33.** $x^4 - 6x^3 + 7x^2 - 46x + 24$
35. $x^7 - 5x^6 - 4x^3 + 18x^2 + 10x$ **37.** $2a^3 - 4a^2 - 10a + 20$ **39.** $2x^3 - x^2 - 9x + 9$
41. $8x^3 - 7x^2 - 9x - 1$ **43.** $2x^4 - 3x^3 - 7x^2 + 5x + 3$
45. $4x^5 - 5x^4 + x^3 - 4x^2 + 5x - 1$ **47.** $4x^2 + 12x + 9$ **49.** $25x^2 - 10x + 1$
51. $64x^2 + 80x + 25$ **53.** $4x^2 - 12xy + 9y^2$ **55.** $4x^2 + 12x + 9$ **57.** $49y^2 + 28y + 4$
59. $4c^2 - 4abc + a^2b^2$ **61.** $a^2 + 2ab + b^2 - 10a - 10b + 25$
63. $x^2 + 6xy + 9y^2 + 10x + 30y + 25$ **65.** $9x^2 - 6x + 1$ **67.** $15x^2y - 6x$
69. $81a^2 + 36a + 4$ **71.** $2x^2 + 3y^2 + 5xy - y - 2$ **73.** $8x^3 + 12x^2y + 6xy^2 + y^3$
75. $a^4 - b^4$ **77.** $x^4 - 81$ **79.** $x^{2n+3} - 3x^3$ **81.** $2x^{2n} - x^{m+n}$ **83.** $3a^{2n} - a^{2n+1}$
85. $2y^{n+6} + 3y^{n+5}$ **87.** $297x^2 - 144x + 15$ **89.** $53.29x^2 - 39.42x + 7.29$

Section 2–4

1. $5(a - b)$ **3.** $x(2a - b)$ **5.** $5a(ab - 2 + a)$ **7.** $3x(3y^2 - 2xy^2 - 1)$
9. $-3a(-3ad + 4d - 6ac)$ **11.** $-(4x^2 + 5yz)$ **13.** $6(x - y)$ **15.** $6a(b + 3)$
17. $a(-5 + 7b)$ **19.** $a^2b^3z^3(a^2b^3z^4 + az - b^2)$ **21.** $13x(x + 3x^3 - 2)$
23. $12x(1 - 2x^2y + 3x^3)$ **25.** $a(7a - 6 + 7a^3)$ **27.** $x(13xy + 15y + 16)$
29. $5ab(10a - 5b^2 + 1)$ **31.** $(2x - 3)(5a - 7b)$ **33.** $(3x - 4)(x^2 - 2a)$
35. $(8a - b)(7ab - 6x + 5y)$ **37.** $(4y - 3)^2(7b - 24ay + 18a)$
39. $(2x - 3)(5x - 4)(5c - 7d)$ **41.** $3x^5(1 - 2x^{c-5})$ **43.** $y^7(5 + 3y^{b-7})$
45. $3p^n(-1 + 2p^{m-n})$ **47.** $x^3(5x^m - 2x^{m-3})$ **49.** $A = \pi(r_2^2 - r_1^2)$ **51.** $V_f = V_b(1 + kt)$

Section 2–5

1. $(x + 4)(x - 3)$ **3.** $(t + 7)(t - 4)$ **5.** $(x + 12)(x - 1)$ **7.** $2(u - 9)(u + 5)$
9. prime **11.** $-(x - 8)(x + 5)$ **13.** $-(2c + 7)(c - 3)$ **15.** $(3x - 1)(x + 5)$
17. $-(7x - 1)(x - 4)$ **19.** $(7a - 2)(a + 1)$ **21.** prime **23.** $(3a - 2b)(a - 5b)$
25. $(7c - 2d)(3c - d)$ **27.** $(5x - 4)(3x + 2)$ **29.** $(x - 10)(x - 4)$
31. $(2a - 2b + 3)(a - b + 2)$ **33.** $(3b + 20)(b + 1)$ **35.** yes **37.** yes **39.** no
41. no **43.** $(2B - 3)^2$ **45.** $(3x + 4)^2$ **47.** $(3x - 5y)^2$ **49.** not a perfect-square
trinomial **51.** $4(4a - 1)^2$ **53.** $3(4a + 1)^2$ **55.** $y(3 - 4s)^2$ **57.** $-b(3x + 4y)^2$
59. $(x^a + 2)(x^a + 3)$ **61.** $(x - 2)(x + 2)(x^2 + 3)$ **63.** $(7a^m - b^n)(a^m + b^n)$

Section 2–6

1. $(x - 4)(x + 4)$ **3.** $4(2x - y)(2x + y)$ **5.** $(xy - 2)(xy + 2)$ **7.** $(a + b - 3)(a + b + 3)$
9. $(2x - 7)(4x + 3)$ **11.** $(x - 3y + 5)(x + 3y + 5)$ **13.** $(-x - 3)(3x + 2y + 1)$
15. $4x^2(y + 2)$ **17.** $(x - y)(x^2 + xy + y^2)$ **19.** $(a + b)(a^2 - ab + b^2)$
21. $(2x + y)(4x^2 - 2xy + y^2)$ **23.** $(x - 3y)(x^2 + 3xy + 9y^2)$

25. $(4x - 3y)(16x^2 + 12xy + 9y^2)$ **27.** $2(2x + y)(4x^2 - 2xy + y^2)$
29. $(x + y - 2)(x^2 + y^2 + 2xy + 2x + 2y + 4)$ **31.** $2(x - 1)(4x^2 - 14x + 13)$
33. $(x - y)(x + y)(x^2 + y^2)$ **35.** $4(a - 2y)(a + 2y)(a^2 + 4y^2)$
37. $(x - y)(x^2 + xy + y^2)(x + y)(x^2 - xy + y^2)$ **39.** $(s - t)(s + t)(s^2 + t^2)(s^4 + t^4)$
41. $(x^5 - y^5)(x^5 + y^5)$ These two factors can be factored again, but the method was not discussed in the text.) **43.** $2(x - 2y)(x^2 + 2xy + 4y^2)$ **45.** $(m^2 + n)(m^4 - m^2n + n^2)$
47. $(x^2 - y)(x^2 + y)$ **49.** $6(a - 2y^2)(a^2 + 2ay^2 + 4y^4)$ **51.** $(c + d)(a + 2b)$
53. $(a - 3)(2b + 1)$ **55.** $(a - c)(a - 2b)$ **57.** $(x + 4)(2y - 3)$ **59.** $(a - 4)(a^2 + 2)$
61. $(x - y)(x - y^2)$ **63.** $(2c^2 - 3)(3c + 4)$ **65.** $(b - 3)(9a - 2)$ **67.** $(cx + 1)(a - 1)$
69. $(pq - 1)(t - 1)$ **71.** $(c - d + 2)(a + 2)$ **73.** $V = \frac{4}{3}\pi(r_2 - r_1)(r_2^2 + r_1r_2 + r_1^2)$

Chapter 2 Review Problems

1. x^4 **3.** $(7y)^4$, or $2,401y^4$ **5.** $10x^2$ **7.** $(5x + 3)^3(9x + 2)$ **9.** $-18x$ **11.** $\frac{25}{144}$
13. $d = 35$ **15.** $7x - 7y$ **17.** $10x^4 - 7x^3 - 7x^2 - 2$ **19.** (a) 10 (b) 10 **21.** -3
23. $x^3 - x^2 - 30x$ **25.** $49x^2 - 42x + 9$ **27.** $2x^3 + x^2 - 5x + 2$ **29.** $x^2(7x^3 - 2x + 7)$
31. $(x + 7)(x - 3)$ **33.** $(7x - 2)(2x + 3)$ **35.** $(3x - 2)(3x + 2)$ **37.** $(x + 10)^2$
39. $2(3x - 4)^2$ **41.** $(x - 3)(x - 2y)$ **43.** $(3x + 2y - 7)(3x + 2y + 1)$ **45.** $\frac{1}{2}b(H - h)$

Cumulative Test for Chapters 1 and 2

1. natural numbers, whole numbers, integers, rational numbers, and irrational numbers
3. false **5.** true **7.** false **9.** 1 **11.** multiplicative identity axiom **13.** -2
15. $-2b^3/a$ **17.** $5x^2 + 25x + 6$ **19.** $5x(3x - 2)(x + 4)$

Chapter 3

Section 3–1

1. -2 **3.** -5 **5.** 2 **7.** 2, -4 **9.** $-4, -2$ **11.** $-3, 3$ **13.** $-3, 0, 3$
15. $-5, 0$ **17.** no **19.** yes **21.** yes **23.** no **25.** no **27.** $\frac{12}{32}$
29. $-18ab/(-12b^2)$ **31.** $(10c + 15)/(5b + 15)$ **33.** $(x^2 - 4x + 3)/(2x^2 + x - 3)$ **35.** $\frac{8}{3}$
37. $-x/2y$ **39.** $\frac{-(a - b)}{2c}$, or $-\frac{a - b}{2c}$ **41.** $(-6y + 2)/(x - y)$ **43.** $(6x - y)/(a + b)$
45. $7x/2$ **47.** $-(a + b)/4$ **49.** $(6x - 5)/a$ **51.** $(2a - 7)/3$ **53.** $-(6)/(9 - c)$ **55.** 0
57. C must be zero; no.

Section 3–2

1. $\frac{5}{8}$ **3.** $2x/y$ **5.** $3x/y^3z^2$ **7.** $8ab^2$ **9.** $-3b^4/a^4$ **11.** $(x - y)/2$ **13.** $(1 + 2x)/5y^2$
15. $5x/3a$ **17.** $(2ab - a^2c + 3)/2$ **19.** $(2x^4y^2 - 1 + 3x^2)/(-2)$ **21.** $h - k$
23. $7x + 2y$ **25.** $3x + 4y$ **27.** $(x - y)/2$ **29.** $3x - 2y$ **31.** $a + b$ **33.** $1/(x + y)$
35. $(x - 3)/(x - 5)$ **37.** $(x - 2y)/(x - y)$ **39.** $c^2 + d^2$ **41.** $6/(x^2 + 4); \frac{3}{4}$
43. $x/(0.5x + 0.2)$

Section 3–3

1. $2x \cdot \frac{1}{3y}$ **3.** $(5v - 2) \cdot \frac{1}{v - 3}$ **5.** $(3x - 2)(x + 4) \cdot \frac{1}{x - 1}$ **7.** $-7x + 2$
9. $7x + 3 - 2y$ **11.** $2 - t$ **13.** $\frac{45}{56}$ **15.** $4a^3b$ **17.** $-y^2z^2/3x^2$ **19.** $-3bx/4a^3$
21. $-6x^4y/z^3$ **23.** $4x^3 + 5x - 3$ **25.** $-xy^2 + 2y^3$ **27.** $2y - a/x + 3/xy$
29. $-3a^2b^3/5 + 6b^2/5 - 4/5b$ **31.** $-9/x$ **33.** $7(x - 2y)/(x + 3y)(x + y)$ **35.** 1

37. $-(x + y)(3x - 2)/(x + 5)(x - y)$ **39.** $x - 2$ **41.** $(a + c)/(x + y)$
43. $(2x - y)/(x + 2y)$ **45.** $1/[3x(x + 2)]$ **47.** $(x - 2)^2(2x + 5)/6x^3$

Section 3-4

1. $2^4 \cdot 3$ **3.** $2 \cdot 3 \cdot 5 \cdot 7$ **5.** $2^4 \cdot 3 \cdot 5 \cdot 7$ **7.** $2 \cdot 3^4 \cdot 5$ **9.** $2 \cdot 3^2 \cdot a \cdot b^2$
11. $2^4 \cdot 11 \cdot a^3 b^2$ **13.** $2 \cdot 3 \cdot 7 \cdot 11 \cdot x^3 y z^2$ **15.** $2^4 \cdot 3^2 \cdot 5 \cdot x^3 y$ **17.** $5x(2x + 3)$
19. $2 \cdot 7(x - 3)$ **21.** $(x - y)(x + y)$ **23.** $3 \cdot 7(4x - 3)(4x + 3)$ **25.** $(3a - 2)(a - 1)^2$
27. $(2x - 3)^2(x - 2)(x + 2)(x + 1)$ **29.** $2^3 \cdot 3 \cdot 7 \cdot x^2(6x - 1)^3(7x - 3)(2x - 1)$
31. $2^3 \cdot 3 \cdot 5 \cdot x^2(x - 3)^2(x + 5)^2(x - 2)^4$
33. $2 \cdot 11 \cdot x^3(8x - 1)^2(2c + 3)^2(3x + 1)(3c - 1)(3x - 1)^3$

Section 3-5

1. $\frac{4}{5}$ **3.** $2/a$ **5.** 1 **7.** $x + 5$ **9.** $-1/(y + 6)$ **11.** 3 **13.** 1 **15.** $-\frac{1}{10}$
17. $(2u - 9)/6$ **19.** $(4z - 4)/[z(z - 4)]$ **21.** $23/[2(m - 2)]$ **23.** $\frac{2x^2 + 2x - 42}{(x + 3)(x - 2)(x + 4)(x - 3)}$
25. $\frac{11x + 19}{(x - 2)(x + 2)(2x + 3)}$ **27.** $\frac{3p - 3q - 5}{(p - q)^2}$ **29.** $\frac{5x - 2}{(x - 2)(x^2 + 2x + 4)}$ **31.** 1
33. $\frac{3(x^2 - 5)}{(x - 3)(x + 3)}$ **35.** $-(a + b)/ab$ **37.** $-6/(x - 3)$ **39.** $(x^{n+2} + x^2 + x^y - 1)/x^4$
41. $(x^{2c-1} - 3 - x^{3c+5} - x^4 + x^{c+3} - 3x^3)/x^5$ **43.** $(4^{3a+1} + 4^{-2a} + 5)/64$ **45.** $38\frac{1}{8}$

Section 3-6

1. $\frac{3}{4}$ **3.** $9/5x$ **5.** $(3b + 2)/(9b - 3)$ **7.** $(2y - x)/6y$ **9.** $9/(x - 3)$ **11.** $4x/(x + 4)$
13. $(3s - 4)/s$ **15.** $(9 - 2y)/y$ **17.** $(n - 3)/(n + 5)$ **19.** $(-x + 3)/(4x + 3)$
21. $(x + 2y)(-x + y)/[-x(x + y)]$ **23.** $(2x + 3)/(x - 3)$ **25.** $(5x - 24)/(5x - 26)$
27. $(a + 6b)/(a + b)$ **29.** $(c^2 V + c^2 v)/(c^2 + Vv)$

Section 3-7

1. $x + 4$ **3.** $x - 4$ **5.** $x^2 + 5x + 3$ **7.** $x^3 + 2 - 2/(2x + 1)$ **9.** $9x^2 - 6x + 4$
11. $x + 2$ **13.** $a + 3$ **15.** $x^3 - 8x^2 + 18x - 36 + 68/(x + 2)$
17. $x^2 - x + 1 - 1/(x + 1)$ **19.** $v^3 - v^2 + v - 1 - 2/(v + 1)$ **21.** $x - 6$ **23.** $x + 4$
25. $x - 1$ **27.** $5a^3 - 6a^2$

Section 3-8

1. $x + 3$ **3.** $x^2 + 2x - 3$ **5.** $2x^3 - 3x$ **7.** $y^2 + 2y + 4$
9. $3x^5 - 6x^4 + 12x^3 - 29x^2 + 58x - 118 + 236/(x + 2)$ **11.** $x^2 + x - 4 + 11/(x - 2)$
13. -181 **15.** -39 **17.** 236 **19.** 157 **21.** yes   **23.** no   **25.** yes
27. $x^4 - x^3 + x^2 - x + 1$

Chapter 3 Review Problems

1. no   **3.** no   **5.** $(9x^2 + 18x)/(2xy^2 + 4y^2)$ **7.** $-7/9$ **9.** $-6/(-m + 7)$
11. $1/(r^2 + 3r + 9)$ **13.** $(x - 6)/(x + 3)$ **15.** $5t(t + 4)$ **17.** $5y^3/2x^2$
19. $2^2 \cdot 3^2 \cdot 5 \cdot 7$ **21.** $(c + 15)/6c$ **23.** $(-w^2 + 3w - 8)/(w + 3)(w + 4)$ **25.** $\frac{3}{4}$
27. $4x/(x + 4)$ **29.** $x^4 + x^3 - 3x^2 - 3x - 1 - 4/(x - 1)$
31. $x^3 + 2x^2 + 4x + 8 + 8/(x - 2)$

Cumulative Test for Chapters 1–3

1. {3, 5}   **3.** A, or {1, 2, 3, 4, 5, 6}   **5.** 190   **7.** $-2/x^3$ **9.** 175
11. $3x^2y^2(3y - 4x^3 - 1)$ **13.** $(a + 2c)(a - 3b)$ **15.** $x/(3x - 1)$ **17.** $-(x + y)/xy$
19. $3x^2 + 6x + 8 + 18/(x - 2)$

Chapter 4

Section 4–1

1. {−1}   **3.** {−1}   **5.** {7}   **7.** {3}   **9.** {5}   **11.** {4}   **13.** {5}   **15.** {3}
17. {−4}   **19.** {5}   **21.** {2}   **23.** {8}   **25.** {−7}   **27.** {−5}   **29.** ∅   **31.** {real numbers}   **33.** {4}   **35.** {10}   **37.** {−2}   **39.** ∅   **41.** {$\frac{26}{5}$}   **43.** $n + 5$
45. $x - 3$ **47.** $x + y$ **49.** $z - (x + y)$ **51.** $2N - 3$ **53.** $S + 5$ **55.** $5y$
57. $\frac{5}{8}z + x^2$ **59.** $z - t$ **61.** $9y + 2 = 20; y = 2$ **63.** $16v - 2 = 6v - 22; v = -2$
65. $\frac{9}{13}$ **67.** {4.64}

Section 4–2

1. {30}   **3.** {6}   **5.** {2}   **7.** {12}   **9.** {$\frac{59}{9}$}   **11.** {2}   **13.** {$-\frac{7}{3}$}   **15.** {$-\frac{12}{11}$}
17. ∅   **19.** {4}   **21.** {5}   **23.** ∅   **25.** {−2}   **27.** {3}   **29.** $\frac{1}{2}n + \frac{1}{3}n = \frac{2}{3}n + 10$; $n = 60$ **31.** $\frac{1}{3}c + \frac{1}{9}c = \$20{,}000; c = \$45{,}000$

Section 4–3

1. The length is 49 feet; the width is 23 feet.   **3.** The length is 10 feet; the width is 5 feet.
5. The length is 20.808 feet.   **7.** The length is 40 feet; the width is 30 feet.   **9.** 30 feet
11. $3\frac{3}{7}$ hours   **13.** 75 minutes   **15.** 8 hours   **17.** 5 liters of the 30% solution, 15 liters of the 70% solution   **19.** 6 liters of the 8% solution, 4 liters of the 13% solution   **21.** 30 quarts   **23.** 24 pounds of $8 tea, 72 pounds of $6 tea   **25.** 20 pounds of the 1:3 alloy, 30 pounds of the 2:3 alloy   **27.** 20, 21, 22   **29.** −18, −16, −14   **31.** The integers can be any three consecutive integers.   **33.** −18, −16, −14   **35.** 720 miles   **37.** Let x, x + 1, x + 2, x + 3, and x + 4 represent the integers. The sum of the integers is x + (x + 1) + (x + 2) + (x + 3) + (x + 4) = 5x + 10; five times the middle integer is 5(x + 2) = 5x + 10.   **39.** $5,000 at 8%; $7,000 at 10%   **41.** $5,000 at 12%; $3,000 at 9%
43. 8.5%   **45.** No, a straight 25% discount gives the greater discount.   **47.** The first car travels at 50 miles per hour; the second car travels at 55 miles per hour.   **49.** 10:30 a.m.
51. The rate of the slower runner is 7 yards per second; the rate of the faster runner is $7\frac{1}{2}$ yards per second; the race was run in 142 seconds.   **53.** 150 miles per hour; 315 miles   **55.** $5,936
57. $80   **59.** $1,800 per month

Section 4–4

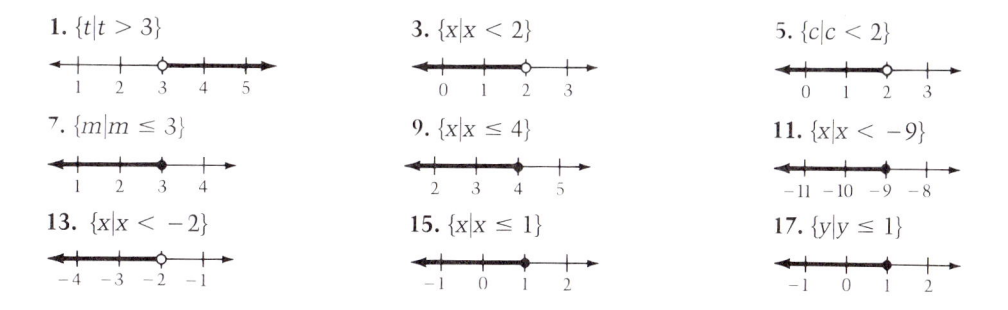

1. {t|t > 3}   **3.** {x|x < 2}   **5.** {c|c < 2}

7. {m|m ≤ 3}   **9.** {x|x ≤ 4}   **11.** {x|x < −9}

13. {x|x < −2}   **15.** {x|x ≤ 1}   **17.** {y|y ≤ 1}

19. $\{x|x < 1\}$

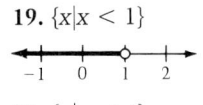

21. $\{m|m > 0\}$

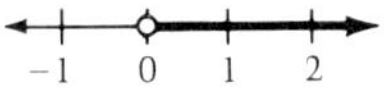

23. $\{m|m < 0\}$

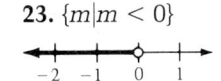

25. $\{y|y < 4\}$

27. $\{x|x \leq 13\}$

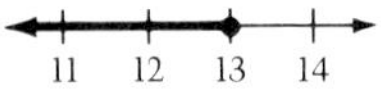

29. $\{d|d \leq 0\}$

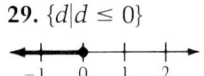

31. $\{y|y > -11\}$

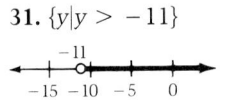

33. $\{p|p \geq 2\}$

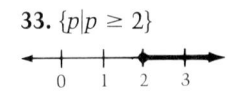

35. $\{d|d > 4\}$ ($d \geq 3$ and $d > 4$ implies $d > 4$)

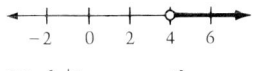

37. $\{x|-8 \leq x < -2\}$

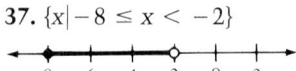

39. $\{x|1 \leq x \leq 4\}$

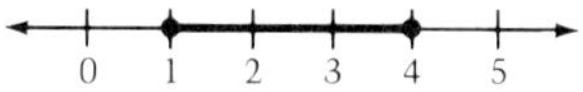

41. $\{x|6 < x < 8\}$

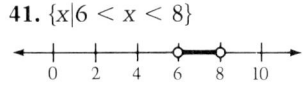

43. $\{x|1 < x < 3\}$

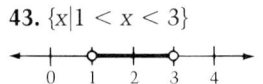

45. $\{x|x > 6 \text{ or } x < 3\}$

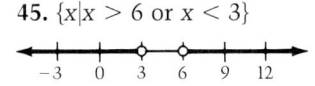

47. $\{x|x < \frac{21}{11} \text{ or } x \geq 5\}$

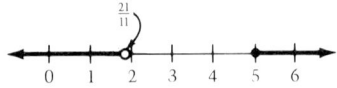

49. $\{a|a \text{ is a real number}\}$

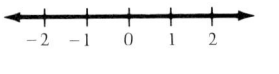

51. $\varnothing$ **53.** $\varnothing$ **55.** $2 - 5x < 3x - 14$; $\{x|x > 2\}$ **57.** $-5y + 8 > 3^2 + 14$; $\{y|y < -3\}$
59. $x/3 > 15$; $\{x|x > 45\}$ **61.** no more than 1,000 calculators   **63.** $\{x|x < 9.05\}$

Section 4–5

1. $\{5, -5\}$ **3.** $\varnothing$ **5.** $\{12, -4\}$ **7.** $\{4, 10\}$ **9.** $\{14, -10\}$ **11.** $\{-\frac{4}{5}, 2\}$
13. $\{-1, 2\}$ **15.** $\{-\frac{8}{7}, 2\}$ **17.** $\{-\frac{17}{3}, 1\}$ **19.** $\{-2, 8\}$ **21.** $\{1, 3\}$ **23.** $\varnothing$
25. $\{-9, 4\}$ **27.** $\{-4, -2\}$ **29.** $\{\frac{2}{5}, 2\}$ **31.** $\{0\}$ **33.** $\{\frac{2}{3}, 8\}$ **35.** $\{7\}$ **37.** $\{-1, 4\}$
39. $\varnothing$ **41.** For example, if $a = 5$ and $b = -6$, $|ab| = |5(-6)| = |-30| = 30$;
$|a| \, |b| = |5| \, |-6| = 5(6) = 30$.   **43.** For example, if $a = -2$ and $b = 5$,
$|a + b| = |-2 + 5| = |3| = 3$, but $|a| + |b| = |-2| + |5| = 2 + 5 = 7$ **45.** $\{-10.79, 2.03\}$

Section 4–6

1. $\{x|-3 < x < 3\}$ **3.** $\{a|a > 8 \text{ or } a < -8\}$ **5.** $\{x|-10 < x < 10\}$ **7.** $\{s|-7 < s < 13\}$
9. $\{x|x > 3 \text{ or } x < -27\}$ **11.** $\{u|-3 < u < 6\}$ **13.** $\{x|-\frac{8}{5} \leq x \leq 2\}$
15. $\{z|z > \frac{65}{14} \text{ or } z < -\frac{71}{14}\}$ **17.** $\{x|-2 \leq x \leq 7\}$ **19.** $\{a|a > \frac{9}{5} \text{ or } a < -3\}$
21. $\{x|-\frac{2}{5} < x < 2\}$ **23.** $\{c|c \geq 3 \text{ or } c \leq 0\}$ **25.** $\{x|x \text{ is a real number}\}$
27. $\{x|x \geq 7 \text{ or } x \leq -5\}$ **29.** $\{x|x > -5 \text{ or } x < -17\}$ **31.** $\{t|-\frac{11}{2} < t < -5\}$
33. $\{r|\frac{2}{5} \leq r \leq 2\}$ **35.** $\{x|x < -4 \text{ or } x > \frac{26}{3}\}$ **37.** $|f(x) - L| < \epsilon$ implies $-\epsilon < f(x) - L < \epsilon$,
which implies $L - \epsilon < f(x) < L + \epsilon$ **39.** $|M - J| \leq 0.5$ **41.** $|x - 3| < 8$
43. $\{x|-0.11 < x < 0.15\}$

Section 4–7

1. $\pi = A/r^2$ **3.** $t = d/r$ **5.** $m = 2E/v^2$ **7.** $x = y^2/4p$ **9.** $x = -3/(a+b)$
11. $x = (3-c)/2a$ **13.** $x = 8y/(y-12)$ **15.** $x = c+d$ **17.** $x = (3a-30)/25$
19. $x = (c-bd)/4$ **21.** $x = s/r$; 10   **23.** $r = mv^2/F$; 600   **25.** $r = (s-P)/Pt$; $\frac{3}{50}$
27. $m = 360/(180-S)$; 5

Chapter 4 Review Problems

1. {10}   **3.** {1}   **5.** 18   **7.** {1}   **9.** $\varnothing$ **11.** $10,000   **13.** $\{u | u < 5\}$
15. $\{x | x < -14\}$ **17.** $\{x | x \geq 13\}$ **19.** $\{-1, \frac{7}{3}\}$ **21.** $\{-\frac{16}{3}, 4\}$ **23.** $\varnothing$
25. $\{x | -4 < x < -1\}$ **27.** $\pi = 3V/4r^3$ **29.** $x_1 = (mx_2 + y_1 - y_2)/m$

Cumulative Test for Chapters 1–4

1. true   **3.** false   **5.** $-x/3y$ **7.** $8y^3 + 10y - 8$ **9.** $2(x+6)(x+5)$
11. $(x-3)^2/(2x-1)(x+5)$ **13.** $9a/(a+9)$ **15.** {6}
17. $\{n | n \geq 3\}$ **19.** $\{x | x \geq 6 \text{ or } x \leq -1\}$ **21.** 17, 19, 21

Chapter 5

Section 5–1

1. $\frac{1}{125}$ **3.** $\frac{1}{2}$ **5.** -18 **7.** $\frac{7}{12}$ **9.** $\frac{1}{16}$ **11.** $\frac{1}{2}$ **13.** $y^3/125x^4$ **15.** $1/w^4$
17. y^8/x^6 **19.** h^2k **21.** $1/x^6y^8$ **23.** $s^3/t^{12}z^2$ **25.** $3/x^2y^4$ **27.** a^3b^3/c **29.** a^6c^6/b^2
31. $a^{-3}b^{-1}/5^{-2}$ **33.** $3^{-2}y^{-8}/x^{-4}$ **35.** $y^{-2}/2^{-2}x^{-4}z^{-4}$ **37.** $x^{-3}/2^{-1}(a+b)^{-2}$ **39.** $m^{-5}t^5$
41. $x^{12}y^{-2}$ **43.** $a^8b^{-4}c^{-2}$ **45.** $(u^2+v)/u^2v$ **47.** $(y^2+y)/x$ **49.** $(c^3d+1)/c^2$
51. $(x^3+y^4)/2x^4y^5$ **53.** $a^2b^2/(a^2+b^2)$ **55.** 6.3×10^2 **57.** 7.3×10^7
59. 6.13×10^{-5} **61.** 0.912   **63.** 0.9   **65.** 120   **67.** (a) 1.32×10^{25} pounds;
(b) 6×10^{24} kilograms

Section 5–2

1. 5   **3.** -6 **5.** 125   **7.** 64   **9.** $\frac{1}{4}$ **11.** $\frac{1}{2}$ **13.** -3 **15.** $\frac{4}{5}$ **17.** $\frac{9}{4}$
19. 3   **21.** -2 **23.** 81   **25.** u **27.** $\sqrt[4]{x^3}\sqrt[6]{y}$ **29.** $\sqrt[5]{b^4}/a^2$ **31.** $1/(\sqrt[4]{x^3}\sqrt[4]{y})$
33. $a\sqrt{c}/\sqrt[3]{b^2}$ **35.** $\sqrt[3]{b^2}/(\sqrt[3]{a}\,c^8)$ **37.** $4\sqrt[4]{y}/(5\sqrt[12]{x^5})$ **39.** $a^{3/4} + a^{5/4}$ **41.** $x^{-5/6} - x^{-1/6}$
43. $h^{-11/15} - h^{-7/12}$ **45.** $x^{9/2} - x^{1/2}$ **47.** $t - 2t^{1/6} + t^{-2/3}$ **49.** $x^{2/3} + 2x^{5/6} + x$
51. $s - 2s^{3/4} + s^{1/2}$ **53.** $x^{2/3} - 2x^{1/3}y^{2/3} + y^{4/3}$ **55.** $4x^{2/7} - 12x^{1/7}y^{2/5} + 9y^{4/5}$
57. $16a^{-2} + 24a^{-1}b^{-2} + 9b^{-4}$ **59.** $4r^{1/2} - 12r^{1/4}s^{2/5} + 9s^{4/5}$
61. $64a^{2/3}b^4 + 80a^{1/3}b^2c^{1/5}d^3 + 25c^{2/5}d^6$ **63.** $x^{4n/15}$ **65.** x^n **67.** $x^{9n/2}/y^{3/2}$ **69.** xy^3z^2

Section 5–3

1. $3\sqrt{3}$ **3.** $6\sqrt{2}$ **5.** $3\sqrt[3]{2}$ **7.** $2|x|\sqrt{2}$ **9.** $2x^2\sqrt{3}$ **11.** $-2\sqrt[3]{2}$
13. $-2p\sqrt[3]{2}$ **15.** $4xy^2\sqrt{xy}$ **17.** $-4c^2d^2\sqrt{2d}$ **19.** $2x\sqrt[3]{xy}$ **21.** $\sqrt{3}/2$ **23.** $2u^2\sqrt{uv}$
25. $x^3y\sqrt[3]{y}$ **27.** $a^2b^3\sqrt{6}$ **29.** $y^2\sqrt[3]{y^2}/2$ **31.** $\sqrt{5}/5$ **33.** $\sqrt{3}/3$ **35.** 2
37. $\sqrt{6pq}/2q$ **39.** $2\sqrt{2xy}/y^2$ **41.** $\sqrt{2u}\sqrt[3]{6}/2u$ **43.** $5\sqrt{2z}/z^2$ **45.** $x\sqrt{2}/10$
47. $\sqrt{a+b}$ **49.** $1/(r+s)$ **51.** $\sqrt[3]{12x^2y}/3y$ **53.** $\sqrt[3]{9b^2}$ **55.** $\sqrt{x}$ **57.** $\sqrt{hk}$

468   Answers to Selected Exercises

59. $\sqrt[6]{3x}$ **61.** $\sqrt[6]{432}$ **63.** $\sqrt[6]{2{,}000}$ **65.** $\sqrt[10]{6{,}075}$ **67.** $\sqrt[6]{108h^5}$ **69.** $\sqrt[12]{a^9 b^6 c^8}$
71. $\sqrt[12]{x^9 y^2 z^{-4}}$ **73.** $\sqrt[6]{108}/2$ **75.** $3\sqrt{2}m_0/4$ **77.** $3\sqrt{7}/14$
79. $\sqrt[n]{a/b} = (a/b)^{1/n} = a^{1/n}/b^{1/n} = \sqrt[n]{a}/\sqrt[n]{b}$

Section 5-4

1. $7\sqrt{5}$ **3.** $13\sqrt{7}$ **5.** $-15\sqrt{3}$ **7.** $32 + \sqrt{7}$ **9.** $4\sqrt{2}$ **11.** $19\sqrt{2}$ **13.** $9\sqrt{5}$
15. $7\sqrt[3]{2}$ **17.** $-14\sqrt[4]{2}$ **19.** $4\sqrt[5]{2}$ **21.** $23x\sqrt{x}$ **23.** $-2m^2 n\sqrt{n}$ **25.** $15x^2\sqrt[3]{x}$
27. $7\sqrt[4]{z^3} - 5\sqrt{z}$ **29.** $2\sqrt{3}/3$ **31.** $17\sqrt{7}/84$ **33.** $5\sqrt{2} - 15$ **35.** $2\sqrt{3} + 2\sqrt{7}$
37. $\sqrt{6} + \sqrt{10}$ **39.** $2\sqrt{3} + 2\sqrt{5}$ **41.** $3 + \sqrt[3]{15}$ **43.** -2 **45.** 68 **47.** $x^2 - y$
49. $7 - 4\sqrt{3}$ **51.** $9 + 2\sqrt{14}$ **53.** -1 **55.** $3b + 3\sqrt[3]{b^2 c} - 2\sqrt[3]{bc^2} - 2c$
57. $(4 - \sqrt{5})/2$ **59.** $(2 - 3\sqrt{u})/3$ **61.** $3 - \sqrt[3]{x}$ **63.** $(3\sqrt{2} + 9)/(-7)$
65. $(\sqrt{6} + 4\sqrt{3})/(-14)$ **67.** $3\sqrt{2} + \sqrt{10} - \sqrt{15} - 2\sqrt{3}$ **69.** $(10 + 5\sqrt{y})/(4 - y)$
71. $(\sqrt{2y} + \sqrt{2x} - \sqrt{3y} - \sqrt{3x})/(y - x)$

Section 5-5

1. $\{29\}$ **3.** $\{26\}$ **5.** $\{\frac{8}{3}\}$ **7.** $\{7\}$ **9.** $\{\frac{13}{3}\}$ **11.** $\emptyset$ **13.** $\{\frac{8}{3}\}$ **15.** $\{\frac{7}{2}\}$ **17.** $\{-5\}$
19. $\{9\}$ **21.** $\emptyset$ **23.** $\{2\}$ **25.** $\{8\}$ **27.** $\emptyset$ **29.** $\{-3\}$ **31.** $\emptyset$ **33.** $\{11\}$
35. $\{-29\}$ **37.** $\{-\frac{3}{4}\}$ **39.** $\{0\}$ **41.** 3.25 feet   **43.** $V = 9.8$ feet per second

Chapter 5 Review Problems

1. $\frac{1}{27}$ **3.** 48 **5.** n^2/m **7.** $y^2 z^2/x^8$ **9.** $r^{-3} s^{-2}/5^{-3}$ **11.** $(x^3 + y^2)/x^3 y^2$ **13.** 12
15. $\frac{1}{7}$ **17.** $\sqrt[8]{t^7}$ **19.** $3\sqrt[12]{xy}/(2\sqrt[20]{z})$ **21.** $k^{1/6} - k^{23/21}$ **23.** $1/x^{7n/6}$ **25.** $4n^2 t\sqrt{nt}$
27. $2x\sqrt[4]{3}$ **29.** $\sqrt{10hk}/2k$ **31.** $\sqrt{x}$ **33.** $\sqrt[5]{3a}$ **35.** $5\sqrt{3}$ **37.** $3\sqrt{2} - 3$
39. $21 - 8\sqrt{5}$ **41.** $(\sqrt{21} - \sqrt{6} - \sqrt{14} + 2)/5$ **43.** $\sqrt[6]{12}$ **45.** $b\sqrt[12]{b^5 c^{10}}$ **47.** $\{110\}$
49. $\emptyset$ **51.** $E = mc^2$

Cumulative Test for Chapters 1-5

1. true   **3.** false   **5.** commutative axiom for multiplication   **7.** $384 x^9 y^5$
9. $9x^n - 4x + 11$ **11.** $3c^2 + 9c$ **13.** $(x - 6)/5$ **15.** $(x^n + 3 + x^{m+1} - x)/x^2$ **17.** $\emptyset$
19. $\{\frac{8}{3}, 4\}$ **21.** 48 inches   **23.** $\$86.40$ **25.** $\frac{17}{72}$ **27.** $x^2 z^{10}/y^8$ **29.** $6xy^3\sqrt{2xy}$
31. $(\sqrt{11} + \sqrt{5})/2$

Chapter 6

Section 6-1

1. $\{-3, 2\}$ **3.** $\{-3, 4\}$ **5.** $\{-\frac{5}{2}, 2\}$ **7.** $\{-\frac{1}{2}, \frac{3}{4}\}$ **9.** $\{\frac{1}{12}, \frac{3}{16}\}$ **11.** $\{-5, 4, 6\}$
13. $\{-\frac{4}{5}, -\frac{2}{3}, \frac{5}{2}\}$ **15.** $\{-4, -3, 7\}$ **17.** $\{-2, -\frac{1}{3}, 5\}$ **19.** $\{-\frac{5}{3}, 0\}$ **21.** $\{-5, 5\}$
23. $\{-\frac{5}{9}, \frac{5}{9}\}$ **25.** $\{-2, 3\}$ **27.** $\{-4, \frac{3}{2}\}$ **29.** $\{-6, \frac{1}{2}\}$ **31.** $\{\frac{1}{2}, \frac{2}{3}\}$ **33.** $\{-\frac{4}{3}, 2\}$
35. $\{\frac{3}{2}, 4\}$ **37.** $\{2, 3\}$ **39.** $\{\frac{1}{2}\}$ **41.** $\{\frac{2}{3}\}$ **43.** $x^2 - 7x - 12 = 0$
45. $x^2 + 10x + 16 = 0$ **47.** $4x^2 + 9x - 9 = 0$ **49.** $4x^2 - 4x + 1 = 0$
51. $\{3a/2, -3a/2\}$ **53.** $\{-a/2, a\}$

Section 6-2

1. $\{-3, 3\}$ **3.** $\{-\frac{5}{4}, \frac{5}{4}\}$ **5.** $\{-3, 3\}$ **7.** $\{-4, 4\}$ **9.** $\{-2\sqrt{3}, 2\sqrt{3}\}$ **11.** $\{1, 5\}$
13. $\{-2 - 2\sqrt{2}, -2 + 2\sqrt{2}\}$ **15.** $\{-\frac{1}{2}, \frac{7}{2}\}$ **17.** $\{(-1 - 2\sqrt{2})/3, (-1 + 2\sqrt{2}/3)\}$

19. {(−2 − 3√2)/3, (−2 + 3√2)/3}   **21.** {−5, 0}   **23.** {−$\frac{2}{3}$, 2}   **25.** 4; $(x + 2)^2$
27. 16; $(x − 4)^2$ **29.** $\frac{49}{4}$; $(x − \frac{7}{2})^2$ **31.** $\frac{1}{9}$; $(x + \frac{1}{3})^2$ **33.** $\frac{25}{36}$; $(x − \frac{5}{6})^2$ **35.** {3, 9}
37. {−5, 4}   **39.** {−2, $\frac{1}{2}$}   **41.** {−3, 1}   **43.** {(−5 − √33)/2, (−5 + √33)/2}
45. {(1 − √41)/4, (1 + √41)/4}   **47.** {(−3 − √15)/4, (−3 + √15)/4}   **49.** {−$\frac{1}{10}$, $\frac{7}{10}$}

Section 6–3

1. $3i$ **3.** $2i\sqrt{3}$ **5.** $−2i$ **7.** $−2i\sqrt{6}$ **9.** $−6$ **11.** $−3\sqrt{2}$ **13.** $4i\sqrt{7}$
15. $−1$ **17.** $−1$ **19.** i **21.** $−1$ **23.** $19i\sqrt{2}$ **25.** $12i\sqrt{2} − 39\sqrt{2}$
27. $−4 − 18i$ **29.** $−18 − 16i$ **31.** $11 − i$ **33.** $11 − i$ **35.** $1 + 19i$ **37.** $4 − 6i$
39. $−1 + 0i$ **41.** $0 + 8i$ **43.** $31 + 12i$ **45.** $3 − 24i$ **47.** $47 + 27i$ **49.** $5 + 12i$
51. $20 − 48i$ **53.** $−\frac{1}{13} + (\frac{18}{13})i$ **55.** $−\frac{2}{13} − (16/13)i$ **57.** $−3 − 4i$ **59.** $−1 + 3i$
61. $x = 3$; $y = −4$ **63.** $x = −2$; $y = −4$ **65.** $x = \pm 3$; $y = 3$ **67.** $x = \frac{28}{5}$; $y = 5$

Section 6–4

1. {−5, 4}   **3.** {−5, $\frac{1}{2}$}   **5.** {−4, $\frac{3}{2}$}   **7.** {−$\frac{1}{2}$, 4}   **9.** {2 − 3i, 2 + 3i}
11. {3 − 7i, 3 + 7i}   **13.** {(2 − i√6)/2, (2 + i√6)/2}   **15.** {(2 − √19)/5, (2 + √19)/5}
17. {(2 − √22)/6, (2 + √22)/6}   **19.** {(−2 − √26)/2, (−2 + √26)/2}   **21.** {−2, 1}
23. {−($\frac{3}{2}$)i, ($\frac{3}{2}$)i}   **25.** {−$\frac{2}{5}$, 0}   **27.** The roots a real, rational, and unequal.   **29.** The roots are real, rational, and equal.   **31.** The roots are complex conjugates.   **33.** The roots are real, irrational, and unequal.   **35.** The roots are complex conjugates.   **37.** $k = \pm 2\sqrt{3}$

Section 6–5

1. {−4, 4}   **3.** {−8, 2}   **5.** {6, 8}   **7.** {$\frac{3}{2}$, 4}   **9.** {$\frac{2}{3}$}   **11.** {3}   **13.** {7}
15. {−2}   **17.** {−2, −1}   **19.** {2}   **21.** {6}   **23.** {3, 9}   **25.** {−8, 2}
27. $L = 1/4\pi^2 f^2 C$ **29.** $\sqrt{2}$

Section 6–6

1. {−2, −1, 1, 2}   **3.** {−3, −2, 2, 3}   **5.** {−√3, √3, −3i, 3i}   **7.** {−3, 1}
9. {−27, 1}   **11.** {2√2, 8}   **13.** {4} (The equation √x = −3 is not true for any x.)
15. {16}   **17.** {−i√2, i√2, −√2, √2}   **19.** {−1, 0, 1}   **21.** {−$\frac{1}{3}$, −$\frac{1}{2}$}   **23.** {−5, −4}
25. $x = −2y \pm \sqrt{2}$

Section 6–7

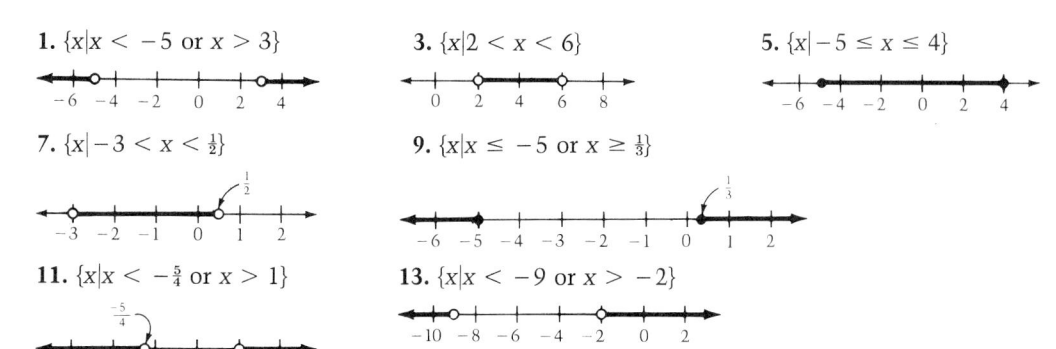

1. {x|x < −5 or x > 3}   **3.** {x|2 < x < 6}   **5.** {x|−5 ≤ x ≤ 4}

7. {x|−3 < x < $\frac{1}{2}$}   **9.** {x|x ≤ −5 or x ≥ $\frac{1}{3}$}

11. {x|x < −$\frac{5}{4}$ or x > 1}   **13.** {x|x < −9 or x > −2}

15. $\{x \mid x < -4\}$

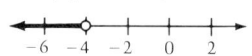

17. $\{x \mid x < -3 \text{ or } x > -2\}$

19. $\{x \mid x < -5 \text{ or } -2 < x < 3\}$

21. $\{x \mid -4 < x < -\tfrac{1}{2} \text{ or } x > 5\}$

23. $\{x \mid x < -3 \text{ or } -1 < x < 1\}$

25. $\{x \mid -2 \le x \le \tfrac{1}{2} \text{ or } x \ge 4\}$

27. $\{x \mid (-2 - \sqrt{7} < x < -2 + \sqrt{7}\}$

29. $\{x \mid x \ge (-1 + \sqrt{5})/2 \text{ or } x \le (-1 - \sqrt{5})/2\}$

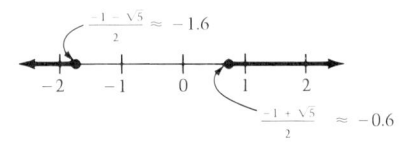

31. $\{x \mid x \le -5 \text{ or } -3 < x < 1 \text{ or } x \ge 3\}$

Section 6–8

1. 5 and 9   **3.** 4 and 7   **5.** 7 and 9   **7.** The integers are either -10 and -8 or 8 and 10.   **9.** 6   **11.** The width is 5 centimeters; the length is 7 centimeters.   **13.** The width is 5 feet; the length is 13 feet.   **15.** The base is 18 centimeters; the altitude is $9\sqrt{3}$ centimeters.   **17.** One-inch squares were cut from each corner.   **19.** 3 feet   **21.** The width is 9 centimeters; the length is 22 centimeters.   **23.** The first part was at 8 miles per hour; the last part of the trip was at 5 miles per hour.   **25.** 6 centimeters   **27.** One leg is 3 inches; the other is 4 inches.   **29.** 10 feet by 70 feet or 35 feet by 20 feet   **31.** 7 and 9

Chapter 6 Review Problems

1. $\{-2, \tfrac{3}{2}\}$ **3.** $\{4\}$ **5.** $\{-3, 5\}$ **7.** $4i\sqrt{2}$ **9.** -1 **11.** $4 + i$ **13.** $14 - 8i$ **15.** $x = -4; y = -6$ **17.** The roots are real, rational, and unequal.   **19.** $k > \tfrac{25}{16}$ **21.** $\{5\}$ **23.** $\{x \mid x \le -3 \text{ or } x \ge 5\}$ **25.** The numbers are 3 and 8.   **27.** The base is 16 centimeters.

Cumulative Test for Chapters 1–6

1. true   **3.** false   **5.** transitive axiom for equality on the substitution axiom   **7.** $x^5 z$ **9.** $7xy(2x^2 + 7x - 12)$ **11.** $10x + 15$ **13.** $2x + 1$ **15.** $11/(x - 3)$ **17.** $\{\tfrac{2}{3}\}$

19. $\{-\frac{28}{3}, 8\}$ **21.** 9 nickels, 13 dimes, 2 quarters   **23.** $y^6/x^{10}z^8$ **25.** $4\sqrt{2}$ **27.** $21x\sqrt{5}$
29. $\varnothing$ **31.** $\{-\frac{5}{2}, 4\}$ **33.** $10i\sqrt{5}$ **35.** $k \geq -\frac{1}{3}$ **37.** $\{x \mid -2 < x < 7\}$

Chapter 7

Section 7–1

1. for example, (2, 2), (0, 4), (1, 3)   **3.** for example, (2, 2), (5, 3), (10, 2);   **5.** (7, 3), (4, 6)
7. (−4, −12), (1, −2)   **9.** (4, 12), (18, −2)   **11.** The domain is {5, −2, 0}; the range is {3, 4, 1}.   **13.** The domain is {0, −1}; the range is {0, −1}.   **15.** $R = \{(-1, 0), (-1, 7), (0, 7)\}$; the domain is {−1, 0}; the range is {0, 7}.   **17.** $R = \{(4, 4), (4, 2), (4, -10), (2, 2), (2, -10), (-10, -10)\}$; the domain is {4, 2, −10}; the range is {4, 2, −10}.
19. $R = \{(0, -3), (1, -1), (2, 1)\}$ **21.** $R = \{(3, -2), (2, 0), (1, 2)\}$

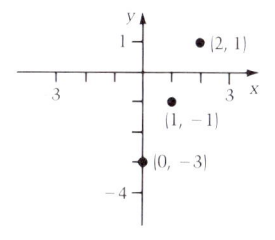

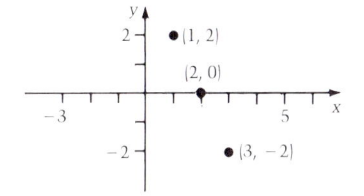

23. $R = \{(-\frac{1}{2}, 4\frac{1}{2}), (-\frac{1}{3}, 4\frac{1}{3}), (\frac{1}{2}, 3\frac{1}{2})\}$

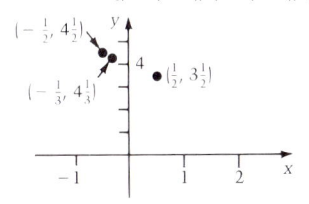

25. true   **27.** false   **29.** true   **31.** true

Section 7–2

1. Yes; the domain is {−4, 2, 0, 4}; the range is {3, 5}.   **3.** No; the domain is {9, 7}; the range is {6, 3, 8}.   **5.** Yes; the domain is {4, −4, −8}; the range is {9, 3, 6}.   **7.** yes   **9.** no
11. $R = \{(-1, -1), (0, 0), (1, 1), (2, 2)\}$; R is a function.

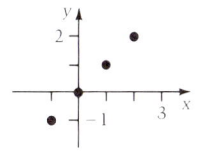

13. $R = \{(2, 1), (2, 0), (2, -1), (1, 0), (1, -1), (0, -1)\}$; R is not a function.

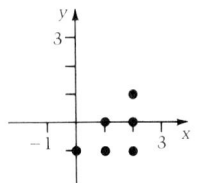

15. $R = \{(0, 0), (1, 2)\}$; R is a function.

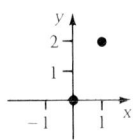

17. $R = \{(-1, 1), (0, 0), (1, 1), (2, 2)\}$; R is a function.

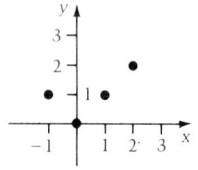

19. $\{x|x$ is a real number and $x \neq 4\}$ **21.** $\{x|x$ is a real number and $x \neq \frac{4}{3}\}$ **23.** $\{x|x$ is a real number and $x \neq 1, -1\}$ **25.** $\{x|x \geq -7\}$ **27.** $\{x|x > 4\}$ **29.** 13   **31.** 85
33. $a^2 + 4$ **35.** $x^2 + 2hx + h^2 + 4$ **37.** 5   **39.** $2hx + h^2$
41 (a) 224 feet per second
  (b) 160 feet per second
  (c) 0
  (d) yes

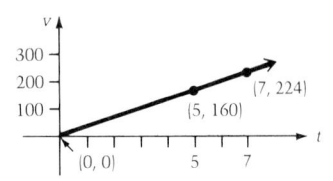

Section 7-3

1. 5   **3.** 13   **5.** $2\sqrt{29}$ **7.** 8   **9.** $\sqrt{a^2 - 2ab + b^2 + 9}$ **11.** 6 or 2
13. 9 or -1 **15.** -13 or 5   **17.** $\frac{3}{2}$ **19.** $\frac{5}{11}$ **21.** $-\frac{5}{11}$ **23.** undefined
25. $(c - d)/9$ **27.** $13 + \sqrt{145}$ **29.** $\sqrt{17} + 2\sqrt{2} + \sqrt{13}$ **31.** Since $3\sqrt{2} = 2\sqrt{2} + \sqrt{2}$, the three points lie on the same line.   **33.** Since $2\sqrt{2} \neq \sqrt{5} + 1$, the three points do not lie on the same line.   **35.** The three slopes are *not* the same; the three points do not lie on the same line.   **37.** parallel   **39.** perpendicular   **41.** neither   **43.** perpendicular
45. perpendicular (one line is horizontal and the other is vertical.)   **47.** $-\frac{1}{9}$ **49.** 0
51. 12

Section 7-4

1. $y = -(\frac{3}{4})x + 3$; $m = -\frac{3}{4}$; $b = 3$ **3.** $y = x - 6$; $m = 1$; $b = -6$ **5.** $y = (\frac{5}{3})x - 2$; $m = \frac{5}{3}$; $b = -2$ **7.** $y = (\frac{3}{4})x - \frac{5}{4}$; $m = \frac{3}{4}$; $b = -\frac{5}{4}$ **9.** $y = (\frac{3}{2})x + 13$; $m = \frac{3}{2}$; $b = 13$
11. $y = (\frac{2}{3})x + \frac{4}{3}$; $m = \frac{2}{3}$; $b = \frac{4}{3}$ **13.** $y = 2x - 2$; $m = 2$; $b = -2$ **15.** $5x - y = 2$
17. $3x - 6y = 2$ **19.** $3x - 5y = -25$
21. $2x - y = -3$ **23.** $2x + 3y = 12$

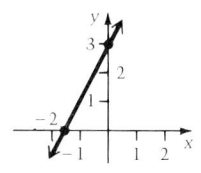

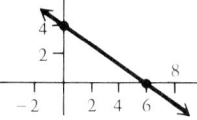

25. $6x + y = 23$ **27.** $y = 3$ **29.** $4x - 3y = -8$ **31.** $x(d - 2) - y(c + 8) = -2c - 8d$
33. parallel   **35.** perpendicular   **37.** neither   **39.** parallel
41.   **43.** 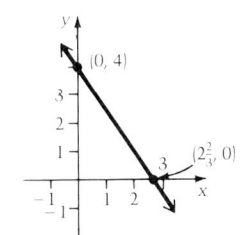   **45.**

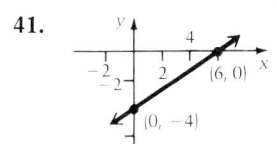

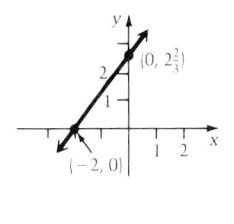

47. $x - 2y = -7$ **49.** line parallel: $x - 3y = -14$; line perpendicular: $3x + y = -2$
51. $37x - 20y = 25$ **53 (a)** $y = 4x$ **(b)** 14   **55 (a)** $V = 10,000 - 2,000t$ **(b)** \$4,000

Section 7-5

1.   **3.** 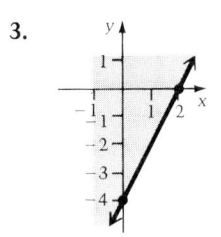   **5.**

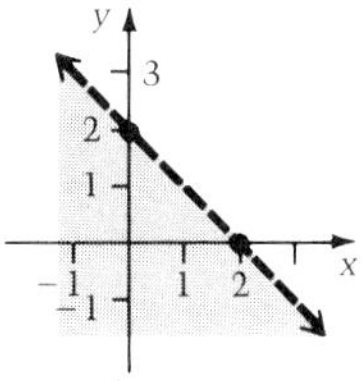

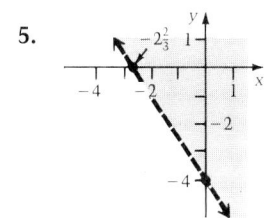

7. 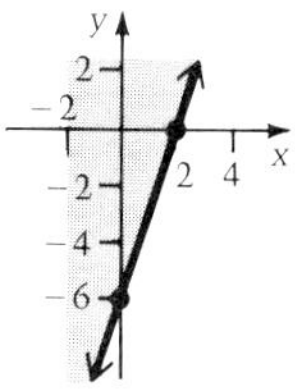   **9.** 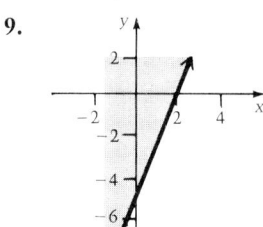   **11.**

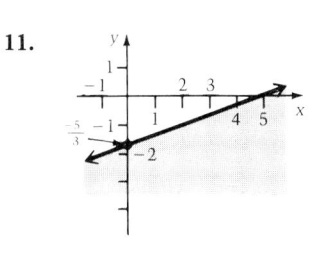

13.

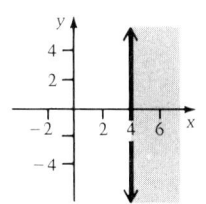

15.

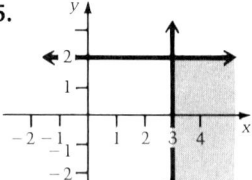

17.

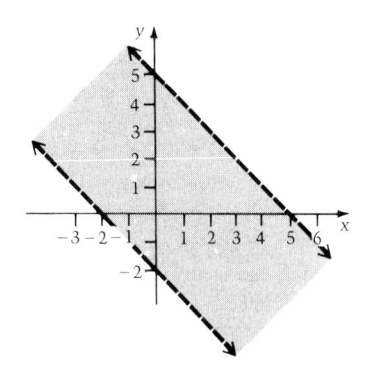

19.

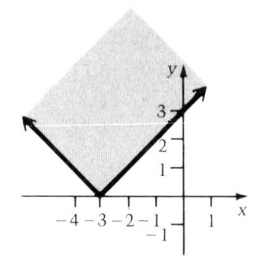

21.

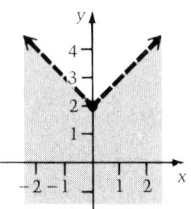

23.

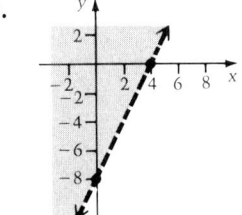

25.

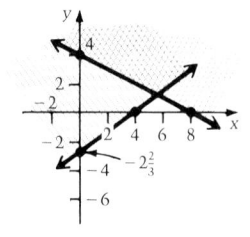

27.

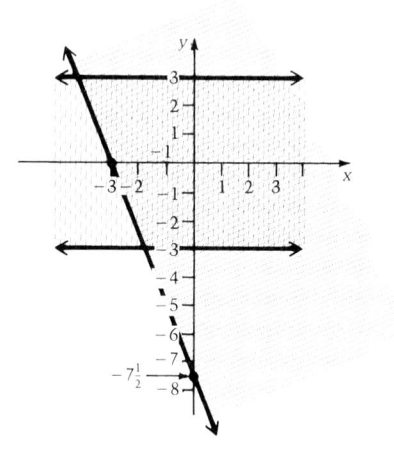

Chapter 7 Review Problems

1. for example: $(8, -12)$, $(-1, 3)$ **3.** false **5.** Yes, this is a function because for every value of x there is one and only one value of y. **7.** $2a^2 - a^4$ **9.** 24 **11.** $\sqrt{13} \neq \sqrt{2} + \sqrt{5}$; the three points do not lie on the same line. **13.** $m = -5$; $b = -3$ **15.** $2x + 11y = 38$ **17.** not perpendicular

19.

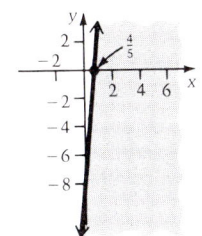

21.

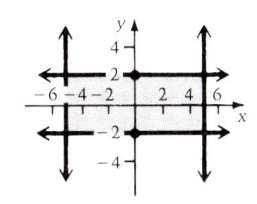

Cumulative Test for Chapters 1–7

1. $3[6 + 2(3 + 4)] = 60$ **3.** -5 **5.** $2x^9/y^5$ **7.** $2x^3 - 15x^2 + 14x + 6$
9. $(-x - y)/3$ **11.** x **13.** $5/3$ **15.** $\{2\}$ **17.** $\{x|x > -4\}$ **19.** $\{v|7/3 < v < 4\}$
21. 12 noon   **23.** m^9/n^{12} **25.** 2.2×10^{-1} **27.** $10\sqrt{6}$ **29.** $\{7\}$
31. $x^2 + 2x - 15 = 0$ **33.** $26i\sqrt{5}$ **35.** 5, 12, 13   **37.** $\{(-3, 6), (0, 4), (3, 2)\}$
39. $\sqrt{34}$ **41.** $2x + 3y = -18$
43.

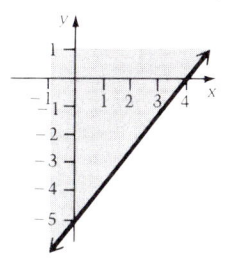

Chapter 8

Section 8–1

1. upward   **3.** downward   **5.** upward   **7.** to the left   **9.** to the left   **11.** $(-3, -12)$
13. $(1, 3)$ **15.** $\left(-\frac{5}{2}, \frac{3}{2}\right)$ **17.** $(-3, -12)$ **19.** $(2, 11)$ **21.** $(-13, 3)$
23. vertex at $(0, 0)$ **25.** vertex at $(3, 0)$

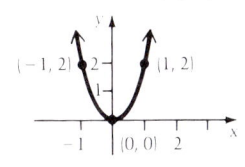

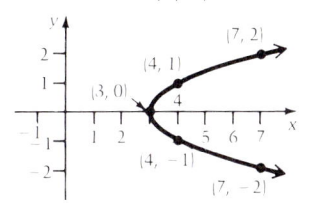

27. vertex at $(3, 2)$ 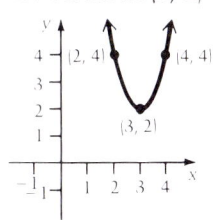   **29.** $y = \sqrt{3x}/3$ or $y = -\sqrt{3x}/3$

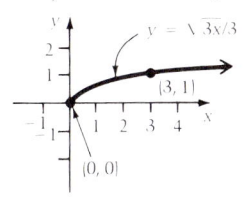

31. $y = \sqrt{2x + 6}/2$ or $y = -\sqrt{2x + 6}/2$ **33.** 8 lawns per day

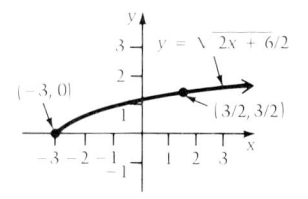

35.
$$d_1 = d_2$$
$$\sqrt{[x - (-p)]^2 + 0^2} = \sqrt{(x - p)^2 + y^2}$$
$$(x + p)^2 = (x - p)^2 + y^2$$
$$x^2 + 2px + p^2 = x^2 - 2px + p^2 + y^2$$
$$2px = -2px + y^2$$
$$4px = y^2$$

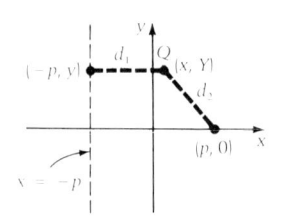

Section 8–2

1. circle   **3.** circle   **5.** parabola   **7.** circle   **9.** (0, 0); 4   **11.** (3, −4); 5
13. (−7, −2); $\sqrt{10}$ **15.** $(x - 3)^2 + (y + 4)^2 = 36$; (3, −4); 6
17. $(x + 5)^2 + (y - 3)^2 = 36$; (−5, 3); 6   **19.** $(x + \frac{3}{2})^2 + (y - \frac{5}{2})^2 = 9$; $(-\frac{3}{2}, \frac{5}{2})$; 3
21. $(x - 2)^2 + (y - 4)^2 = 58$; $x^2 + y^2 - 4x - 8y - 38 = 0$ **23.** $(x + 3)^2 + y^2 = 52$; $x^2 + y^2 + 6x - 43 = 0$ **25.** $x^2 + (y + 6)^2 = 81$; $x^2 + y^2 + 12y - 45 = 0$
27.   **29.** $(x + 4)^2 + (y - 3)^2 = 36$

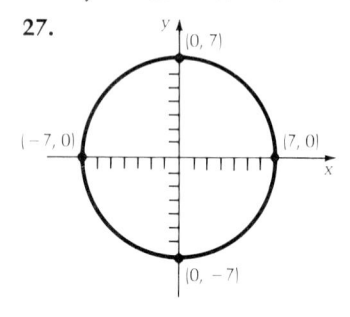

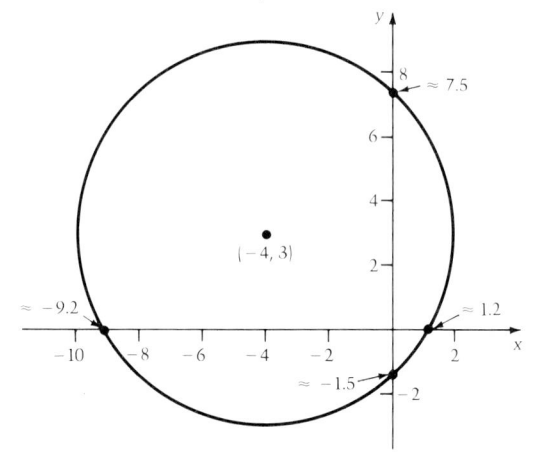

31. $(x - 9)^2 + (y - 5)^2 = 4$ **33.** $(x + 4)^2 + (y + 3)^2 = 20$ **35.** $x^2 + y^2 = 64$
37. $x^2 + (y + 6)^2 = 256$ **39.** $(x - \frac{3}{4})^2 + (y - 5)^2 = 36$ **41.** The standard form of the given equation is $(x + 2)^2 + (y - 3)^2 = 0$. Since $r^2 = 0$, $r = 0$, and so the graph is a point circle.

Section 8–3

1. hyperbola   **3.** circle   **5.** straight line   **7.** parabola   **9.** straight line
11. hyperbola with vertices at (3, 0) and (−3, 0)   **13.** ellipse with vertices at (6, 0), (−6, 0), (0, 2), and (0, −2)   **15.** hyperbola with vertices at $(\sqrt{5}, 0)$, $(-\sqrt{5}, 0)$

17.

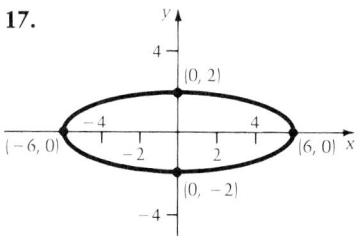

19.

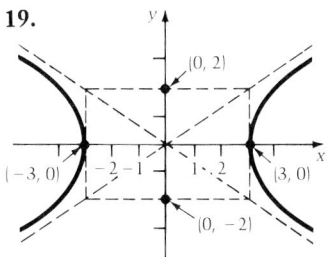

21.

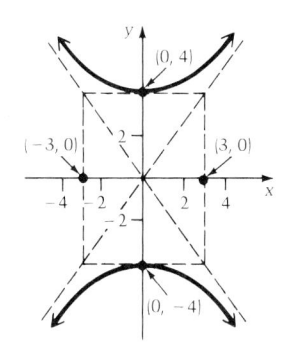

23. Set the left side of the equation equal to zero.
$$\frac{y^2}{a^2} - \frac{x^2}{b^2} = 0$$
$$\frac{y^2}{a^2} = \frac{x^2}{b^2}$$
$$y^2 = \frac{a^2 x^2}{b^2}$$
$y = ax/b$ or $y = -ax/b$.

Section 8–4

1.

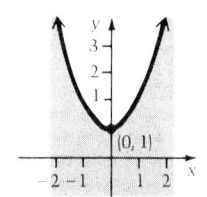

3.

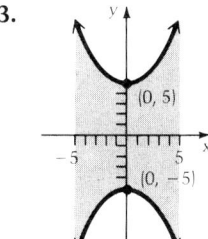

5.

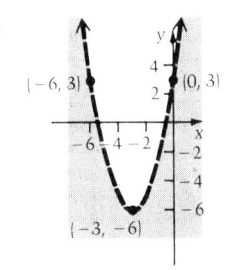

7.

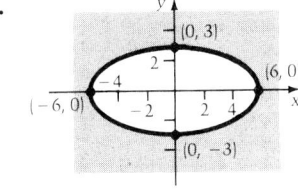

9.

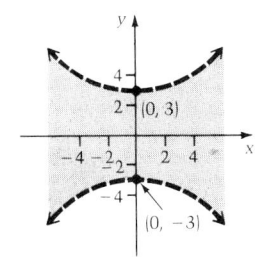

11.

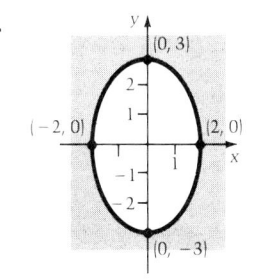

13.

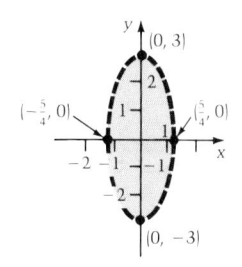

15.

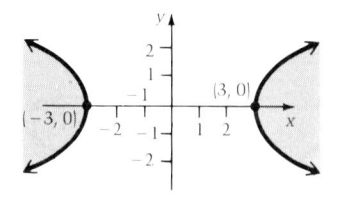

17.

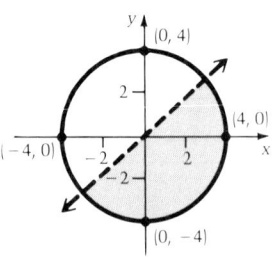

19.

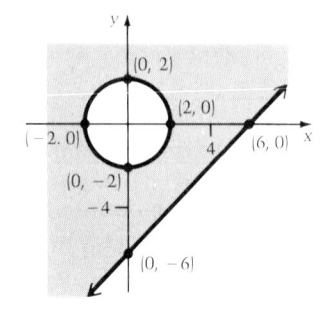

21.

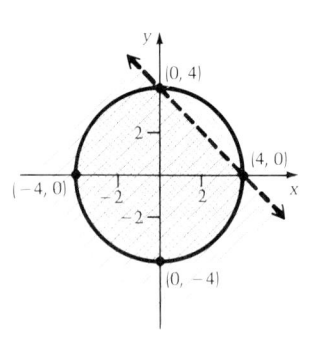

23.

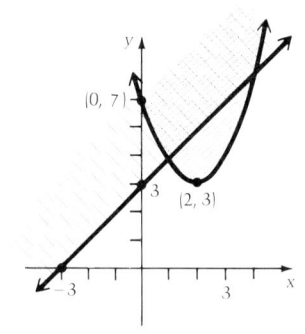

Section 8–5

1. 4 feet, 14 feet **3.** 39 of one kind, 9 of the other **5.** 9 60-watt bulbs, 12 100-watt bulbs **7.** 4:5 **9.** 4:21 **11.** $x:y::1:5$ **13.** 20 **15.** $25b$ **17.** $1/2a$ **19.** 9 **21.** 1 **23.** −8 or 8 **25.** $69.93 **27.** 39 loads **29.** The team must win all 5 of the remaining games.

31. $\dfrac{a}{b} = \dfrac{c}{d}$

$ad = bc$

$\dfrac{ad}{ac} = \dfrac{bc}{ac}$

$\dfrac{d}{c} = \dfrac{b}{a}$

Section 8–6

1. 35 **3.** 1 **5.** 6 **7.** $\sqrt[3]{36}$ **9.** no effect **11.** y is half as large. **13.** 128 feet per second **15.** $\frac{1}{2}$ ampere **17.** 18 pounds **19.** 2,198 cubic inches **21.** 2,000 ohms **23.** $4,800 **25.** 36 square feet **27.** 162 foot-pounds **29.** 96 pounds

Section 8–7

1. (a) R is a function.
   (b) The domain is $\{2, 4, 5\}$; the range is $\{3, -1, 2\}$.
   (c) $R^{-1} = \{(3, 2), (-1, 4), (2, 5)\}$; R^{-1} is a function.
   (d) The domain is $\{3, -1, 2\}$; the range is $\{2, 4, 5\}$.
3. (a) R is a function.
   (b) The domain is $\{4, 5, -1\}$; the range is $\{3, 0\}$.
   (c) $R^{-1} = \{(3, 4), (3, 5), (0, -1)\}$; R^{-1} is not a function.
   (d) The domain is $\{3, 0\}$; the range is $\{4, 5, -1\}$.
5. (a) R is not a function.
   (b) The domain is $\{2, 0\}$; the range is $\{3, 5, -1\}$.
   (c) $R^{-1} = \{(3, 2), (5, 0), (-1, 2)\}$; R^{-1} is a function.
   (d) The domain is $\{3, 5, -1\}$; the range is $\{2, 0\}$.
7. (a) R is not a function.
   (b) The domain is $\{2, 7\}$; the range is $\{3, 7\}$.
   (c) $R^{-1} = \{(3, 2), (3, 7), (7, 2)\}$; R^{-1} is not a function.
   (d) The domain is $\{3, 7\}$; the range $\{2, 7\}$.
9. (a) $f^{-1} = \{(x, y) | y = (3x + 6)/2\}$ (c)
   (b) f^{-1} is a function.

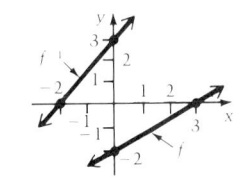

11. (a) $f^{-1} = \{(x, y) | y = \sqrt[3]{x}\}$ (c)
    (b) f^{-1} is a function.

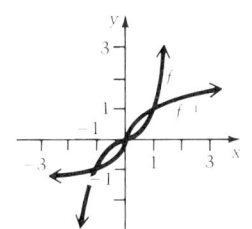

13. (a) $f^{-1} = \{(x, y) | y = \pm\sqrt{x}\}$ (c)
    (b) f^{-1} is not a function.

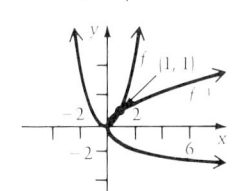

15. (a) $f^{-1} = \{(x, y) | y = \sqrt{x + 1}\}$ (c)
    (b) f^{-1} is a function.

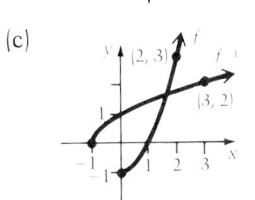

17. (a) $f^{-1}(x) = (x + 1)/3$
    (b) $f(2) = 5$
    (c) $f^{-1}(f(2)) = f^{-1}(5) = 2$

19. (a) $f^{-1}(x) = \sqrt[3]{x}$
    (b) $f(2) = 8$
    (c) $f^{-1}(f(2)) = f^{-1}(8) = 2$

21. (a) $f^{-1}(x) = \sqrt{x-1}$
(b) $f(2) = 5$
(c) $f^{-1}(f(2)) = f^{-1}(5) = 2$
23. 3   **25.** The inverse of the identity function is itself.
27. Write $f(x) = ax + b$ as $y = ax + b$. The inverse of $y = ax + b$ is $x = ay + b$. When $x = ay + b$ is solved for y, we get $y = (x - b)/a$. If $a = 0$, $y = (x - b)/a$ is not defined. Also, if $a = 0$, $f(x) = ax + b$ becomes $f(x) = b$, or $y = b$, which has a horizontal line as a graph. The inverse of $y = b$ is $x = b$, which has a vertical line as a graph. A vertical line does not represent a function.

Chapter 8 Review Problems

1. downward   **3.** to the left   **5.** $(1, -1)$
7. vertex at $(2, 1)$

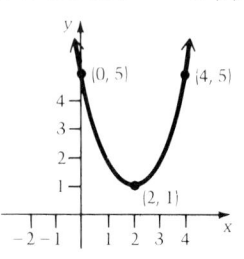

9. Center at $(0, 0)$; $r = 10$ **11.** center at $(-1, 0)$; $r = 2\sqrt{6}$ **13.** $(x + 2)^2 + (y + 3)^2 = 125$
15. ellipse   **17.** hyperbola   **19.** circle
21.

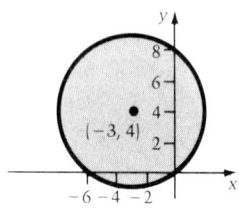

23. $6\frac{2}{3}$ **25.** $R^{-1} = \{(3, 6), (5, 7), (-4, -2)\}$; the domain is $\{3, 5, -4\}$; the range is $\{6, 7, -2\}$.
27. 2

Cumulative Test for Chapters 1–8

1. a nonterminating, nonrepeating decimal   **3.** 980   **5.** $(2x - 3)^3$
7. $3ab^2(x - 3)[a^2b^2(x - 3)^2 - 5]$ **9.** 1 and -3 **11.** $x + 4y$ **13.** $x^2 + 6x - 3$
15. $\{-\frac{1}{2}\}$ **17.** $\{\frac{22}{3}, -\frac{20}{3}\}$ **19.** $13{,}000$ **21.** $xy/(x + y)$ **23.** $\sqrt{6y}/3x^3y$
25. $(2 - \sqrt{3x})/3$ **27.** $\{-6, \frac{4}{3}\}$ **29.** $-6/13 + (17/13)$
31. $\{x \mid x < -4 \text{ or } x > 5\}$ **33.** 9   **35.** $-\frac{2}{5}$ **37.** $x + 2y = 1$

39. 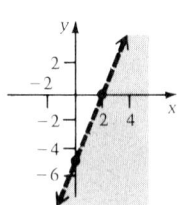   **41.** Center at $(-3, 4)$; $r = 5$ **43.** 8   **45.** 400 feet

Chapter 9

Section 9–1

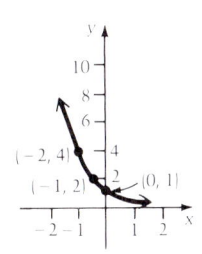

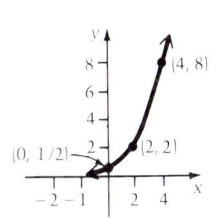

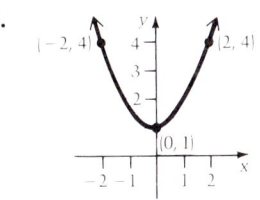

1. [graph: points (0,1), (2,9)]

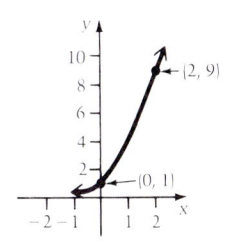

3. [graph: points (−2,4), (−1,2), (0,1)]

5. [graph: points (0,1/2), (2,2), (4,8)]

7. [graph: points (−2,4), (0,1), (2,4)]

9. [graph: points (−1,4), (0,2), (1,4)]

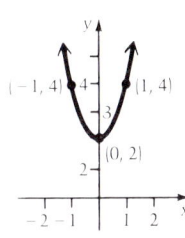

11. 5 **13.** $\frac{5}{4}$ **15.** $\frac{1}{3}$ **17.** 1 **19.** $-\frac{4}{3}$ **21.** 5 **23.** 2 **25.** −1 **29.** 0
29. 1 **31.** 3 **33.** 3 **35.** 27 **37.** 1,225 **39.** 5 **41.** 22,000

Section 9–2

1. The domain of $y = 2^x$ is {real numbers}; the range of $y = 2^x$ is $\{y \mid y > 0\}$; the domain of $x = 2^y$ is $\{x \mid x > 0\}$; the range of $x = 2^y$ is {real numbers}. The two equations are inverses.

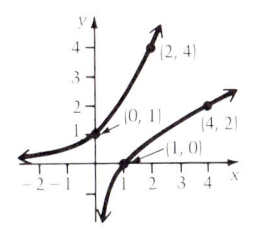

[graph: points (0,1), (2,4), (1,0), (4,2)]

3. The domain is {real numbers}; the range is $\{y \mid y > 0\}$.

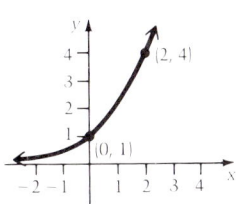

[graph: points (0,1), (2,4)]

5. $3^2 = 9$ **7.** $8^{-1/3} = \frac{1}{2}$ **9.** $4^3 = 64$ **11.** $(\sqrt[3]{11})^3 = 11$ **13.** $8^{2/3} = 4$
15. $10^{-2} = 0.01$ **17.** $27^{2/3} = 9$ **19.** $(0.02)^2 = 0.004$ **21.** $(1.7)^3 = 4.913$ **23.** $a^c = b$
25. $a^{3/2} = b$ **27.** $\log_2 32 = 5$ **29.** $\log_4 2 = \frac{1}{2}$ **31.** $\log_{27} \frac{1}{3} = -\frac{1}{3}$ **33.** $\log_{64} \frac{1}{8} = -\frac{1}{2}$

35. $\log_{1.6} 2.56 = 2$ **37.** $\log_{5/8} \frac{25}{64} = 2$ **39.** $\log_4 \frac{1}{64} = -3$ **41.** $\log_{0.01}(0.0001) = 2$
43. $\log_e e = 1$ **45.** $\log_a c = \frac{1}{2}$ **47.** 2 **49.** 2 **51.** 5 **53.** $\frac{1}{3}$ **55.** -2 **57.** 2
59. 32 **61.** 10 **63.** 9 **65.** b represents any real number such that $b \neq 0$. Also, for logarithmic work, $b > 0$ and $b \neq 1$. See definition 9–2. **67.** 9.63 **69.** 1 **71.** 0
73. 0
75. If the base is zero, then we have two cases:
1. 0^x, where $x \neq 0$, always has a value of zero.
2. 0^x, where $x = 0$, is undefined.

Section 9–3

1. $\log_b 2 + \log_b x$ **3.** $\log_b 3 + \log_b m - \log_b 4$ **5.** $\log_b 5 + 2\log_b t$ **7.** $\frac{1}{2}[\log_b 3 + \log_b x]$
9. $\frac{1}{5}[\log_b 4 + \log_b p] - 2\log_b r$ **11.** $\frac{1}{3}[\log_{10} 4 + 2\log_{10} r] - \frac{1}{2}\log_{10} s$
13. $\frac{1}{2}[\log_3 5 + \frac{1}{2}(\log_3 2 + \log_3 u)]$ **15.** $\log_b 15$ **17.** $\log_b (m/2)$ **19.** $\log_b x^2 y^3$
21. $\log_b \frac{\sqrt{tu}}{t^3}$ **23.** $\log_b \frac{\sqrt{xy}}{\sqrt[3]{3x}}$ **25.** 2.30 **27.** 3.22 **29.** 4.37 **31.** 5.98
33. -0.92 **35.** -2.07 **39.** true

Section 9–4

1. 2 **3.** 4 **5.** -3 **7.** -4 **9.** -6 **11.** 0 **13.** 2.8075 **15.** 1.9425
17. $9.3345 - 10$ **19.** $6.8751 - 10$ **21.** 6.9410 **23.** 9,400 **25.** 800 **27.** 0.0347
29. 0.00298 **31.** 3.57 **33.** 0.000519 **35.** 1.7499 **37.** 0.8699 **39.** 3.9848
41. $7.7872 - 10$ **43.** $6.8829 - 10$ **45.** 219.1 **47.** 80.17 **49.** 5.922
51. 0.002553 **53.** 0.0003045 **55.** 52.81 **57.** 11.33 **59.** 316.8 **61.** 38.03
63. 1.761 **65.** 16.52 **67.** 113.0 **69.** 0.8010 **71.** 7,842 centimeters

Section 9–5

1. 2.696 **3.** 1.309 **5.** -0.1990 **7.** -1.930 **9.** 5.429 **11.** 2.048 **13.** 1
15. 3.1 **17.** 5 **19.** 9 **21.** 3 **23.** 6 **25.** 2 **27.** $\frac{1}{49}$ **29.** 9 or 1 **31.** 6%
33. $171.30

Section 9–6

1. 4.2 **3.** 9.1 **5.** 2.5×10^{-4} **7.** $72,450 **9.** $5,599 **11.** $1,450 more
13. 0.101 **15.** (a) 40,000; (b) 64,660 **17.** 11.53 years or approximately 11 years, 6 months
19. 9.01 years or 9 years, 4 days (approximately)

Chapter 9 Review Problems

1. $\frac{3}{2}$ **3.** 0 **5.** $(\frac{1}{2})^{-4} = 16$ **7.** $\log_{49}(\frac{1}{7}) = -\frac{1}{2}$ **9.** $\frac{1}{49}$ **11.** $\frac{1}{3}\log_b 2.3 + \frac{1}{2}\log_b 19.6$
13. $\frac{1}{2}(\log_b x + \frac{1}{2}\log_b y) - \log_b z$ **15.** $\log_b (5x)^{-2/3}$ **17.** $8.7666 - 10$ **19.** 0.02584
21. 0.7344 **23.** $10,450

Cumulative Test for Chapters 1–9

1. multiplicative inverse axiom **3.** $-\frac{36}{5}$ **5.** $1/(9x - 3)^2(7x - 3)^3$ **7.** $4(a - 2)(a + 2)$
9. $(x + 4)/(5x - 3)$ **11.** $2(2x + 5)$ **13.** $(k - 4)/k$ **15.** $\{3\}$ **17.** $\{-\frac{13}{3}, 7\}$ **19.** $\{13\}$
21. 13 centimeters **23.** $\frac{7}{12}$ **25.** 9.3×10^{-4} **27.** $u^{1/3} - 2u^{2/3} + u$ **29.** $\{\frac{1}{2}\}$
31. $\{1 + i\sqrt{3}, 1 - i\sqrt{3}\}$

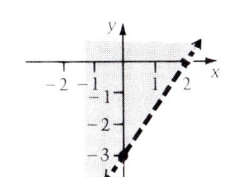

33. $\{x \mid -6 < x < 5\}$ **35.** (0, 4)

37. **39.** $r = 2\sqrt{10}$ **41.** 3 **43.** $9^{1/2} = 3$
45. $2 \log x + (\frac{1}{2}) \log y - \log 2 - \log z$ **47.** {0.0973}

Chapter 10

Section 10-1

1. inconsistent **3.** one solution **5.** dependent **7.** {(3, -2)} **9.** {(0, -1)}
11. {(3, 4)} **13.** {(2, -2)} **15.** ∅ **17.** {(13, 3)} **19.** {(2, 20)} **21.** {(-1, 0)}
23. $\{(-\frac{41}{22}, \frac{35}{22})\}$ **25.** {(-1, -2)} **27.** {(1, -1)} **29.** {(-2, 2)} **31.** $\{(-\frac{3}{2}, 4)\}$
33. {(5, 10)} **35.** {(2, 6)} **37.** $\{(2, -\frac{12}{5})\}$ **39.** $5,000 at 8%; $3,500 at 6%
41. They both crossed the finish line at the same time. **43.** 800 gallons for the first pump, 1200 gallons for the second **45.** $\frac{5}{24}$

Section 10-2

1. {(0, -1, 5)} **3.** {(2, 7, 4)} **5.** {(2, -4, -1)} **7.** dependent **9.** {(0, -2, -2)}
11. {(0, 1, 0)} **13.** {(3, 2, -8)} **15.** 8 nickels, 7 dimes, 3 quarters **17.** aluminum, 103 pounds; tin, 57 pounds; iron, 2,100 pounds

Section 10-3

1. -14 **3.** -13 **5.** 8 **7.** -13a **9.** $a_1 b_2 - a_2 b_1$ **11.** {(1, 1)} **13.** {(2, -3)}
15. {(3, 1)} **17.** {(5, -8)} **19.** {(3, -2)} **21.** inconsistent **23.** {(a, a²)} **25.** 0
27. The two determinants have opposite values.

Section 10-4

1. 44 **3.** -8 **5.** $6x - 6y$ **7.** 2 **9.** 3 **11.** $-x^3$ **13.** $-\frac{1}{2}$ or 1 **15.** -2
17. 0 **19.** The values of the determinants differ by a factor corresponding to the value of the multiple.

Section 10-5

1. {(-2, 2, 3)} **3.** {(3, -2, 0)} **5.** {(1, 2, 0)} **7.** {(1, 1, 2)} **9.** Cramer's rule does not apply. **11.** Cramer's rule does not apply. **13.** {(1, 1, -2, -2)} (arranged in the order w, x, y, z.) **15.** $x^2 + y^2 - 8x - 6y + 9 = 0$; $(x - 4)^2 + (y - 3)^2 = 16$

Section 10-6

1. {(2, 5), (5, 2)} **3.** {(2, 3), (-2, -3), (3, 2), (-3, -2)} **5.** {(1, 2), (1, -2), (-1, 2), (-1, -2)}
7. {(2, 0), (5, 3)} **9.** $\{(0, -2), (\frac{9}{5}, \frac{8}{5})\}$ **11.** {(1, 2), (-1, -2), (2i, -i), (-2i, i)}
13. {(1, 1), (-1, -1)}

15a.

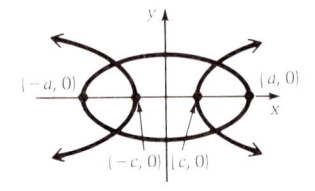

15b.

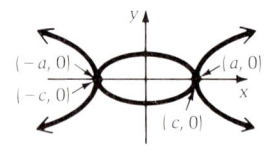

15c.

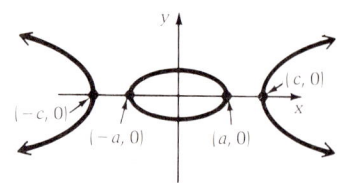

17. 3 feet by 5 feet

Section 10–7

1. 40, 35   **3.** 50, 35   **5.** 8, 12   **7.** 3, 4   **9.** The fuel costs $1.20 per gallon; the oil costs 90¢ per quart.   **11.** The shirts cost $18 each; the pants $22 each; the socks $2 per pair.
13. The electrician earns $425 per week; the assitant $350 per week; the helper $275 per week.
15. Michele worked $2\frac{1}{2}$ hours; Debbie worked 6 hours; Greg worked 1 hour.
17. 2 subcompacts; 3 compacts; 4 full size   **19.** The rate of the boat is 6 mile per hour; the rate of the current is 3 mile per hour.

Chapter 10 Review Problems

1. $\{(-2, 3)\}$ **3.** $\{(60, 150)\}$ **5.** 44   **7.** 8   **9.** $\{(-2, 1, 5)\}$ **11.** $\{(8, -11), (1, -4)\}$
13. Gasoline costs $1.30 per gallon; oil costs $1.10 per quart.

Cumulative Test for Chapters 1–10

1. $\{5\}$ **3.** 41   **5.** 5   **7.** $64x^2 - 48x + 9$ **9.** $(x + 6)/4a$ **11.** $\frac{1}{3}$
13. $\{x | x > \frac{11}{4}\}$ **15.** 11:42 A.M.   **17.** $y^6/4z^4$ **19.** $19x\sqrt{2x}$ **21.** $\emptyset$
23. $\{(3 + \sqrt{21})/3, (3 - \sqrt{21})/3\}$ **25.** $7i\sqrt{6}$ **27.** $\{6\}$ **29.** true   **31.** $m = -\frac{5}{6}$
33.

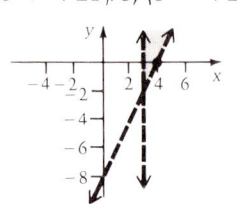

35. $(\sqrt{3}, 0)$ and $(-\sqrt{3}, 0)$ **37.** $(x + 8)^2 + (y - 5)^2 = 5$ **39.** f^{-1} is a function also.
41. $\{5\}$ **43.** $\{-2\}$ **45.** $\{(-\frac{7}{26}, \frac{25}{26})\}$ **47.** $\{(-1, 0)\}$ **49.** 21, 49

Chapter 11

Section 11–1

1. 2, 4, 6, 8   **3.** 1, 3, 5, 7   **5.** $-2, 11, -26, 47$ **7.** $-4, 8, -16, 32$
9. $2, -8, 32, -128$ **11.** $-1, 0, 3, 8$ **13.** $2, 2\frac{1}{2}, 3\frac{1}{3}, 4\frac{1}{4}$ **15.** $\frac{1}{2}, \frac{1}{4}, \frac{1}{8}, \frac{1}{16}$ **17.** $\frac{1}{2}, \frac{1}{6}, \frac{1}{12}, \frac{1}{20}$
19. $-2, 8, -24, 64$ **21.** $1, \frac{1}{2}, \frac{1}{3}, \frac{1}{4}$ **23.** $2, \frac{33}{4}, \frac{244}{9}, 1,025/16$ **25.** $3n + 1$ **27.** $1/2n$
29. $(n + 1)/n$ **31.** $3^n/n$ **33.** $\sqrt{n}/(n + 1)$ **35.** $(-1)^n n^2/(n + 1)$ **37.** x^n/n^2
39. $\sqrt[n+1]{x}/3^n$ **41.** $\frac{5}{2}$ kilogram; $\frac{5}{4}$ kilogram

Section 11-2

1. 32   3. 20   5. -72 7. 88   9. $\frac{115}{36}$ 11. -8 13. $-\frac{17}{35}$ 15. 340
17. $3 + 5 + 7 + \cdots$ 19. $\frac{4}{3} + \frac{3}{2} + \frac{18}{11} + \cdots$ 21. $\sum_{i=1}^{5} 2i$ 23. $\sum_{i=1}^{4} \frac{i}{i+1}$
25. $\sum_{i=1}^{4}(-1)^i \frac{i}{i+2}$ 27. $\sum_{i=4}^{\infty}(-1)^i(i)^2$ 29. $\sum_{i=1}^{3}(i^2 + 2i) = 26;$ $\sum_{i=1}^{3} i^2 + \sum_{i=1}^{3} 2i = 14 + 12 = 26$
31. Write $\sum_{i=1}^{4} 3$ as $\sum_{i=1}^{4}(3 + 0i)$ and expand to get 12.   33. For example, $\sum_{i=3}^{6} i = 3 + 4 + 5 + 6$, which has four terms, and $(6 - 3) + 1 = 3 + 1 = 4$.   35. By exercise 33, the expansion will have $(n - 0) + 1 = n + 1$ terms.

Section 11-3

1. yes; $d = 2$ 3. yes; $d = \sqrt{3}/3$ 5. no   7. 48   9. $-\frac{17}{4}$ 11. 2, 10, 18
13. $\frac{9}{20}, \frac{7}{10}, \frac{19}{20}, \frac{6}{5}$ 15. $a = \frac{1}{3}; \frac{1}{3}, 1, \frac{5}{3}, \frac{7}{3}$ 17. 33   19. 2,850   21. 1,785
23. 0.12 (or $\frac{3}{25}$)   25. $42,000   27a. 5 years   b. $900 interest   29. 8, 5, 2, or 2, 5, 8
31. 9 years   33. Show that the right side of $S_n = (n/2)[2a_1 + (n-1)d]$ will simplify to the desired result when a_1 and d are replaced with 3 and 2, respectively.

Section 11-4

1. yes; $r = 3$ 3. yes; $r = 2\sqrt{5}$ 5. no   7. $\frac{1}{8}$ 9. $-\frac{3}{16}$ 11. $\frac{1}{162}$ 13. 5, $\frac{5}{2}, \frac{5}{4}$ or $-5, \frac{5}{2}, -\frac{5}{4}$ 15. $r = -\frac{1}{2}; a_6 = \frac{1}{4}$ 17. $r = -2; -2, 4, -8, 16, \cdots$ 19. $\frac{121}{3}$ 21. 40
23. $a_1 = 12, S_6 = \frac{189}{8}$ 25. $17\frac{2}{3}$ feet
27. $S_n = \frac{a_1(r^n - 1)}{r - 1} = \frac{a_1 r^n - a_1}{r - 1}$. We know $a_n = a_1 r^{n-1}$; multiply both sides of this equation by r, which gives $a_n r = a_1 r^n$. Now substitute this value into the first equation:
$S_n = \frac{a_n r - a_1}{r - 1}$.

Section 11-5

1. $\frac{3}{16}$ 3. $-\frac{160}{13}$ 5. no sum; $r = \frac{4}{3} \geq 1$ 7. $(5 + \sqrt{5})/20$ 9. $\frac{5}{98}$ 11. $\frac{9}{28}$
13. $-\frac{27}{10}$ 15. $\frac{2}{11}$ 17. 2,293/990   19. $r = \frac{1}{2}; \frac{5}{8} + \frac{5}{16} + \frac{5}{32} + \frac{5}{64}$
21. If $r = \frac{1}{3}$, $a_1 = 3$ and the series is $3 + 1 + \frac{1}{3} + \cdots$
    If $r = \frac{2}{3}$, $a_1 = \frac{3}{2}$ and the series is $\frac{3}{2} + 1 + \frac{2}{3} + \cdots$
23. $r = -\frac{2}{3}; \frac{5}{9} - \frac{10}{27} + \frac{20}{81} - \frac{40}{243} + \cdots$ 25. $33\frac{1}{3}$ feet

Section 11-6

1. 20   3. 35   5. 25   7. $a^4 + 4a^3b + 6a^2b^2 + 4ab^3 + b^4$
9. $x^4 - 4x^3y + 6x^2y^2 - 4xy^3 + y^4$ 11. $a^5 - 10a^4b + 40a^3b^2 - 80a^2b^3 + 80ab^4 - 32b^5$
13. $x^4 - \frac{4}{3}x^3 + \frac{2}{3}x^2 - \frac{4}{27}x + \frac{1}{81}$ 15. $16x^4 + \frac{32}{3}x^3y + \frac{8}{3}x^2y^2 + \frac{8}{27}xy^3 + \frac{1}{81}y^4$
17. $x^{18} + 12x^{15} + 60x^{12} + 160x^9 + 240x^6 + 192x^3 + 64$
19. $x^2 + 2x^{3/2}y + \frac{3}{2}xy^2 + \frac{1}{2}x^{1/2}y^3 + y^4/16$
21. $9 - 6\sqrt{3}y + \frac{9}{2}y^2 - \frac{1}{2}\sqrt{3}y^3 + y^4/16$ 23. $1{,}024x^{10} + 15{,}360x^9y + 103{,}680x^8y^2$
25. $x^{12} - 8x^{10}\sqrt{x}\sqrt[4]{y} + 28x^9\sqrt{y}$ 27. $60x^2y^4$ 29. $160x^6$ 31. $10x^3y^3\sqrt{y}$ 33. 1.06
35. 0.93   37. $\frac{1}{x^2} - \frac{2y}{x^3} + \frac{3y^2}{x^4} - \frac{4y^3}{x^5}$ 39. $x^{1/3} + \frac{y}{3x^{2/3}} - \frac{y^2}{9x^{5/3}} + \frac{5y^3}{81x^{8/3}}$ 41. $\frac{35}{128}$

Chapter 11 Review Problems

1. $2, \frac{5}{2}, \frac{8}{3}, \frac{11}{4}$ 3. $1 + 2 + 3 + 4$ 5. $\frac{19}{14}$ 7. $-8, -3, 2, 7$ 9. $\frac{1}{288}$ 11. $16{,}000$
13. $\frac{214}{99}$ 15. $x^4 - 12x^3y + 54x^2y^2 - 108xy^3 + 81y^4$ 17. 0.89 19. 1.194

Cumulative Test for Chapters 1–11

1. true   3. 759 5. $-10x^4 - 23x^3 + 15x^2 - 2x$ 7. $(x-3)/[x(5x-1)]$ 9. $(7-3k)/k$
11. $\{\frac{37}{3}\}$ 13. $c = 2ab/(4b - 3a)$ 15. $5a^4/2b^3c^2$ 17. $17\sqrt{6}/6$ 19. $\{1\}$
21. $\{-2, \frac{4}{3}\}$ 23. $\{5, 13\}$ 25. $\{2, 4, 7\}$ 27. 10
29.

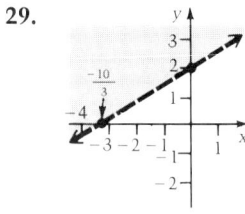

31. $(x-4)^2 + (y+3)^2 = 17$ 33. 139 feet   35. $\{\frac{7}{15}\}$ 37. 1 39. $\{0.55\}$ 41. $\{4\}$
43. 150 miles per hour   45. 13 47. 185 49. 640 51. $\frac{15}{2}$
53. $32a^5 + 80a^4b + 80a^3b^2 + 40a^2b^3 + 10ab^4 + b^5$

Index

A

Abscissa, 256
Absolute value
  definition of, 29, 160
  equations, 160
  inequalities, 164
Absolute value inequalities, graph of, 164
Addition
  axioms of, 22, 23
  of complex numbers, 221
  of like terms, 54
  of polynomials, 52
  of radicals, 194
  of rational expressions, 106
  of real numbers, 30
Addition method of solving a system of equations, 381
Addition theorem of equality, 24
Additive identity, 23
Additive inverse, 23
Additive inverse axiom, 23
Algebraic expression, 52, 128
Algorithm, 115
Antilogarithm, 362
Applications, 138, 241, 369, 409
Arithmetic means, 433
Arithmetic progression
  arithmetic means of, 433
  common difference, 432
  definition of, 432
  nth term of, 433
  sum of first n terms, 434
Arithmetic sequence (see arithmetic progressions)
Arithmetic series
  definition of, 433
  sum of first n terms, 434
Array of signs for a determinant, 399
Associative axioms, 22
Asymptote, 313
Axiom(s)
  associative, 22
  closure, 22
  commutative, 22
  distributive, 24
  identity, 23
  inverse, 23
  of order, 19
  reflexive, 15
  substitution, 16
  symmetric, 15
  transitive, 15, 19
  trichotomy, 19
Axis of symmetry, 298

B

Base
  of an exponent, 45
  of a logarithm, 350
Binary operation, 21
Binomial(s)
  definition of, 52
  expansion of, 448
  product of, 59, 60
  square of, 61
  theorem, 448
Boundary line, 282

C

Center of a circle, 304
Characteristic of a logarithm, 359
Circle
  center of, 304
  definition of, 304
  equations for, 306, 307
  general form of, 307
  radius of, 304
  standard form of, 306
Closure axioms, 22
Coefficient, 53
Combining like terms, 54
Common difference, 432
Common factor, 64
Common logarithms, 357
Common ratio, 437
Commutative axioms, 22
Completing the square, 214, 215
Complex fraction
  definition of, 97
  simplification of, 111
Complex number(s)
  addition of, 221
  conjugate of, 222
  definition of, 220
  equality of, 220
  imaginary part, 220
  product of, 222
  quotient of, 222
  real part, 220
  subtraction of, 221
Components of an ordered pair, 256
Composite number, 103

487

Conditional equation, 129
Conic section, 304
Conjugate
  axis, 313
  of a complex number, 222
  definition of, 196
Consecutive integer problems, 143, 150
Constant of variation, 324
Continued inequality, 157
Coordinate, 7
Coordinate system, rectangular, 258
Cramer's rule
  for a system of three equations, 402
  for a system of two equations, 394
Critical numbers of a quadratic inequality, 236
Cube root, 181, 182

D

Decimal(s)
  nonterminating, nonrepeating, 8
  repeating, 7, 444
  terminating, 7
Degree
  of a monomial, 53
  of a polynomial, 53
Denominator
  least common, 103
  rationalizing the, 190, 196
Dependent equations, 381
Determinant(s)
  array of signs for, 399
  Cramer's rule, 394, 402
  definition of, 392
  evaluation of, 393, 397, 398
  higher-ordered, 397
  minor of, 398
  order of, 392
  second-order, 393
  third-order, 397
Difference
  of complex numbers, 221
  of rational expressions, 106
  of real numbers, 31
  of two cubes, 76
  of two squares, 74
Direct variation, 324
Directrix, 295
Discriminant, 228
Disjoint sets, 8
Distance formula, 266
Distributive axiom, 24
Dividend, 120
Division
  of complex numbers, 222
  of polynomials, 115
  of rational expressions, 97

of real numbers, 35
synthetic, 118
Divisor, 120
Domain
  of a function, 260, 262
  of a relation, 256
Double-negative theorem, 25

E

Elimination method of solving a system of equations, 381
Ellipse
  center of, 311
  definition of, 310
  focus of, 310
  major axis of, 310
  minor axis of, 310
  standard form of, 311
Empty set, 3
Equal sets, 4
Equality
  addition theorem, 24
  axioms of, 15 16
  of complex numbers, 220
  multiplication theorem, 25
  reflexive axiom of, 15
  of sets, 4
  substitution axiom of, 16
  symmetric axiom of, 15
  transitive axiom of, 15
Equation(s)
  absolute value, 160
  of a circle, 306, 307
  circle
    general form of, 307
    standard form of, 306
  conditional, 129
  definition of, 128
  dependent, 381
  of an ellipse, 311
  equivalent, 129
  exponential, 366
  first-degree, 128
  fractional, 134
  of a hyperbola, 312
  inconsistent, 381
  of line
    point-slope form, 277
    slope-intercept form, 275
    standard form, 274
  linear systems
    in three variables, 388
    in two variables, 381
  linear in three variables, 388
  linear in two variables, 381
  logarithmic, 366, 367
  of a parabola, 296
  quadratic, 209
  quadratic in form, 233
  radical, 198, 230

solution, 130
solution set of, 130
solving, 129
subject of, 167
systems of linear, 381, 388
systems of quadratic, 405
Equivalent equations, 129
Equivalent fractions, 88
Evaluation of determinants, 393, 397, 398
Expansion of determinants, 393, 397, 398
Exponent(s)
  definition of, 45
  integral, 175
  natural number, 45
  negative, 175
  rational, 180
  rules of, 46–49, 175
  zero, 49, 175
Exponential equations, 366
Exponential form of an equation, 350
Exponential function, 344, 345
Extraneous root, 198
Extraneous solution, 136
Extremes of a proportion, 320

F

Factor, common monomial, 64
Factorial notation, 447
Factoring
  definition of, 45
  the difference of two cubes, 76
  the difference of two squares, 75
  by grouping, 77
  perfect-square trinomials, 72
  solving quadratic equations by, 209
  the sum of two cubes, 76
  by removing the greatest common factor, 64
  trinomials, 68–70
Factor, 45
Finite sequence, 424, 425
First-degree equation
  in one variable, 128
  in two variables (see linear functions, standard form of)
First-degree inequalities, 153
Focus, 295, 310
Formula(s)
  distance, 266
  quadratic, 226
Fraction(s)
  complex, 97
  equivalent, 88
  fundamental theorem of, 89
  least common denominator for, 103
  raising to higher terms, 88

Index

reducing (*see* rational expressions, reducing)
signs of, 90
standard form of, 91
Fractional equation, 134
Function(s)
  definition of, 260
  domain of, 260, 262
  exponential, 344, 345
  inverse of, 330
  linear, 273
  logarithmic, 349
  one-to-one, 331
  parts of, 262
  range of, 260, 262
  sequence, 424
Fundamental theorem of fractions, 89

G

General term of a sequence, 425
Geometric means, 439
Geometric progression
  common ratio, 437
  definition of, 437
  geometric means of, 439
  nth term of, 438
Geometric sequence (*see* geometric progressions)
Geometric series
  infinite, 442
  sum of first n terms, 439
  sum of infinite, 443
Geometry problems, 139, 148
Graph(s)
  of absolute value inequalities, 164
  of an ellipse, 312
  of exponential functions, 346
  of first-degree inequalities, 155
  of a hyperbola, 313–316
  of the inverse of a function, 332–334
  of linear inequalities, 283–286
  of ordered pairs, 258
  of a parabola, 296, 297
  of quadratic inequalities
    in one variable, 237, 238
    in two variables, 317, 318
  of a relation, 258
Greatest common factor, 64
Grouping symbols, 12

H

Half-plane, 282
Hyperbola
  asymptotes of, 313
  conjugate axis of, 313
  definition of, 312
  transverse axis of, 313

I

i
  definition of, 219
  powers of, 220
Identity, 129
Identity axioms, 23
Identity element
  of addition, 23
  of multiplication, 23
Imaginary numbers, 220
Imaginary part of a complex number, 220
Imaginary unit, 219
Inconsistent equations, 381
Index of a radical, 181
Index of summation, 429
Inequalities
  absolute value, 164
  addition property, 154
  division properties, 154
  first-degree, 153
  graph of, 155
  linear, 282
  multiplication properties, 154
  quadratic, 236, 317
  sense of, 153
  solution set for, 153
  subtraction property, 154
  in two variables, 282
Inequality
  azioms of, 17
  definition of, 153
Infinite geometric series, 442
Infinite sequence, 424, 425
Integers, set of, 6
Integral exponents, 175
Intercepts, 279
Interest problems, 144, 151
Interpolation, linear, 360
Intersection of sets, 3
Inverse
  axioms, 23
  of a function, 330
  multiplicative, 23
Inverse variation, 325
Irrational numbers, set of, 6

J

Joint variation, 326

L

Least common denominator, 103
Least common multiple
  of natural numbers, 103
  of polynomials, 104

Less than, 18
Like radicals, 195
Like terms
  combining, 54
  definition of, 54
Length of a line segment, 266
Linear equations
  system of
    in three variables, 388
    in two variables, 381
Linear function(s)
  point-slope form, 277
  slope-intercept form, 275
  standard form of, 273
Linear inequalities, 282
Linear interpolation, 360
Line(s)
  boundary of an inequality, 282
  parallel, 270
  perpendicular, 270
  slope of, 267, 268
Logarithm(s)
  base of, 350
  characteristic of, 359
  common, 357
  definition of, 350
  mantissa of, 359
  properties of, 351, 354
Logarithmic equations, 366
Logarithmic functions, 349
Lowest terms (*see* rational expression, reducing)

M

Mantissa of a logarithm, 359
Means of a proportion, 320
Minor, 398
Miscellaneous problems, 147, 152
Mixture problems, 141, 149
Monomial, 52
Multiplication
  of complex numbers, 222
  equality multiplication theorem, 25
  of polynomials, 58
  of radicals, 195
  of rational expressions, 97
  of real numbers, 34, 35
Multiplication by zero theorem, 25
Multiplicative identity, 23
Multiplicative inverse, 23

N

nth root, 182
Natural number(s)
  composite, 103
  prime, 103
  set of, 6

Notation
  factorial, 447
  functional value, 263
  scientific, 177
  set-builder, 2
  sigma, 428
  summation, 428
Number line, 7
Number(s)
  absolute value, 29
  complex, 218, 220
  composite, 103
  graph of a real, 7
  integers, 6
  irrational, 6
  natural, 6
  prime, 103
  pure imaginary, 220
  rational, 6
  real, 6
  whole, 6
Numerical coefficient, 53

O

One-to-one correspondence, 8
One-to-one function, 331
Open sentence, 128
Operations, order of, 11
Opposite of a real number, 23
Order
  axioms of, 19
  of a determinant, 392
Ordered pair of real numbers
  abscissa of, 256
  components of, 256
  definition of, 256
  graph of, 258
  ordinate of, 256
Ordered triple, 388
Ordinate, 256
Origin, 258

P

Parabola
  definition of, 295
  vertex of, 296
Perfect-square trinomial, 72
Point-slope form, 277
Polynomial(s)
  addition of, 52
  definition of, 52
  degree of, 53
  division of, 115
  factoring, 64, 68, 74
  multiplication of, 58
  subtraction of, 52
  value of a, 55
  written in descending powers, 54
Powers of i, 220

Prime
  factorization, 103
  factors, 103
  number, 103
Principal nth root of a real
  number, 182
Principal square root, 181
Principal square root of a negative
  number, 219
Product(s)
  of complex numbers, 222
  definition of, 45
  of polynomials, 58
  of radicals, 195
  of rational expressions, 97
  of real numbers, 35
Progression(s)
  arithmetic, 432
  geometric, 437
Proper subset, 4
Proportion, 319
Pure imaginary number, 220
Pythagorean theorem, 247

Q

Quadrant, 258
Quadratic equation(s)
  definition of, 209
  discriminant, 228
  incomplete, 209
  nature of the roots of, 228
  solution of
    by completing the square, 214
    by factoring, 209
    by the quadratic formula, 225
  standard form of, 209
  systems of, 405
Quadratic formula, 225, 226
Quadratic inequalities
  definition of, 236
  in one variable, 236
  in two variables, 317
Quotient(s)
  of complex numbers, 222
  of radicals, 188, 196
  of rational expressions, 97
  of real numbers, 35

R

Radicals
  addition of, 194
  equations containing, 198, 230
  index of, 181
  like, 195
  multiplication of, 195
  multiplying with different
    indices, 191
  operations on, 193
  properties of, 188, 189

  rationalizing the denominator of,
    190, 196
  simplified form of, 190
  subtraction of, 194
Radical equations, 198, 230
Radical sign, 181
Radicand, 181
Radius of a circle, 304
Raising a fraction to higher terms
  (see terms of a fraction)
Range of a function, 260, 262
Range of a relation, 256
Ratio, 319
Rational exponents, 180, 183, 184
Rational expression(s)
  addition of, 106
  definition of, 52
  multiplication and division of,
    97
  reducing, 94
  subtraction of, 106
Rational numbers, set of, 6
Rationalizing the denominator,
  190, 196
Real number system, 6
Real part of a complex number,
  220
Reciprocal, 23
Rectangular coordinate system,
  258
Reflexive axiom of equality, 15
Relation
  definition of, 256
  domain of, 256
  graph of, 258
  inverse of, 330
  range of, 256
Remainder, 117
Repeating decimal, 7, 444
Replacement set of a variable, 128
Root(s)
  cube, 181
  multiplicity of, 211
  nth, of a real number, 182
  of a quadratic equation, 209, 212
  square, 180

S

Scientific notation, 177
Second-degree equations (see
  quadratic equations)
Sequence(s)
  arithmetic (see arithmetic
    progressions)
  definition of, 425
  finite, 424, 425
  function, 424
  gereral term of, 425
  geometric (see geometric
    progressions)
  infinite, 424, 425

Series
  arithmetic, 433
  arithmetic, sum of first n terms, 434
  definition of, 428
  geometric, 439
  geometric, nth term of, 438
  geometric, sum of, 439
  infinite geometric, 442
  sum of infinite geometric, 443
Set(s)
  definition of, 2
  disjoint, 8
  element of, 2
  empty, 3
  equal, 4
  of integers, 6
  intersection of, 3
  of irrational numbers, 6
  member of, 2
  of natural numbers, 6
  proper subset, 4
  of rational numbers, 6
  replacement, 128
  solution, 128
  subset, 4
  union of, 3
  well-defined, 2
  of whole numbers, 6
Set-builder notation, 2
Sigma notation, 428
Signed number, 29
Slope, 267, 268
Slope-intercept form, 275
Solution(s)
  extraneous, 136
  of first-degree equations, 130
  of inequalities, 153, 237
  of quadratic equations, 209
  of quadratic inequalities, 236–240
  of systems of equations, 382, 389
Solution set
  of an absolute value inequality, 164
  definition of, 128
  of an equation, 130
  of an inequality, 153, 237
  of quadratic equations, 209
  of systems of equations, 382, 389
Special products, 44
Square root, 180, 182
Standard form
  of the equation of a circle, 306
  of the equation of an ellipse, 311
  of the equation of a hyperbola, 312
  of a fraction, 91
  of a linear equation, 274
  of a quadratic equation, 209
Statement, 128
Steve and Ross, 140, 149, 247, 252, 303, 373, 414
Subject of an equation, 167
Subset, 4
Subset, proper, 4
Substitution axiom, 16
Substitution method of solving a system of equations, 384
Subtraction
  of complex numbers, 221
  of fraction (see rational expressions, subtraction of)
  of like terms, 54
  of polynomials, 52
  of radicals, 194
  of rational expression, 106
  or real numbers, 31
  of signed numbers, 29
Sum
  of an arithmetic series, 434
  of complex numbers, 221
  of finite geometric series, 439
  of infinite geometric series, 443
  of radicals, 193
  of rational expressions, 106
  of real numbers, 29
  of two cubes, 76
Summation
  index of, 429
  notation, 428
Symmetric axiom of equality, 15
Symmetry with respect to a line, 298
Synthetic division, 118
System(s)
  of linear equation
    in three variables, 388
    in two variables, 381
  of quadratic equations, 405

T

Terminating decimal, 7
Term(s)
  of an algebraic expression, 52
  general, of a progression, 425
  of a fraction, 88
  of a progression, 424
Theorem(s)
  definition, 24
  double-negative, 25
  equality addition, 24
  equality multiplication, 25
  multiplication by zero, 25
Transitive axiom of equality, 15
Transitive axiom of inequality, 19
Transverse axis, 313
Trichotomy axiom, 19
Trinomial(s)
  definition of, 52
  factoring of, 68–70
  perfect-square, 72

U

Uniform-motion problems, 145, 151
Union of sets, 3

V

Variable, replacement set for, 128
Variation
  constant of, 324
  direct, 324
  inverse, 325
  joint, 326
Vertex of a parabola, 296

W

Well-defined set, 2
Whole numbers, set of, 6
Work problems, 140, 149
Writing a quadratic equation from its roots, 212

X

x-axis, 258
x-intercept, 279

Y

y-axis, 258
y-intercept, 279

Z

Zero-factor rule, 209